Solid Mechanics and Its Applications

Volume 247

Aims and Scope of the Series

The fundamental questions arising in mechanics are: *Why*?, *How*?, and *How much*? The aim of this series is to provide lucid accounts written by authoritative researchers giving vision and insight in answering these questions on the subject of mechanics as it relates to solids.

The scope of the series covers the entire spectrum of solid mechanics. Thus it includes the foundation of mechanics; variational formulations; computational mechanics; statics, kinematics and dynamics of rigid and elastic bodies: vibrations of solids and structures; dynamical systems and chaos; the theories of elasticity, plasticity and viscoelasticity; composite materials; rods, beams, shells and membranes; structural control and stability; soils, rocks and geomechanics; fracture; tribology; experimental mechanics; biomechanics and machine design.

The median level of presentation is to the first year graduate student. Some texts are monographs defining the current state of the field; others are accessible to final year undergraduates; but essentially the emphasis is on readability and clarity.

More information about this series at http://www.springer.com/series/6557

Hornsen (HS) Tzou

Piezoelectric Shells

Sensing, Energy Harvesting, and Distributed Control—Second Edition

Springer

Hornsen (HS) Tzou
State Key Laboratory of Mechanics
and Control of Mechanical Structures;
Interdisciplinary Research Institute,
College of Aerospace Engineering
Nanjing University of Aeronautics
and Astronautics
Nanjing, China

ISSN 0925-0042 ISSN 2214-7764 (electronic)
Solid Mechanics and Its Applications
ISBN 978-94-024-1256-7 ISBN 978-94-024-1258-1 (eBook)
https://doi.org/10.1007/978-94-024-1258-1

Library of Congress Control Number: 2018935210

This Springer imprint is published by the registered company Springer Nature B.V.
The registered company address is: Van Godewijckstraat 30, 3311 GX Dordrecht, The Netherlands

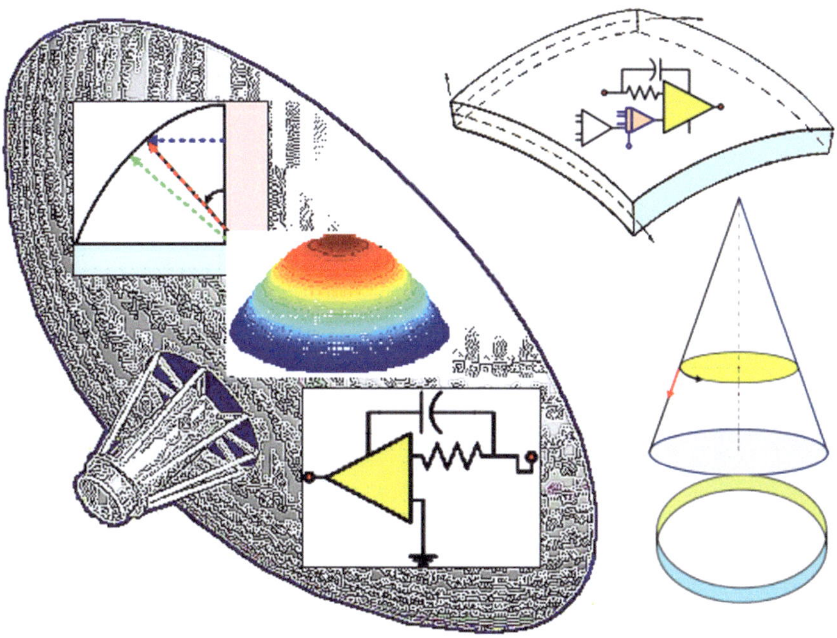

Preface (2nd Edition)

It has been over three decades since the beginning of vigorous R&D activities of smart structures and structronic systems in the mid-1980s. The area started from a couple of papers a year to many technical sessions and symposia a year, to today's more than fifteen major technical conferences, symposia, and workshops a year. The discipline of smart structures and structronic systems was also recognized as one of the key technologies in mechanical engineering in the twenty-first century.

Among the smart materials, e.g., piezoelectric, flexoelectric, shape-memory, electro-/magneto-/photo-strictive materials, electro- and magneto-rheological fluids, polyelectrolytes, pyroelectric materials, superconducting materials, magneto-optical materials, optical fibers, etc., piezoelectric materials are very versatile in both sensor and actuator applications. Accordingly, it becomes the most utilized smart material in many sensors, actuators, energy harvesters, and precision devices, ranging from micro-/nano-electromechanical systems, nano-energy generators, thin-film sensors/actuators, accelerometers, force/pressure transducers, ultrasonic motors, precision machines, medical devices, automobile components, etc., to helicopters, airplanes, and large-scale space structures and systems. There have been a few books on piezoelectricity published in the past; however, a unified presentation of piezoelectric shells and distributed sensing, energy harvesting and control applications is still lacking. This book is intended to fill the gap and to provide researchers and practicing engineers with an introduction to advanced piezoelectric shell theories, energy harvesting and distributed sensor/actuator technologies in structural identification, energy generation and control.

This book represents a collection of the author's three-decade research and development on piezoelectric shells and related applications to distributed measurement, energy harvesting and control of continua; it reflects six Best Paper Awards and NASA Technical Disclosures, including two ASME and one AIAA Best Paper Awards and three NASA Technical Disclosures over the years. To provide an "umbrella" theory to most common piezoelectric continua, a generic double-curvature piezoelectric deep shell defined in a tri-orthogonal curvilinear coordinate system is used as the fundamental continuum to derive the piezoelectric/elasticity and distributed sensing/control/energy generation theories. There are a number of shell continua and related theories discussed in the book: (1) three piezoelectric shell continua, (2) one elastic shell continuum laminated with distributed shell layers, (3) one multi-layered shell actuator, (4) one thin piezoelectric shell structure, and (5) one generic shell energy generator. All of these shells are based on the generic double-curvature deep shell continuum. The major advantage of the generic shell theories is their *generality*. All theories developed based on the generic shell continuum can be easily simplified to account for a large number of shell and non-shell continua and structures (piezoelectric or elastic), such as shells of revolution, parabolic shells, toroidal shells, spherical shells, cylindrical shells, conical shells, plates (circular, elliptical, or rectangular), arches, rings, beams, rods, with a variety of mechanical or electric boundary conditions. Therefore, the important aspects of the book are to introduce (1) the generic approach of piezoelectric shells, (2) the fundamental electromechanics and sensing/control effectiveness of distributed piezoelectric sensors/actuators in structural controls, (3) the energy generation or harvesting based on piezoelectric laminated shell structures, and (4) the development of piezoelectric finite element and its numerical techniques.

Continued from the 1st edition of the book, this 2nd edition is divided into four major subjects. The first subject focuses on piezoelectric shell continua, the second on distributed sensing and control of elastic continua and structures, the third on energy generation and harvesting, and the fourth on the finite element development. To introduce the essential background in the area, Chap. 1 provides a general introduction to piezoelectric phenomena, piezoelectric materials, piezoelectric continua, and distributed sensing, energy generation and control of continua or structures. Detailed derivations of three piezoelectric shell theories are presented in Chap. 2. Applications of the generic piezoelectric shell theories to common piezoelectric continua based on a reduction procedure are given in Chap. 3. Distributed convolving sensor and actuator theories in distributed identification and control are also discussed. In addition, frequency and damping controls of piezoelectric structures are also demonstrated. Chap. 4 focuses on generic distributed shell sensor/actuator theories for elastic shell continua laminated with distributed piezoelectric thin sensor/actuator layers. Applications of the generic theories to other shell and non-shell common geometries are also demonstrated in case studies. Two generic sensor/actuator design principles, namely *the segmentation technique* and *the shaping technique*, are also presented. Chap. 5 presents a multi-layered shell actuator theory based on equivalent induced control actions. Distributed measurements (sensing or identifications) and controls of typical elastic continua or structures, e.g., beams, plates, rings, cylindrical shells, and paraboloidal shells, are demonstrated in Chaps. 6–10.

Boundary controls of an elastic cantilever beam are investigated in Chap. 6. Theoretical, finite element and experimental data are compared. Distributed sensing and control of plates with fully distributed and segmented sensors/actuators, i.e., based on *the segmentation technique*, are studied in Chap. 7. Detailed parametric studies of segmented sensors/actuators are emphasized. In Chap. 8, distributed shell sensor electromechanics of a generic piezoelectric shell layer are first presented and followed by generic theories of distributed shaped and convolved sensors/actuators, i.e., *the shaping technique*. These theories are then applied to an elastic circular ring. Distributed convolving modal sensors/actuators are designed, and their performances are evaluated. Following the same procedures, distributed sensing and control of cylindrical shells are presented in Chap. 9. Spatially distributed microscopic modal actuation characteristics and optimal control of parabolic shells are evaluated in Chap. 10.

Piezoelectricity-based energy generation and harvesting of generic shell harvesters are presented in Chap. 11. Electromechanics and power generation of a tubular shell are discussed in Chap. 12. Analytical results are compared with experimental data. Note that laboratory experiments were carried out over time. Thus, piezoelectric materials with various dielectric constants from different vendors were used in experimental models presented in newly added Chaps. 10–12.

Furthermore, it is well known that theoretical models and solutions are limited to relatively ideal geometries and boundary conditions. In order to broaden the practical applications of piezoelectric technology, piezoelectric finite elements, especially a "thin" solid element, and sensing/control algorithms are formulated in Chap. 13. Validations and applications of the piezoelectric thin finite element to micro-manipulations, distributed identification (modal voltages), and distributed control of continua are presented in case studies. Note that minor typos and mistakes are corrected; specifically, laboratory experimental components are added to many chapters in the revised book.

This book can be used as a professional reference for scientists and practicing engineers who would like to understand the new piezoelectric technology in distributed sensing, identification, energy generation and control. The book can also be used in classroom teaching as a textbook for graduate students. Technical backgrounds in vibration and solid mechanics should be adequate to master the contents.

Nanjing, China Hornsen (HS) Tzou

Acknowledgements (2nd Edition)

This research was supported by grants from the National Science Foundation (No. RII-8610671), the Commonwealth of Kentucky (1986–1991), the Army Research Office (DAAL03-91-G-0065, 1991–1994, Technical Monitor: Dr. Gary L. Anderson), the Center for Robotics and Manufacturing Systems (1987–1988 and 1992–1993) at the University of Kentucky, etc. (Note that contents of the information do not necessarily reflect the position or the policy of the government, nor should official endorsement be inferred.) A fellowship supported by the Institute of Space and Astronautical Science (ISAS), Japanese Ministry of Education, is gratefully acknowledged. (The author is particularly thankful to his hosts, Prof. M. C. Natori and Prof. K. Miura, in the Division of Spacecraft Engineering, for their warm hospitalities during that period.) The later chapters on energy generation and harvesting were supported by grants from the National Natural Science Foundation of China (#11872206, #11472241 and #11172262), the Nanjing University of Aeronautics and Astronautics Foundation (NUAA-NP2016203; 56SYQR18002), the State Key Laboratory of Mechanics and Control of Mechanical Structures (NUAA-MCMS-0516G01), the State Key Laboratory of Fluid Power Transmission and Control (No. SKLoFP_CX_001), and also the 985 (2nd phase) grant at Zhejiang University.

The author would like to express his sincere gratitude to a number of people who were influential to his career development: Prof. A. J. Schiff (advisor), Prof. H. T. Y. Yang, Prof. R. A. Altenkirch, Prof. J. T. P. Yao, etc. He also likes to thank Dr. D. J. Segalman and Dr. R. D. Skocypec for encouraging his professional services on the ASME Board on Technical Knowledge Dissemination (BTKD), the Technical Communities Operating Board (TCOB), and the Interdisciplinary Councils (ICs). The author would like to thank all his past graduate students, post-doctors, and visiting scholars who actually executed the research. In addition, he likes to thank his young wife Amy Jianhua Lin for her kindness, patience, and love. His two lovely children Jessica and Austin also paid the full price for the book over the years.

Nanjing, China Hornsen (HS) Tzou

Contents

Chapter 1
Introduction

The concept of smart, intelligent, and adaptive materials and structures originated in mid-1980s in an attempt to address the new emerging activities of synergistically integrating active electro-/magneto-/photo- and other functional materials with large-scale structures and systems, as in situ sensors, actuators and control components (Tzou, Lee and Arnold, 2004). Previously, electro- and other active materials had been mostly used in small and micro-scale transducers and precision mechatronic (mechanical + electronic) control systems. The general perception of smart, intelligent, and adaptive materials or structures implies an ability of being clever, sharp, active, fashionable, and sophisticated. However, in reality, materials or structures themselves can never achieve true intelligence or reasoning capability without the artificial intelligence (AI) through microprocessors, computers, control electronics, control logics/algorithms, etc. Accordingly, the materials can only be active and the structures could ultimately be smart and intelligent in practical applications. Furthermore, the synergistic integration of smart materials, structures, sensors, actuators, and control electronics has redefined the concept of structures from a conventional passive elastic system to an active or adaptive (life-like) multi-functional *structronic* (structure + electronic) *system* with inherent self-sensing, diagnosis, and control capabilities (Tzou and Gabbert, 1997; Gabbert and Tzou, 2001; Tzou, et al., 1998; Tzou and Anderson, 1992). The common smart materials include piezoelectrics, flexoelectrics (Tzou and Zhang, 2016; Hu, Li and Tzou, 2013a, b, 2014, 2015), shape memory materials, electrostrictive materials, magnetostrictive materials, photostrictive materials, electrorheological fluids, magnetorheological fluids, polyelectrolyte gels, pyroelectrics, photoferroelectric materials, magneto-optical materials, light-actuated shape memory polymers (LaSMP) (Li, et al., 2015), superconducting materials, etc. (Tzou, Lee and Arnold, 2004), Table 1.1. The multi-field opto-thermo-electro-magneto-mechanical systems and structures applied to complicated multi-field controls coupling elastic, temperature, electric, magnetic, light, etc. and control interactions are discussed later.

Mass, stiffness, static and dynamic characteristics of elastic continua, machines, and structures (e.g., shells and plates) are generally *distributed* in nature, i.e., their

H. (HS) Tzou, *Piezoelectric Shells*, Solid Mechanics and Its Applications 247,
https://doi.org/10.1007/978-94-024-1258-1_1

Table 1.1 Smart materials and their milestone years

Piezoelectrics (1880), Flexoelectrics, Pyroelectrics (315BC)	Shape-memory materials (1932), Light-actuated shape memory polymer	Electro- and Magneto-strictive Materials (1954/ 1840)	Electro- and Magneto-rheological Fluids (1784/1947)
Polyelectrolyte gels (pH Muscles) (1949)	Photostrictive Materials (1974), Photovoltaic Materials	Superconductors (1911)	Liquid Crystals
Optical fibers	Electro-luminescence Particles	Magneto-optical, Electromagnets, Magneto-elastic Materials	Electro-static Materials

properties or behaviors are functions of *time* and *spatial* coordinates. Accordingly, these systems are usually classified as *distributed parameter systems* (DPSs) and they are usually modeled by partial differential equations (PDEs). In practical applications, however, due to their complexity, discretization techniques, e.g., the finite difference or finite element method, etc., are often employed and, for convenience, the derived simplified *lumped* (or *discrete*) *parameter systems* (LPSs) modeled by ordinary differential equations (ODEs) are analyzed and evaluated. Accordingly, simplified LPS only reveals partial system characteristics of the original DPS. Similarly, sensing and control of DPSs using conventional *discrete* (or point) sensors and actuators at spatially discrete locations also pose many observation/control deficiencies, especially they are placed at (or near) modal nodes or lines of DPSs. Eventually, these modes are neither observable nor controllable by these discretely located transducers. To remedy the discrete sensor/actuator deficiencies and to observe and control these elastic DPSs, spatially distributed sensors and actuators are highly desirable. As discussed, the complexity of multi-field photo-thermo-electro-magneto-mechanical and control couplings of multifunctional DPSs poses many challenging research issues over the years. Recent development of smart structures and structronic systems emphasizing synergistic integration of controllable smart materials, e.g., piezoelectrics, shape memory materials, electro- and magneto-strictive materials, electro- and magneto-rheological fluids, photostrictives, magneto-optical materials, flexoelectric materials, light-actuated shape memory polymers (LaSMP), etc., with high-performance structures, precision structronic and mechatronic systems represents a highlight of classical multi-field coupling and interactions, Fig. 1.1.

In general, these smart materials respond to certain stimuli, e.g., strain, force, pressure, light, etc., and also be actively controlled via electrical field, magnetic field, currents, high-energy lights, etc. (Tzou, Lee and Arnold, 2004; Shih, et al., 2005; Chai, et al., 2008; Shih and Tzou, 2005 and 2007; Tzou, Chai and Arnold, 2004; Chai and Tzou, 2007). Accordingly, structures and structronic systems incorporating these smart materials are capable of responding to certain stimuli in a

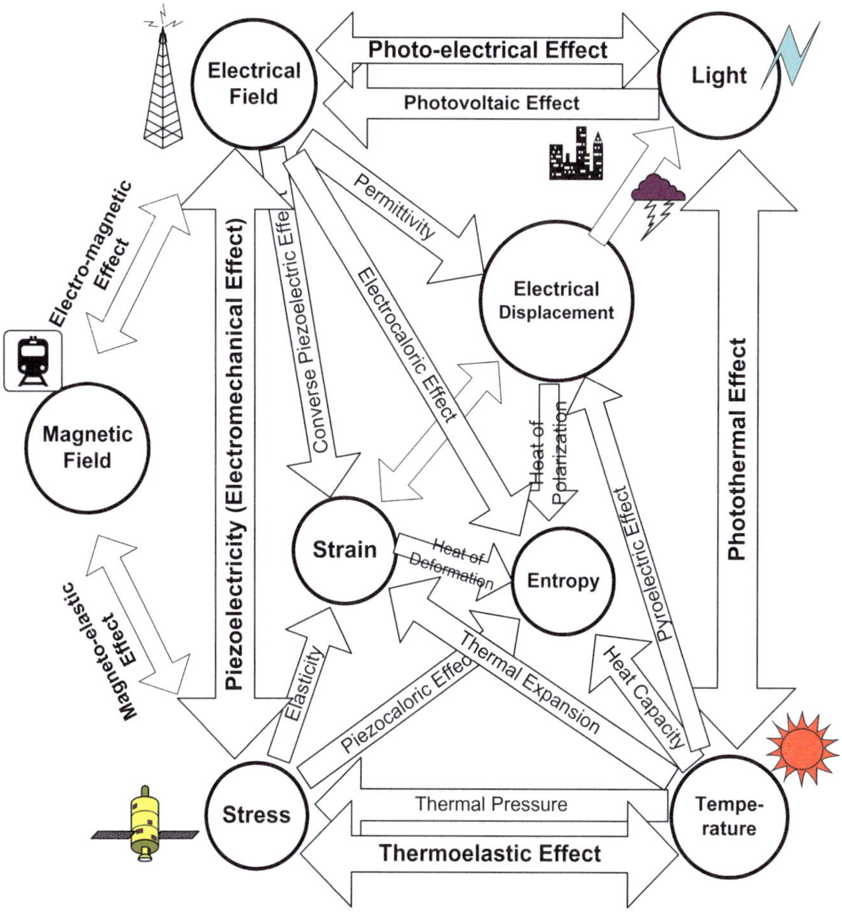

Fig. 1.1 Multi-field coupling among elastic, electric, magnetic, temperature and light fields

prescribed and controllable manner based on specified performance or control requirements. Furthermore, the multi-field photo-piezo(electric)-magneto-thermo-elastic-control coupling and interactions of distributed smart structures and structronic systems have been becoming much more complicated and they deserve in-depth studies to establish profound database and design guidelines for the next-generation high-performance smart structures, machines and structronic systems. This book focuses on electromechanical/control coupling of piezoelectric shells, and sensing, energy harvesting and distributed control of structronic systems based on distributed piezoelectric devices.

1.1 Piezoelectricity

The discovery of piezoelectricity by Curie brothers (Jacques and Pierre) in 1880 had led the science and technology into a new dimension. It has been nearly one hundred and forty years (1880–2017) since the first observation of piezoelectric phenomena. Over the years, sophisticated piezoelectricity theories have been proposed and refined; new piezoelectric materials have been discovered or synthesized (Mason, 1950; Cady, 1964; Mindlin, 1961 & 1972; Tiersten, 1969; Sesseler, 1981; Dökmeci, 1983; Tzou & Zhong, 1990; etc.). Furthermore, novel piezoelectric devices, e.g., transducers and actuators, have been invented and applied to a variety of engineering applications and systems (Mason, 1981; Sessler, 1981; Dökmeci, 1983; Tzou, 1990 & 1992b; Tzou & Fukuda, 1992).

Piezoelectricity is an electromechanical phenomenon which couples the elastic (*dynamic coupling*) and electric (*static coupling*) fields. In general, a piezoelectric material responds to mechanical forces/pressures and generates an electric charge/voltage. This phenomenon is called the *direct piezoelectric effect*. Conversely, an electric charge/field applied to the material induces mechanical stresses or strains, and this phenomenon is called the *converse piezoelectric effect*. However, temperature fluctuation also influences the thermo-elecromechanical behavior and the two-field coupling extends to the three-field coupling—the piezo(electric)-thermoelastic coupling, illustrated in Fig. 1.2.

Engineering applications of piezoelectric materials started with a depth-sounding device, based on Rochelle salts, invented by Langevin in 1917. Many other sensors (e.g., accelerometers, pressure transducers, force transducers, microphones, etc.), actuators (precision manipulators, robot manipulators, ultrasonic motors, driving mechanisms for scanning tunneling microscopes, etc.), smart structures and structronic systems (e.g., active truss system, antenna surface and vibration control, etc.) have been documented over the years. Energy harvesting mechanisms and devices have also been proposed and evaluated (Erturk and Inman, 2008; Kim, 2010; Zhang, Hu and Tzou, 2014). New devices are being invented and patented constantly. In the recent development of active, adaptive or smart structures and structronic systems, active piezoelectric and elastic/piezoelectric structures (elastic materials integrated with piezoelectric sensors/actuators and control electronics) and thin-layer piezoelectric devices are very promising in both precision static and dynamic applications, e.g., precision drivers, robot manipulators, helicopter rotor blades, airplane wings, precision trusses, aerospace/aircraft structures, vibration controls and isolations, high-precision positioners, micro-sensors/actuators, thin-film devices/systems, nano-electromechanical systems, etc. (Tzou & Fukuda, 1991 & 1992). All these activities have driven a renewed and widely spread interests in piezoelectricity related researches and developments. In active piezoelectric structures or smart structronic structures, as discussed previously, the *direct* effect is used for structural sensing, measurements and energy harvesting; the *converse* effect is applied to actuation and active vibration controls of structures and systems. Table 1.2 summarizes natural and synthetic piezoelectric materials (Dökmeci, 1983).

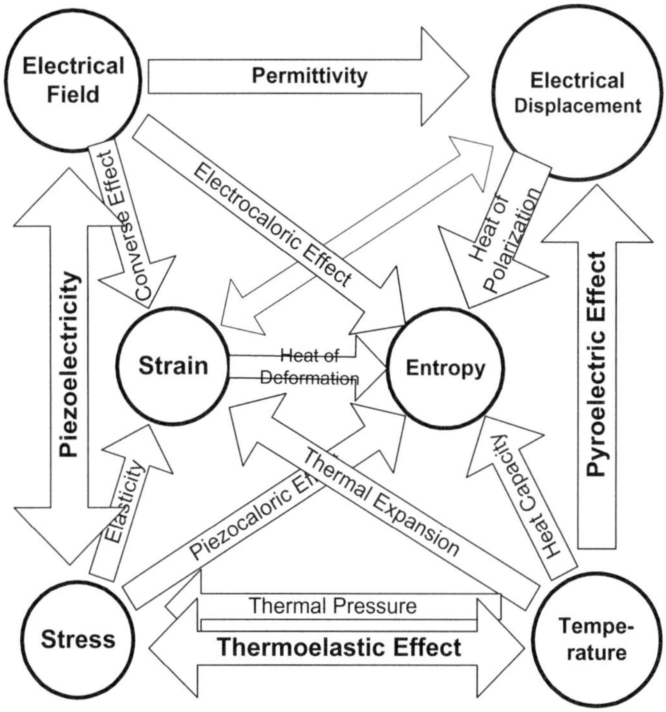

Fig. 1.2 Piezo(electric)-thermoelasticity—a piezothermoelectromechanical coupling

Table 1.2 Common piezoelectric materials

Natural crystals	Quartz, Rochelle salt, ammonium phosphate, etc.
Liquid crystals	
Noncrystalline materials	Glass rubber, paraffin, etc.
Textures	Bone, wood, etc.
Synthetic piezoelectric materials	1) Piezoceramics: lead zirconate titanate (PZT), barium titanate, lead niobrate, lead lanthinum zirconate titanate (PLZT), etc.
	2) Crystallines: ammonium dihydrogen phosphate, lithium sulfate, etc.
	3) Piezoelectric polymer: polyvinylidene fluoride (PVDF or PVF_2), etc.

In general, synthetic materials can be designed and fabricated into arbitrary geometries and shapes. Thus, synthetic piezoelectric ceramics (e.g., PZT & PLZT) and polymers (e.g., PVDF) are very popular in many sensor and actuator applications. Two subject areas are briefly reviewed next. The first subject is concerned with general theoretical developments and applications of active piezoelectric shell continua; the second subject is on distributed sensing, identification, energy harvesting and control of elastic shell continua (e.g., shells and plates) using distributed piezoelectric sensor/actuator layers. Detailed reviews, theories, derivations, formulations, distributed sensing/control, energy harvesting and case studies are presented in later chapters.

1.2 Piezoelectric Continua

Studies of piezoelectricity were initiated in 1880, and significant advances have been continuously made over the years (Mason; 1950; Cady, 1964; Mindlin, 1961 & 1972; Tiersten, 1969; Sesseler, 1981; Dökmeci, 1983; Tzou and Zhong, 1990; Tzou, 1993, etc.). Earlier work mostly concentrated on vibrations and waves of specific geometries with finite and infinite dimensions, such as thin rods, plates, rings, disks, circular cylindrical shells, etc. (Dökmeci, 1980). Toupin (1959) derived an equilibrium equation for a polarized elastic spherical shell. Haskins and Walsh (1957) presented vibration theories for piezoelectric shells of revolution and a hollow ceramic cylinder. Drumhaller and Kalnins (1970) presented a coupled theory for the vibration analysis of piezoelectric shells of revolution, and it satisfied all electrostatic and elastic equations. Adelman and Stavsky studied axisymmetric vibrations of radially polarized piezoelectric ceramic cylinders (1975a) and vibrations of radially polarized composite piezoceramic cylinders and disks (1975b). Paul studied 1) asymptotic analyses of wave propagation in a piezoelectric solid cylinder (1982), 2) vibrations of a hollow piezoceramic circular cylinder (1978), 3) axisymmetric vibrations of a piezoelectric solid cylinder guided by a thin film (1986), and 4) wave propagations in a piezoelectric solid cylinder of arbitrary cross sections (1987).

Generic theories on piezoelectric shells of arbitrary shape are of importance in many applications. Dökmeci (1978) proposed a theory on coated thermopiezoelectric laminae. Senik and Kudriavtsev (1980) formulated the equations of motion for piezoelectric shells polarized along the normal of the shell middle surface. Chau (1986) proposed a variation formulation to solve the equilibrium problem of anisotropic piezoelectric shells. Rogacheva (1982, 1984,1986) studied state equations and boundary conditions of piezoelectric shells polarized along coordinate directions. Tzou and Gadre (1989) derived a generic theory for multi-layered piezoelectric shell actuators using equivalent induced strains; they also studied micro-excitations and isolations (1988, 1990). Tzou proposed generic distributed sensing and control theories for a generic shell continuum laminated with piezoelectric thin layers (1988, 1991a), and the concept was extended to distributed neurons and muscles for elastic continua (1991b). Tzou and Tseng formulated a new thin piezoelectric finite element and applied it to distributed identification and control of continua using mono-axial actuators (1990). Later, they derived a finite element control formulation and compared the control effects of distributed mono- and bi-axial actuators in distributed vibration controls (1991). Although studies of piezoelectric shells of generic shapes have advanced in the last few decades, most of them have primarily been concerned with wave propagations, i.e., in-plane motions. Electromechanics and vibration behaviors encompassing all three principal directions, three translational and two rotatory coordinates, of generic piezoelectric shell continua and applications to active and adaptive smart structures and structronic systems still need to be further explored and studied.

Thus, a piezo(electric)elastic vibration theory of a generic piezoelectric (deep) shell continuum with a symmetrical hexagonal structure is derived based on the linear piezoelectricity theory and Hamilton's principle, Chap. 2. Generic equations of mechanical motion and mechanical boundary conditions, as well as a charge equation of electrostatics and electric boundary conditions are formulated. The derived system equations are very generic; they can be simplified to apply to a broad class of piezoelectric structures, e.g., beams, arches, rings, plates, cylinders, spheres, etc. The simplification procedures are proposed and demonstrated in case studies, Chap. 3. There are three piezoelectric shell vibration theories respectively for 1) a thick generic shell continuum, 2) a thick generic piezoelectric shell continuum with transverse shears and rotatory inertias, and 3) a thin piezoelectric shell without shear deformation and rotatory inertia effects, i.e., based on the Kirchhoff-Love thin shell assumptions, Chap. 2. Applications of the theories to active adaptive structures and energy harvestings are also demonstrated in later chapters.

1.3 Distributed Sensing, Energy Harvesting and Vibration Controls

As discussed previously, structures are generally "distributed" in nature, i.e., structural behaviors are functions of time and space; they are classified as *distributed parameter systems (DPSs)*. In practical applications, however, discretization techniques are employed to lead to simplified *discrete* (or *lumped*) *parameter systems*. There are many fruitful research literatures on identification and control of lumped parameter systems available today, but relatively few on distributed parameter systems. Wang (1966) surveyed earlier researches on control of distributed parameter systems. Robinson (1971) summarized a variety of problems on theory and application of distributed parameter control. Stavroulakis (1983) reviewed various topics on control and estimation of distributed parameter systems. In this book, the distributed parameter systems are **continuous structures** (continua), e.g., shells, plates, etc. Structural sensing, identification, energy harvesting and vibration control of continua (e.g., shells, plates, etc.) are investigated.

Precision structural identification and control needs an effective coordination of sensors and actuators. Conventional transducers and sensors are generally "discrete"; they measure spatially discrete locations of a continuum or structure. In order to observe and control the continuum, distributed sensors and actuators are highly desirable. In this book, distributed piezoelectric layers are integrated with a generic deep shell continuum; one layer serves as a distributed sensor and the other a distributed actuator for structural identification and vibration control of the continuum, Chap. 4. Applications of the generic theory to distributed sensing, control and energy harvesting of continuous structures, e.g., beams, plates, ring shells, cylindrical shells, parabolic shells, etc., are presented in Chaps. 6–12.

Distributed active vibration controls of flexible beams using piezoelectric materials have been studied (Plumb, Hubbard, and Bailey, 1987; Crawley & deLuis, 1987; Baz and Poh, 1988, Hanagud and Obal, 1988; Tzou, 1987; Tzou, Johnson and Liu, 1999). Tzou and Gadre derived a multi-layered shell actuator theory for distributed vibration control of flexible shell structures (1989). Polla (1992) developed thin-film piezoceramic micro-electromechanical systems (MEMS). Lee and Moon (1988) proposed modal sensors and actuators for beams. Tzou, et al. studies orthogonal sensors/actuators for rings and shells (Tzou and Zhong, 1996; Tzou, Zhong and Natori, 1993; Tzou, Zhong and Hollkamp, 1994). An integrated distributed sensing and control theory for thin shells was also proposed (Tzou, 1988, 1991a). Piezoelectric vibration exciters and isolators were theoretically studied and experimentally verified (Tzou and Gadre, 1988, 1990). A similar technique was applied to a rotordynamic vibration control (Palazzolo et al., 1989). Sirlin (1987) used a flexible piezoelectric polymer in spacecraft isolations. Fason and Gabar (1988) designed an active piezoelectric member in a space truss structure. Other robotic applications included: 1) distributed vibration control of flexible robots (Tzou, 1989b), 2) a robot wrist actuator (Lee & Arjunan, 1989), 3) a micro-displacement robot gripper (Tzou, 1989a), etc. Additional practical applications include piezoelectric damped skis, control of aeroelastic wing flutter, optical metering truss control, helicopter blade control, noise reduction, active constrained layer damping, biomedical applications, ultrasonic motors, self-sensing actuators, etc. (Lynch, 1996; Ashley, 1995; Bank and Smith, 1998; Baz and Ro, 1994; Ueha, et al., 1993; Sashida and Kenjo, 1993; Uchino, 1997; Dosch, Inman, Garcia, 1992; Anderson, Hagood, Goodliffe, 1992; Tzou and Hollkamp, 1994) and energy harvesters (Erturk and Inman, 2008; Kim, 2010; Zhang, Hu and Tzou, 2014), etc. Tzou and Tseng also developed a piezoelectric finite element for 1) distributed sensing and control of shells and plates and 2) evaluation of mono- and bi-axial actuators (1990, 1991). Generic theories on structural identification and vibration control of continua using electroded piezoelectric layers were proposed (Tzou, 1992a; Tzou, 1993). In this book, generic distributed structural identification, energy harvesting, and vibration control theories of a generic deep shell continuum are presented. Open and closed-loop dynamic system equations and state equations of the continuum are formulated. Simple reduction procedures are proposed and applications to other common geometries are demonstrated in case studies.

As the function requirement increases, the geometry of the piezoelectric devices also extends from the one-dimensional (1D) DPSs (i.e., rods, beams, etc.) to two-dimensional (2D) zero-curvature DPSs (e.g., flat rectangular (Tzou and Fu, 1994), circular (Tzou and Zhou, 1995) and elliptical plates), and 2D curved DPSs (e.g., cylindrical shells, conical shells, spherical shells, etc.) (Tzou, 1993). Linear piezoelastic phenomena and theories have been studied for years (Chau, 1986; Dökmeci, 1983; Drumheller and Kalnins, 1970; Mindlin, 1972; Rogacheva, 1994; Senik and Kudriavtsev, 1980; Tzou and Gadre, 1989). An umbrella linear piezo (electric)-thermo-elastic shell theory based on a double-curvature generic shell was proposed and the derived governing equations can be easily simplified to account for many standard piezoelectric continua. Linear thermo electromechanical behaviors and precision position control of distributed piezoelectric sensors and actuators were also investigated (Koppe, et al., 1998; Ye and Tzou, 2000). As the deformation goes beyond the linear elastic range, the large deformation geometric nonlinearity also comes into consideration. Geometrical nonlinearity of elastic shells was investigated over the years. Pai, et al. (1993) proposed a nonlinear model for a piezoelectric plate. Geometrical nonlinearity of anisotropic piezothermoelastic shells simultaneously exposed to mechanical, electric, and thermal fields was investigated (Tzou and Bao, 1998; Tzou and Zhou, 1995, 1997). Tzou and Yang (2000) presented an advanced nonlinear piezo(electric)-thermoelastic shell theory, distributed sensing and control of structronic shells, and their applications to common or standard distributed structronic structures and systems, e.g., isolators, reflectors, nozzles, mirrors, antennas, disks, wing boxes, rocket fairings, flexible robots, active struts, etc. Detailed micro-thermo-electromechanics and sensing/control effectiveness of various 1D, 2D plate DPSs, and advanced shell DPSs are discussed (Tzou, Zhu and Hagiwara, 2002; Howard, et al, 2001; Tzou and Zhou, 1997). Distributed sensing and control of ring shells (Hu, Li and Tzou, 2013a; Smithmaitrie and Tzou, 2005; Tzou, Zhong and Natori, 1993), cylindrical shells (Li, et al., 2012; DeHaven, et al., 2007; Tzou, Bao and Venkayya, 1996a,b; Tzou and Zhong, 1996), spherical shells (Zhou and Tzou, 2000; Smithmaitrie and Tzou, 2004; Tzou, Smithmaitrie and Ding, 2002), conical shells (Tzou, Chai and Wang, 2003, Wang, et al, 2006; Chai, Tzou and Higuchi, 2006; Chai, DeHaven and Tzou, 2005), paraboloidal shells (Hu, Li and Tzou, 2013b; Tzou and Ding 2004; Tzou, Ding and Hagiwara, 2002; Yue, Deng and Tzou, 2008), toroidal shells (Tzou, Chai and Wang, 2004; Tzou and Wang, 2002, 2003), etc. were also evaluated. Figure 1.3 illustrates three generic double-curvature shells, i.e., 1) a multi-functional shell, 2) a smart shell structure (an elastic shell coupled with sensor and actuator layers) and 3) a composite shell with multi-sensor/actuator layers, their derivatives, e.g., shells of revolution, paraboloidal shells, toroidal shells, spherical shells, conical shells, cylindrical shells, arches, rings, plates (circular and rectangular), beams, etc., and engineering applications, e.g., antennas, reflectors, UFOs, rockets, automobiles, space structures, micro- and nano-structures, etc. Detailed discussions are presented in Chaps. 3, 4 and later chapters on sensing, energy harvesting and/or distributed control of plate and shell structures.

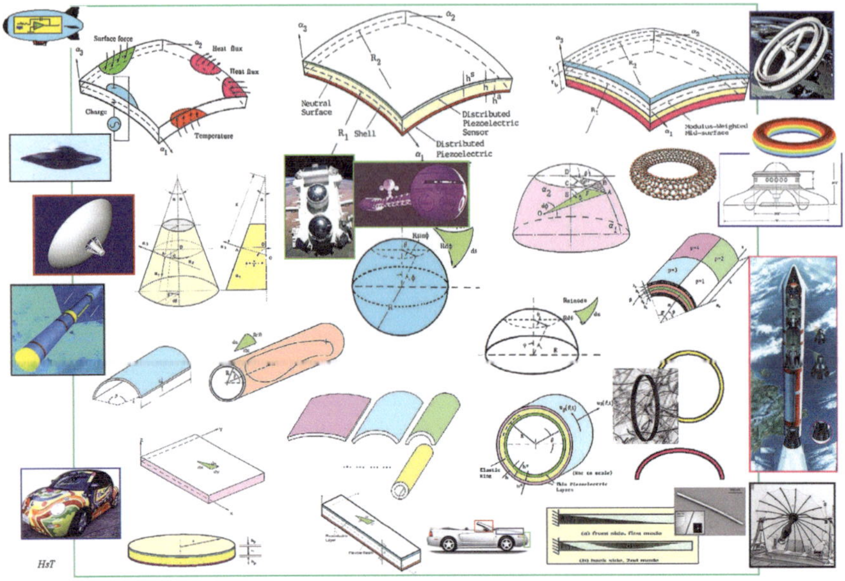

Fig. 1.3 Generic double-curvature shells, their derivatives and engineering applications

1.4 Finite Element Development

In general, experimental models are limited by size, cost, noise, and many other laboratory uncertainties. Theoretical models can be more generic; however, analytical solutions are restricted to relatively simple geometries and/or boundary conditions. Thus, finite element development becomes essential in the modeling and analysis of continua or struatures with complicated distributed piezoelectric sensors and/or actuators. Isoparametric piezoelectric finite elements have been developed and used in piezoceramic transducer designs (Allik & Hughes, 1979; Nailon et al., 1983). However, the derived isoparametric hexahedron and tetrahedral elements are too thick for thin shell/plate applications. In general, the active adaptive shell structures or structronic shells are composed of an elastic shell with coupled or embedded piezoelectric sensor/actuator layers. The thickness of the elastic structure can be about two to three orders thicker than that of the piezoelectric layers. It would be very inefficient and time-consuming if the entire active structure were modeled by the isoparametric hexahedron or tetrahedral solid elements. Distributed vibration control of a beam modeled by piezoelectric beam finite elements was investigated by Obal (1986). Distributed sensing and active vibration control of a layered plate with top and bottom piezoelectric layers was studied by Tzou and Tseng (1990). Piezoelastic and pyroelectric behaviors of laminated plates and shells were evaluated by piezo(electric)-thermo-elastic triangle shell elements (Tzou, Ye and Ding, 2001; Ye and Tzou, 2000; Tzou and Ye, 1996a, b). Detailed formulation and development of a "thin" solid piezoelectric finite element with applications to distributed sensing and control of continua is presented in Chap. 13. High-precision manipulation, modal voltages, distributed modal identification, sensing, energy harvesting and vibration control are also investigated in various chapters.

1.5 Remarks

In this book, generic double-curvature piezoelectric shell theories are derived. Generic distributed structural sensing, identification, energy harvesting and vibration control theories of a generic deep shell continuum are presented. Open and closed-loop dynamic system equations and state equations of piezoelectric structronic systems are formulated. Simple reduction procedures are proposed and applications to other common geometries are demonstrated in case studies. The revised book not only corrected typos and minor mistakes, but also added new chapters on optimal control of parabolic shells and energy harvesting of shells, including both theoretical and experimental aspects. Furthermore, laboratory and experimental components are added to, almost, all chapters on distributed sensing, energy harvesting and control of shell and non-shell structures and structronic systems. Note that performances of piezoelectric sensors/harvesters/actuators are restricted by breakdown voltages, hysteresis effects, limited strain rates, etc. These material properties need to be further improved in order to enhance the sensor/actuator performance and efficiency. Also, laboratory experiments were carried out over time; different materials with various dielectric constants from different venders were used in various studies presented in newly added Chaps. 10–12. Extreme care should be taken when repeating those studies.

It should be pointed out that all piezoelectric shell theories and distributed sensing/control and energy harvesting theories are based on a symmetrical hexagonal piezoelectric structure—class C_{6v} = 6 mm. Extension of these theories to more generic piezoelectric materials, such as a triclinic structure, would make them even more comprehensive and versatile. Besides, the temperature effect, e.g., the pyroelectricity and thermal induced stress/strains, is not considered in all studies; it should be considered when a working environment has significant temperature variations. Applications in active adaptive structures and other engineering applications can be found in references (Tzou & Anderson, 1992; Tzou & Fukuda, 1991 & 1992; Tzou, 1990 & 1992b; Mason, 1981; Sessler, 1981; Dökmeci, 1983; Tzou, Lee and Arnold, 2004; Tzou, 1993). New novel applications still need to be further explored in the near future. Furthermore, the piezoelectricity is recently extended to the *flexoelectricity* and its distributed sensing, energy harvesting and control are investigated (Tzou and Zhang, 2016; Hu, Li and Tzou, 2013a, 2014, 2015; Zhang, Li and Tzou, 2016; Tzou, Deng and Li., 2017) in recent years.

References

Adelman, N.T. and Stavsky, Y., 1975a, "Axisymmetric Vibrations of Radially Polarized Piezoelectric Ceramic Cylinders," *J. Sound and Vibration,* 38(2), pp.245-254.

Adelman, N.T. and Stavsky, Y., 1975b, "Vibrations of Radially Polarized Composite Piezoelectric Cylinders and Disks," *J. Sound and Vibration,* 43(1), pp.37-44.

Allik, H. and Hughes, T.J.R., "Finite Element Method for Piezoelectric Vibration", *Int. J. of Numerical Methods Eng.*, Vol. 2, 1979, pp.151-168.

Anderson, E.H., Hagood, N.W., and Goodliffe, J.M., 1992, "Self-sensing Piezoelectric Actuation: Analysis and Application to Controlled Structures," *AIAA Paper*: AIAA-92-2465-CP at 33 SDM Conference.

Ashley, S., 1995, "Smart Skis and Other Adaptive Structures, *Mechanical Engineering*, Vol.117, No.11, pp.76-81.

Banks, H.T. and Smith, R.C., 1998, "Numerical Techniques for Simulation, Parameter Estimation, and Noise Control in Structural Acoustic Systems," in *Dynamics and Control of Distributed Systems*, Tzou, H.S. and Bergman, L.A., (Ed), Cambridge University Press, New York, pp.202-263.

Baz A. and Poh S., 1988, "Performance of an Active Control System with Piezoelectric Actuators," *Journal of Sound and Vibration*, Vol.126, No.2, pp.327-343.

Baz A. and Ro, J., 1994, "The Concept and Performance of Active Constrained Layer Damping Treatments," *J. of Sound and Vibration*, Vol.28, No.3, pp.18-21.

Cady, W.G., *Piezoelectricity*, Dover Pub., New York, 1964.

Chai, W.K., DeHaven, J.G. and Tzou, H.S., 2005, "Distributed Micro-control Characteristics of Shallow Conical Shell Sections," *Journal of Vibration and Control*, Vol.11, No.11, pp.1397-1411.

Chai, W.K., Tzou, H.S., 2007, "Design and Testing of a Hybrid Electrostrictive/Piezoelectric Polymeric Beam with Bang-bang Control," *Mechanical Systems and Signal Processing (Journal of)*, Vol.21, pp.417-429.

Chai, W.K., Tzou, H.S., Arnold, S.M. and Lee, H.J., 2008, "Magnetostrictive Micro-actuations and Modal Sensitivities of Thin Cylindrical Magnetoelastic Shells," ASME Transactions, *Journal of Pressure Vessel Technology*, Vol.130, pp.011206-011206-5.

Chai, W.K., Tzou, H.S. and Higuchi, K., 2006, "Micro-actuation Characteristics of Conical Shell Sections," *Journal of Sound & Vibration*. Vol.293, No1-2, pp.286-298.

Chau, L.K., 1986, "The Theory of Piezoelectric Shells," *PMM U.S.S.R.*, 50(1), pp.98-105.

Crawley, E.F. and de Luis, J., 1987, "Use of Piezoelectric Actuator as Elements of Intelligent Structures," *AIAA Journal*, 25(10), pp.1373-1385.

DeHaven, J.G., Han, Y. and Tzou, 2007, "Transition of Membrane/bending Neural Signals on Transforming Adaptive Shells," *Journal of Vibration and Control*, Vol.13, No.7, pp.1007-1029.

Dökmeci, M.C., 1978, "Theory of Vibrations of Coated, Thermopiezoelectric Laminae," *J. Math. Phys.*, 19(1), January.

Dökmeci, M.C., 1980, "Vibrations of Piezoelectric Crystals," *J. Eng. Soc.*, 18, pp.431-448.

Dökmeci, M.C., 1983, "Dynamic Applications of Piezoelectric Crystals," *The Shock and Vibration Digest*, 15(3), pp.9-22.

Dosch, J.J., Inman, D.J., and Garcia, E., 1992, "A Self-sensing Piezoelectric Actuator for Collocated Control," *J. of Intell. Materl. Systems and Structures*, Vol.3, pp.166-185.

Drumheller, D.S. and Kalnins, A., 1970, "Dynamic Shell Theroy for Ferroelectric Ceramics," *J. Acoust. Soc. Am.*,47(5), pp.1343-1349.

Erturk, A. and D. J. Inman, D.J., "A Distributed Parameter Electromechanical Model for Cantilevered Piezoelectric Energy Harvesters," *Journal of Vibration and Acoustics* 130 (2008) 0410021-04100215.

Fason J.L. and Gabra, J.A., 1988, "Experimental Studies of Active Members in Control of Large Space Structures," AIAA Paper 88-2207.

Gabbert, U. and Tzou, H.S. (Edtr.), 2001, *Smart Structures and Structronic Systems,* IUTAM Symposium on Smart Structures and Structronic Systems, (ISBN 0-7923-6968-8), Kluwer Academic Publishers, Dordrecht/Boston/London.

Hanagud S. and Obal, M.W., 1988, "Identification of Dynamic Coupling Coefficients in a Structure with Piezoelectric Sensors and Actuators," AIAA paper No.88-2418.

Haskins, J.F. and Walsh, 1957, "Vibration of Ferroelectric Cylindrical Shells with Transverse Isotropy," *J. Acoust. Soc. Am.*, 29, pp.729-734.

Howard, R.V., Chai, W.K. and Tzou, H.S., 2001, "Modal Voltages of Linear and Non-Linear Structures Using Distributed Artificial Neurons," *Mechanical Systems and Signal Processing* 15, 629-640.

Hu, S.D., Li, H. and Tzou, H.S., 2013a, "Flexoelectric responses of circular rings," *Journal of Vibration and Acoustics* -Transactions of the ASME, 2013, 135(2): 021003.

Hu, S.D., Li, H. and Tzou, H.S., 2013b, "Distributed Neural Signals on Parabolic Cylindrical Shells," *Journal of Sound and Vibration* 332, 2984-3001.

Hu, S.D., H. Li, H. and Tzou, H.S., 2014, "Comparison of Flexoelectric and Piezoelectric Dynamic Signal Responses on Flexible Rings," *Journal of Intelligent Material Systems and Structures* 25, 832-844.

Hu, S.D., Li, H. and Tzou, H.S., 2015, "Distributed flexoelectric structural sensing: Theory and experiment," *Journal of Sound and Vibration*, 2015, 348(2015): 126-136.

Koppe, H., Gabbert, U., and Tzou, H.S., 1998, "On Three-Dimensional Layered Piezoelectric Shell Elements for Design Simluation of Adaptive Structures, Fortschritt-Berichte VDI, *Smart Mechanical Systems – Adaptronics*, 1998.

Lee, C.K. and Moon, F., 1988, "Modal Sensors/Actuators," IBM Report, RJ 6306 (61975), Research Division, IBM.

Lee K.M. and Arjunan, S., 1989, "A Three Degree of Freedom Micro-motion in Parallel Actuated Manipulator," *Proceedings of 1989 IEEE Intl. Conf. on Robotics and Automation*, Vol.(3), pp.1698-1703.

Li, H.Y., Li, H. and Tzou, H.S., 2015, "Frequency Control of Beams and Cylinrical Shells with Light-Activated Shape Memory Polymers," ASME *Journal of Vibration and Acoustics*, 37, p.011006.

Li, H., Hu, S. D., Tzou, H. S., and Chen, Z. B., 2012, "Optimal Vibration Control of Conical Shells with Collocated Helical Sensor/Actuator Pairs," *Journal of Theoretical and Applied Mechanics*, 50(3), pp.769-784.

Lynch, T., 1996, "Piezoelectric Damper Hones Ski Performance," Reprint from *Design News*, February 5.

Mason, W. P., 1950, *Piezoelectric Crystals and Their Application to Ultrasonics*, Nostrand, NY.

Mason, W.P., 1981, "Piezoelectricity, its History and Applications," *J. Acoust. Soc. Am.*, 70(6), Dec. 1981, pp.1561-1566.

Mindlin, R.D., 1961, "On the Equations of Motion of Piezoelectric Crystals, *Problems on Continuum Mechanics*, Radok, J., Editor, Soc. Ind. Appl. Math., Philadelphia, pp.282-290.

Mindlin, R.D., 1972, "High Frequency Vibrations of Piezoelectric Crystal Plates," Inl. J. Solid Struc., Vol.(8), pp.895-906.

Nailon, M., Coursant, R.H., and Besnier, F., 1983, "Analysis of Piezoelectric Structures by a Finite Element Method", *ACTA Electronica*, Vol.25, No.4, pp.341-362.

Obal, M.W., 1986, *Vibration Control of Flexible Structures Using Piezoelectric Devices as Sensors and Actuators*, Ph.D. Thesis, Georgia Institute of Technology.

Palazzolo, A.B., Lin, R.R., Kascak, R.R., and Alexander, R.M., 1989, "Active Control of Transient Rotordynamic Vibration by Optimal Control methods," *ASME Journal of Engineering for Gas Turbines and Power*, Vol.(111), p.265.

Paul, H.S., 1978, "Vibrations of a Hollow Circular Cylinder of Piezoelectric Ceramics," *J. Acoust. Soc. Am.*, 82(3), pp.952-956.

Paul, H.S., 1982, "Asymptotic Analysis of the Modes of Wave Propagation in a Piezoelectric Solid Cylinder," *J. Acoust. Soc. Am.*, 71(2), pp.255-263.

Paul, H.S., 1986, "Axisymmetric Vibration of a Piezoelectric Solid Cylinder Guided by a Thin Film," *J. Acoust. Soc. Am.*, 80(4), pp.1091-1096.

Paul, H.S., 1987, "Wave Propagation in a Piezoelectric Solid Cylinder of Arbitrary Cross Section," *J. Acoust. Soc. Am.* 82(6), pp.2013-2020.

Plumb, J.M., Hubbard, J.E., and Bailey, T., 1987, "Nonlinear Control of a Distributed System: Simulation and Experimental Results," ASME *Journal of Dynamic Systems, Measurements, and Control*, 109(2), pp.133-139.

Polla, D.L., 1992, "Micromachining of Piezoelectric Microsensors and Microactuators for Robotics applications," *Precision Sensors, Actuators and Systems*, Tzou and Fukuda (Ed.), Kluwer Pub., pp.139-174.

Robinson, A.C., 1971, "A Survey of Optimal Control of Distributed Parameter Systems", *Automatica* 7, pp.371-388.

Rogacheva, N.N., 1982, "Equations of State of Piezoceramic Shells," *PMM U.S.S.R.*, 45(5), pp.677-684.

Rogacheva, N.N., 1984, "On Boundary Conditions in the Theory of Piezoceramic Shells Polarized Along Coordinate Lines," *PMM U.S.S.R.*, 47(2), pp.220-226.

Rogacheva, N.N., 1986, "Classification of Free Piezoceramic Shell Vibrations," *PMM U.S.S.R.*, 50 (1), pp.106-111.

Rogacheva, N.N., 1994, *The Theory of Piezoelectric Shells and Plates*, CRC Press, Inc.

Sashida, T. and Kenjo, T., 1993, *An Introduction to Ultrasonic Motors*, Clarendon Press, Oxford.

Senik, N.A. and Kudriavtsev, B.A., 1980, "Equations on the Theory of Piezoceramic Shells," In: *Mechanics of a solid deformable body and related analytical problems. Moscow, Izd. mosk. Inst. Chim. Mashinostroeniia*, U.S.S.R.

Sessler, G.M., 1981, "Piezoelectricity in Polyvinylidene Fluoride," *J. Acoust. Soc. Am.*, 70(6), pp.1596-1608.

Shih H.R. and Tzou, H.S., 2005, "Structural Vibration Control Using Spatially Configured Opto-electromechanical Actuators," *Journal of Sound & Vibration*, Vol.284, pp.361-378.

Shih H.R. and Tzou, H.S., 2007, Photostrictive actuators for photonic control of shallow spherical shells," *Smart Materials and Structures*, Vol.16, 1712–1717.

Shih, H.R., Watkin. J. and Tzou, H.S., 2005, "Shape and Displacement Control of Beams via Photostrictive Optical Actuators," *Journal of Intelligent Structures and Material Systems*. Vol.16, pp.355-363.

Sirlin, S.W., 1987, "Vibration Isolation for Spacecraft Using the Piezoelectric Polymer PVF_2," Proc. of the 114th Meeting of the Acoustic Soc. Am., Nov.

Stavroulakis, P., 1983, *Distributed Parameter System Theory, Part 1 _ 2*, Hutchinson Ross Pub. Co. Stroudsburg, PA.

Smithmaitrie, P. and Tzou, H.S., 2004, Micro-control actions of actuator patches laminated on hemispherical shells, *Journal of Sound and Vibration*, 277, 691-710.

Smithmaitrie, P. and Tzou, H.S., 2005, "Electro-dynamics, Micro-actuation and Design of Ultrasonic Curvilinear Arc Stators," *Journal of Sound & Vibration*, Vol.284, pp.635-650.

Tiersten, H.F., 1969, *Linear Piezoelectric Plate Vibrations*, Plenum, New York.

Toupin, R.A., 1959, "Piezoelectric Relation and the Rachial Deformation of Polarized Spherical Shell," *J. Acoust. Soc. Am.*, 31(3), pp.315-318.

Tzou, H.S., 1987, "Active Vibration Control of Flexible Structures via Converse Piezoelectricity," *Development in Mechanics*, 14(b), 20th Midwest Mechanical Conference, pp.1201-1206.

Tzou, H.S., 1988, "Integrated Sensing and Adaptive Vibration Suppression of Distributed Systems," *Recent Development in Control of Nonlinear and Distributed Parameter Systems*, ASME-DSC-Vol.(10), December 1988, 51-58.

Tzou, H.S., 1989a, "Development of a Light-weight Robot End-effector using Polymeric Piezoelectric Bimorph," *Proceedings of 1989 IEEE Intl. Conf. on Robotics and Automation*, Vol.(3), pp.1704-1709.

Tzou, H.S., 1989b, "Integrated Distributed Sensing and Active Vibration Suppression of Flexible Manipulators using Distributed Piezoelectrics," *Journal of Robotic Systems*, Vol.(6.6), pp.745-767. December 1989.

Tzou, H.S., 1990, *Intelligent Piezoelectric Systems*, Industrial Technology Research Institute, Mechanical Industry Research Laboratories, Hsinchu, Taiwan, ROC.

Tzou, H.S., 1991a, "Distributed Modal Identification and Vibration Control of Continua: Theory and Applications," ASME *Journal of Dynamic Systems, Measurements, and Control*, 113(3), pp.494-499, September 1991.

Tzou, H.S., 1991b, "Distributed Piezoelectric Neurons and Muscles for Shell Continua," *Structural Vibration and Acoustics*, Ed. Huang, Tzou, et al., ASME-DE-Vol.34, pp.1-6, 1991 ASME 13th Biennial Conference on Mechanical Vibration and Noise, Symposium on Intelligent Structures and Systems, Miami, FL, September 22-25, 1991.

Tzou, H.S., 1992a, "A New Distributed Sensation and Control Theory for "Intelligent" Shells," *Journal of Sound _ Vibration*, Vol.(152), No.(3), pp.335-350, March 1992.

Tzou, H.S., 1992b, *Distributed Sensors and Actuators*, Tutorial Notes (300 pages), 1992 IEEE International Conference on Intelligent Robots and Systems (IROS'92), Raleigh, NC, July 7-10, 1992.

Tzou, H.S., 1993, *Piezoelectric Shells (Distributed Sensing and Control of Continua)*, Kluwer Academic Publishers.

Tzou, H.S. and Anderson, G.L. (Editors), 1992, *Intelligent Structural Systems*, ISBN No.0-7923-1920-6, 488 pages, Book, Kluwer Academic Publishers, Dordrecht/Boston/ London, August 1992.

Tzou, H.S., and Bao, Y., 1998, "Analysis of Nonlinear Piezothermoelastic Laminated Beams with Electric and Temperature Effects," *Journal of Sound & Vibration*, Vol.209, No.3, pp.505-518.

Tzou, H.S., Bao, Y., Venkayya, V.B., 1996a, "Study of Segmented Transducers Laminated on Cylindrical Shells, Part 1: Sensor Patches," *Journal of Sound & Vibration*, Vol.197, No.2, pp.207 224.

Tzou, H.S., Bao, Y., Venkayya, V.B., 1996b, "Study of Segmented Transducers Laminated on Cylindrical Shells, Part 2: Actuator Patches," *Journal of Sound & Vibration*, Vol.197, No.2, pp.225 249, October 1996.

Tzou, H.S., Chai, W.K. and Arnold, S.M., 2004, "Structronics and Actuation of Hybrid Electrostrictive/piezoelectric Thin Shells," ASME *Journal of Vibration and Acoustics*. Vol.128, pp.79-87.

Tzou, H.S., Chai, W.K. and Wang, D.W., 2003, "Modal Voltages and Micro-Signal Analysis of Conical Shells of Revolution," *Journal of Sound and Vibration* 260, 589-609.

Tzou, H. S., Chai, W. K., and Wang, D. W., 2004, "Micro-Control Actions and Location Sensitivity of Actuator Patches Laminated on Toroidal Shells," *Journal of Vibration and Acoustics-Transactions of the ASME*, 126(2), pp.284-297.

Tzou, H.S., Deng, B.L. and Li, H.Y., 2017, "Flexoelectric Actuation and Vibration Control of Ring Shells," ASME Transactions, *Journal of Vibration & Acoustics*, Vol.139 -031014.

Tzou, H. S., Ding, J. H., and Hagiwara, I., 2002, "Micro-Control Actions of Segmented Actuator Patches Laminated on Deep Paraboloidal Shells," *JSME International Journal Series C-Mechanical Systems Machine Elements and Manufacturing*, 45(1), pp. 8-15.

Tzou, H. S. and Ding, J. H., 2004, "Optimal Control of Precision Paraboloidal Shell Structronic Systems," *Journal of Sound and Vibration*, 276(1-2), pp.273-291.

Tzou, H.S. and Fu, H., 1994, "A Study of Segmentation of Distributed Sensors and Actuators, Part 2, Parametric Study and Vibration Controls," *Journal of Sound & Vibration*, Vol.172, No.2, pp.261 276, April 1994.

Tzou, H.S. and Fukuda, T., 1991, *Piezoelectric Smart Systems Applied to Robotics, Micro-Systems, Identification, and Control*, Workshop Notes, IEEE Robotics and Automation Society, 1991 IEEE International Conference on Robotics and Automation, Sacramento, CA, April 7-12, 1991.

Tzou, H.S. and Fukuda, T. (Editors), 1992, *Precision Sensors, Actuators, and Systems*, Book (470 pages), Kluwer Academic Publishers, December 1992.

Tzou H.S. and Gadre, M., 1988, "Active Vibration Isolation by Piezoelectric Polymer with Variable Feedback Gain," *AIAA Journal*, Vol.(26), No.8, 1014-1017.

Tzou, H.S., and Gadre, M, 1989, "Theoretical Analysis of a Multi-Layered Thin Shell Coupled with Piezoelectric Shell Actuators for Distributed Vibration Control," *Journal of Sound and Vibration*, 132(3), pp.433-450.

Tzou H.S., and Gadre, M., 1990, "Active Vibration Isolation and Excitation by a Piezoelectric Slab with Constant Feedback Gains," *Journal of Sound and Vibration*, Vol.136, No.3, pp.477-490.

Tzou H.S. and Gabbert, U., 1997, STRUCTRONICS - A New Discipline and Its Challenging Issues, Fortschritt-Berichte VDI, *Smart Mechanical Systems – Adaptronics*, Reihe 11: Schwingungstechnik Nr.244, pp.245-250, Düsseldorf, Germany.

Tzou, H.S., Guran, A. (Editors), Gabbert, U., Tani, J. and Breitbach, A., (Associate Editors), 1998, *Structronic Systems - Smart Structures, Devices, and Systems*, Volume-2: *Systems and Control*, World Scientific Publishing Co., New Jersey/Singapore.

Tzou, H.S., Zhu, Y.T. and Hagiwara, I, 2002, "Distributed Precision Control of Structronic Shells and Common Shapes– A New Approach," *Recent Research Developments in Sound & Vibration*, ISBN: 81-7895-031-6, Vol.1, Part-2, pp.613-646, TransWorld Research, 2002.

Tzou, H.S. and Hollkamp, J.J., 1994, "Collocated Independent Modal Control with Self Sensing Orthogonal Piezoelectric Actuators (Theory and Experiment)," *Journal of Smart Materials and Structures*, Vol.3, pp.277 284.

Tzou, H.S., Johnson, D., and Liu, K.J., 1999, "Damping Behavior of Cantilever Structronic Systems with Boundary Control," ASME Transactions, *Journal of Vibration & Acoustics*, pp.402-407, July 1999.

Tzou, H.S., Lee, H.J. and Arnold, S.M., 2004, "Smart Materials, Precision Sensors/Actuators, Smart Structures and Structronic Systems," *Mechanics of Advanced Materials and Structures*, Vol.11, pp.367-393.

Tzou, H. S., Smithmaitrie, P. and Ding, J. H., 2002, "Sensor Electromechanics and Distributed Signal Analysis of Piezo(Electric)-Elastic Spherical Shells," *Mechanical Systems and Signal Processing (Journal of)*, Vol.16(2-3), pp.185-199.

Tzou H.S. and Tseng, C.I., 1990, "Distributed piezoelectric sensor/actuator design for dynamic measurement/control of distributed parameter systems: a finite element approach," *Journal of Sound and Vibration*, Vol.(138), No.(1), pp.17-34.

Tzou, H.S., and Tseng, C.I., 1991, "Distributed Modal Identification and Vibration Control of Continua: Piezoelectric Finite Element Formulation and Analysis," ASME *Journal of Dynamic Systems, Measurements, and Control*, 113(3), pp.500-505, September 1991.

Tzou, H.S. and Wang, D.W., 2002, "Micro-Sensing Characteristics and Modal Voltages Of Linear/Non-Linear Toroidal Shells," *Journal of Sound and Vibration*, 2002, 254(2): 203-218.

Tzou, H.S. and Wang, D.W., 2003, "Vibration Control of Toroidal Shells with Parallel Aand Diagonal Piezoelectric Actuators," *Journal of Pressure Vessel Technology* -Transactions of the ASME, 2003, 125(2): 171-176.

Tzou H.S. and Yang, R.J., 2000, "Nonlinear Piezo-thermoelastic Shell Theory applied to Control of Variable-geometry Shells," *Journal of Theoretical and Applied Mechanics*, No.3, Vol.38, pp.623-644.

Tzou, H.S. and Ye, R., 1996a, "Analysis of Piezoelastic Systems with Laminated Piezoelectric Triangle Shell Elements," *AIAA Journal*, Vol.34, No.1, pp.110 115.

Tzou, H.S. and Ye, R., 1996b, "Pyroelectric and Thermal Strain Effects in Piezoelectric (PVDF and PZT) Devices," *Mechanical Systems and Signal Processing* (Journal of), Vol.10, No.4, pp.459 479.

Tzou, H.S., Ye, R., and J. H. Ding 2001, "A New X-Actuator Design For Controlling Wing Bending And Twisting Modes," *Journal of Sound & Vibration*, 241(2), pp.271-281.

Tzou, H.S. and Zhang, X.F., 2016, "A flexoelectric double-curvature nonlinear shell energy harvester," *Journal of Vibration and Acoustics* -Transactions of the ASME, 2016, 138(3): 031006.

Tzou, H.S., and Zhong, J.P., 1990, *"Electromechanical Dynamics of Piezoelectric Shell Distributed Systems, Part-1: Theory," Robotics Research-1990*, ASME-DSC-Vol.26, pp.207-211, 1990 ASME Winter Annual Meetings, Dallas, Texas, Nov. 25-30, 1990;

"Electromechanics and Vibrations of Piezoelectric Shell Distributed Systems," *ASME Journal of Dynamic Systems, Measurements, and Control*, 1993.
Tzou, H.S., and Zhong, J.P., 1991, "Sensor Mechanics of Distributed Shell Convolving Sensors Applied to Flexible Rings," *Structural Vibration and Acoustics*, Edrs. Huang, Tzou, et al., ASME-DE-Vol.34, pp.67-74, Symposium on Intelligent Structural Systems, 1991 ASME 13th Biennial Conference on Mechanical Vibration and Noise, Miami, Florida, September 22-25, 1991.
Tzou, H.S. and Zhong, J., 1996, "Spatially Filtered Vibration Control of Cylindrical Shells," *Shock and Vibration Journal*, Vol.3, No.4, pp.269 278.
Tzou, H.S., Zhong, J.P., and Natori, M.C., 1993, "Sensor Mechanics of Distributed Shell Convolving Sensors Applied to Flexible Rings," ASME Transactions, *Journal of Vibration & Acoustics*, Vol.115, No.1, pp.40-46.
Tzou, H.S., Zhong, J.P., and Hollkamp, J.J., 1994, "Spatially Distributed Orthogonal Piezoelectric Shell Actuators: Theory and Applications," *Journal of Sound & Vibration*, Vol.177, No.3, pp.363-378.
Tzou, H.S. and Zhou, Y., 1995, "Dynamics and Control of Nonlinear Circular Plates with Piezoelectric Actuators," *Journal of Sound & Vibration*, Vol.188, No.2, pp.189-207.
Tzou, H.S., and Zhou, Y., 1997, "Nonlinear Piezothermoelasticity and Multi field Actuations, Part 2: Control of Nonlinear Buckling and Dynamics," ASME Transactions, *Journal of Vibration & Acoustics*, Vol.119, pp.382-389.
Uchino, K., 1997, *Piezoelectric Actuators and Ultrasonic Motors*, Kluwer Academic Pub., Dordrecht/Boston/London.
Ueha, S., Tomikawa, Y., Kurosawa, M., and Nakamura, N., 1993, *Ultrasonic Motors*, Clarendon Press, Oxford.
Wang, P.K.C., 1966, "On the Feedback Control of Distributed Parameter Systems", *Int. J. on Control*, Vol. 3, No. 3, pp.255-273.
Wang, D.W., Tzou, H.S., Arnold, S.M. and Lee, H.J., 2006, "Control of Static Shape, Dynamic Oscillation and Thermally Induced Vibration of Nozzles," ASME Transactions, *Journal of Pressure Vessel Technology*, Vol.128, pp.357-363.
Ye, R. and Tzou, H.S., 2000, "Control of Adaptive Shells with Thermal and Mechanical Excitations," *Journal of Sound & Vibration*, Vol.231(5), pp.1321-1338.
Yue, H. H., Deng, Z. Q., and Tzou, H.S., 2008, "Optimal Actuator Locations and Precision Micro-Control Actions on Free Paraboloidal Membrane Shells," *Communications in Nonlinear Science and Numerical Simulation*, 13(10), pp.2298-2307.
Zhou, Y.H., and Tzou, H.S., 2000, "Active Control of Nonlinear Piezoelectric Circular Shallow Spherical Shells," *International Journal of Solids and Structures*, 37(12), pp.1663-1677.
Zhang, X.F., Hu, S.D. and Tzou, H.S., 2014, A generic double-curvature piezoelectric shell energy harvester: Linear/nonlinear theory and applications, *Journal of Sound and Vibration,* 333, 7286-7298
Zhang, X.F., Li, H.Y., and Tzou, H.S., 2016, "Analytical and Experimental Studies of Flexoelectric Beam Control," ASME 2016 International Mechanical Engineering Congress & Exposition, IMECE2016, Paper No. IMECE2016-66527, Phoenix, Arizona, USA.

Chapter 2
Piezoelectric Shell Vibrations

Active piezoelectric structures capable of self-adaptation (Tzou & Anderson, 1992) and high-precision manipulations (Tzou & Fukuda, 1992) have drawn much attention in recent years. In this chapter, generic electromechanically coupled vibration theories of deep piezoelectric shell continua are derived. The system equations include both mechanical and electric components. The mechanical components are related to conventional elastic vibrations of shells; the electric components are electromechanical coupling effects induced by the piezoelectricity. Eliminating these electromechanical coupling terms from the generic piezoelectric system equations yields a set of conventional vibration equations for elastic shell continua. Note that the electromechanical coupling terms can also be used in sensor and actuator applications applied to distributed sensing, identification, energy harvesting and precision control of shell and plate structures. (Detailed discussions will be presented in later chapters.) A background introduction and a brief review of the subject area are presented first. Detailed derivations of the piezoelectric shell theory using Hamilton's principle are presented next. Simplification of the piezoelectric shell vibration theory to the conventional elastic shell vibration theory is also discussed. Applications of the generic theories to commonly occurring geometries and structures, e.g., spherical shells, cylindrical shells, plates, etc., and distributed control of piezoelectric shells are demonstrated in Chap. 3. Energy generation and harvesting of piezoelectric laminated linear and nonlinear structures is presented in Chaps. 11 and 12.

Piezoelectricity is an electromechanical phenomenon coupling the electric and elastic fields in piezoelectric materials, which was first investigated by the Curie brothers, Jacques and Pierre, in 1880 (Cady, 1946; Manson, 1950). Due to the distinct electromechanical characteristics, engineering applications are constantly being explored and detailed electromechanical coupling behaviors investigated. Earlier studies were mostly concerned with the waves and vibrations of specific geometries, such as thin rods and plates with either finite or infinite dimensions. Mindlin (1955), Tiersten and Mindlin (1962) proposed a three-dimensional theory of elastodynamics and applied it to vibrations of piezoelectric crystal plates and rods. Eer Nisse (1966), Holland (1968), Holland and Eer Nisse (1969) proposed a

H. (HS) Tzou, *Piezoelectric Shells*, Solid Mechanics and Its Applications 247,
https://doi.org/10.1007/978-94-024-1258-1_2

variation technique and applied the piezoelectrics to engineering devices. Paul (1968), Bleustain (1969), Kagawa and Yamabuchi (1976), Schmidt (1972) and Keuning (1972) also studied the electromechanics and waves of (thin or thick) piezoelectric crystal plates.

Electromechanics and wave propagations of piezoelectric shells with well defined geometries and boundary conditions have been investigated. Haskins and Walsh (1957) studied vibrations of piezoelectric cylindrical shells with transverse isotropy. Toupin (1959) derived the equilibrium relations for a polarized spherical shell. Based on the classical shell theory, Drumheller and Kalnins (1970) proposed a vibration theory for piezoelectric shells of revolution and derived electrostatic and elasticity equations. Adelman and Stavsky studied vibrations of radially polarized composite piezoceramic cylinders and disks (1975a), axisymmetric vibrations of radially polarized piezoelectric ceramic cylinders (1975b), etc. Paul investigated axisymmetric vibrations of a piezoelectric solid cylinder guided by a thin film (Paul 1986), modes of wave propagation in a piezoelectric solid cylinder (1982), wave propagation in a piezoelectric solid cylinder of arbitrary cross section (1987), vibrations of a hollow circular cylinder of piezoelectric ceramics (1978), etc. Dökmeci (1980) studied 1) radial vibrations induced by asymmetrical conditions in either infinitesimally thin or infinitely long transducers, 2) radial, torsional, axisymmetric and flexural vibrations of spherical and circular cylindrical shells polarized in either axial or radial direction, 3) radial oscillations, both extensional and shear modes of piezoelectric crystals. Dökmeci (1978) also proposed a vibration theory of coated thermopiezoelectric laminae.

An electromechanical theory of a generic deep piezoelectric shell continuum is essential to provide an "umbrella" theory encompassing a variety of commonly occurring geometries, such as beams, circular plates, rectangular plates, rings, cylindrical shells, spherical shells, conical shells, paraboloidal shells, etc. (Tzou, 1992a, b; Tzou, 2000; Tzou, Zhu and Hagiwara, 2002). Rogacheva (1982, 1984a & b, 1986) derived state equations and boundary conditions for piezoelectric shells polarized along coordinate directions. Senik and Kudriavtsev (1980) formulated equations of motion for piezoelectric shells polarized normal to the shell middle surface using the variation principle of electroelasticity and the quadratic approximation of the electric field potential along the thickness coordinate. Chau (1986) proposed a variation formulation to study the equilibrium problem of completely anisotropic piezoelectric shells. Tzou (1991, 1992a, b) derived a general distributed sensing and control theory for a generic elastic shell continuum coupled with piezoelectric thin layers. Although studies of generic piezoelectric shells have been advanced in the last decade, wave propagations were the primary emphases. Electromechanics and vibration behaviors encompassing all three principal directions of generic piezoelectric shell continua and applications to distributed sensing, identification, energy harvesting and control still need to be explored.

In this chapter, electromechanical vibration theories of generic piezoelectric shell continua are derived based on the linear piezoelectricity theory and Hamilton's principle (Tzou and Zhong, 1990 & 1991b). The generic equations of mechanical motion, mechanical boundary conditions, the charge equation of electrostatics, and electric boundary conditions are derived. The system equations are very general; they can be directly simplified to apply to a broad class of piezoelectric shell/non-shell continua, e.g., cylinders, spheres, plates, arches, rings, beams, rods, etc., and conventional elastic shell/non-shell continua. The system equations can be easily reduced to conventional shell vibration equations if all electric coupling terms are removed. This general subject area is divided into two chapters: Chap. 2—derivations of generic piezoelectric shell theories and Chap. 3—applications to common piezoelectric continua and active piezoelectric structures.

In Chap. 2, derivations of the electromechanical vibration theories for generic piezoelectric shell continua with a symmetrical hexagonal structure (Class C_{6v} = 6 mm) are presented. Three piezoelectric shell continua are considered here:

1) A thick generic piezoelectric shell continuum (in which deflections U_js are defined in a generic form),
2) A thick piezoelectric shell with the effects of transverse shear deformations and rotatory inertias, and
3) A thin piezoelectric shell based on the Kirchhoff-Love hypothesis in which the effects resulting from transverse shear deformations and rotatory inertias are neglected.

Fundamental physics and definitions are reviewed first. Detailed derivations and system equations are presented afterwards.

2.1 Fundamentals

In this section, system definitions, fundamental physics and theories are briefly reviewed. Note that the derivation is based on the linear piezoelectricity theory and Hamilton's principle. Both mechanical and electric energies are considered in the system energy expressions. Taking the variation calculation of the total energy and work, one can derive the electromechanical equations of mechanical motion, mechanical boundary conditions, charge equations of electrostatics, and electric boundary conditions.

2.1.1 Definition of a Generic Piezoelectric Shell Continuum

Figure 2.1 shows a generic piezoelectric shell continuum defined in a tri-orthogonal curvilinear coordinate system with α_1 and α_2 defining the shell neutral surface and α_3 the normal direction. The piezoelectric shell continuum has a constant thickness h which is relatively thin with respect to its radii of curvature (R_1 and R_2). Generic deflections (U_1,U_2, and U_3) in three principal directions (α_1, α_2, and α_3) are assumed to be small and linear. It is also assumed that the piezoelectric shell continuum has a symmetrical hexagonal structure (Class $\mathbf{C}_{6v}$ = 6 mm).

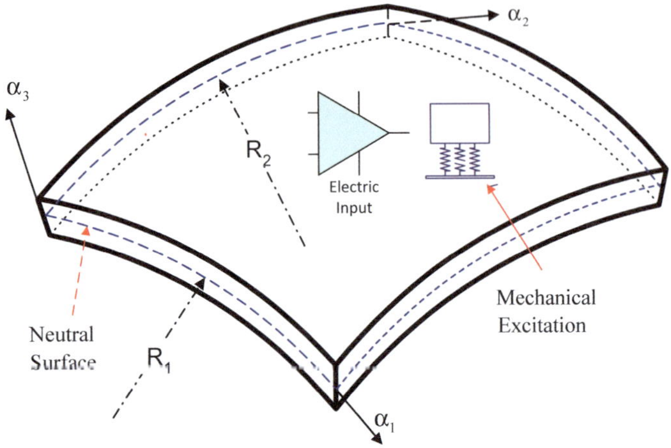

Fig. 2.1 A generic piezoelectric shell continuum

2.1.2 Physical Laws

In this section, a number of fundamental physical laws used in derivations are discussed. Hamilton's principle is the fundamental in theoretical derivations and it is written as (Tzou and Zhong, 1990 & 1991b)

$$\delta \int_{t_0}^{t_1} dt[\mathcal{K} - \mathcal{U}] = 0, \tag{2.1.1}$$

where $\mathcal{K}$ is the kinetic energy; $\mathcal{U}$ is the total potential energy (including the mechanical energy, electric energy, and work done by externally applied forces and charges); δ denotes the variation. For a piezoelectric continuum subjected to a prescribed surface traction t and a surface charge per unit area Q, Hamilton's principle states:

$$\begin{aligned} &\delta \int_{t0}^{t1} dt \int_V \left[\frac{1}{2}\rho \dot{U}_j \dot{U}_j\right] dV - \left\{\delta \int_{t0}^{t1} dt \int_V [\mathcal{H}(S_{ij}, E_j)] dV \right. \\ &\left. - \int_{t0}^{t1} dt \int_S (t_j \delta U_j - Q_j \delta\varphi) dS \right\} = 0 \end{aligned}, \tag{2.1.2}$$

where $\mathcal{H}(S_{ij}, E_j)$ is the electric enthalpy; ρ is the mass density; U_j is the deflection in the α_j direction; S_{ij} is the mechanical strain of the i-th surface and the j-th direction; E_j is the electric field strength in the α_j direction; t_j is the surface traction in the α_j direction; Q_j is the surface charge; φ is the electric potential; V is the piezoelectric volume considered; S is the surface over the volume. Electric field E_j and potential

φ relations in the curvilinear coordinate system are defined by the Maxwell equation:

$$E_1 = -\frac{1}{A_1(1+\frac{\alpha_3}{R_1})}\frac{\partial \varphi}{\partial \alpha_1}, \tag{2.1.3}$$

$$E_2 = -\frac{1}{A_2(1+\frac{\alpha_3}{R_2})}\frac{\partial \varphi}{\partial \alpha_2}, \tag{2.1.4}$$

$$E_3 = -\frac{\partial \varphi}{\partial \alpha_3}, \tag{2.1.5}$$

where A_1 and A_2 are the Lamé parameters and R_1 and R_2 are the radii of curvature of the α_1 and α_2 axes, respectively. (These will be defined later.) The electric enthalpy $\mathcal{H}(\mathbf{S}_{ij}, E_j)$ for the piezoelectric continuum is written as

$$\mathcal{H} = \frac{1}{2}\{S_{ij}\}^t[c]\{S_{ij}\} - \{S_{ij}\}^t[e]^t\{E_j\} - \frac{1}{2}\{E_j\}^t[\varepsilon]\{E_j\} \tag{2.1.6}$$

where [c] is the elastic constant matrix; [e] is the piezoelectric constant matrix; $[\varepsilon]$ is the dielectric constant matrix. In general, the linear piezoelectric relations of a piezoelectric continuum can be described as

$$\{\mathbf{T}_{ij}\} = [c]\{\mathbf{S}_{ij}\} - [e]^t\{\mathbf{E}_j\}, \tag{2.1.7}$$

$$\{\mathbf{D}_j\} = [e]\{\mathbf{S}_{ij}\} + [\varepsilon]\{\mathbf{E}_j\}, \tag{2.1.8}$$

where $\{\mathbf{T}_{ij}\}$ is the stress vector induced by two effects: 1) mechanical and 2) electric effects; $\{\mathbf{D}_j\}$ is the electric displacement vector. (Note that $\{\sigma_{ij}\}$ denotes conventional mechanical stress defined later.) Eq. (2.1.7) denotes the ***converse*** piezoelectric effect and Eq. (2.1.8) the ***direct*** piezoelectric effect. A piezoelectric material with a symmetrical hexagonal structure (Class C_{6v} = 6 mm) is isotropic in the transverse direction α_3, but is anisotropic in the α_1 and α_2 directions. If it is polarized in the thickness direction, [c], [e], and $[\varepsilon]$ matrices are defined as (Tiersten, 1969)

$$[c] = \begin{bmatrix} c_{11} & c_{12} & c_{13} & 0 & 0 & 0 \\ c_{12} & c_{11} & c_{13} & 0 & 0 & 0 \\ c_{13} & c_{13} & c_{33} & 0 & 0 & 0 \\ 0 & 0 & 0 & c_{44} & 0 & 0 \\ 0 & 0 & 0 & 0 & c_{44} & 0 \\ 0 & 0 & 0 & 0 & 0 & c_{66} \end{bmatrix}, \tag{2.1.9}$$

$$[e] = \begin{bmatrix} 0 & 0 & 0 & 0 & e_{15} & 0 \\ 0 & 0 & 0 & e_{24} & 0 & 0 \\ e_{31} & e_{32} & e_{33} & 0 & 0 & 0 \end{bmatrix}, \tag{2.1.10}$$

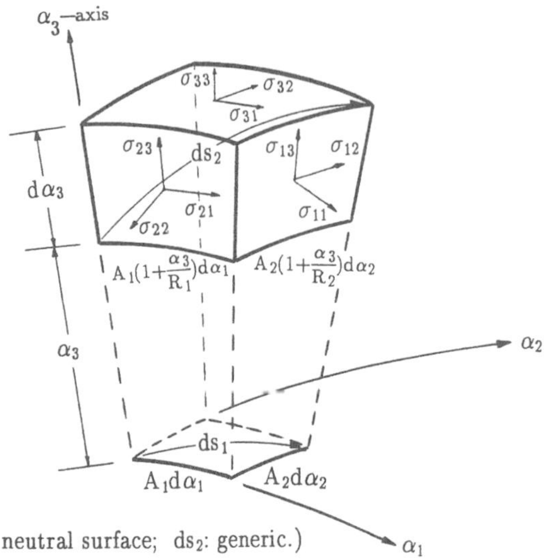

Fig. 2.2 Definition of mechanical stresses and strains

$$[\varepsilon] = \begin{bmatrix} \varepsilon_{11} & 0 & 0 \\ 0 & \varepsilon_{11} & 0 \\ 0 & 0 & \varepsilon_{S33} \end{bmatrix}, \tag{2.1.11}$$

where $c_{66} = (c_{11} - c_{12})/2$. Note that $e_{31} = e_{32}$ for a 6 mm structure. If a piezoelectric material is electrically polarized, but is not mechanical stretched in the process, $e_{24} = e_{15}$. Based on these matrices, the enthalpy $\mathcal{H}(\mathbf{S}_{ij}, E_j)$ can be written as

$$\begin{aligned} \mathcal{H} = & \frac{1}{2}(\sigma_{11}S_{11} + \sigma_{22}S_{22} + \sigma_{12}S_{12} + \sigma_{13}S_{13} + \sigma_{23}S_{23} + \sigma_{33}S_{33}) \\ & - (e_{15}E_1S_{13} + e_{15}E_2S_{23} + e_{31}E_3S_{11} + e_{31}E_3S_{22} + e_{33}E_3S_{33}) \\ & - \frac{1}{2}\left(\varepsilon_{11}E_1^2 + \varepsilon_{11}E_2^2 + \varepsilon_{33}E_3^2\right) \end{aligned} \tag{2.1.12}$$

Figure 2.2 illustrates the components of mechanical stresses and strains in a curvilinear coordinate system.

The mechanical stress $\boldsymbol{\sigma}_{ij}$ and mechanical strain $\mathbf{S}_{ij}$ relations are

1) Normal Stresses:

$$\sigma_{11} = c_{11}S_{11} + c_{12}S_{22} + c_{13}S_{33}, \tag{2.1.13}$$

$$\sigma_{22} = c_{12}S_{11} + c_{11}S_{22} + c_{13}S_{33}, \tag{2.1.14}$$

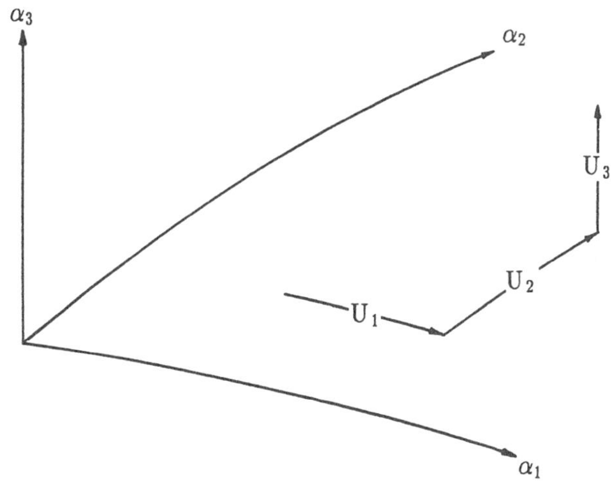

Fig. 2.3 Definition of generic displacements U_j

$$\sigma_{33} = c_{13}S_{11} + c_{13}S_{22} + c_{33}S_{33}, \qquad (2.1.15)$$

2) Shear Stresses:

$$\sigma_{12} = c_{66}S_{12}, \qquad (2.1.16)$$

$$\sigma_{13} = c_{44}S_{13}, \qquad (2.1.17)$$

$$\sigma_{23} = c_{44}S_{23}. \qquad (2.1.18)$$

The strain-displacement relations defined in terms of deflections U_1, U_2, and U_3, Fig. 2.3, are (Soedel, 1981):

1) Normal Strains:

$$S_{11} = \frac{1}{A_1(1+\frac{\alpha_3}{R_1})}[\frac{\partial U_1}{\partial \alpha_1} + \frac{U_2}{A_2}\frac{\partial A_1}{\partial \alpha_2} + U_3\frac{A_1}{R_1}], \qquad (2.1.19)$$

$$S_{22} = \frac{1}{A_2(1+\frac{\alpha_3}{R_2})}[\frac{\partial U_2}{\partial \alpha_2} + \frac{U_1}{A_1}\frac{\partial A_2}{\partial \alpha_1} + U_3\frac{A_2}{R_2}], \qquad (2.1.20)$$

$$S_{33} = \frac{\partial U_3}{\partial \alpha_3}, \qquad (2.1.21)$$

2) Shear Strains:

$$S_{12} = \frac{A_1(1+\frac{\alpha_3}{R_1})}{A_2(1+\frac{\alpha_3}{R_2})}\frac{\partial}{\partial\alpha_2}\left[\frac{U_1}{A_1(1+\frac{\alpha_3}{R_1})}\right] + \frac{A_2(1+\frac{\alpha_3}{R_2})}{A_1(1+\frac{\alpha_3}{R_1})}\frac{\partial}{\partial\alpha_1}\left[\frac{U_2}{A_2(1+\frac{\alpha_3}{R_2})}\right], \quad (2.1.22)$$

$$S_{13} = A_1(1+\frac{\alpha_3}{R_1})\frac{\partial}{\partial\alpha_3}\left[\frac{U_1}{A_1(1+\frac{\alpha_3}{R_1})}\right] + \frac{1}{A_1(1+\frac{\alpha_3}{R_1})}\frac{\partial U_3}{\partial\alpha_1}, \quad (2.1.23)$$

$$S_{23} = A_2(1+\frac{\alpha_3}{R_2})\frac{\partial}{\partial\alpha_3}\left[\frac{U_2}{A_2(1+\frac{\alpha_3}{R_2})}\right] + \frac{1}{A_2(1+\frac{\alpha_3}{R_2})}\frac{\partial U_3}{\partial\alpha_2}. \quad (2.1.24)$$

Note that ***Codazzi's relations*** were used in the strain derivations (Soedel, 1981); these are defined as:

$$\frac{\partial}{\partial\alpha_2}\left[A_1(1+\frac{\alpha_3}{R_1})\right] = (1+\frac{\alpha_3}{R_2})\frac{\partial A_1}{\partial\alpha_2}, \quad (2.1.25a)$$

$$\frac{\partial}{\partial\alpha_1}\left[A_2(1+\frac{\alpha_3}{R_2})\right] = (1+\frac{\alpha_3}{R_1})\frac{\partial A_2}{\partial\alpha_1}. \quad (2.1.25b)$$

Substituting Eqs. (2.1.10) & (2.1.11) into Eq. (2.1.8) gives

$$E_1 = \frac{1}{\varepsilon_{11}}D_1 - \frac{e_{15}}{\varepsilon_{11}}S_{13} = E_1^e + E_1^d, \quad (2.1.26)$$

$$E_2 = \frac{1}{\varepsilon_{11}}D_2 - \frac{e_{15}}{\varepsilon_{11}}S_{23} = E_2^e + E_2^d, \quad (2.1.27)$$

$$E_3 = \frac{1}{\varepsilon_{33}}D_3 - \frac{e_{31}S_{11} + e_{31}S_{22} + e_{33}S_{33}}{\varepsilon_{33}} = E_3^e + E_3^d, \quad (2.1.28)$$

where E_j^e denotes the electric field induced by an electric displacement; E_j^d denotes the electric field induced by the direct piezoelectric effect (j = 1,2,3). These two separate effects are further defined as

$$E_j^e = \frac{D_j}{\varepsilon_{jj}}, (j = 1, 2, 3), \quad (2.1.29)$$

$$E_1^d = -\frac{e_{15}}{\varepsilon_{11}}S_{13}, \quad (2.1.30)$$

$$E_2^d = -\frac{e_{15}}{\varepsilon_{11}}S_{23}, \quad (2.1.31)$$

$$E_3^d = -\frac{e_{31}S_{11} + e_{31}S_{22} + e_{33}S_{33}}{\varepsilon_{33}}. \tag{2.1.32}$$

These fundamental definitions and mechanical/electric relations will be used in derivations of piezoelectric shell theories.

2.2 Thick Piezoelectric Shell Continuum

When a "thick" piezoelectric deep shell continuum is considered, transverse shear deformation and rotatory inertia effects need to be considered in the derivation. Note that the generic displacement U_j (j = 1,2,3) includes both in-plane deformations (membrane components) and out-of-plane rotations (bending components). These two components will be further separated in later derivations. The electric fields E_j's in all three directions are considered in this case.

In order to derive the electromechanical system equations and mechanical/electric boundary conditions of the piezoelectric shell continuum, one needs to carry out all variations in Eq. (2.1.2) and collect coefficients of similar terms. Because of tedious and complicated procedures, only crucial steps in the derivations are outlined. The variation of kinetic energy is presented first and followed by energies associated with electric enthalpy $\mathcal{H}(\mathbf{S}_{ij}, E_j)$, traction forces t_j, and electric charges Q_j. A final variation equation is derived, which leads to all electromechanical system equations and boundary conditions (Tzou and Zhong, 1991b).

2.2.1 *Variation of Kinetic Energy*

The variation of kinetic energy $\mathcal{K}$ is

$$\begin{aligned}
&\delta \int_{t_0}^{t_1} dt \int_{\alpha_1}\int_{\alpha_2}\int_{\alpha_3} \mathcal{K} dV \\
&= \delta \int_{t_0}^{t_1} dt \int_{\alpha_1}\int_{\alpha_2}\int_{\alpha_3} \left(\tfrac{1}{2}\rho \dot{U}_j \dot{U}_j\right) A_1 A_2 \left(1 + \tfrac{\alpha_3}{R_1}\right)\left(1 + \tfrac{\alpha_3}{R_2}\right) d\alpha_1 d\alpha_2 d\alpha_3 \\
&= -\rho \int_{t_0}^{t_1} dt \int_V \left(\ddot{U}_1 \delta U_1 + \ddot{U}_2 \delta U_2 + \ddot{U}_3 \delta U_3\right) A_1 A_2 \left(1 + \tfrac{\alpha_3}{R_1}\right)\left(1 + \tfrac{\alpha_3}{R_2}\right) d\alpha_1 d\alpha_2 d\alpha_3
\end{aligned} \tag{2.2.1}$$

Note that integration by parts was used in the kinetic energy variation.

2.2.2 Variation of Electric Enthalpy

This variation includes two components as described in Eq. (2.1.6), i.e., a) the mechanical strains S_{ij} and b) the electric fields E_j which are presented separately.

a) Variation of Mechanical Strain Energy

There are three normal strains and three shear strains needed to be considered in the derivation. Carrying out all variations of mechanical strain energy is very tedious and time consuming. Using the variation of S_{11} as an example, one can derive

$$\begin{aligned}
&\frac{\partial \mathcal{H}}{\partial S_{11}}\delta S_{11}\\
&=\frac{1}{2}[(c_{11}S_{11}+c_{12}S_{22}+c_{13}S_{33})\delta S_{11}+c_{11}S_{11}\delta S_{11}+c_{12}S_{22}\delta S_{11}\\
&\quad+c_{13}S_{33}\delta S_{11}]-e_{31}E_3\delta S_{11}\\
&=\frac{1}{2}[(c_{11}S_{11}+c_{12}S_{22}+c_{13}S_{33})\delta S_{11}+(c_{11}S_{11}+c_{12}S_{22}+c_{13}S_{33})\delta S_{11}]\\
&\quad-e_{31}E_3\delta S_{11}\\
&=(c_{11}S_{11}+c_{12}S_{22}+c_{13}S_{33})\delta S_{11}-e_{31}E_3\delta S_{11}\\
&=(\sigma_{11}-e_{31}E_3)\delta S_{11}
\end{aligned} \quad . \quad (2.2.2)$$

Proceeding each term in Eq. (2.1.16) in the same fashion gives

$$\begin{aligned}
&\frac{\partial \mathcal{H}}{\partial S_{ij}}\delta S_{ij}\\
&\quad=(\sigma_{11}-e_{31}E_3)\delta S_{11}+(\sigma_{22}-e_{31}E_3)\delta S_{22}+(\sigma_{33}-e_{33}E_3)\delta S_{33}\\
&\qquad+\sigma_{12}\delta S_{12}+(\sigma_{13}-e_{15}E_1)\delta S_{13}+(\sigma_{23}-e_{15}E_2)\delta S_{23}
\end{aligned} \quad . \quad (2.2.3)$$

$$\begin{aligned}
&\int_{t_0}^{t_1}dt\int_V\frac{\partial \mathcal{H}}{\partial S_{ij}}\delta S_{ij}dV\\
&=\int_{t_0}^{t_1}dt\int_V[(\sigma_{11}-e_{31}E_3)\delta S_{11}+(\sigma_{22}-e_{31}E_3)\delta S_{22}+(\sigma_{33}-e_{33}E_3)\delta S_{33}\\
&\quad+\sigma_{12}\delta S_{12}+(\sigma_{13}-e_{15}E_1)\delta S_{13}+(\sigma_{23}-e_{15}E_2)\delta S_{23}]\\
&\quad\cdot A_1A_2\left(1+\frac{\alpha_3}{R_1}\right)\left(1+\frac{\alpha_3}{R_2}\right)d\alpha_1 d\alpha_2 d\alpha_3
\end{aligned} \quad . \quad (2.2.4)$$

Variation of each strain component, six in total, also needs to be carried out respectively. These strain variations are presented as follows.

1) δS_{11} Term:

$$
\begin{aligned}
&\int_{t_0}^{t_1} dt \int_V (\sigma_{11} - e_{31}E_3)\delta S_{11} A_1 A_2 (1+\frac{\alpha_3}{R_1})(1+\frac{\alpha_3}{R_2}) d\alpha_1 d\alpha_2 d\alpha_3 \\
&= \int_{t_0}^{t_1} dt \int_V (\sigma_{11} - e_{31}E_3) A_2 (1+\frac{\alpha_3}{R_2}) \left[\frac{\partial(\delta U_1)}{\partial \alpha_1} + \frac{\delta U_2}{A_2}\frac{\partial A_1}{\partial \alpha_2} + \delta U_3 \frac{A_1}{R_1}\right] d\alpha_1 d\alpha_2 d\alpha_3 \\
&= \int_{t_0}^{t_1} dt \int_V \left\{ -\frac{\partial[(\sigma_{11} - e_{31}E_3) A_2 (1+\frac{\alpha_3}{R_2})]}{\partial \alpha_1} \delta U_1 + (\sigma_{11} - e_{31}E_3)(1+\frac{\alpha_3}{R_2})\frac{\partial A_1}{\partial \alpha_2}\delta U_2 \right. . \\
&\left. + (\sigma_{11} - e_{31}E_3)\frac{A_1 A_2 (1+\frac{\alpha_3}{R_2})}{R_1}\delta U_3 \right\} d\alpha_1 d\alpha_2 d\alpha_3 \\
&+ \int_{t_0}^{t_1} dt \int_{S_1} (\sigma_{11} - e_{31}E_3) A_2 (1+\frac{\alpha_3}{R_2}) \delta U_1 d\alpha_2 d\alpha_3
\end{aligned}
\tag{2.2.5}
$$

2) δS_{22} Term:

$$
\begin{aligned}
&\int_{t_0}^{t_1} dt \int_V (\sigma_{22} - e_{31}E_3)\delta S_{22} A_1 A_2 (1+\frac{\alpha_3}{R_1})(1+\frac{\alpha_3}{R_2}) d\alpha_1 d\alpha_2 d\alpha_3 \\
&= \int_{t_0}^{t_1} dt \int_V (\sigma_{22} - e_{31}E_3) A_1 (1+\frac{\alpha_3}{R_1}) \left[\frac{\partial(\delta U_2)}{\partial \alpha_2} + \frac{\delta U_1}{A_1}\frac{\partial A_2}{\partial \alpha_1} \right. \\
&\left. + \delta U_3 \frac{A_2}{R_2}\right] d\alpha_1 d\alpha_2 d\alpha_3 \\
&= \int_{t_0}^{t_1} dt \int_V \left\{ -\frac{\partial[(\sigma_{22} - e_{31}E_3) A_1 (1+\frac{\alpha_3}{R_1})]}{\partial \alpha_2} \delta U_2 + (\sigma_{22} - e_{31}E_3)(1+\frac{\alpha_3}{R_1})\frac{\partial A_2}{\partial \alpha_1}\delta U_1 \right. \\
&\left. + (\sigma_{22} - e_{31}E_3)\frac{A_1 A_2 (1+\frac{\alpha_3}{R_1})}{R_2}\delta U_3 \right\} d\alpha_1 d\alpha_2 d\alpha_3 \\
&+ \int_{t_0}^{t_1} dt \int_{S_2} (\sigma_{22} - e_{31}E_3) A_1 (1+\frac{\alpha_3}{R_1}) \delta U_2 d\alpha_1 d\alpha_3 .
\end{aligned}
\tag{2.2.6}
$$

3) δS_{33} Term:

$$
\begin{aligned}
&\int_{t_0}^{t_1} dt \int_V (\sigma_{33} - e_{33}E_3)\delta S_{33} A_1 A_2 (1+\frac{\alpha_3}{R_1})(1+\frac{\alpha_3}{R_2}) d\alpha_1 d\alpha_2 d\alpha_3 \\
&= \int_{t_0}^{t_1} dt \int_V (\sigma_{33} - e_{33}E_3) A_1 A_2 (1+\frac{\alpha_3}{R_1})(1+\frac{\alpha_3}{R_2}) \frac{\partial(\delta U_3)}{\partial \alpha_3} d\alpha_1 d\alpha_2 d\alpha_3 \\
&= \int_{t_0}^{t_1} dt \int_V \left\{ -\frac{\partial[(\sigma_{33} - e_{33}E_3) A_1 A_2 (1+\frac{\alpha_3}{R_1})(1+\frac{\alpha_3}{R_2})]}{\partial \alpha_3} \right\} \delta U_3 d\alpha_1 d\alpha_2 d\alpha_3 \\
&\quad + \int_{t_0}^{t_1} dt \int_{S_3} (\sigma_{33} - e_{33}E_3) A_1 A_2 (1+\frac{\alpha_3}{R_1})(1+\frac{\alpha_3}{R_2}) \delta U_3 d\alpha_1 d\alpha_2
\end{aligned}
\tag{2.2.7}
$$

4) δS_{12} Term:

$$
\begin{aligned}
&\int_{t_0}^{t_1} dt \int_V \sigma_{12}\delta S_{12} A_1 A_2 (1+\frac{\alpha_3}{R_1})(1+\frac{\alpha_3}{R_2}) d\alpha_1 d\alpha_2 d\alpha_3 \\
&= \int_{t_0}^{t_1} dt \int_V \sigma_{12} \left\{ A_1^2 (1+\frac{\alpha_3}{R_1})^2 \frac{\partial}{\partial \alpha_2} [\frac{\delta U_1}{A_1(1+\frac{\alpha_3}{R_1})}] \right. \\
&\quad \left. + A_2^2 (1+\frac{\alpha_3}{R_2})^2 \frac{\partial}{\partial \alpha_1} [\frac{\delta U_2}{A_2(1+\frac{\alpha_3}{R_2})}] \right\} d\alpha_1 d\alpha_2 d\alpha_3 \\
&= \int_{t_0}^{t_1} dt \int_V \left[\sigma_{12} A_1 (1+\frac{\alpha_3}{R_1}) \frac{\partial(\delta U_1)}{\partial \alpha_2} - \sigma_{12}(1+\frac{\alpha_3}{R_2}) \frac{\partial A_1}{\partial \alpha_2} \delta U_1 \right. \\
&\quad \left. + \sigma_{12} A_2 (1+\frac{\alpha_3}{R_2}) \frac{\partial(\delta U_2)}{\partial \alpha_1} - \sigma_{12}(1+\frac{\alpha_3}{R_1}) \frac{\partial A_2}{\partial \alpha_1} \delta U_2 \right] d\alpha_1 d\alpha_2 d\alpha_3 \\
&= \int_{t_0}^{t_1} dt \int_V \left\{ \left[-\frac{\partial[\sigma_{12} A_1 (1+\frac{\alpha_3}{R_1})]}{\partial \alpha_2} - \sigma_{12}(1+\frac{\alpha_3}{R_2}) \frac{\partial A_1}{\partial \alpha_2} \right] \delta U_1 \right. \\
&\quad \left. + \left[-\frac{\partial[\sigma_{12} A_2 (1+\frac{\alpha_3}{R_2})]}{\partial \alpha_1} - \sigma_{12}(1+\frac{\alpha_3}{R_1}) \frac{\partial A_2}{\partial \alpha_1} \right] \delta U_2 \right\} d\alpha_1 d\alpha_2 d\alpha_3 \\
&\quad + \int_{t_0}^{t_1} dt \int_{S_1} \sigma_{12} A_2 (1+\frac{\alpha_3}{R_2}) \delta U_2 d\alpha_2 d\alpha_3 + \int_{t_0}^{t_1} dt \int_{S_2} \sigma_{12} A_1 (1+\frac{\alpha_3}{R_1}) \delta U_1 d\alpha_1 d\alpha_3
\end{aligned}
\tag{2.2.8}
$$

Note that ***Codazzi's relations***, Eqs. (2.1.25a & 2.1.25b), were used in the above derivations (Soedel, 1981).

5) δS_{13} Term:

$$
\begin{aligned}
&\int_{t0}^{t1} dt \int_V (\sigma_{13} - e_{15}E_1)\delta S_{13} A_1 A_2 (1+\frac{\alpha_3}{R_1})(1+\frac{\alpha_3}{R_2}) d\alpha_1 d\alpha_2 d\alpha_3 \\
&= \int_{t0}^{t1} dt \int_V (\sigma_{13} - e_{15}E_1)\Big[-A_1 A_2 (1+\frac{\alpha_3}{R_2})\frac{\delta U_1}{R_1} \\
&\quad + A_1 A_2 (1+\frac{\alpha_3}{R_1})(1+\frac{\alpha_3}{R_2})\frac{\partial(\delta U_1)}{\partial\alpha_3} + A_2(1+\frac{\alpha_3}{R_2})\frac{\partial(\delta U_3)}{\partial \alpha_1}\Big] d\alpha_1 d\alpha_2 d\alpha_3 \\
&= \int_{t0}^{t1} dt \int_V \Big\{\Big[-(\sigma_{13} - e_{15}E_1) A_1 A_2 (1+\frac{\alpha_3}{R_2})\frac{1}{R_1} \\
&\quad - \frac{\partial[(\sigma_{13} - e_{15}E_1) A_1 A_2 (1+\frac{\alpha_3}{R_1})(1+\frac{\alpha_3}{R_2})]}{\partial\alpha_3}\Big]\delta U_1 \\
&\quad - \frac{\partial[(\sigma_{13} - e_{15}E_1) A_2 (1+\frac{\alpha_3}{R_2})]}{\partial\alpha_1}\delta U_3\Big\} d\alpha_1 d\alpha_2 d\alpha_3 \\
&\quad + \int_{t0}^{t1} dt \int_{S_3} (\sigma_{13} - e_{15}E_1) A_1 A_2 (1+\frac{\alpha_3}{R_1})(1+\frac{\alpha_3}{R_2})\delta U_1 d\alpha_1 d\alpha_2 \\
&\quad + \int_{t0}^{t1} dt \int_{S_1} (\sigma_{13} - e_{15}E_1) A_2 (1+\frac{\alpha_3}{R_2})\delta U_3 d\alpha_2 d\alpha_3
\end{aligned}
\tag{2.2.9}
$$

6) δS_{23} Term:

$$
\begin{aligned}
&\int_{t0}^{t1} dt \int_V (\sigma_{23} - e_{15}E_2)\delta S_{23} A_1 A_2 (1+\frac{\alpha_3}{R_1})(1+\frac{\alpha_3}{R_2}) d\alpha_1 d\alpha_2 d\alpha_3 \\
&= \int_{t0}^{t1} dt \int_V (\sigma_{23} - e_{15}E_2)\Big[-A_1 A_2 (1+\frac{\alpha_3}{R_1})\frac{\delta U_2}{R_2} \\
&\quad + A_1 A_2 (1+\frac{\alpha_3}{R_1})(1+\frac{\alpha_3}{R_2})\frac{\partial(\delta U_2)}{\partial\alpha_3} + A_1(1+\frac{\alpha_3}{R_1})\frac{\partial(\delta U_3)}{\partial \alpha_2}\Big] d\alpha_1 d\alpha_2 d\alpha_3
\end{aligned}
$$

$$
\begin{aligned}
&= \int_{t0}^{t1} dt \int_V \Bigg\{ \Bigg[-(\sigma_{23} - e_{15}E_2)A_1A_2(1+\frac{\alpha_3}{R_1})\frac{1}{R_2} \\
&\quad - \frac{\partial[(\sigma_{23} - e_{15}E_2)A_1A_2(1+\frac{\alpha_3}{R_1})(1+\frac{\alpha_3}{R_2})]}{\partial\alpha_3} \Bigg] \delta U_2 \\
&\quad - \frac{\partial[(\sigma_{23} - e_{15}E_2)A_1(1+\frac{\alpha_3}{R_1})]}{\partial\alpha_2} \delta U_3 \Bigg\} d\alpha_1 d\alpha_2 d\alpha_3 \\
&\quad + \int_{t0}^{t1} dt \int_{S_3} (\sigma_{23} - e_{15}E_2)A_1A_2(1+\frac{\alpha_3}{R_1})(1+\frac{\alpha_3}{R_2})\delta U_2 d\alpha_1 d\alpha_2 \\
&\quad + \int_{t0}^{t1} dt \int_{S_2} (\sigma_{23} - e_{15}E_2)A_1(1+\frac{\alpha_3}{R_1})\delta U_3 d\alpha_1 d\alpha_3
\end{aligned} \quad . \tag{2.2.10}
$$

Variations of all strains were derived above. Variation of electric field energies is considered next.

b) Variation of Electric-Field Energy

$$
\begin{aligned}
\frac{\partial \mathcal{H}}{\partial E_j}\delta E_j &= (e_{15}S_{13} + \varepsilon_{11}E_1)\frac{1}{A_1(1+\frac{\alpha_3}{R_1})}\frac{\partial(\delta\varphi)}{\partial\alpha_1} \\
&\quad + (e_{15}S_{23} + \varepsilon_{11}E_2)\frac{1}{A_2(1+\frac{\alpha_3}{R_2})}\frac{\partial(\delta\varphi)}{\partial\alpha_2} \\
&\quad + (e_{31}S_{11} + e_{31}S_{22} + e_{33}S_{33} + \varepsilon_{33}E_3)\frac{\partial(\delta\varphi)}{\partial\alpha_3}
\end{aligned} \quad . \tag{2.2.11}
$$

Using the integration by parts, one can derive the first variation term as

$$
\begin{aligned}
&\int_{t_0}^{t_1} dt \int_{\alpha_1}\int_{\alpha_2}\int_{\alpha_3} (e_{15}S_{13} + \varepsilon_{11}E_1)\frac{1}{A_1(1+\frac{\alpha_3}{R_1})}\frac{\partial(\delta\varphi)}{\partial\alpha_1} \\
&\quad \cdot A_1A_2(1+\frac{\alpha_3}{R_1})(1+\frac{\alpha_3}{R_2})d\alpha_1 d\alpha_2 d\alpha_3 \\
&= \int_{t_0}^{t_1} dt \int_{\alpha_2}\int_{\alpha_3} (e_{15}S_{13} + \varepsilon_{11}E_1)A_2(1+\frac{\alpha_3}{R_2})\delta\varphi d\alpha_2 d\alpha_3 \\
&\quad - \int_{t_0}^{t_1} dt \int_{\alpha_1}\int_{\alpha_2}\int_{\alpha_3} \frac{\partial\left[(e_{15}S_{13} + \varepsilon_{11}E_1)A_2(1+\frac{\alpha_3}{R_2})\right]}{\partial\alpha_1}\delta\varphi d\alpha_1 d\alpha_2 d\alpha_3
\end{aligned} \quad . \tag{2.2.12}
$$

Proceeding with all terms in Eq. (2.2.11) yields

$$
\begin{aligned}
&\int_{t_0}^{t_1} dt \int_V \frac{\partial \mathcal{H}}{\partial E_j} \delta E_j dV \\
&= \int_{t_0}^{t_1} dt \int_{\alpha_2} \int_{\alpha_3} (e_{15} S_{13} + \varepsilon_{11} E_1) A_2 (1 + \frac{\alpha_3}{R_2}) \delta\varphi d\alpha_2 d\alpha_3 \\
&+ \int_{t_0}^{t_1} dt \int_{\alpha_1} \int_{\alpha_3} (e_{15} S_{23} + \varepsilon_{11} E_2) A_1 (1 + \frac{\alpha_3}{R_1}) \delta\varphi d\alpha_1 d\alpha_3 \\
&+ \int_{t_0}^{t_1} dt \int_{\alpha_1} \int_{\alpha_2} (e_{31} S_{11} + e_{31} S_{22} + e_{33} S_{33} + \varepsilon_{33} E_3) \\
&\cdot A_1 A_2 (1 + \frac{\alpha_3}{R_1})(1 + \frac{\alpha_3}{R_2}) \delta\varphi d\alpha_1 d\alpha_2 \\
&- \int_{t_0}^{t_1} dt \int_{\alpha_1} \int_{\alpha_2} \int_{\alpha_3} \left\{ \frac{\partial \left[(e_{15} S_{13} + \varepsilon_{11} E_1) A_2 (1 + \frac{\alpha_3}{R_2}) \right]}{\partial \alpha_1} \right. \\
&+ \frac{\partial \left[(e_{15} S_{23} + \varepsilon_{11} E_2) A_1 (1 + \frac{\alpha_3}{R_1}) \right]}{\partial \alpha_2} \\
&\left. + \frac{\partial \left[(e_{31} S_{11} + e_{31} S_{22} + e_{33} S_{33} + \varepsilon_{33} E_3) A_1 A_2 (1 + \frac{\alpha_3}{R_1})(1 + \frac{\alpha_3}{R_2}) \right]}{\partial \alpha_3} \right\} \delta\varphi d\alpha_1 d\alpha_2 d\alpha_3
\end{aligned}
\tag{2.2.13}
$$

2.2.3 *Variation of Traction Force Energy*

The work done by the traction forces t_{ij}'s can be written as

$$
\begin{aligned}
&\int_{t0}^{t1} dt \int_S t_{ij} \delta U_j dS \\
&= \int_{t_0}^{t_1} dt \int_{S_1} (t_{11} \delta U_1 + t_{12} \delta U_2 + t_{13} \delta U_3) A_2 (1 + \frac{\alpha_3}{R_2}) d\alpha_2 d\alpha_3 \\
&+ \int_{t_0}^{t_1} dt \int_{S_2} (t_{21} \delta U_1 + t_{22} \delta U_2 + t_{23} \delta U_3) A_1 (1 + \frac{\alpha_3}{R_1}) d\alpha_1 d\alpha_3 \\
&+ \int_{t_0}^{t_1} dt \int_{S_3} (t_{31} \delta U_1 + t_{32} \delta U_2 + t_{33} \delta U_3) A_1 A_2 (1 + \frac{\alpha_3}{R_1})(1 + \frac{\alpha_3}{R_2}) d\alpha_1 d\alpha_2
\end{aligned}
\tag{2.2.14}
$$

Note that for a traction force t_{ij}, the subscript i indicates the surface on which t_{ij} acts and j indicates the direction in which t_{ij} acts. The surface traction forces can be used to define mechanical boundary conditions.

2.2.4 Variation of Electric Potential Energy

Carrying out the variation of electric potential energy in the variation equation gives

$$\begin{aligned}\int_{t_0}^{t_1} dt \int_S Q\delta\varphi dS &= \int_{t_0}^{t_1} dt \int_{\alpha_2}\int_{\alpha_3} Q_1 A_2(1+\frac{\alpha_3}{R_2})\delta\varphi d\alpha_2 d\alpha_3 \\ &+ \int_{t_0}^{t_1} dt \int_{\alpha_1}\int_{\alpha_3} Q_2 A_1(1+\frac{\alpha_3}{R_1})\delta\varphi d\alpha_1 d\alpha_2 d\alpha_3 \\ &+ \int_{t_0}^{t_1} dt \int_{\alpha_1}\int_{\alpha_2} Q_3 A_1 A_2(1+\frac{\alpha_3}{R_1})(1+\frac{\alpha_3}{R_2})\delta\varphi d\alpha_1 d\alpha_2\end{aligned} \quad . \tag{2.2.15}$$

Thus, the variations of all energies were carried out. Derivations of system equations of the piezoelectric shell continuum can be proceeded.

2.2.5 Electromechanical Equations by Hamilton's Principle

Substituting all energy variation terms into Hamilton's equation yields the variation equation

$$\begin{aligned}\int_{t_0}^{t_1} dt \int_V \Bigg\{ \Bigg[&-\frac{\partial[(\sigma_{11}-e_{31}E_3)A_2(1+\frac{\alpha_3}{R_2})]}{\partial\alpha_1} - \frac{\partial[\sigma_{12}A_1(1+\frac{\alpha_3}{R_1})]}{\partial\alpha_2} - \sigma_{12}(1+\frac{\alpha_3}{R_2})\frac{\partial A_1}{\partial\alpha_2} \\ &+ (\sigma_{22}-e_{31}E_3)(1+\frac{\alpha_3}{R_1})\frac{\partial A_2}{\partial\alpha_1} - (\sigma_{13}-e_{15}E_1)A_1A_2(1+\frac{\alpha_3}{R_2})\frac{1}{R_1} \\ &- \frac{\partial[(\sigma_{13}-e_{15}E_1)A_1A_2(1+\frac{\alpha_3}{R_1})(1+\frac{\alpha_3}{R_2})]}{\partial\alpha_3} + \rho A_1A_2(1+\frac{\alpha_3}{R_1})(1+\frac{\alpha_3}{R_2})\ddot{U}_1 \Bigg] \delta U_1\end{aligned}$$

$$
\begin{aligned}
&+\Bigg[-\frac{\partial[\sigma_{12}A_2(1+\frac{\alpha_3}{R_2})]}{\partial\alpha_1}-\frac{\partial[(\sigma_{22}-e_{31}E_3)A_1(1+\frac{\alpha_3}{R_1})]}{\partial\alpha_2}\\
&-\sigma_{12}(1+\frac{\alpha_3}{R_1})\frac{\partial A_2}{\partial\alpha_1}+(\sigma_{11}-e_{31}E_3)(1+\frac{\alpha_3}{R_2})\frac{\partial A_1}{\partial\alpha_2}\\
&-(\sigma_{23}-e_{15}E_2)A_1A_2(1+\frac{\alpha_3}{R_1})\frac{1}{R_2}\\
&-\frac{\partial[(\sigma_{23}-e_{15}E_2)A_1A_2(1+\frac{\alpha_3}{R_1})(1+\frac{\alpha_3}{R_2})]}{\partial\alpha_3}+\rho A_1A_2(1+\frac{\alpha_3}{R_1})(1+\frac{\alpha_3}{R_2})\ddot{U}_2\Bigg]\delta U_2\\
&\quad+\Bigg[-\frac{\partial[(\sigma_{13}-e_{15}E_1)A_2(1+\frac{\alpha_3}{R_2})]}{\partial\alpha_1}-\frac{\partial[(\sigma_{23}-e_{15}E_2)A_1(1+\frac{\alpha_3}{R_1})]}{\partial\alpha_2}\\
&\quad+(\sigma_{11}-e_{31}E_3)\frac{A_1A_2(1+\frac{\alpha_3}{R_2})}{R_1}+(\sigma_{22}-e_{31}E_3)\frac{A_1A_2(1+\frac{\alpha_3}{R_1})}{R_2}\\
&\quad-\frac{\partial[(\sigma_{33}-e_{33}E_3)A_1A_2(1+\frac{\alpha_3}{R_1})(1+\frac{\alpha_3}{R_2})]}{\partial\alpha_3}\\
&\quad+\rho A_1A_2(1+\frac{\alpha_3}{R_1})(1+\frac{\alpha_3}{R_2})\ddot{U}_3\Bigg]\delta U_3\}d\alpha_1d\alpha_2d\alpha_3\\
&+\int_{t_0}^{t_1}dt\int_{S_1}[(\sigma_{11}-e_{31}E_3-t_{11})\delta U_1+(\sigma_{12}-t_{12})\delta U_2\\
&+(\sigma_{13}-e_{15}E_1-t_{13})\delta U_3]A_2(1+\frac{\alpha_3}{R_2})d\alpha_2d\alpha_3\\
&+\int_{t_0}^{t_1}dt\int_{S_2}[(\sigma_{12}-t_{21})\delta U_1+(\sigma_{22}-e_{31}E_3-t_{22})\delta U_2\\
&+(\sigma_{23}-e_{15}E_2-t_{23})\delta U_3]A_1(1+\frac{\alpha_3}{R_1})d\alpha_1d\alpha_3\\
&+\int_{t_0}^{t_1}dt\int_{S_3}[(\sigma_{13}-e_{15}E_1-t_{31})\delta U_1+(\sigma_{23}-e_{15}E_2-t_{32})\delta U_2\\
&+(\sigma_{33}-e_{33}E_3-t_{33})\delta U_3]A_1A_2(1+\frac{\alpha_3}{R_1})(1+\frac{\alpha_3}{R_2})d\alpha_1d\alpha_2\\
&+\int_{t_0}^{t_1}dt\int_V\Bigg\{\frac{\partial[(e_{15}S_{13}+\varepsilon_{11}E_1)A_2(1+\frac{\alpha_3}{R_2})]}{\partial\alpha_1}+\frac{\partial[(e_{15}S_{23}+\varepsilon_{11}E_2)A_1(1+\frac{\alpha_3}{R_1})]}{\partial\alpha_2}\\
&+\frac{\partial[(e_{31}S_{11}+e_{31}S_{22}+e_{33}S_{33}+\varepsilon_{33}E_3)A_1A_2(1+\frac{\alpha_3}{R_1})(1+\frac{\alpha_3}{R_2})]}{\partial\alpha_3}\Bigg\}\delta\varphi d\alpha_1d\alpha_2d\alpha_3
\end{aligned}
$$

$$
\begin{aligned}
&+\int_{t_0}^{t_1} dt \int_{S_1} (e_{15}S_{13}+\varepsilon_{11}E_1+Q_1)A_2(1+\frac{\alpha_3}{R_2})\delta\varphi d\alpha_2 d\alpha_3 \\
&+\int_{t_0}^{t_1} dt \int_{S_2} (e_{15}S_{23}+\varepsilon_{11}E_2+Q_2)A_1(1+\frac{\alpha_3}{R_1})\delta\varphi d\alpha_1 d\alpha_3 \\
&+\int_{t_0}^{t_1} dt \int_{S_3} (e_{31}S_{11}+e_{31}S_{22}+e_{33}S_{33}+\varepsilon_{33}E_3+Q_3) \\
&\cdot A_1A_2(1+\frac{\alpha_3}{R_1})(1+\frac{\alpha_3}{R_2})\delta\varphi d\alpha_1 d\alpha_2 = 0
\end{aligned}
\quad . \tag{2.2.16}
$$

Based on Hamilton's principle, Eq. (2.2.16) can only be satisfied if each of integral parts are zero respectively. Moreover, since the variation displacements and electric potentials are arbitrary, each integral equation can only be satisfied if the coefficients of the variation displacements and electric potentials are zero. Thus, the electromechanical system equations and boundary conditions (mechanical and electric) can be derived accordingly.

2.2.6 Electromechanical Equations of Mechanical Motion

Three electromechanical equations in three principal directions can be derived from the three displacement variations δU_j.

$$
\begin{aligned}
&\frac{\partial[(\sigma_{11}-e_{31}E_3)A_2(1+\frac{\alpha_3}{R_2})]}{\partial\alpha_1}+\frac{\partial[\sigma_{12}A_1(1+\frac{\alpha_3}{R_1})]}{\partial\alpha_2} \\
&\quad+\sigma_{12}(1+\frac{\alpha_3}{R_2})\frac{\partial A_1}{\partial\alpha_2}-(\sigma_{22}-e_{31}E_3)(1+\frac{\alpha_3}{R_1})\frac{\partial A_2}{\partial\alpha_1} \\
&\quad+(\sigma_{13}-e_{15}E_1)A_1A_2(1+\frac{\alpha_3}{R_2})\frac{1}{R_1}+\frac{\partial[(\sigma_{13}-e_{15}E_1)A_1A_2(1+\frac{\alpha_3}{R_1})(1+\frac{\alpha_3}{R_2})]}{\partial\alpha_3}, \\
&=\rho A_1A_2(1+\frac{\alpha_3}{R_1})(1+\frac{\alpha_3}{R_2})\ddot{U}_1
\end{aligned}
\tag{2.2.17}
$$

$$
\begin{aligned}
&\frac{\partial[\sigma_{12}A_2(1+\frac{\alpha_3}{R_2})]}{\partial\alpha_1}+\frac{\partial[(\sigma_{22}-e_{31}E_3)A_1(1+\frac{\alpha_3}{R_1})]}{\partial\alpha_2} \\
&\quad+\sigma_{12}(1+\frac{\alpha_3}{R_1})\frac{\partial A_2}{\partial\alpha_1}-(\sigma_{11}-e_{31}E_3)(1+\frac{\alpha_3}{R_2})\frac{\partial A_1}{\partial\alpha_2} \\
&\quad+(\sigma_{23}-e_{15}E_2)A_1A_2(1+\frac{\alpha_3}{R_1})\frac{1}{R_2}+\frac{\partial[(\sigma_{23}-e_{15}E_2)A_1A_2(1+\frac{\alpha_3}{R_1})(1+\frac{\alpha_3}{R_2})]}{\partial\alpha_3}, \\
&=\rho A_1A_2(1+\frac{\alpha_3}{R_1})(1+\frac{\alpha_3}{R_2})\ddot{U}_2
\end{aligned}
\tag{2.2.18}
$$

$$\frac{\partial[(\sigma_{13}-e_{15}E_1)A_2(1+\frac{\alpha_3}{R_2})]}{\partial\alpha_1}+\frac{\partial[(\sigma_{23}-e_{15}E_2)A_1(1+\frac{\alpha_3}{R_1})]}{\partial\alpha_2}$$
$$-(\sigma_{11}-e_{31}E_3)\frac{A_1A_2(1+\frac{\alpha_3}{R_2})}{R_1}-(\sigma_{22}-e_{31}E_3)\frac{A_1A_2(1+\frac{\alpha_3}{R_1})}{R_2} \quad . \quad (2.2.19)$$
$$+\frac{\partial[(\sigma_{33}-e_{33}E_3)A_1A_2(1+\frac{\alpha_3}{R_1})(1+\frac{\alpha_3}{R_2})]}{\partial\alpha_3}$$
$$=\rho A_1A_2(1+\frac{\alpha_3}{R_1})(1+\frac{\alpha_3}{R_2})\ddot{U}_3$$

Note that the deflections U_j's are generic displacements, which includes both membrane and bending components. This general definition will become more explicit in later derivations.

2.2.7 *Mechanical Boundary Conditions*

Mechanical boundary conditions are derived from the surface traction forces and the electric fields; they can be defined in three principal directions.

1) α_1 direction:

$$\sigma_{11}-e_{31}E_3=t_{11}, \quad (2.2.20a)$$
$$\sigma_{12}=t_{12}, \quad (2.2.20b)$$
$$\sigma_{13}-e_{15}E_1=t_{13}. \quad (2.2.20c)$$

2) α_2 direction:

$$\sigma_{22}-e_{31}E_3=t_{22}, \quad (2.2.21a)$$
$$\sigma_{12}=t_{12}, \quad (2.2.21b)$$
$$\sigma_{23}-e_{15}E_2=t_{23}. \quad (2.2.21c)$$

3) α_3 direction:

$$\sigma_{13}-e_{15}E_1=t_{13}, \quad (2.2.22a)$$
$$\sigma_{23}-e_{15}E_2=t_{23}, \quad (2.2.22b)$$
$$\sigma_{33}-e_{33}E_3=t_{33}. \quad (2.2.22c)$$

Note that there are three electromechanical equations governing the mechanical motion in the α_1, α_2, and α_3 directions, respectively. The electric field strengths in all three directions are included in the equations; the mechanical forces and electric voltages both contribute the mechanical boundary conditions of the piezoelectric shell continuum. The electric terms can be used in conjunction with control systems in structural controls (Tzou & Zhong, 1991a & d).

2.2.8 *Charge Equation of Electrostatics*

Charge equation can be derived in a similar way, i.e.,

$$\frac{\partial[(e_{15}S_{13}+\varepsilon_{11}E_1)A_2(1+\frac{\alpha_3}{R_2})]}{\partial\alpha_1}+\frac{\partial[(e_{15}S_{23}+\varepsilon_{11}E_2)A_1(1+\frac{\alpha_3}{R_1})]}{\partial\alpha_2} +\frac{\partial[(e_{31}S_{11}+e_{31}S_{22}+e_{33}S_{33}+\varepsilon_{33}E_3)A_1A_2(1+\frac{\alpha_3}{R_1})(1+\frac{\alpha_3}{R_2})]}{\partial\alpha_3}=0 . \quad (2.2.23)$$

The charge equation of electrostatics can be used in sensor applications in Chaps. 4 and 8 (Tzou, Zhong and Natori, 1993).

2.2.9 *Electric Boundary Conditions*

Electric boundary conditions are defined on the outside surfaces of the piezoelectric shell.

$$(e_{15}S_{13}+\varepsilon_{11}E_1+Q_1)A_2(1+\frac{\alpha_3}{R_2})=0, \quad (2.2.24a)$$

$$(e_{15}S_{23}+\varepsilon_{11}E_2+Q_2)A_1(1+\frac{\alpha_3}{R_1})=0, \quad (2.2.24b)$$

$$(e_{31}S_{11}+e_{31}S_{22}+e_{33}S_{33}+\varepsilon_{33}E_3+Q_3)A_1A_2(1+\frac{\alpha_3}{R_1})(1+\frac{\alpha_3}{R_2})=0. \quad (2.2.24c)$$

Note that the electric boundary conditions in Eqs. (2.2.24a,b,c) indicate that the electric displacements on the surfaces are equal to the densities of surface charges. All $A_i(1+\frac{\alpha_3}{R_i})$ and $A_1A_2(1+\frac{\alpha_3}{R_1})(1+\frac{\alpha_3}{R_2})$ terms can be removed since $A_i(1+\frac{\alpha_3}{R_i})$ / $=0$ and $A_1A_2(1+\frac{\alpha_3}{R_1})(1+\frac{\alpha_3}{R_2})\neq 0$. Thus,

$$(e_{15}S_{13}+\varepsilon_{11}E_1+Q_1)=0, \quad (2.2.25a)$$

$$(e_{15}S_{23}+\varepsilon_{11}E_2+Q_2)=0, \quad (2.2.25b)$$

$$(e_{31}S_{11}+e_{31}S_{22}+e_{33}S_{33}+\varepsilon_{33}E_3+Q_3)=0. \quad (2.2.25c)$$

The electric boundary conditions can also be used in structural controls, i.e., ***boundary controls***.

2.3 Thick Piezoelectric Shell

If the shell thickness is relatively thick compared with either the wavelength of the highest frequency of interest or the physical dimensions, in this case, transverse shear deformation and rotatory inertia effects need to be considered. For a generic shell continuum, the resultant deflection U_j includes two components: a membrane component (u_i) and a bending component (β_i). In this section, these two components are separated from the generalized deflection coordinate U_j; the substitutions lead to five electromechanical equations in three displacement coordinates $u_i(\alpha_1, \alpha_2)$, i = 1,2,3, and two rotary coordinates $\beta_i(\alpha_1, \alpha_2)$, i = 1,2.

2.3.1 *Assumptions*

It is assumed that the transverse displacement U_3 is independent of thickness, and in-plane displacements U_1 and U_2 vary linearly through the shell thickness α_3, Fig. 2.4. Thus, the resultant deflections can be written as

$$U_1(\alpha_1, \alpha_2, \alpha_3) = u_1(\alpha_1, \alpha_2) + \alpha_3 \beta_1(\alpha_1, \alpha_2). \tag{2.3.1a}$$

$$U_2(\alpha_1, \alpha_2, \alpha_3) = u_2(\alpha_1, \alpha_2) + \alpha_3 \beta_2(\alpha_1, \alpha_2), \tag{2.3.1b}$$

$$U_3(\alpha_1, \alpha_2, \alpha_3) = u_3(\alpha_1, \alpha_2), \tag{2.3.1c}$$

where $\beta_1(\alpha_1, \alpha_2)$ and $\beta_2(\alpha_1, \alpha_2)$ represent rotation angles in the positive sense of α_1 and α_2 axes, respectively. (Note that the shear deformation and rotary inertia effects

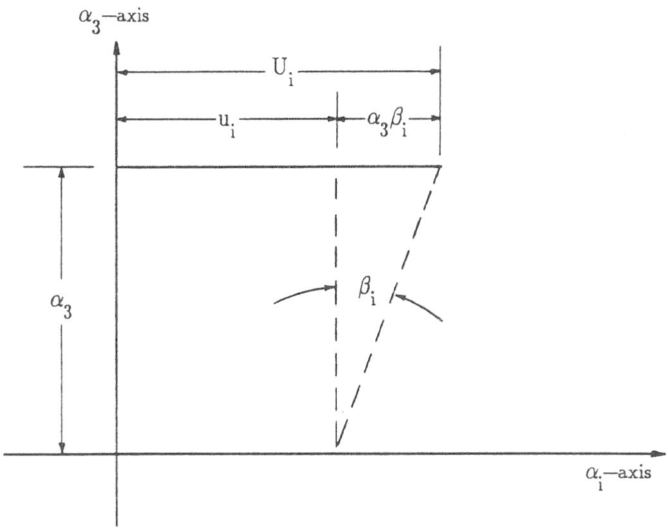

Fig. 2.4 Relation between deflection $\mathbf{U_i}$ and membrane displacement $\mathbf{u_i}$ & rotation angle $\beta_i(\alpha_1, \alpha_2)$

are still preserved in this derivation; they will be neglected in the next case when the Kirchhoff–Love thin-shell assumptions are employed.)

The linear acceleration terms can be redefined as

$$\int_{\alpha_3} \ddot{U}_1 d\alpha_3 = \int_{-h/2}^{h/2} \left(\ddot{u}_1 + \alpha_3 \ddot{\beta}_1\right) d\alpha_3 = h\ddot{u}_1, \tag{2.3.2a}$$

$$\int_{\alpha_3} \ddot{U}_2 d\alpha_3 = \int_{-h/2}^{h/2} \left(\ddot{u}_2 + \alpha_3 \ddot{\beta}_2\right) d\alpha_3 = h\ddot{u}_2, \tag{2.3.2b}$$

$$\int_{\alpha_3} \ddot{U}_3 d\alpha_3 = \int_{-h/2}^{h/2} \ddot{u}_3 d\alpha_3 = h\ddot{u}_3. \tag{2.3.2c}$$

In general, $\alpha_3 \ll R_1$ and $\alpha_3 \ll R_2$; thus, the ratio of finite distance to radius of curvature is negligible, i.e.,

$$1 + \frac{\alpha_3}{R_1} \cong 1 \text{ and } 1 + \frac{\alpha_3}{R_2} \cong 1. \tag{2.3.3}$$

2.3.2 *Electromechanical Equations in Translations*

Substituting Eqs. (2.3.1a,b,c) into Eqs. (2.2.17)–(2.2.19) and integrating these equations along the α_3 direction, one can derive a new set of electromechanical equations in three axial directions of the piezoelectric shell:

$$\begin{aligned} &\frac{\partial[(N_{11}^m - N_{11}^e)A_2]}{\partial\alpha_1} + \frac{\partial(N_{21}^m A_1)}{\partial\alpha_2} + N_{12}^m \frac{\partial A_1}{\partial\alpha_2} - (N_{22}^m - N_{22}^e)\frac{\partial A_2}{\partial\alpha_1} \\ &+ (Q_{13}^m - Q_{13}^e)A_1A_2\frac{1}{R_1} + A_1A_2(\sigma_{13} - e_{15}E_1)\Big|_{-h/2}^{h/2} = \rho h A_1 A_2 \ddot{u}_1 \end{aligned}, \tag{2.3.4}$$

$$\begin{aligned} &\frac{\partial(N_{12}^m A_2)}{\partial\alpha_1} + \frac{\partial[(N_{22}^m - N_{22}^e)A_1]}{\partial\alpha_2} + N_{21}^m \frac{\partial A_2}{\partial\alpha_1} - (N_{11}^m - N_{11}^e)\frac{\partial A_1}{\partial\alpha_2} \\ &+ (Q_{23}^m - Q_{23}^e)A_1A_2\frac{1}{R_2} + A_1A_2(\sigma_{23} - e_{15}E_2)\Big|_{-h/2}^{h/2} = \rho h A_1 A_2 \ddot{u}_2 \end{aligned}, \tag{2.3.5}$$

$$\begin{aligned} &-(N_{11}^m - N_{11}^e)\frac{A_1A_2}{R_1} - (N_{22}^m - N_{22}^e)\frac{A_1A_2}{R_2} + \frac{\partial[(Q_{13}^m - Q_{13}^e)A_2]}{\partial\alpha_1} \\ &+ \frac{\partial[(Q_{23}^m - Q_{23}^e)A_1]}{\partial\alpha_2} + A_1A_2(\sigma_{33} - e_{33}E_3)\Big|_{-h/2}^{h/2} = \rho h A_1 A_2 \ddot{u}_3 \end{aligned}. \tag{2.3.6}$$

Note that the shear stress components, $(\sigma_{i3} - e_{ij}E_i)\Big|_{-h/2}^{h/2}$, can be treated as boundary forces on the top and bottom surfaces; or set to zero if both surfaces are force free surfaces. $(\sigma_{33} - e_{33}E_3)\Big|_{-h/2}^{h/2}$ denotes the transverse normal stress effect.

2.3.3 Electromechanical Equations in Rotations

The two rotatory effects can be derived as

$$\int_{\alpha_3} \alpha_3 \ddot{U}_1 d\alpha_3 = \int_{-h/2}^{h/2} \alpha_3 \left(\ddot{u}_1 + \alpha_3 \ddot{\beta}_1 \right) d\alpha_3 = \frac{h^3}{12} \ddot{\beta}_1, \tag{2.3.7a}$$

$$\int_{\alpha_3} \alpha_3 \ddot{U}_2 d\alpha_3 = \int_{-h/2}^{h/2} \alpha_3 \left(\ddot{u}_2 + \alpha_3 \ddot{\beta}_2 \right) d\alpha_3 = \frac{h^3}{12} \ddot{\beta}_2. \tag{2.3.7b}$$

Multiplying Eqs. (2.2.17)–(2.2.18) by α_3 and integrating along the α_3 direction, one can derive two rotational electromechanical equationsEquations of the piezoelectric shell:

$$\begin{aligned}
&\frac{\partial[(M_{11}^m - M_{11}^e)A_2]}{\partial \alpha_1} + \frac{\partial(M_{21}^m A_1)}{\partial \alpha_2} + M_{12}^m \frac{\partial A_1}{\partial \alpha_2} - (M_{22}^m - M_{22}^e) \frac{\partial A_2}{\partial \alpha_1} \\
&+ A_1 A_2 \frac{1}{R_1} \int_{\alpha_3} (\sigma_{13} - e_{15} E_1) \alpha_3 d\alpha_3 + A_1 A_2 [\alpha_3 (\sigma_{13} - e_{15} E_1)] \Big|_{-h/2}^{h/2}, \\
&- A_1 A_2 (Q_{13}^m - Q_{13}^e) = A_1 A_2 \frac{\rho h^3}{12} \ddot{\beta}_1
\end{aligned} \tag{2.3.8}$$

$$\begin{aligned}
&\frac{\partial(M_{12}^m A_2)}{\partial \alpha_1} + \frac{\partial[(M_{22}^m - M_{22}^e)A_1]}{\partial \alpha_2} + M_{21}^m \frac{\partial A_2}{\partial \alpha_1} - (M_{11}^m - M_{11}^e) \frac{\partial A_1}{\partial \alpha_2} \\
&+ A_1 A_2 \frac{1}{R_2} \int_{\alpha_3} (\sigma_{23} - e_{15} E_2) \alpha_3 d\alpha_3 + A_1 A_2 [\alpha_3 (\sigma_{23} - e_{15} E_2)] \Big|_{-h/2}^{h/2}. \\
&- A_1 A_2 (Q_{23}^m - Q_{23}^e) = A_1 A_2 \frac{\rho h^3}{12} \ddot{\beta}_2
\end{aligned} \tag{2.3.9}$$

where superscript "m" denotes the mechanically induced component and "e" the electrically induced component. The resultant forces and moments, both mechanical and electric, are defined next.

2.3.4 Definition of Resultant Forces and Moments

Resultant forces and moments derived from mechanical and electric components are defined in this section. (Note that these components are defined as force or moment per unit length.) It is assumed that there is no in-plane twisting effect in the piezoelectric shell continuum.

1) Mechanical Membrane Forces

$$N_{11}^{m} = \int_{\alpha_3} \sigma_{11} d\alpha_3, \tag{2.3.10a}$$

$$N_{22}^{m} = \int_{\alpha_3} \sigma_{22} d\alpha_3, \tag{2.3.10b}$$

$$N_{12}^{m} = \int_{\alpha_3} \sigma_{12} d\alpha_3. \tag{2.3.10c}$$

2) Mechanical Bending Moments

$$M_{11}^{m} = \int_{\alpha_3} \sigma_{11} \alpha_3 d\alpha_3, \tag{2.3.11a}$$

$$M_{22}^{m} = \int_{\alpha_3} \sigma_{22} \alpha_3 d\alpha_3, \tag{2.3.11b}$$

$$M_{12}^{m} = \int_{\alpha_3} \sigma_{12} \alpha_3 d\alpha_3. \tag{2.3.11c}$$

3) Mechanical Transverse Shear Forces

$$Q_{13}^{m} = \int_{\alpha_3} \sigma_{13} d\alpha_3, \tag{2.3.12a}$$

$$Q_{23}^{m} = \int_{\alpha_3} \sigma_{23} d\alpha_3. \tag{2.3.12b}$$

4) Electric Membrane Forces

$$\begin{aligned} N_{11}^{e} &= \int_{\alpha_3} e_{31} E_3 d\alpha_3 \\ &= \int_{\alpha_3} e_{31} E_3^{e} d\alpha_3 + \int_{\alpha_3} e_{31} E_3^{d} d\alpha_3, \\ &= N_{11}^{c} + N_{11}^{d} \end{aligned} \tag{2.3.13a}$$

$$\begin{aligned} N_{22}^{e} &= \int_{\alpha_3} e_{31} E_3 d\alpha_3 \\ &= \int_{\alpha_3} e_{31} E_3^{e} d\alpha_3 + \int_{\alpha_3} e_{31} E_3^{d} d\alpha_3, \\ &= N_{22}^{c} + N_{22}^{d} \end{aligned} \tag{2.3.13b}$$

$$N_{12}^{e} = 0. \tag{2.3.13c}$$

5) Electric Bending Moments

$$\begin{aligned} M_{11}^{e} &= \int_{\alpha_3} e_{31} E_3 \alpha_3 d\alpha_3 \\ &= \int_{\alpha_3} e_{31} E_3^{e} \alpha_3 d\alpha_3 + \int_{\alpha_3} e_{31} E_3^{d} \alpha_3 d\alpha_3 , \\ &= M_{11}^{c} + M_{11}^{d} \end{aligned} \tag{2.3.14a}$$

$$\begin{aligned} M_{22}^{e} &= \int_{\alpha_3} e_{31} E_3 \alpha_3 d\alpha_3 \\ &= \int_{\alpha_3} e_{31} E_3^{e} \alpha_3 d\alpha_3 + \int_{\alpha_3} e_{31} E_3^{d} \alpha_3 d\alpha_3 , \\ &= M_{22}^{c} + M_{22}^{d} \end{aligned} \tag{2.3.14b}$$

$$M_{12}^{e} = 0. \tag{2.3.14c}$$

6) Electric Transverse Shear Forces

$$\begin{aligned} Q_{13}^{e} &= \int_{\alpha_3} e_{15} E_1 d\alpha_3 \\ &= \int_{\alpha_3} e_{15} E_1^{e} d\alpha_3 + \int_{\alpha_3} e_{15} E_1^{d} d\alpha_3 , \\ &= Q_{13}^{c} + Q_{13}^{d} \end{aligned} \tag{2.3.15a}$$

$$\begin{aligned} Q_{23}^{e} &= \int_{\alpha_3} e_{15} E_2 d\alpha_3 \\ &= \int_{\alpha_3} e_{15} E_2^{e} d\alpha_3 + \int_{\alpha_3} e_{15} E_2^{d} d\alpha_3 , \\ &= Q_{23}^{c} + Q_{23}^{d} \end{aligned} \tag{2.3.15b}$$

where superscript "c" indicates the converse piezoelectric effect and "d" the direct piezoelectric effect. E^e and E^d were defined in Sect. 2.1. Note that three electric fields E_i's are still preserved in the electromechanical equations and boundary conditions. Mechanical boundary conditions are derived next.

2.3.5 Mechanical Boundary Conditions

Mechanical boundary conditions can also be directly derived from the variation equation,

$$\begin{aligned}&\int_{t_0}^{t_1} dt \int_{S_1} [(\sigma_{11} - e_{31}E_3 - t_{11})\delta(u_1 + \alpha_3\beta_1) + (\sigma_{12} - t_{12})\delta(u_2 + \alpha_3\beta_2)\\ &\quad + (\sigma_{13} - e_{15}E_1 - t_{13})\delta u_3]A_2 d\alpha_2 d\alpha_3\\ &= \int_{t_0}^{t_1} dt \int_{\alpha_2} [(N_{11}^m - N_{11}^e - N_{11}^*)\delta u_1 + (N_{12}^m - N_{12}^*)\delta u_2 + (Q_{13}^m - Q_{13}^e - Q_{13}^*)\delta u_3\\ &\quad + (M_{11}^m - M_{11}^e - M_{11}^*)\delta\beta_1 + (M_{12}^m - M_{12}^*)\delta\beta_2]A_2 d\alpha_2 = 0\end{aligned}, \tag{2.3.16}$$

$$\begin{aligned}&\int_{t_0}^{t_1} dt \int_{S_2} [(\sigma_{12} - t_{21})\delta(u_1 + \alpha_3\beta_1) + (\sigma_{22} - e_{31}E_3 - t_{22})\delta(u_2 + \alpha_3\beta_2)\\ &\quad + (\sigma_{23} - e_{15}E_2 - t_{23})\delta u_3]A_1 d\alpha_1 d\alpha_3\\ &= \int_{t_0}^{t_1} dt \int_{\alpha_1} [(N_{21}^m - N_{21}^*)\delta u_1 + (N_{22}^m - N_{22}^e - N_{22}^*)\delta u_2 + (Q_{23}^m - Q_{23}^e - Q_{23}^*)\delta u_3\\ &\quad + (M_{21}^m - M_{21}^*)\delta\beta_1 + (M_{22}^m - M_{22}^e - M_{22}^*)\delta\beta_2]A_1 d\alpha_1 = 0\end{aligned}, \tag{2.3.17}$$

where boundary membrane forces, bending moments, and transverse shear forces (denoted by a super-scribe *) are

$$N_{ij}^* = \int_{\alpha_3} t_{ij} d\alpha_3 \quad (i = 1, 2 \text{ and } j = 1, 2), \tag{2.3.18a}$$

$$M_{ij}^* = \int_{\alpha_3} t_{ij}\alpha_3 d\alpha_3 \quad (i = 1, 2 \text{ and } j = 1, 2), \tag{2.3.18b}$$

$$Q_{13}^* = \int_{\alpha_3} t_{13} d\alpha_3, \tag{2.3.18c}$$

$$Q_{23}^* = \int_{\alpha_3} t_{23} d\alpha_3. \tag{2.3.18d}$$

Equations (2.3.16) & (2.3.17) are satisfied by either the virtual displacements vanishing or the coefficients of the virtual displacements vanishing. Thus, mechanical boundary conditions are:

1) On the Surface Defined by α_1:

$$N_{11}^m - N_{11}^e = N_{11}^* \quad or \quad u_1 = u_1^*, \tag{2.3.19a}$$

$$N_{12}^m = N_{12}^* \quad or \quad u_2 = u_2^*, \tag{2.3.19b}$$

$$Q_{13}^m - Q_{13}^e = Q_{13}^* \quad or \quad u_3 = u_3^*, \tag{2.3.19c}$$

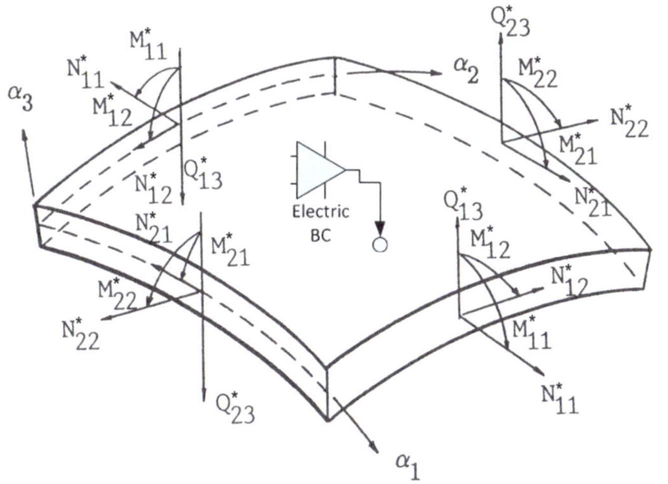

Fig. 2.5 Mechanical boundary conditions

$$M^m_{11} - M^e_{11} = M^*_{11} \quad or \quad \beta_1 = \beta^*_1, \tag{2.3.19d}$$

$$M^m_{12} = M^*_{12} \quad or \quad \beta_2 = \beta^*_2; \tag{2.3.19e}$$

2) On the Surface Defined by $\boldsymbol{\alpha}_2$:

$$N^m_{21} = N^*_{21} \quad or \quad u_1 = u^*_1, \tag{2.3.20a}$$

$$N^m_{22} - N^e_{22} = N^*_{22} \quad or \quad u_2 = u^*_2, \tag{2.3.20b}$$

$$Q^m_{23} - Q^e_{23} = Q^*_{23} \quad or \quad u_3 = u^*_3, \tag{2.3.20c}$$

$$M^m_{21} = M^*_{21} \quad or \quad \beta_1 = \beta^*_1, \tag{2.3.20d}$$

$$M^m_{22} - M^e_{22} = M^*_{22} \quad or \quad \beta_2 = \beta^*_2. \tag{2.3.20e}$$

Note that all terms with "*" denote the resultant components on the boundaries of the piezoelectric shell continuum, Fig. 2.5. Both electric and mechanical components contribute the resultant boundary forces and moments.

2.3.6 Electrostatic Charge Equation and Boundary Conditions

A charge equation of electrostatics of the piezoelectric shell continuum becomes

$$\frac{\partial[(e_{15}S_{13}+\varepsilon_{11}E_1)A_2(1+\frac{\alpha_3}{R_2})]}{\partial\alpha_1}+\frac{\partial[(e_{15}S_{23}+\varepsilon_{11}E_2)A_1(1+\frac{\alpha_3}{R_1})]}{\partial\alpha_2}$$
$$+\frac{\partial[(e_{31}S_{11}+e_{31}S_{22}+e_{33}S_{33}+\varepsilon_{33}E_3)A_1A_2(1+\frac{\alpha_3}{R_1})(1+\frac{\alpha_3}{R_2})]}{\partial\alpha_3}=0 \quad (2.3.21)$$

Note that the transverse strain S_{33} is usually small. i.e., $S_{33} \simeq 0$, due to its relatively small dimension, unless the location where the concentrated load is. Electric boundary conditions on three surfaces are

$$(e_{15}S_{13}+\varepsilon_{11}E_1+Q_1)A_2(1+\frac{\alpha_3}{R_2})=0, \quad (2.3.22a)$$

$$(e_{15}S_{23}+\varepsilon_{11}E_2+Q_2)A_1(1+\frac{\alpha_3}{R_1})=0, \quad (2.3.22b)$$

$$(e_{31}S_{11}+e_{31}S_{22}+e_{33}S_{33}+\varepsilon_{33}E_3+Q_3)A_1A_2(1+\frac{\alpha_3}{R_1})(1+\frac{\alpha_3}{R_2})=0. \quad (2.3.22c)$$

where $A_i(1+\frac{\alpha_3}{R_i})$ and $A_1A_2(1+\frac{\alpha_3}{R_1})(1+\frac{\alpha_3}{R_2})$ terms can be removed since $A_i(1+\frac{\alpha_3}{R_i})\neq 0$ and $A_1A_2(1+\frac{\alpha_3}{R_1})(1+\frac{\alpha_3}{R_2})\neq 0$. Thus,

$$(e_{15}S_{13}+\varepsilon_{11}E_1+Q_1)=0, \quad (2.3.23a)$$

$$(e_{15}S_{23}+\varepsilon_{11}E_2+Q_2)=0, \quad (2.3.23b)$$

$$(e_{31}S_{11}+e_{31}S_{22}+e_{33}S_{33}+\varepsilon_{33}E_3+Q_3)=0. \quad (2.3.23c)$$

Again, $S_{33} \simeq 0$ and the transverse equation becomes $(e_{31}S_{11}+e_{31}S_{22}+\varepsilon_{33}E_3+Q_3)=0$. Note that this set of electromechanical system equations includes the effects of all three electric fields, transverse shear deformations, and rotary inertias. In general, this theory can be applied to piezoelectric structures that involve large motions, translations and rotations, e.g., space structures, robot manipulators, etc.

2.4 Thin Piezoelectric Shell

If the piezoelectric shell continuum is thin, the transverse shear deformations and rotatory inertias are neglected. The transverse shear strains are also negligible, i.e., $S_{13}=0$ and $S_{23}=0$. In addition, the in-plane electric fields E_1 and E_2 are neglected and only the transverse electric field E_3 is considered.

2.4.1 Assumptions and Fundamentals

Using the Kirchhoff–Love thin shell assumptions and linear variations of in-plane displacements, Sect. 2.3.3, one can derive the strain–displacement relations for thin shell continuum as

$$S_{11}=\frac{1}{A_1}\frac{\partial}{\partial\alpha_1}(u_1+\alpha_3\beta_1)+\frac{1}{A_1A_2}\frac{\partial A_1}{\partial\alpha_2}(u_2+\alpha_3\beta_2)+\frac{u_3}{R_1}, \tag{2.4.1a}$$

$$S_{22}=\frac{1}{A_2}\frac{\partial}{\partial\alpha_2}(u_2+\alpha_3\beta_2)+\frac{1}{A_1A_2}\frac{\partial A_2}{\partial\alpha_1}(u_1+\alpha_3\beta_1)+\frac{u_3}{R_2}, \tag{2.4.1b}$$

$$S_{33}=0, \tag{2.4.1c}$$

$$S_{13}=0, \tag{2.4.1d}$$

$$S_{23}=0, \tag{2.4.1e}$$

$$S_{12}=\frac{A_2}{A_1}\frac{\partial}{\partial\alpha_1}\left(\frac{u_2+\alpha_3\beta_2}{A_2}\right)+\frac{A_1}{A_2}\frac{\partial}{\partial\alpha_2}\left(\frac{u_1+\alpha_3\beta_1}{A_1}\right). \tag{2.4.1f}$$

Based on assumptions that transverse shear strains are negligible, i.e., $S_{13}=0$ and $S_{23}=0$, one can further derive the two rotations β_1 and β_2 (Soedel, 1981; Tzou & Gadre, 1989):

$$\beta_1=\frac{u_1}{R_1}-\frac{1}{A_1}\frac{\partial u_3}{\partial\alpha_1}, \tag{2.4.2a}$$

$$\beta_2=\frac{u_2}{R_2}-\frac{1}{A_2}\frac{\partial u_3}{\partial\alpha_2}. \tag{2.4.2b}$$

Thus, the stresses introduced by mechanical strains become

$$\sigma_{11}=\frac{Y}{1-\mu^2}(S_{11}+\mu S_{22}), \tag{2.4.3a}$$

$$\sigma_{22}=\frac{Y}{1-\mu^2}(S_{22}+\mu S_{11}), \tag{2.4.3b}$$

$$\sigma_{12}=\frac{Y}{2(1+\mu)}S_{12}, \tag{2.4.3c}$$

where Y is Young's Modulus and μ is Poisson's ratio. Thus, $c_{11}=Y/(1-\mu^2)$, $c_{22}=Y\mu/(1-\mu^2)$, and $c_{66}=Y/2(1+\mu)$. (Note that σ_{33} is neglected since the shell is thin, except in the neighborhood of concentrated loads.)

Introducing the ***membrane strains*** S_{ij}^o and the ***bending strains*** k_{ij}, one can separate strain expressions, Eq. (2.4.1) into two parts:

$$S_{11} = S_{11}^{o} + \alpha_3 k_{11}, \tag{2.4.4a}$$

$$S_{22} = S_{22}^{o} + \alpha_3 k_{22}, \tag{2.4.4b}$$

$$S_{12} = S_{12}^{o} + \alpha_3 k_{12}. \tag{2.4.4c}$$

The ***membrane strains*** S_{ij}^{o} are defined as

$$S_{11}^{o} = \frac{1}{A_1}\frac{\partial u_1}{\partial \alpha_1} + \frac{u_2}{A_1 A_2}\frac{\partial A_1}{\partial \alpha_2} + \frac{u_3}{R_1}, \tag{2.4.5a}$$

$$S_{22}^{o} = \frac{1}{A_2}\frac{\partial u_2}{\partial \alpha_2} + \frac{u_1}{A_1 A_2}\frac{\partial A_2}{\partial \alpha_1} + \frac{u_3}{R_2}, \tag{2.4.5b}$$

$$S_{12}^{o} = \frac{A_2}{A_1}\frac{\partial}{\partial \alpha_1}\left(\frac{u_2}{A_2}\right) + \frac{A_1}{A_2}\frac{\partial}{\partial \alpha_2}\left(\frac{u_1}{A_1}\right). \tag{2.4.5c}$$

The ***bending strains*** k_{ij} are defined as

$$k_{11} = \frac{1}{A_1}\frac{\partial \beta_1}{\partial \alpha_1} + \frac{\beta_2}{A_1 A_2}\frac{\partial A_1}{\partial \alpha_2}, \tag{2.4.6a}$$

$$k_{22} = \frac{1}{A_2}\frac{\partial \beta_2}{\partial \alpha_2} + \frac{\beta_1}{A_1 A_2}\frac{\partial A_2}{\partial \alpha_1}, \tag{2.4.6b}$$

$$k_{12} = \frac{A_2}{A_1}\frac{\partial}{\partial \alpha_1}\left(\frac{\beta_2}{A_2}\right) + \frac{A_1}{A_2}\frac{\partial}{\partial \alpha_2}\left(\frac{\beta_1}{A_1}\right). \tag{2.4.6c}$$

The mechanical stresses can be further defined in terms of membrane and bending strains.

$$\sigma_{11} = \frac{Y}{1-\mu^2}\left[S_{11}^{o} + \mu S_{22}^{o} + \alpha_3(k_{11} + \mu k_{22})\right], \tag{2.4.7a}$$

$$\sigma_{22} = \frac{Y}{1-\mu^2}\left[S_{22}^{o} + \mu S_{11}^{o} + \alpha_3(k_{22} + \mu k_{11})\right], \tag{2.4.7b}$$

$$\sigma_{12} = \frac{Y}{2(1+\mu)}\left(S_{12}^{o} + \alpha_3 k_{12}\right). \tag{2.4.7c}$$

(Note that σ_{33} is neglected here.) Substituting Eqs. (2.4.7a,b,c) into Eqs. (2.3.10a,b,c) and finishing the integrations, one can derive the mechanical membrane forces (per unit length) for thin shells:

$$N_{11}^{m} = K\left(S_{11}^{o} + \mu S_{22}^{o}\right), \tag{2.4.8a}$$

$$N_{22}^{m} = K\left(S_{22}^{o} + \mu S_{11}^{o}\right), \tag{2.4.8b}$$

$$N_{12}^m = \frac{K(1-\mu)}{2} S_{12}^o, \tag{2.4.8c}$$

where $K = [Yh/(1-\mu^2)]$-the ***membrane stiffness***. Substituting Eqs. (2.4.7a,b,c) into Eqs. (2.3.11a,b,c) and finishing the integrations, one can further derive mechanical bending moments:

$$M_{11}^m = D(k_{11} + \mu k_{22}), \tag{2.4.9a}$$

$$M_{22}^m = D(k_{22} + \mu k_{11}), \tag{2.4.9b}$$

$$M_{12}^m = \frac{D(1-\mu)}{2} k_{12}, \tag{2.4.9c}$$

where $D = [Yh^3/12(1-\mu^2)]$-the ***bending stiffness***.

2.4.2 Thin Shell Electromechanical Equations

Using Eqs. (2.3.4)–(2.3.6) and considering electric field E_3 only, i.e., $E_1 = 0$ and $E_2 = 0$, one can derive the electromechanical equations for thin piezoelectric shells:

$$\begin{aligned}&\frac{\partial[(N_{11}^m - N_{11}^e)A_2]}{\partial\alpha_1} + \frac{\partial(N_{21}^m A_1)}{\partial\alpha_2} + N_{12}^m \frac{\partial A_1}{\partial\alpha_2} - (N_{22}^m - N_{22}^e)\frac{\partial A_2}{\partial\alpha_1} \\ &+ A_1 A_2 \frac{Q_{13}^m}{R_1} = \rho h A_1 A_2 \ddot{u}_1\end{aligned}, \tag{2.4.10}$$

$$\begin{aligned}&\frac{\partial(N_{12}^m A_2)}{\partial\alpha_1} + \frac{\partial[(N_{22}^m - N_{22}^e)A_1]}{\partial\alpha_2} + N_{21}^m \frac{\partial A_2}{\partial\alpha_1} - (N_{11}^m - N_{11}^e)\frac{\partial A_1}{\partial\alpha_2} \\ &+ A_1 A_2 \frac{Q_{23}^m}{R_2} = \rho h A_1 A_2 \ddot{u}_2\end{aligned}, \tag{2.4.11}$$

$$\begin{aligned}&\frac{\partial(Q_{13}^m A_2)}{\partial\alpha_1} + \frac{\partial(Q_{23}^m A_1)}{\partial\alpha_2} - A_1 A_2\left(\frac{N_{11}^m - N_{11}^e}{R_1} + \frac{N_{22}^m - N_{22}^e}{R_2}\right) \\ &+ A_1 A_2 (\sigma_{33} - e_{33}E_3)\Big|_{-h/2}^{h/2} = \rho h A_1 A_2 \ddot{u}_3\end{aligned}. \tag{2.4.12}$$

Note that the normal stress $(\sigma_{33} - e_{33}E_3)\Big|_{-h/2}^{h/2}$ can be neglected as discussed previously. Q_{13}^m and Q_{23}^m can be derived using the two rotatory equations $\ddot{\beta}_1$ and $\ddot{\beta}_2$, i.e., Equations (2.3.8)–(2.3.9), for the thick shell continuum.

$$Q_{13}^m A_1 A_2 = \frac{\partial[(M_{11}^m - M_{11}^e)A_2]}{\partial\alpha_1} + \frac{\partial(M_{21}^m A_1)}{\partial\alpha_2} + M_{12}^m \frac{\partial A_1}{\partial\alpha_2} - (M_{22}^m - M_{22}^e)\frac{\partial A_2}{\partial\alpha_1}, \tag{2.4.13}$$

$$Q_{23}^{m}A_1A_2 = \frac{\partial(M_{12}^{m}A_2)}{\partial\alpha_1} + \frac{\partial[(M_{22}^{m} - M_{22}^{e})A_1]}{\partial\alpha_2} + M_{21}^{m}\frac{\partial A_2}{\partial\alpha_1} - (M_{11}^{m} - M_{11}^{e})\frac{\partial A_1}{\partial\alpha_2}, \tag{2.4.14}$$

where ρ is the mass density and h is the thickness of piezoelectric shell. Equations (2.4.10)–(2.4.12) describe the translational electromechanical vibratory motions in the α_1, α_2 and α_3 directions, respectively.

2.4.3 Mechanical Boundary Conditions

Substituting Eqs. (2.4.2a & b) into Eqs. (2.3.16) & (2.3.17) yields

$$\begin{aligned}\int_{t_0}^{t_1} dt \int_{\alpha_2} \Big[&\left(N_{11}^{m} - N_{11}^{e} - N_{11}^{*}\right)\delta u_1 + \left(N_{12}^{m} - N_{12}^{*}\right)\delta u_2 \\ &+ \left(Q_{13}^{m} - Q_{13}^{e} - Q_{13}^{*}\right)\delta u_3 + \left(M_{11}^{m} - M_{11}^{e} - M_{11}^{*}\right)\delta\beta_1, \\ &+ \left(M_{12}^{m} - M_{12}^{*}\right)\left(\frac{\delta u_2}{R_2} - \frac{1}{A_2}\frac{\partial(\delta u_3)}{\partial\alpha_2}\right)\Big]A_2 d\alpha_2 = 0\end{aligned} \tag{2.4.15}$$

$$\begin{aligned}\int_{t_0}^{t_1} dt \int_{\alpha_1} \Big[&\left(N_{21}^{m} - N_{21}^{*}\right)\delta u_1 + \left(N_{22}^{m} - N_{22}^{e} - N_{22}^{*}\right)\delta u_2 \\ &+ \left(Q_{23}^{m} - Q_{23}^{e} - Q_{23}^{*}\right)\delta u_3 + \left(M_{21}^{m} - M_{21}^{*}\right)\left(\frac{\delta u_1}{R_1} - \frac{1}{A_1}\frac{\partial(\delta u_3)}{\partial\alpha_1}\right). \\ &+ \left(M_{22}^{m} - M_{22}^{e} - M_{22}^{*}\right)\delta\beta_2\Big]A_1 d\alpha_1 = 0\end{aligned} \tag{2.4.16}$$

Note that Q_{13}^{e} and Q_{23}^{e} are zero because E_1 and E_2 are 0. Performing the integration by parts of δu_3 terms and then collecting the coefficients, one can derive

$$\begin{aligned}\int_{t_0}^{t_1} dt \int_{\alpha_2} \Big\{&\left(N_{11}^{m} - N_{11}^{e} - N_{11}^{*}\right)\delta u_1 + \left[\left(N_{12}^{m} + \frac{M_{12}^{m}}{R_2}\right) - \left(N_{12}^{*} + \frac{M_{12}^{*}}{R_2}\right)\right]\delta u_2 \\ &+ \left(M_{11}^{m} - M_{11}^{e} - M_{11}^{*}\right)\delta\beta_1 + \left[\left(Q_{13}^{m} + \frac{1}{A_2}\frac{\partial M_{12}^{m}}{\partial\alpha_2}\right)\right. \\ &\left. - \left(Q_{13}^{*} + \frac{1}{A_2}\frac{\partial M_{12}^{*}}{\partial\alpha_2}\right)\right]\delta u_3\Big\}A_2 d\alpha_2 = 0\end{aligned}, \tag{2.4.17}$$

$$\int_{t_0}^{t_1} dt \int_{\alpha_1} \left\{ \left[\left(N_{21}^m + \frac{M_{21}^m}{R_1} \right) - \left(N_{21}^* + \frac{M_{21}^*}{R_1} \right) \right] \delta u_1 + \left(N_{22}^m - N_{22}^e - N_{22}^* \right) \delta u_2 \right.$$
$$+ \left(M_{22}^m - M_{22}^e - M_{22}^* \right) \delta \beta_2 + \left[\left(Q_{23}^m + \frac{1}{A_1} \frac{\partial M_{21}^m}{\partial \alpha_1} \right) \right. \quad .$$
$$\left. \left. - \left(Q_{23}^* + \frac{1}{A_1} \frac{\partial M_{21}^*}{\partial \alpha_1} \right) \right] \delta u_3 \right\} A_1 d\alpha_1 = 0 \tag{2.4.18}$$

These equations are satisfied if either the virtual displacements vanishing or the coefficients of the virtual displacements vanishing. The shear stress resultants are defined as follows:

$$V_{13} = Q_{13}^m + \frac{1}{A_2} \frac{\partial M_{12}^m}{\partial \alpha_2}, \tag{2.4.19a}$$

$$V_{23} = Q_{23}^m + \frac{1}{A_1} \frac{\partial M_{21}^m}{\partial \alpha_1}, \tag{2.4.19b}$$

$$T_{12} = N_{12}^m + \frac{M_{12}^m}{R_2}, \tag{2.4.19c}$$

$$T_{21} = N_{21}^m + \frac{M_{21}^m}{R_1}. \tag{2.4.19d}$$

Thus, mechanical boundary conditions are defined accordingly and they are summarized in Tables 2.1 and 2.2.

Note that usually only either force boundary conditions (B.C.'s) or displacement (Disp.) boundary conditions are selected for a given physical boundary condition. Force/moment boundary conditions are shown in Fig. 2.5. For a totally fixed edge at $\alpha_1 = \alpha_1^*$ (i.e., no motion allowed), the boundary conditions are: $u_1 = 0$, $\beta_1 = 0$,

Table 2.1 Mechanical boundary conditions on the surface defined by $\alpha_1 = \alpha_1^*$

Force B.C.	Disp. B.C.
$N_{11}^m - N_{11}^e = N_{11}^*$	$u_1 = u_1^*$
$M_{11}^m - M_{11}^e = M_{11}^*$	$\beta_1 = \beta_1^*$
$V_{13} = V_{13}^*$	$u_3 = u_3^*$
$T_{12} = T_{12}^*$	$u_2 = u_2^*$

Table 2.2 Mechanical boundary conditions on the surface defined by $\alpha_2 = \alpha_2^*$

Force B.C.	Disp. B.C.
$N_{22}^m - N_{22}^e = N_{22}^*$	$u_2 = u_2^*$
$M_{22}^m - M_{22}^e = M_{22}^*$	$\beta_2 = \beta_2^*$
$V_{23} = V_{23}^*$	$u_3 = u_3^*$
$T_{21} = T_{21}^*$	$u_1 = u_1^*$

$u_3 = 0$, and $u_2 = 0$. For a totally free edge at $\alpha_2 = \alpha_2^*$, i.e., no external forces and moments. The boundary conditions at $\alpha_2 = \alpha_2^*$ are: $N_{22}^m - N_{22}^e = 0$, $M_{22}^m - M_{22}^e = 0$, $V_{23} = 0$, and $T_{21} = 0$.

2.4.4 *Electrostatic Charge Equation and Electric Boundary Conditions*

According to the assumptions discussed earlier, one can further simplify the transverse charge equation of electrostatics as

$$\frac{\partial\left[(e_{31}S_{11} + e_{31}S_{22} + \varepsilon_{33}E_3)A_1A_2(1 + \frac{\alpha_3}{R_1})(1 + \frac{\alpha_3}{R_2})\right]}{\partial\alpha_3} = 0. \tag{2.4.20}$$

Electric boundary condition can be simplified to

$$(e_{31}S_{11} + e_{31}S_{22} + \varepsilon_{33}E_3 + Q_3)A_1A_2(1 + \frac{\alpha_3}{R_1})(1 + \frac{\alpha_3}{R_2}) = 0. \tag{2.4.21}$$

Considering $\frac{\alpha_3}{R_1} \ll 1, \frac{\alpha_3}{R_2} \ll 1$ and $A_1A_2 \neq 0$, one can derive

$$e_{31}S_{11} + e_{31}S_{22} + \varepsilon_{33}E_3 + Q_3 = 0. \tag{2.4.22}$$

Note that Eqs. (2.4.10)–(2.4.12) are the electromechanical vibration equations for a generic piezoelectric thin shell without the effects contributed by the transverse shear deflections and rotatory inertias. The system equations and boundary conditions are all influenced by electromechanical coupling terms introduced by the piezoelectricity. In addition to inherent electromechanical properties, these electric terms can also be used as control forces/moments in distributed structural control of shells (Tzou and Zhong, 1991a & d). The charge equation of electrostatics can be used to estimate output signals of a piezoelectric shell in sensor applications (Tzou, Zhong and Natori, 1993). Substituting generic force/moment and strain expressions into the system equations yields a set of complete system equations ultimately defined by deflections u_k's in the α_1, α_2, and α_3 directions, respectively.

The differences between these electromechanical equations of the piezoelectric thin shell and Love's equations of an elastic thin shell are the electromechanical coupling terms in three governing equations, membrane forces, bending moments, and boundary conditions respectively. (This will be discussed next.)

2.5 Thin Elastic Shell

For a thin elastic shell continuum, all electromechanical coupling terms in the piezoelectric system equations and boundary conditions derived in Sect. 2.4 should be eliminated. (Note that it is assumed that the elastic thin shell follows all the Kirchhoff–Love thin shell assumptions.) Thus, the vibration equations for an elastic thin shell are

$$\begin{aligned}&\frac{\partial(N_{11}^m A_2)}{\partial\alpha_1}+\frac{\partial(N_{21}^m A_1)}{\partial\alpha_2}+N_{12}^m\frac{\partial A_1}{\partial\alpha_2}-N_{22}^m\frac{\partial A_2}{\partial\alpha_1}\\&+A_1A_2\frac{Q_{13}^m}{R_1}=\rho hA_1A_2\ddot{u}_1\end{aligned}, \tag{2.5.1}$$

$$\begin{aligned}&\frac{\partial(N_{12}^m A_2)}{\partial\alpha_1}+\frac{\partial(N_{22}^m A_1)}{\partial\alpha_2}+N_{21}^m\frac{\partial A_2}{\partial\alpha_1}-N_{11}^m\frac{\partial A_1}{\partial\alpha_2}\\&+A_1A_2\frac{Q_{23}^m}{R_2}=\rho hA_1A_2\ddot{u}_2\end{aligned}, \tag{2.5.2}$$

$$\frac{\partial(Q_{13}^m A_2)}{\partial\alpha_1}+\frac{\partial(Q_{23}^m A_1)}{\partial\alpha_2}-A_1A_2\left(\frac{N_{11}^m}{R_1}+\frac{N_{22}^m}{R_{22}}\right)=\rho hA_1A_2\ddot{u}_3. \tag{2.5.3}$$

Transverse shear effects Q_{13}^m and Q_{23}^m are defined as

$$Q_{13}^m A_1A_2=\frac{\partial(M_{11}^m A_2)}{\partial\alpha_1}+\frac{\partial(M_{21}^m A_1)}{\partial\alpha_2}+M_{12}^m\frac{\partial A_1}{\partial\alpha_2}-M_{22}^m\frac{\partial A_2}{\partial\alpha_1}, \tag{2.5.4}$$

$$Q_{23}^m A_1A_2=\frac{\partial(M_{12}^m A_2)}{\partial\alpha_1}+\frac{\partial(M_{22}^m A_1)}{\partial\alpha_2}+M_{21}^m\frac{\partial A_2}{\partial\alpha_1}-M_{11}^m\frac{\partial A_1}{\partial\alpha_2}. \tag{2.5.5}$$

This set of shell system equations is identical to that of conventional elastic shell vibrations (Soedel, 1981). Mechanical boundary conditions of the elastic thin shell are defined accordingly, Tables 2.3 and 2.4.

Table 2.3 Mechanical boundary conditions on the surface defined by $\alpha_1 = \alpha_1^*$

Force B.C.	Disp. B.C.
$N_{11}^m = N_{11}^*$	$u_1 = u_1^*$
$M_{11}^m = M_{11}^*$	$\beta_1 = \beta_1^*$
$V_{13} = V_{13}^*$	$u_3 = u_3^*$
$T_{12} = T_{12}^*$	$u_2 = u_2^*$

Table 2.4 Mechanical boundary conditions on the surface defined by $\alpha_2 = \alpha_2^*$

Force B.C.	Disp. B.C.
$N_{22}^m = N_{22}^*$	$u_2 = u_2^*$
$M_{22}^m = M_{22}^*$	$\beta_2 = \beta_2^*$
$V_{23} = V_{23}^*$	$u_3 = u_3^*$
$T_{21} = T_{21}^*$	$u_1 = u_1^*$

The shear stress resultants become

$$V_{13} = Q_{13}^{m} + \frac{1}{A_2}\frac{\partial M_{12}^{m}}{\partial \alpha_2}, \tag{2.5.6a}$$

$$V_{23} = Q_{23}^{m} + \frac{1}{A_1}\frac{\partial M_{21}^{m}}{\partial \alpha_1}, \tag{2.5.6b}$$

$$T_{12} = N_{12}^{m} + \frac{M_{12}^{m}}{R_2}, \tag{2.5.6c}$$

$$T_{21} = N_{21}^{m} + \frac{M_{21}^{m}}{R_1}. \tag{2.5.6d}$$

These equations are all identical to the derived formulations in (Soedel 1981).

2.6 Summary

In the recent development of active smart structures and structronic systems (integrating sensor, actuator, and control electronics with elastic structures), piezoelectric materials are widely accepted as sensor and actuator materials because of their distinct electromechanical characteristics: the *direct* and *converse piezoelectric effects*. Based on the direct effect, a variety of piezoelectric sensors, e.g., accelerometers, pressure transducers, resonators, acoustic pickups, etc., can be developed. On the other hand, precision piezoelectric actuators and manipulators are based on the converse effect.

This chapter is devoted to a fundamental theoretical development of electromechanical equations and boundary conditions for generic piezoelectric thick and thin shell continua with a symmetrical hexagonal structure (Class $C_{6v} = 6mm$). The effects arising from the transverse shears and rotary inertias, as well as three electric fields were all included. Hamilton's principle and the linear piezoelectricity theory were used to formulate the variation equation of a thick shell continuum. Three electromechanical equations expressed in three generalized coordinates were derived first; three electric fields, transverse shear deformations and rotary inertia effects were all considered. Mechanical and electric boundary conditions were also derived accordingly. (Note that generic mechanical excitations were not included in the system equations.) Generic electromechanical vibration theories of the thick piezoelectric continuum were further simplified to a thick piezoelectric shell expressed in three principal coordinates and two rotary coordinates. In this case, the effects of transverse shear deflections and rotary inertias were still preserved. Piezoelectric shell vibration equations of thin piezoelectric shell continua were also derived by using the thin shell assumptions and Kirchhoff–Love theory. The thin shell equations were further simplified to generic vibration equations of thin elastic shell continua. Applications of the theories to other piezoelectric shell and non-shell continua are demonstrated in Chap. 3.

References

Adelman, N. T. and Stavsky, Y., 1975a, "Vibrations of Radially Polarized Composite Piezoelectric Cylinders and Disks," *J. Sound and Vibration,* 37-44, 1975.

Adelman, N. T. and Stavsky, Y., 1975b, "Axisymmetric Vibrations of Radially Polarized Piezoelectric Ceramic Cylinders," *J. Sound and Vibration,* 38(2), 245-254, 1975.

Bleustain, J. L., 1969, "Some Simple Modes of Wave Propagation in an Infinite Piezoelectric Plate," *J. Acoust. Soc. Am.* 45, 00.614-620, 1969.

Cady, W. G., 1946, *Piezoelectricity*, McGraw?Hill, New York, 1946.

Chau, L. K., 1986, "The Theory of Piezoelectric Shells," *PMM U.S.S.R.,* Vol. 50, No. 1, 98-105, 1986.

Dökmeci, M. C., 1978, "Theory of Vibrations of Coated, Thermopiezoelectric Laminae," *J. Math. Phys.,* Vol. 19, No. 1, January 1978.

Dökmeci, M. C., 1980, "Vibrations of Piezoelectric Crystals," *J. Eng. Soc.* Vol. 18, 431-448, 1980.

Drumheller, D. S. and Kalnins, A., 1970, "Dynamic Shell Theroy for Ferroelectric Ceramics," *J. Acoust. Soc. Am.* 47, 1343, 1970.

Eer Nisse, E. P., 1966, "Coupled?Mode Approach to Elastic Vibration Analysis," *J. Acoust. Soc. Am.* 40, pp.1045-1055, 1966.

Haskins, J. F. and Walsh, 1957, "Vibration of Ferroelectric Cylindrical Shells with Transverse Isotropy," *J. L., J. Acoust. Soc. Am.* 29, 729-734, 1957.

Holland, R. and Eer Nisse, E. P., 1969, *Design of Resonant Piezoelectric Devices*, M.I.T. Press, Cambridge, Massachusetts, 1969.

Holland, R, "Resonant Properties of Piezoelectric Ceramic Rectangular Parallelepipeds," *J. Acoust. Soc. Am.* 43, pp.988-997, 1968.

Kagawa, Y. and Yamabuchi, T., 1976, "A Finite Element Appraoch to Electromechanical Problems with an Application to Energy?Trapped and Suface?Wave Divices," *IEEE Trans. Sonics Ultrason., SU-*23, pp.263-272, 1976.

Keuning, D. H., 1972, "Exact Resonant Frequencies for the Thickness?Twist Trapped Energy Mode in a Piezoceramic Plate," *J. Eng. Math.* 6, pp.143-154, 1972.

Mason, W. P., *Piezoelectric Crystals and Their Applications to Ultrasonics*, Van Nostrand, New York, 1950.

Mindlin, R. D., 1955, *An Introduction to the Mathematical Theory of Vibrations of Elastic Plates*, U.S. Army Signal Corps Engineering Laboratories, Fort Monmouth, New Jersey.

Paul, H. S., 1968, "Vibrational Waves in a Thick Infinite Plate of Piezoelectric Crystal," *J. Acoust. Soc. Am.* 44, pp.478-482, 1968.

Paul, H. S., 1978, "Vibrations of a Hollow Circular Cylinder of Piezoelectric Ceramics," *J. Acoust. Soc. Am. 82(3), September 1978.*

Paul, H. S., 1982, "Asymptotic Analysis of the Modes of Wave Propagation in a Piezoelectric Solid Cylinder," *J. Acoust. Soc. Am.* 71(2), February 1982.

Paul, H. S., 1986, "Axisymmetric Vibration of a Piezoelectric Solid Cylinder Guided by a Thin Film," *J. Acoust. Soc. Am.* 80(4), October 1986.

Paul, H. S., 1987, "Wave Propagation in a Piezoelectric Solid Cylinder of Arbitrary Cross Section," *J. Acoust. Soc. Am.* 82(6), December 1987.

Rogacheva, N. N., 1982, "Equations of State of Piezoceramic Shells," *PMM U.S.S.R.,* Vol. 45, 677-684, 1982.

Rogacheva, N. N., 1984a, "On Boundary conditions in the Theory of Piezoceramic Shells Polarized Along Coordinate Lines," *PMM U.S.S.R.,* Vol. 47, No. 2, 220-226, 1984.

Rogacheva, N. N., 1984b, "On Stain?Venant Type Conditions in the Theory of Piezoelastic Shells," *PMM U.S.S.R.,* Vol. 48, No. 2, 213-216, 1984.

Rogacheva, N. N., 1986, "Classification of Free Piezoceramic Shell Vibrations," *PMM U.S.S.R.,* Vol. 50, No. 1,106-111, 1986.

Schmidt, G. H., 1972, "Extensional Vibrations of Piezoelectric Plates," *J. Eng. Math.* 6, pp.133-142, 1972.

Senik, N. A. and Kudriavtsev, B. A., 1980, "Equations on the Theory of Piezoceramic Shells," In: *Mechanics of a solid deformable body and related analytical problems. Moscow, Izd. mosk. Inst. Chim. Mashinostroeniia, 1980.*

Soedel, W., 1981, *Vibrations of Shells and Plates*, Marcel Dekker, Inc., New York, 1981.

Tiersten, H. F. and Mindlin, R.D., 1962, *Appl. Math.* 20, 107-109, 1962.

Tiersten, H. F., 1969, *Linear Piezoelectric Plate Vibrations*, Plenum, New York, 1969.

Toupin, R. A., 1959, "Piezoelectric Relation and the Rachial Deformation of Polarized Spherical Shell," *J. Acoust. Soc. Am.* 31, 315-318, 1959.

Tzou, H.S., 1991, "Distributed Modal Identification and Vibration Control of Continua: Theory and Applications," ASME *Journal of Dynamic Systems, Measurements, and Control*, Vol. (133), No.(3), 1991.

Tzou, H.S., 1992, "A New Distributed Sensation and Control Theory for "Intelligent" Shells," *Journal of Sound and Vibration*, Vol.152, No.3, pp.335-350, March 1992.

Tzou, H.S. and Anderson, G.L. (Editors), *Intelligent Structural Systems*, ISBN No.0-7923-1920-6, 488 pages, Book, Kluwer Academic Publishers, August 1992.

Tzou, H.S., and Gadre, M, 1989, "Theoretical Analysis of a Multi-Layered Thin Shell Coupled with Piezoelectric Shell Actuators for Distributed Vibration Control," *Journal of Sound and Vibration*, Vol.132, No.3, pp.433-450, August 1989.

Tzou, H.S. and Zhong, J.P., 1990, "Electromechanical Dynamics of Piezoelectric Shell Distributed Systems, Part-1&2," *Robotics Research-1990*, ASME-DSC-Vol.26, pp.207-211, 1990 ASME Winter Annual Meetings, Dallas, Texas, Nov. 25-30, 1990; *ASME Journal of Dynamic Systems, Measurements, and Control.*

Tzou, H.S. and Zhong, J.P., 1991a, "Adaptive Piezoelectric Shell Structures: Theory and Experiments," *AIAA/ASME/ASCE/AHA/ASC 32nd Structures, Structural Dynamics and Materials Conference*, pp.2290-2296, Paper No. AIAA-91-1238-CP, Baltimore, Maryland, April 8-10, 1991. *Mechanical Systems and Signal Processing*, Vol.(7), No.(3), May 1993.

Tzou, H.S. and Zhong, J.P., 1991b, "Theory on Hexagonal Symmetrical Piezoelectric Thick Shells Applied to Smart Structures," *Structural Vibration and Acoustics*, Edrs. Huang, Tzou, et al., ASME-DE-Vol.34, pp.7-15, Symposium on Intelligent Structural Systems, 1991 ASME 13th Biennial Conference on Mechanical Vibration and Noise, Miami, Florida, September 22-25, 1991.

Tzou, H.S., Zhong, J.P., and Natori, M.C., 1993, "Sensor Mechanics of Distributed Shell Convolving Sensors Applied to Flexible Rings," ASME Transactions, Journal of Vibration & Acoustics, Vol.115, No.1, pp.40 46, January 1993.

Tzou, H.S. and Zhong, J.P., 1991d, "Control of Piezoelectric Cylindrical Shells via Distributed In-Plane Membrane Forces," *Controls for Aerospace Systems*, DSC-Vol.35, pp.15-20, Distributed Control of Flexible Structures, Aerospace Panel, Dynamic Systems and Control Division, 1991 ASME WAM, Atlanta, GA, December 1-6, 1991.

Tzou, H.S., 1992, "Active Piezoelectric Shell Continua," *Intelligent Structural Systems*, H.S. Tzou & G.L. Anderson, (Editors), pp.9-74, Kluwer Academic Publishers, Aug. 1992.

Tzou, H.S., 2000, "Integrated Piezoelectric Sensor Actuator Design for Distributed Identification and Control of "Smart" Machines and Flexible Robots: Part 1: Theory and Experiments," *Intelligent Systems and Robotics*, G.W. Zobrist and C.Y Ho, (Editors), pp.92-122, Gordon and Breach Science Publishers (2000OPA), Netherlands, 2000.

Tzou, H.S., Zhu, Y.T. and Hagiwara, I., 2002, "Distributed Precision Control of Structronic Shells and Common Shapes – A New Approach," *Recent Research Developments in Sound & Vibration*, ISBN: 81-7895-031-6, Vol.1, Part-2, pp.613-646, TransWorld Research, 2002.

Tzou, H.S. and Fukuda, T. (Editors), Precision Sensors, Actuators, and Systems, ISBN 0-7923-2015-8, 478 pages, Book, Kluwer Academic Publishers, Dordrecht/Boston/London, December 1992; April 2002 (2nd print).

Chapter 3
Common Piezoelectric Continua and Active Piezoelectric Structures

In this chapter, a simple reduction procedure is developed to apply the three generic double-curvature piezoelectric shell continuum theories, derived in Chap. 2, to other common piezoelectric continua or structures, such as shells of revolutions, paraboloidal shells, toroidal shells, spherical shells, cylindrical shells, conical shells, plates, rings, etc. (Tzou and Zhong, 1990; Tzou, 1992a, b; Tzou, 2000; Tzou, Zhu and Hagiwara, 2002). Note that the generic piezoelectric thin shell theory is used as the fundamental theory in later derivations and analyses of distributed sensing, energy harvesting and precision control. Equations of motion of corresponding thin elastic shells can be easily derived by eliminating all electromechanical coupling terms in the piezoelectric shell equations.

As a practical application of the piezoelectric continua, active vibration control of piezoelectric shell structures via the converse piezoelectric effect is proposed. Theories of active piezoelectric continua are derived. Generic feedback schemes, e.g., displacement, velocity, and acceleration, are proposed and corresponding governing equations formulated. Frequency control of a piezoelectric bimorph beam is investigated (Tzou & Zhong, 1991a). Damping control of a piezoelectric cylindrical shell via the in-plane membrane forces is also studied and effects of shell deepness evaluated (Tzou & Zhong, 1991b).

3.1 Reduction Procedures

There are four essential geometric parameters required to carry out the reductions, namely, two Lamé parameters (A_1 and A_2) and two radii (R_1 and R_2) of curvature (Soedel, 1981; Tzou, 1992a, b). The general procedures, a step-by-step approach, are presented in this section; demonstrations of the procedures are presented in case studies (Tzou and Zhong 1990, 1991c).

H. (HS) Tzou, *Piezoelectric Shells*, Solid Mechanics and Its Applications 247,
https://doi.org/10.1007/978-94-024-1258-1_3

1) Select a Coordinate System:

The original generic shell continuum was defined in a generic tri-orthogonal coordinate system (i.e., $\alpha_1, \alpha_2, \& \alpha_3$). Depending on a given piezoelectric continuum, these coordinates can be redefined to best describe its coordinate system, e.g., $\alpha_1 = x$ and $\alpha_2 = y$ for a rectangular plate; $\alpha_1 = x$ and $\alpha_2 = \theta$ for a cylindrical shell and thin cylindrical tube shell, etc.

2) Determine the Radii of Curvature:

The radii of curvature R_1 and R_2 of the two in-plane coordinate axes α_1 and α_2 defining the neutral surface can be easily observed from the coordinate system defined in Step 1. For example, the radii of x and y axes in a rectangular plate are $R_1 = \infty$ and $R_2 = \infty$; the radii are $R_1 = \infty$ and $R_2 = R$ in a cylindrical shell.

3) Derive a Fundamental Form:

A fundamental form represents an infinitesimal distance ds on the neutral surface of the shell continuum; the distance is the hypotenuse of a (or an approximate) right-angle triangle defined by the infinitesimal distances ($d\alpha_1$ and $d\alpha_2$) of the two in-plane coordinates. From the fundamental form, two Lamé parameters A_1 and A_2 and the two selected coordinates α_1 and α_2 can be defined, i.e.,

$$(ds)^2 = (A_1)^2 (d\alpha_1)^2 + (A_2)^2 (d\alpha_2)^2. \tag{3.1.1}$$

For example, the fundamental form for a rectangular plate (defined in a Cartesian coordinate system x and y) is

$$(ds)^2 = (1)^2 (dx)^2 + (1)^2 (dy)^2, \tag{3.1.2}$$

where the Lamé parameters are $A_1 = 1$ and $A_2 = 1$. For a cylindrical shell whose neutral surface defined by the x and θ axes, the fundamental form is defined as

$$(ds)^2 = (1)^2 (dx)^2 + (R)^2 (d\theta)^2, \tag{3.1.3}$$

where $A_1 = 1$ and $A_2 = R$ (radius of the cylinder). These two cases are further explained in details next. Other geometries can be defined accordingly (Tzou, 1992a, b; Tzou, 2000; Tzou, Zhu and Hagiwara, 2002).

4) Simplify the Generic System Equations:

By substituting the four geometric parameters A_1, A_2, R_1, and R_2 into the generic electromechanical shell equations and simplifying, one can easily derive the governing equations for that specific piezoelectric shell or non-shell continuum. System equations for the corresponding elastic shell can be easily obtained by neglecting all electric coupling terms in the piezoelectric equations and boundary conditions. (This was discussed in Sect. 2.5, Chap. 2.)

The reduction procedures described above are demonstrated in two examples: 1) a rectangular piezoelectric plate and 2) a piezoelectric shell of revolution. The equations for the first, the plate, will be compared with published results. The second, the piezoelectric shell of revolution, represents another broad class of piezoelectric shell continua, e.g., paraboloidal shells, cylindrical shells, conical shells, spherical shells, etc. Non-shell-type piezoelectric continua, e.g., piezoelectric arches, beams, rods, are not discussed here; in general, their reduction procedures are identical. However, the derivations are relatively straightforward for non-shell-type piezoelectric continua because usually there is only one dimension involved.

3.2 Piezoelectric Plates

The coordinate system for a piezoelectric rectangular plate, Fig. 3.1, is an ordinary Cartesian coordinate system, x, y, & z, in which the x and y axes define the neutral surface and z the transverse direction. Thus, $\alpha_1 = x$ and $\alpha_2 = y$.

The *fundamental form* for a rectangular plate is

$$(ds)^2 = (1)^2(dx)^2 + (1)^2(dy)^2. \tag{3.2.1}$$

Thus, the Lamé parameters $A_1 = 1$ and $A_2 = 1$, and the radii of curvature of the two in-plane (i.e., the neutral surface) coordinate axes are both infinite, i.e., $R_1 = \infty$ and $R_2 = \infty$. There are two plate cases presented here: 1) a thick piezoelectric plate and 2) a thin piezoelectric plate. The first case will be verified by comparing with published results.

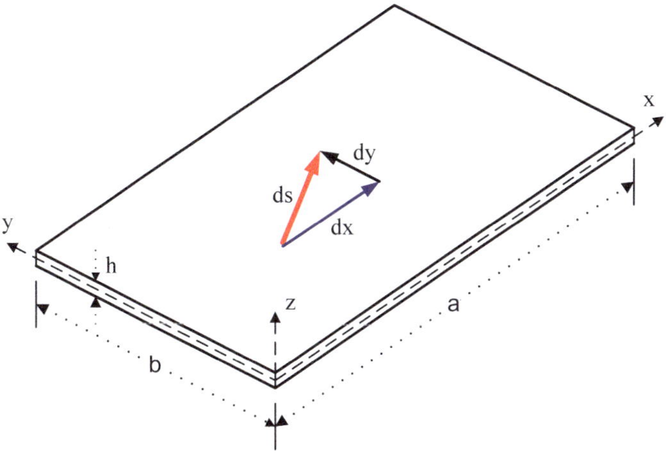

Fig. 3.1 A piezoelectric plate

3.2.1 Thick Piezoelectric Rectangular Plate

For a thick piezoelectric plate made of symmetrical hexagonal piezoelectrics, the plate electromechanical equations can be derived using the equations, Sect. 2.3, of thick piezoelectric shell continua. Substituting the system parameters $A_1, A_2, R_1, \text{and } R_2$ into the shell equations yields

$$\frac{\partial}{\partial \alpha_1}(\sigma_{11} - e_{31}E_3) + \frac{\partial \sigma_{12}}{\partial \alpha_2} + \frac{\partial}{\partial \alpha_3}(\sigma_{13} - e_{15}E_1) = \rho \ddot{U}_1, \tag{3.2.2}$$

$$\frac{\partial \sigma_{12}}{\partial \alpha_1} + \frac{\partial}{\partial \alpha_2}(\sigma_{22} - e_{31}E_3) + \frac{\partial}{\partial \alpha_3}(\sigma_{23} - e_{15}E_2) = \rho \ddot{U}_2, \tag{3.2.3}$$

$$\frac{\partial}{\partial \alpha_1}(\sigma_{13} - e_{15}E_1) + \frac{\partial}{\partial \alpha_2}(\sigma_{23} - e_{15}E_2) + \frac{\partial}{\partial \alpha_3}(\sigma_{33} - e_{33}E_3) = \rho \ddot{U}_3. \tag{3.2.4}$$

Substituting general stress–strain equations, Sect. 2.1, into the above dynamic equations gives

$$\begin{aligned} c_{11}\frac{\partial S_{11}}{\partial \alpha_1} + c_{12}\frac{\partial S_{22}}{\partial \alpha_1} + c_{13}\frac{\partial S_{33}}{\partial \alpha_1} - e_{31}\frac{\partial E_3}{\partial \alpha_1} + c_{66}\frac{\partial S_{12}}{\partial \alpha_2} \\ + c_{44}\frac{\partial S_{13}}{\partial \alpha_3} - e_{15}\frac{\partial E_1}{\partial \alpha_3} = \rho \ddot{U}_1, \end{aligned} \tag{3.2.5}$$

$$\begin{aligned} c_{66}\frac{\partial S_{12}}{\partial \alpha_1} + c_{12}\frac{\partial S_{11}}{\partial \alpha_2} + c_{11}\frac{\partial S_{22}}{\partial \alpha_2} + c_{13}\frac{\partial S_{33}}{\partial \alpha_2} - e_{31}\frac{\partial E_3}{\partial \alpha_2} \\ + c_{44}\frac{\partial S_{23}}{\partial \alpha_3} - e_{15}\frac{\partial E_2}{\partial \alpha_3} = \rho \ddot{U}_2, \end{aligned} \tag{3.2.6}$$

$$\begin{aligned} c_{44}\frac{\partial S_{13}}{\partial \alpha_1} - e_{15}\frac{\partial E_1}{\partial \alpha_1} + c_{44}\frac{\partial S_{23}}{\partial \alpha_2} - e_{15}\frac{\partial E_2}{\partial \alpha_2} + c_{13}\frac{\partial S_{11}}{\partial \alpha_3} \\ + c_{12}\frac{\partial S_{22}}{\partial \alpha_3} + c_{33}\frac{\partial S_{33}}{\partial \alpha_3} - e_{33}\frac{\partial E_3}{\partial \alpha_3} = \rho \ddot{U}_3. \end{aligned} \tag{3.2.7}$$

The strain-displacement relations, defined in Sect. 2.1, are simplified to

$$S_{11} = \frac{\partial U_1}{\partial \alpha_1}, \tag{3.2.8a}$$

$$S_{22} = \frac{\partial U_2}{\partial \alpha_2}, \tag{3.2.8b}$$

$$S_{33} = \frac{\partial U_3}{\partial \alpha_3}, \tag{3.2.8c}$$

$$S_{12} = \frac{\partial U_1}{\partial \alpha_2} + \frac{\partial U_2}{\partial \alpha_1}, \tag{3.2.8d}$$

$$S_{13} = \frac{\partial U_1}{\partial \alpha_3} + \frac{\partial U_3}{\partial \alpha_1}, \tag{3.2.8e}$$

$$S_{23} = \frac{\partial U_2}{\partial \alpha_3} + \frac{\partial U_3}{\partial \alpha_2}. \tag{3.2.8f}$$

The electric field relations between the electric-field and the electric-potential, defined in Sect. 2.1, are also simplified to

$$E_1 = -\frac{\partial \varphi}{\partial \alpha_1}, \tag{3.2.9a}$$

$$E_2 = -\frac{\partial \varphi}{\partial \alpha_2}, \tag{3.2.9b}$$

$$E_3 = -\frac{\partial \varphi}{\partial \alpha_3}. \tag{3.2.9c}$$

Substituting all these reduced expressions into Eqs. (3.2.5)–(3.2.7) and regrouping the terms yields

$$\begin{aligned} &c_{11}\frac{\partial^2 U_1}{\partial \alpha_1^2} + (c_{12}+c_{66})\frac{\partial^2 U_2}{\partial \alpha_1 \partial \alpha_2} + (c_{13}+c_{44})\frac{\partial^2 U_3}{\partial \alpha_1 \partial \alpha_3} + c_{66}\frac{\partial^2 U_1}{\partial \alpha_2^2} \\ &+ c_{44}\frac{\partial^2 U_1}{\partial \alpha_3^2} + (e_{31}+e_{15})\frac{\partial^2 \varphi}{\partial \alpha_1 \partial \alpha_3} = \rho \ddot{U}_1 \end{aligned}, \tag{3.2.10}$$

$$\begin{aligned} &c_{66}\frac{\partial^2 U_2}{\partial \alpha_1^2} + (c_{66}+c_{12})\frac{\partial^2 U_1}{\partial \alpha_1 \partial \alpha_2} + c_{11}\frac{\partial^2 U_2}{\partial \alpha_2^2} + (c_{13}+c_{44})\frac{\partial^2 U_3}{\partial \alpha_2 \partial \alpha_3} \\ &+ c_{44}\frac{\partial^2 U_2}{\partial \alpha_3^2} + (e_{31}+e_{15})\frac{\partial^2 \varphi}{\partial \alpha_2 \partial \alpha_3} = \rho \ddot{U}_2 \end{aligned}, \tag{3.2.11}$$

$$\begin{aligned} &c_{44}\frac{\partial^2 U_3}{\partial \alpha_1^2} + (c_{44}+c_{13})\frac{\partial^2 U_1}{\partial \alpha_1 \partial \alpha_3} + c_{44}\frac{\partial^2 U_3}{\partial \alpha_2^2} + (c_{44}+c_{13})\frac{\partial^2 U_2}{\partial \alpha_2 \partial \alpha_3} \\ &+ c_{33}\frac{\partial^2 U_3}{\partial \alpha_3^2} + e_{15}\frac{\partial^2 \varphi}{\partial \alpha_1^2} + e_{15}\frac{\partial^2 \varphi}{\partial \alpha_2^2} + e_{33}\frac{\partial^2 \varphi}{\partial \alpha_3^2} = \rho \ddot{U}_3 \end{aligned}. \tag{3.2.12}$$

This set of electromechanical equations is identical to that published in a book (Tiersten, 1969, p. 58). Note that Tiersten's piezoelectric plate equations were derived directly from Hamilton's principle. The equations were in tensor expressions and are provided in Appendix for comparison.

3.2.2 Thin Piezoelectric Plate

In this case, the electromechanical equations of thin piezoelectric shell continua, Sect. 2.4, are used in the derivation. For thin shells, transverse shear deformations and rotary inertias are neglected. Note that only the transverse electric field E_3 is considered and $E_1 = E_2 = 0$. Substituting these parameters into the thin shell piezoelectric equations, Sect. 2.4, yields the piezoelectric plate governing equations.

$$\frac{\partial(N_{xx}^m - N_{xx}^e)}{\partial x} + \frac{\partial N_{yx}^m}{\partial y} = \rho h \ddot{u}_x, \tag{3.2.13}$$

$$\frac{\partial N_{xy}^m}{\partial x} + \frac{\partial\left(N_{yy}^m - N_{yy}^e\right)}{\partial y} = \rho h \ddot{u}_y, \tag{3.2.14}$$

$$\frac{\partial Q_{x3}^m}{\partial x} + \frac{\partial Q_{y3}^m}{\partial y} + (\sigma_{33} - e_{33}E_3)|_{-h/2}^{h/2} = \rho h \ddot{u}_3, \tag{3.2.15}$$

where

$$Q_{x3}^m = \frac{\partial(M_{xx}^m - M_{xx}^e)}{\partial x} + \frac{\partial M_{yx}^m}{\partial y}, \tag{3.2.16}$$

$$Q_{y3}^m = \frac{\partial M_{xy}^m}{\partial x} + \frac{\partial\left(M_{yy}^m - M_{yy}^e\right)}{\partial y}. \tag{3.2.17}$$

Note that Q_{x3}^e and Q_{y3}^e are zero because E_1 and E_2 are zero. Transverse stress effects $(\sigma_{33} - e_{33}E_3)|_{-h/2}^{h/2}$ in Eq. (3.2.15) can also be neglected. The charge equation of electrostatics is defined as

$$\frac{\partial(e_{31}S_{11} + e_{31}S_{22} + \varepsilon_{33}E_3)}{\partial \alpha_3} = 0. \tag{3.2.18}$$

In order to relate the resultant forces and moments to the general displacement coordinates u_i's, one must use: 1) force/moment–strain relations, 2) strain–displacement relations, and 3) definitions of rotation angles, Sect. 2.4. For example, the strain-displacement relations for a rectangular plate can be simplified to

$$S_{xx} = \frac{\partial u_x}{\partial x} + \alpha_3 \frac{\partial \beta_x}{\partial x}, \tag{3.2.19a}$$

$$S_{yy} = \frac{\partial u_y}{\partial y} + \alpha_3 \frac{\partial \beta_y}{\partial y}, \tag{3.2.19b}$$

$$S_{xy} = \frac{\partial u_y}{\partial x} + \frac{\partial u_x}{\partial y} + \alpha_3 \left(\frac{\partial \beta_y}{\partial x} + \frac{\partial \beta_x}{\partial y} \right); \tag{3.2.19c}$$

where α_3 defines the distance measured from the neutral surface; the rotation angles β_x and β_y are defined as

$$\beta_x = -\frac{\partial u_3}{\partial x}, \tag{3.2.20a}$$

$$\beta_y = -\frac{\partial u_3}{\partial y}. \tag{3.2.20b}$$

3.3 Thin Piezoelectric Shell of Revolution

In this case, it is assumed that the shell of revolution continuum is subjected to a transverse electric field E_3 which is perpendicular to the surface. The transverse shear deformation and rotary inertia are neglected. Figure 3.2 illustrates the geometry in which coordinates (ψ, θ, u_3) and infinitesimal distance ds are also defined. Note that $u_3(\alpha_3)$ is normal to the surface defined by $\alpha_1 = \psi$ and $\alpha_2 = \theta$.

Recall that the coordinates are selected as $\alpha_1 = \psi$ and $\alpha_2 = \theta$. The radii of curvatures of the two axes are: $R_1 = R_\psi$ and $R_2 = R_\theta$. The infinitesimal distances or fundamental form ds is defined as

$$(ds)^2 = R_\psi^2 (d\psi)^2 + R_\theta^2 \sin^2 \psi (d\theta)^2. \tag{3.3.1}$$

Thus, the two Lamé parameters are

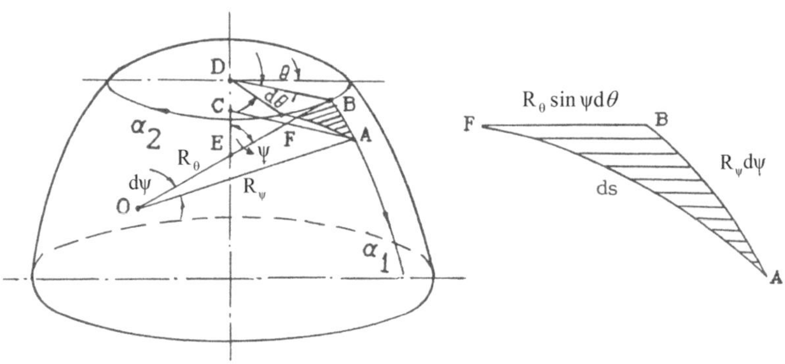

Fig. 3.2 A piezoelectric shell of revolution

$$A_1 = R_\psi, \tag{3.3.2a}$$

$$A_2 = R_\theta \sin\psi. \tag{3.3.2b}$$

Substituting these parameters into Eqs. (2.4.10)–(2.4.14), Sect. 2.4, with subscripts 1 and 2 replaced by ψ and θ, and with $R_\psi \cos\psi d\psi = d(R_\theta \sin\psi)$ yields a set of electromechanical equations.

$$\begin{aligned} &\frac{\partial}{\partial\psi}\left[\left(N^m_{\psi\psi} - N^e_{\psi\psi}\right)R_\theta \sin\psi\right] + R_\psi \frac{\partial N^m_{\theta\psi}}{\partial\theta} - \left(N^m_{\theta\theta} - N^e_{\theta\theta}\right)R_\psi \cos\psi \\ &+ R_\psi R_\theta \sin\psi\left(\frac{Q^m_{\psi 3}}{R_\psi} + F_\psi\right) = R_\psi R_\theta \sin\psi \rho h \frac{\partial^2 u_\psi}{\partial t^2}, \end{aligned} \tag{3.3.3}$$

$$\begin{aligned} &\frac{\partial}{\partial\psi}\left(N^m_{\psi\theta} R_\theta \sin\psi\right) + R_\psi \frac{\partial\left(N^m_{\theta\theta} - N^e_{\theta\theta}\right)}{\partial\theta} + N^m_{\theta\psi} R_\psi \cos\psi \\ &+ R_\psi R_\theta \sin\psi\left(\frac{Q^m_{\theta 3}}{R_\theta} + F_\theta\right) = R_\psi R_\theta \sin\psi \rho h \frac{\partial^2 u_\theta}{\partial t^2} \end{aligned}, \tag{3.3.4}$$

$$\begin{aligned} &\frac{\partial}{\partial\psi}\left(Q^m_{\psi 3} R_\theta \sin\psi\right) + R_\psi \frac{\partial Q^m_{\theta 3}}{\partial\theta} - \left(\frac{N^m_{\psi\psi} - N^e_{\psi\psi}}{R_\psi} + \frac{N^m_{\theta\theta} - N^e_{\theta\theta}}{R_\theta}\right) R_\psi R_\theta \sin\psi \\ &+ F_3 R_\psi R_\theta \sin\psi + R_\psi R_\theta \sin\psi (\sigma_{33} - e_{33} E_3)\Big|_{-h/2}^{h/2} \\ &= R_\psi R_\theta \sin\psi \rho h \frac{\partial^2 u_3}{\partial t^2} \end{aligned}, \tag{3.3.5}$$

where

$$\begin{aligned} Q^m_{\psi 3} = \frac{1}{R_\psi R_\theta \sin\psi}\Bigg(&\frac{\partial}{\partial\psi}\left[\left(M^m_{\psi\psi} - M^e_{\psi\psi}\right)R_\theta \sin\psi\right] + R_\psi \frac{\partial M^m_{\theta\psi}}{\partial\theta} \\ &- \left(M^m_{\theta\theta} - M^e_{\theta\theta}\right)R_\psi \cos\psi\Bigg) \end{aligned}, \tag{3.3.6}$$

$$\begin{aligned} Q^m_{\theta 3} = \frac{1}{R_\psi R_\theta \sin\psi}\Bigg(&\frac{\partial}{\partial\psi}\left(M^m_{\psi\theta} R_\theta \sin\psi\right) + R_\psi \frac{\partial\left(M^m_{\theta\theta} - M^e_{\theta\theta}\right)}{\partial\theta} \\ &+ M^m_{\theta\psi} R_\psi \cos\psi\Bigg) \end{aligned}. \tag{3.3.7}$$

Note that $Q^e_{\psi 3}$ and $Q^e_{\theta 3}$ are zero because E_1 and E_2 are zero. Transverse stress effects $R_\psi R_\theta \sin\psi(\sigma_{33} - e_{33}E_3)\Big|_{-h/2}^{h/2}$ can also be neglected. Note that the mechanical excitations (F_θ, F_ψ, and F_3) are included in this set of electromechanical equations. The charge equation of electrostatics becomes

$$\frac{\partial\left[(e_{31}S_{\psi\psi}+e_{31}S_{\theta\theta}+\varepsilon_{33}E_3)R_\psi R_\theta\sin\psi\left(1+\frac{\alpha_3}{R_\psi}\right)\left(1+\frac{\alpha_3}{R_\theta}\right)\right]}{\partial\alpha_3}=0. \qquad (3.3.8)$$

The strain-displacement relations for the shell of revolution are simplified to

$$S_{\psi\psi}=\frac{1}{R_\psi}\left(\frac{\partial u_\psi}{\partial\psi}+u_3\right)+\frac{\alpha_3}{R_\psi}\frac{\partial\beta_\psi}{\partial\psi}, \qquad (3.3.9a)$$

$$\begin{aligned}S_{\theta\theta}=&\frac{1}{R_\theta\sin\psi}\left(\frac{\partial u_\theta}{\partial\theta}+u_\psi\cos\psi+u_3\sin\psi\right)\\&+\frac{\alpha_3}{R_\theta\sin\psi}\left(\frac{\partial\beta_\theta}{\partial\theta}+\beta_\psi\cos\psi\right)\end{aligned}, \qquad (3.3.9b)$$

$$\begin{aligned}S_{\psi\theta}=&\frac{R_\theta}{R_\psi}\sin\psi\frac{\partial}{\partial\psi}\left(\frac{u_\theta}{R_\theta\sin\psi}\right)+\frac{1}{R_\theta\sin\psi}\frac{\partial u_\psi}{\partial\theta}\\&+\alpha_3\left\{\frac{R_\theta}{R_\psi}\sin\psi\frac{\partial}{\partial\psi}\left(\frac{\beta_\theta}{R_\theta\sin\psi}\right)+\frac{1}{R_\theta\sin\psi}\frac{\partial\beta_\psi}{\partial\theta}\right\}\end{aligned}, \qquad (3.3.9c)$$

where the rotation angles are defined as

$$\beta_\psi=\frac{1}{R_\psi}\left(u_\psi-\frac{\partial u_3}{\partial\psi}\right), \qquad (3.3.10a)$$

$$\beta_\theta=\frac{1}{R_\theta\sin\psi}\left(u_\theta\sin\psi-\frac{\partial u_3}{\partial\theta}\right). \qquad (3.3.10b)$$

Note that all mechanical forces/moments and electromechanical coupling terms were defined in Sect. 2.4 of Chap. 2. Substituting all force/moment and stress relations, stress–strain relations, and strain–displacement relations into the governing equations yields a set of complete electromechanical governing equations in terms of generic displacements u_ψ, u_θ and u_3. Similar procedures can also be applied to the charge equation of electrostatics.

3.3.1 Applications (Thin Piezoelectric Shell of Revolution)

As discussed previously, the thin piezoelectric shell of revolution also represents a class of piezoelectric shell continua, e.g., spheres, cylinders, cones, etc. With appropriate boundary conditions, all related shell panel continua can be defined accordingly. Simplifications of the shell of revolution equations to these engineering structures are demonstrated next.

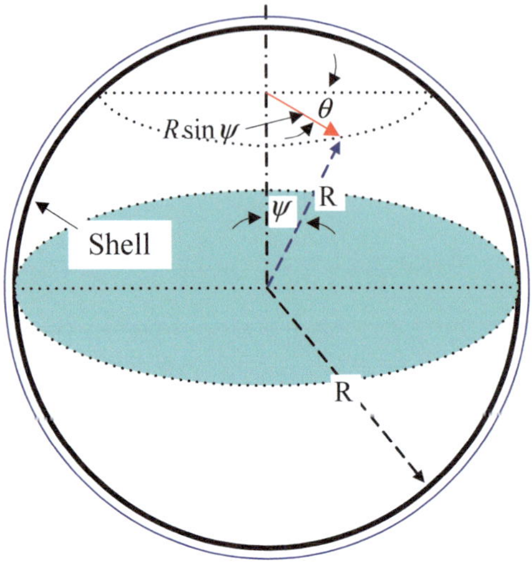

Fig. 3.3 A piezoelectric spherical Shell

3.3.1.1 Piezoelectric Spherical Shell

For a piezoelectric spherical shell, Fig. 3.3, the radii of curvature are constants, i.e., $R_\theta = R$ and $R_\psi = R$; the fundamental form is defined as

$$(ds)^2 = R^2(d\psi)^2 + R^2 \sin^2 \psi (d\theta)^2. \quad (3.3.11)$$

Thus, the Lamé parameters are $A_1 = R$ and $A_2 = R \sin \psi$. Using the parameters defined above, one can easily simplify the original piezoelectric shell equations and derive a new set of electromechanical equations for the piezoelectric spherical shell. Note that one can also substitute these parameters into the equations for the piezoelectric shell of revolution and simplify them accordingly.

3.3.1.2 Piezoelectric Cylindrical Shell

For a piezoelectric cylindrical shell, Fig. 3.4, it is assumed that α_1 is aligned with the cylinder height, i.e., $\alpha_1 = x$; α_2 is the rotational angle θ defining the circumferential direction. The radii of the cylinder are $R_\psi = R_x = \infty$ and $R_\theta = R$ (constant); the fundamental form is defined as

$$(ds)^2 = (1)^2(dx)^2 + (R)(d\theta)^2. \quad (3.3.12)$$

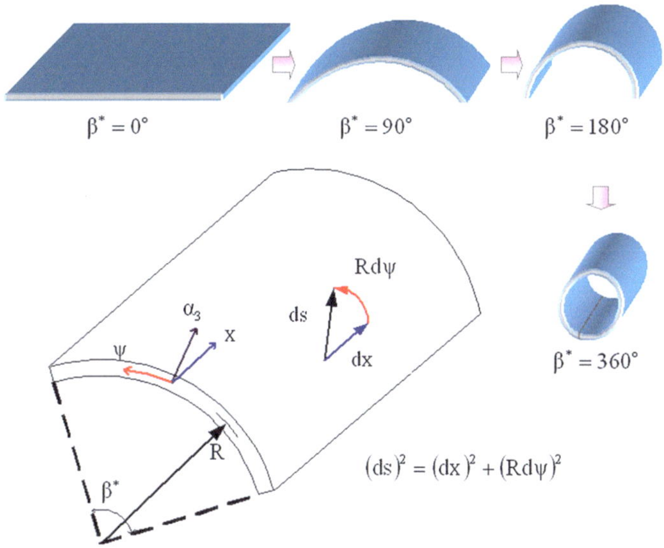

Fig. 3.4 A piezoelectric cylindrical Shell

The Lamé parameters are $A_1 = 1$ and $A_2 = R$. Thus, with the new Lamé parameters and radii of curvature, a new set of electromechanical equations for the piezoelectric cylinder is defined.

3.3.1.3 Piezoelectric Conical Shell

For a piezoelectric cone shell continuum, it is assumed that α_1 is aligned with the height direction on the shell surface, i.e., $\alpha_1 = x$; α_2 is the rotational angle θ, Fig. 3.5. The radius for the α_1 axis is ∞ and that for the α_2 axis is $(x \tan \beta^*)$, i.e., $R_\psi = R_x = \infty$ and $R_\theta = x \tan \beta^*$ where β^* is an angle measured from the center line perpendicular to the cone base at its maximum height to the shell surface. Thus, the fundamental form is defined as

$$(ds)^2 = (1)^2 (dx)^2 + (x \tan \beta^*)^2 (d\theta)^2. \tag{3.3.13}$$

Thus, the Lamé parameters are $A_1 = 1$ and $A_2 = x \tan \beta^*$. Again, a new set of governing equations for the piezoelectric cone is defined.

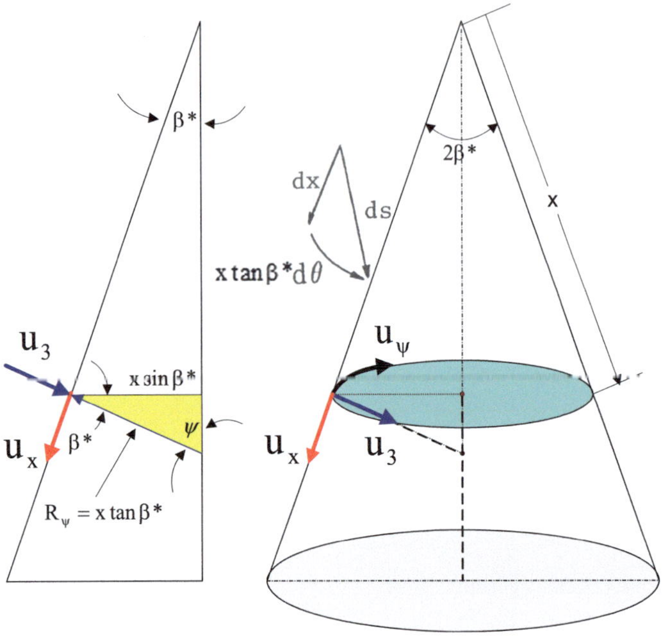

Fig. 3.5 A piezoelectric conical Shell

3.4 Active Piezoelectric Shell Structures

As discussed in Chap. 2, electric forces and moments in piezoelectric shell equations can be used to control the dynamic characteristics, e.g., frequency and damping, of piezoelectric shell continua and structures. In this section, application of the generic piezoelectric shell theory to active thin shell continua is proposed and the governing equations derived. Figure 3.6 illustrates a feedback control system for active piezoelectric structures, in which sensor signals are processed in a central processing unit (C.P.U.) and then fed back to the piezoelectric structures. The C.P.U. decides the control algorithms, control gains, signal filtering, etc. For demonstrations of active piezoelectric structures, two case studies are given: 1) frequency control (*displacement feedback*) of a piezoelectric bimorph beam and 2) damping control (*velocity feedback*) of a piezoelectric cylindrical shell; these are presented in Sects. 3.7 and 3.8, respectively. It is assumed that the active piezoelectric shell continua are made of symmetrical hexagonal piezoelectric materials and subjected to a transverse electric field E_3.

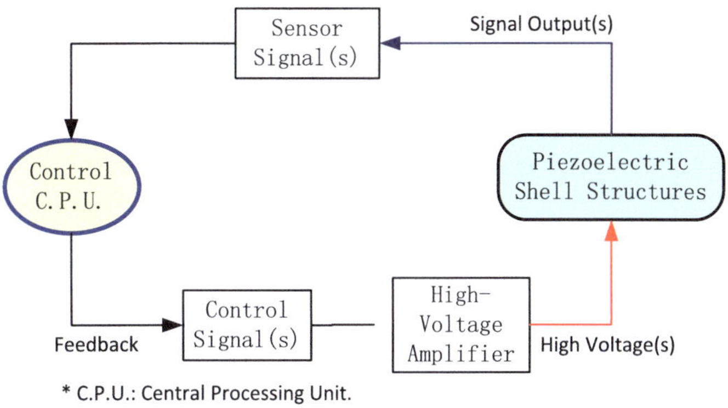

Fig. 3.6 A feedback control system for active piezoelectric shells

3.4.1 Active Piezoelastic Shell Equations

A set of governing equations, in three principal directions, for piezoelectric thin shell continua were derived in Sect. 2.4 of Chap. 2:

$$\begin{gathered}\frac{\partial\left[\left(N_{11}^m - N_{11}^e\right)A_2\right]}{\partial\alpha_1} + \frac{\partial\left(N_{21}^m A_1\right)}{\partial\alpha_2} + N_{12}^m\frac{\partial A_1}{\partial\alpha_2} - \left(N_{22}^m - N_{22}^e\right)\frac{\partial A_2}{\partial\alpha_1} \\ + A_1A_2\frac{Q_{13}^m}{R_1} = \rho h A_1 A_2 \ddot{u}_1\end{gathered}, \tag{3.4.1}$$

$$\begin{gathered}\frac{\partial\left(N_{12}^m A_2\right)}{\partial\alpha_1} + \frac{\partial\left[\left(N_{22}^m - N_{22}^e\right)A_1\right]}{\partial\alpha_2} + N_{21}^m\frac{\partial A_2}{\partial\alpha_1} - \left(N_{11}^m - N_{11}^e\right)\frac{\partial A_1}{\partial\alpha_2} \\ + A_1A_2\frac{Q_{23}^m}{R_2} = \rho h A_1 A_2 \ddot{u}_2\end{gathered}, \tag{3.4.2}$$

$$\begin{gathered}\frac{\partial\left(Q_{13}^m A_2\right)}{\partial\alpha_1} + \frac{\partial\left(Q_{23}^m A_1\right)}{\partial\alpha_2} - A_1A_2\left(\frac{N_{11}^m - N_{11}^e}{R_1} + \frac{N_{22}^m - N_{22}^e}{R_2}\right) \\ + A_1A_2\left(\sigma_{33} - e_{33}E_3\right)\Big|_{-h/2}^{h/2} = \rho h A_1 A_2 \ddot{u}_3\end{gathered}. \tag{3.4.3}$$

Q_{13}^m and Q_{23}^m are defined as

$$Q_{13}^{m} = \frac{1}{A_1A_2}\left(\frac{\partial\left[\left(M_{11}^{m} - M_{11}^{e}\right)A_2\right]}{\partial\alpha_1} + \frac{\partial\left(M_{21}^{m}A_1\right)}{\partial\alpha_2} + M_{12}^{m}\frac{\partial A_1}{\partial\alpha_2} - \left(M_{22}^{m} - M_{22}^{e}\right)\frac{\partial A_2}{\partial\alpha_1}\right), \tag{3.4.4}$$

$$Q_{23}^{m} = \frac{1}{A_1A_2}\left(\frac{\partial\left(M_{12}^{m}A_2\right)}{\partial\alpha_1} + \frac{\partial\left[\left(M_{22}^{m} - M_{22}^{e}\right)A_1\right]}{\partial\alpha_2} + M_{21}^{m}\frac{\partial A_2}{\partial\alpha_1} - \left(M_{11}^{m} - M_{11}^{e}\right)\frac{\partial A_1}{\partial\alpha_2}\right); \tag{3.4.5}$$

where ρ is the mass density and h is the thickness of piezoelectric shell. The superscripts m and e respectively denote the mechanical and electric components. Q_{i3}^{e} (i = 1,2) are zero since $E_1 = E_2 = 0$. σ_{33} can be neglected except the neighborhood where concentrated loads are applied. Note that this set of equations can be reduced to the conventional elastic shell equations by neglecting all electric coupling terms. Transverse shear deformation and rotatory inertia effects were not considered.

In active structure applications, the electric forces and moments induced by the converse piezoelectric effect can be used to control the shell characteristics, such as frequency and damping, in open or closed loop control systems. Moving all electric membrane forces to the right side of system equation yields

$$\begin{aligned}&\frac{\partial\left(N_{11}^{m}A_2\right)}{\partial\alpha_1} + \frac{\partial\left(N_{21}^{m}A_1\right)}{\partial\alpha_2} + N_{12}^{m}\frac{\partial A_1}{\partial\alpha_2} - N_{22}^{m}\frac{\partial A_2}{\partial\alpha_1} + A_1A_2\frac{Q_{13}^{m}}{R_1} - \rho hA_1A_2\ddot{u}_1\\ &= \frac{\partial\left(N_{11}^{e}A_2\right)}{\partial\alpha_1} - N_{22}^{e}\frac{\partial A_2}{\partial\alpha_1}\end{aligned}, \tag{3.4.6}$$

$$\begin{aligned}&\frac{\partial\left(N_{12}^{m}A_2\right)}{\partial\alpha_1} + \frac{\partial\left(N_{22}^{m}A_1\right)}{\partial\alpha_2} + N_{21}^{m}\frac{\partial A_2}{\partial\alpha_1} - N_{11}^{m}\frac{\partial A_1}{\partial\alpha_2} + A_1A_2\frac{Q_{23}^{m}}{R_2} - \rho hA_1A_2\ddot{u}_2\\ &= \frac{\partial\left(N_{22}^{e}A_1\right)}{\partial\alpha_2} - N_{11}^{e}\frac{\partial A_1}{\partial\alpha_2}\end{aligned}, \tag{3.4.7}$$

$$\begin{aligned}&\frac{\partial\left(Q_{13}^{m}A_2\right)}{\partial\alpha_1} + \frac{\partial\left(Q_{23}^{m}A_1\right)}{\partial\alpha_2} - A_1A_2\left(\frac{N_{11}^{m}}{R_1} + \frac{N_{22}^{m}}{R_2}\right) + A_1A_2(\sigma_{33} - e_{33}E_3)\big|_{-h/2}^{h/2}\\ &- \rho hA_1A_2\ddot{u}_3 = -A_1A_2\left(\frac{N_{11}^{e}}{R_1} + \frac{N_{22}^{e}}{R_2}\right).\end{aligned} \tag{3.4.8}$$

Note that the electric membrane forces and bending moments can be used for structural control of piezoelectric shell continua. In the equation of transverse oscillation, the electric membrane forces are divided by a radius such that these terms vanish when the radius of curvature is infinite, e.g., plates, beams, etc. This membrane force control scheme can be used only in shell type structures with non-zero curvatures.

3.4.2 Simplified Representation and Structural Damping

Separating the electric components into the direct (with a superscript d) and converse components (with a superscript c) and introducing a viscous damping term through the force term, one can write a simplified form of the generic equations of motion:

$$L_i^m\{u_1,u_2,u_3\} + L_i^d\{u_1,u_2,u_3\} + L_i^c(\phi_3) - c\dot{u}_i - \rho h\ddot{u}_i = -F_i, \quad i = 1,2,3. \tag{3.4.9}$$

Here F_i is the distributed mechanical load acting on the neutral surface of the piezoelectric shell; c is an equivalent viscous damping factor; ϕ_3 is the electric potential in the α_3 direction. (Note that $\phi_3 = -\int E_3 d\alpha_3$.) It is assumed that the control forces and moments, denoted by a superscript c, are contributed by a transverse voltage ϕ_3, although feedback voltages in the other two directions could also be possible. The viscous damping term is introduced through the distributed mechanical force term (Soedel, 1981). $L_i^m\{u_1,u_2,u_3\}$ is the mechanical Love's operator:

$$\begin{aligned} L_1^m\{u_1,u_2,u_3\} = \frac{1}{A_1A_2}\bigg(&\frac{\partial(N_{11}^m A_2)}{\partial\alpha_1} + \frac{\partial(N_{21}^m A_1)}{\partial\alpha_2} + N_{12}^m\frac{\partial A_1}{\partial\alpha_2} \\ &- N_{22}^m\frac{\partial A_2}{\partial\alpha_1} + A_1A_2\frac{Q_{13}^m}{R_1}\bigg), \end{aligned} \tag{3.4.10a}$$

$$\begin{aligned} L_2^m\{u_1,u_2,u_3\} = \frac{1}{A_1A_2}\bigg(&\frac{\partial(N_{12}^m A_2)}{\partial\alpha_1} + \frac{\partial(N_{22}^m A_1)}{\partial\alpha_2} + N_{21}^m\frac{\partial A_2}{\partial\alpha_1} \\ &- N_{11}^m\frac{\partial A_1}{\partial\alpha_2} + A_1A_2\frac{Q_{23}^m}{R_2}\bigg), \end{aligned} \tag{3.4.10b}$$

$$\begin{aligned} L_3^m\{u_1,u_2,u_3\} = \frac{1}{A_1A_2}\bigg(&\frac{\partial(Q_{13}^m A_2)}{\partial\alpha_1} + \frac{\partial(Q_{23}^m A_1)}{\partial\alpha_2} \\ &- A_1A_2\left(\frac{N_{11}^m}{R_1} + \frac{N_{22}^m}{R_2}\right)\bigg). \end{aligned} \tag{3.4.10c}$$

$L_i^d\{u_1,u_2,u_3\}$ is an operator for the direct piezoelectric effect and is defined as:

$$\begin{aligned} L_1^d\{u_1,u_2,u_3\} = -\frac{1}{A_1A_2}\Bigg\{&\frac{\partial(N_{11}^dA_2)}{\partial\alpha_1} - N_{22}^d\frac{\partial A_2}{\partial\alpha_1} \\ &+\frac{1}{R_1}\left(\frac{\partial(M_{11}^dA_2)}{\partial\alpha_1} - M_{22}^d\frac{\partial A_2}{\partial\alpha_1}\right)\Bigg\}, \end{aligned} \tag{3.4.11a}$$

$$\begin{aligned} L_2^d\{u_1,u_2,u_3\} = -\frac{1}{A_1A_2}\Bigg\{&\frac{\partial(N_{22}^dA_1)}{\partial\alpha_2} - N_{11}^d\frac{\partial A_1}{\partial\alpha_2} \\ &+\frac{1}{R_2}\left(\frac{\partial(M_{22}^dA_1)}{\partial\alpha_2} - M_{11}^d\frac{\partial A_1}{\partial\alpha_2}\right)\Bigg\}, \end{aligned} \tag{3.4.11b}$$

$$\begin{aligned} L_3^d\{u_1,u_2,u_3\} = -\frac{1}{A_1A_2}\Bigg\{&\frac{\partial}{\partial\alpha_1}\left(\frac{1}{A_1}\frac{\partial(M_{11}^dA_2)}{\partial\alpha_1} - \frac{M_{22}^d}{A_1}\frac{\partial A_2}{\partial\alpha_1}\right) \\ &+\frac{\partial}{\partial\alpha_2}\left(\frac{1}{A_2}\frac{\partial(M_{22}^dA_1)}{\partial\alpha_2} - \frac{M_{11}^d}{A_2}\frac{\partial A_1}{\partial\alpha_2}\right) \\ &-A_1A_2\left(\frac{N_{11}^d}{R_1} + \frac{N_{22}^d}{R_2}\right)\Bigg\}. \end{aligned} \tag{3.4.11c}$$

$L_i^c\{\phi_1,\phi_2,\phi_3\}$ is an operator for the converse piezoelectric effect and it is assumed that only a transverse voltage ϕ_3 is considered, as discussed previously. Thus, $L_i^c\{\phi_1,\phi_2,\phi_3\} = L_i^c\{\phi_3\}$, since ϕ_1 and ϕ_2 are set zero.

$$L_1^c\{\phi_3\} = -\frac{1}{A_1A_2}\left\{\frac{\partial(N_{11}^cA_2)}{\partial\alpha_1} - N_{22}^c\frac{\partial A_2}{\partial\alpha_1} + \frac{1}{R_1}\left(\frac{\partial(M_{11}^cA_2)}{\partial\alpha_1} - M_{22}^c\frac{\partial A_2}{\partial\alpha_1}\right)\right\}, \tag{3.4.12a}$$

$$L_2^c\{\phi_3\} = -\frac{1}{A_1A_2}\left\{\frac{\partial(N_{22}^cA_1)}{\partial\alpha_2} - N_{11}^c\frac{\partial A_1}{\partial\alpha_2} + \frac{1}{R_2}\left(\frac{\partial(M_{22}^cA_1)}{\partial\alpha_2} - M_{11}^c\frac{\partial A_1}{\partial\alpha_2}\right)\right\}, \tag{3.4.12b}$$

$$\begin{aligned} L_3^c\{\phi_3\} = -\frac{1}{A_1A_2}\Bigg\{&\frac{\partial}{\partial\alpha_1}\left(\frac{1}{A_1}\frac{\partial(M_{11}^cA_2)}{\partial\alpha_1} - \frac{M_{22}^c}{A_1}\frac{\partial A_2}{\partial\alpha_1}\right) \\ &+\frac{\partial}{\partial\alpha_2}\left(\frac{1}{A_2}\frac{\partial(M_{22}^cA_1)}{\partial\alpha_2} - \frac{M_{11}^c}{A_2}\frac{\partial A_1}{\partial\alpha_2}\right) - A_1A_2\left(\frac{N_{11}^c}{R_1} + \frac{N_{22}^c}{R_2}\right)\Bigg\}. \end{aligned} \tag{3.4.12c}$$

$N_{11}^c, N_{11}^d, N_{22}^c, N_{22}^d, M_{11}^c, M_{11}^d, M_{22}^c, M_{22}^d$ are defined in Sect. 2.3 of Chap. 2. Note that, in general, the electric force and moment components induced by the converse piezoelectric effect can be expressed as spatial functions from which distributed structural controls can be achieved. In addition, the voltage can also be set up as a function of either displacements, velocities, or accelerations; theses are discussed later.

3.5 Control of Active Piezoelectric Structures

The purpose of this section is to investigate distributed controls of active piezoelectric structures. Forced vibrations of piezoelectric structures with applications to active distributed structural controls are derived. There are two kinds of external forces that active piezoelectric and combined elastic/piezoelectric structures could experience: one is a mechanical force and the other an electric force. The ***modal expansion method*** is used to analyze the forced vibration of structures. Transient responses and steady-state responses of the forced vibration are discussed. Three generic feedback algorithms (i.e., displacement, velocity, and acceleration) are proposed and their corresponding governing equations formulated.

3.5.1 The Modal Expansion Method

In forced vibration analysis, a *spectral representation* or *modal expansion* is used to synthesize the dynamic response based on known mode shape functions. Natural modes of the piezoelectric or elastic/piezoelectric shells can be excited by either mechanical means or electric voltage loads. The amount of each modal participation in the total dynamic response is defined by a *modal participation factor* (*or modal coordinate*). The total dynamic response can be represented by the summation of all participating natural modes and their respective modal participation factors:

$$u_i(\alpha_1, \alpha_2, t) = \sum_{k=1}^{\infty} \eta_k(t) U_{ik}(\alpha_1, \alpha_2), \quad i = 1, 2, 3 \tag{3.5.1}$$

where $\eta_k(t)$ is the modal participation factor (i.e., modal coordinate); $U_{ik}(\alpha_1, \alpha_2)$ is the mode shape function; k denotes the k-th mode; k = 1,2,3,...,∞ of a continuum (a distributed parameter system). Substituting the modal expansion equation, Eq. (3.5.1), into the simplified piezoelectric shell equation, Eq. (3.4.9), gives the expanded modal equation (Zhong, 1991):

$$\sum_{k=1}^{\infty} \left\{ \eta_k \left(L_i^m\{U_{1k}, U_{2k}, U_{3k}\} + L_i^d\{U_{1k}, U_{2k}, U_{3k}\} \right) - c\dot{\eta}_k U_{ik} - \rho h \ddot{\eta}_k U_{ik} \right\} \\ = -F_i - L_i^c(\phi_3), \quad i = 1, 2, 3 \tag{3.5.2}$$

Note that it is assumed that a transverse electric voltage ϕ_3 is applied. From the eigenvalue analysis, it is known that

$$L_i^m\{U_{1k}, U_{2k}, U_{3k}\} + L_i^d\{U_{1k}, U_{2k}, U_{3k}\} = -\rho h \omega_k^2 U_{ik}, \quad (3.5.3)$$

where ω_k is the k-th natural frequency. Substituting Eq. (3.5.3) into Eq. (3.5.2) yields

$$\sum_{k=1}^{\infty} (\rho h \ddot{\eta}_k + c\dot{\eta}_k + \rho h \omega_k^2 \eta_k) U_{ik} = F_i + L_i^c(\phi_3), \quad i = 1, 2, 3. \quad (3.5.4)$$

Multiplying Eq. (3.5.4) by a mode shape U_{ip} on both sides and integrating it over the shell surface, one can derive

$$\begin{aligned} &\sum_{k=1}^{\infty} (\rho h \ddot{\eta}_k + c\dot{\eta}_k + \rho h \omega_k^2 \eta_k) \int_{\alpha_1}\int_{\alpha_2} \sum_{i=1}^{3} U_{ik} U_{ip} A_1 A_2 d\alpha_1 d\alpha_2 \\ &= \int_{\alpha_1}\int_{\alpha_2} \left\{ \sum_{i=1}^{3} \left[F_i + L_i^c(\phi_3)\right] U_{ip} \right\} A_1 A_2 d\alpha_1 d\alpha_2 \end{aligned} \quad . \quad (3.5.5)$$

Integrating over the shell surface and imposing the modal orthogonality condition, one can remove the summation since all p ≠ k modes vanish and only the p = k mode left. Thus,

$$\ddot{\eta}_k + \frac{c}{\rho h}\dot{\eta}_k + \omega_k^2 \eta_k = \hat{F}_k, \quad (3.5.6)$$

where the modal force $\hat{F}_k$ is defined as

$$\hat{F}_k = \frac{1}{\rho h N_k} \int_{\alpha_1}\int_{\alpha_2} \left\{ \sum_{i=1}^{3} \left[F_i + L_i^c(\phi_3)\right] U_{ik} \right\} A_1 A_2 d\alpha_1 d\alpha_2, \quad (3.5.7a)$$

and

$$N_k = \int_{\alpha_1}\int_{\alpha_2} \left(\sum_{i=1}^{3} U_{ik}^2 \right) A_1 A_2 d\alpha_1 d\alpha_2. \quad (3.5.7b)$$

Introducing a modal damping ratio ζ_k, one can write the modal equation as:

$$\ddot{\eta}_k + 2\zeta_k \omega_k \dot{\eta}_k + \omega_k^2 \eta_k = \hat{F}_k(t), \quad (3.5.8)$$

where

$$\zeta_k = \frac{c}{2\rho h \omega_k}. \tag{3.5.9}$$

In an under-damped forced oscillation with initial conditions $\eta_k(0)$ and $\dot{\eta}_k(0)$, the solution, i.e., free response, of the modal participation factor equation is

$$\begin{aligned}\eta_k(t) &= e^{-\zeta_k \omega_k t}\left\{\eta_k(0)\cos(\gamma_k t) + [\eta_k(0)\zeta_k\omega_k + \dot{\eta}_k(0)]\frac{\sin(\gamma_k t)}{\gamma_k}\right\} \\ &+ \frac{1}{\gamma_k}\int_0^t \hat{F}_k(\tau) e^{-\zeta_k\omega_k(t-\tau)}\sin[\gamma_k(t-\tau)]d\tau\end{aligned}, \tag{3.5.10}$$

where γ_k is the damped natural frequency and

$$\gamma_k = \omega_k\sqrt{1-\zeta_k^2}. \tag{3.5.11}$$

The initial conditions of the modal participation factor are defined by the initial displacement $u_i(\alpha_1, \alpha_2, 0)$ and velocity $\dot{u}_i(\alpha_1, \alpha_2, 0)$ of the system:

$$\eta_k(0) = \frac{1}{N_k}\int_{\alpha_1}\int_{\alpha_2}\left(\sum_{i=1}^{3} u_i(\alpha_1, \alpha_2, 0) U_{ik}\right) A_1 A_2 d\alpha_1 d\alpha_2, \tag{3.5.12a}$$

$$\dot{\eta}_k(0) = \frac{1}{N_k}\int_{\alpha_1}\int_{\alpha_2}\left(\sum_{i=1}^{3} \dot{u}_i(\alpha_1, \alpha_2, 0) U_{ik}\right) A_1 A_2 d\alpha_1 d\alpha_2. \tag{3.5.12b}$$

In the general modal response of $\eta_k(t)$, Eq. (3.5.10), the first part, determined by initial conditions, denotes the *zero-input* response which decays with respect to time; the second part, represented by a convolution integral, denotes the *zero-state* response induced by the external excitation $\hat{F}_k(\tau)$. $u_i(\alpha_1, \alpha_2, 0)$ and $\dot{u}_i(\alpha_1, \alpha_2, 0)$ are respectively the initial displacement and velocity of the system. The subscript i denotes the i-th direction, i = 1,2,3. The modal damping ratio can be estimated from the snap-back response, with an initial displacement only, using the logarithmic decrement method (Thomson, 1981; Tzou & Pandita, 1987; Tzou, 1989).

$$\zeta \cong \frac{\ln\left(\frac{u_m}{u_{m+n}}\right)}{2\pi(n)}, \tag{3.5.13}$$

where "ln" denotes the natural log; u_m and u_{m+n} denotes the oscillation amplitudes at the m-th and (m + n)-th peaks, respectively. This estimation is valid only for lightly damped systems.

3.5.2 Steady-State Harmonic Response

For a steady-state harmonic response, the excitation can be defined as

$$\mathrm{F_i}(\alpha_1, \alpha_2, t) + \mathrm{L_i^c}(\phi_3) = \left(\mathrm{F_i'}(\alpha_1, \alpha_2) + \mathrm{L_i^{c\prime}}\{\phi_3(\alpha_1, \alpha_2)\}\right)e^{j\omega t}, \tag{3.5.14}$$

where $\mathrm{F_i'}(\alpha_1, \alpha_2)$ is the spatial part of the mechanical excitation; $\mathrm{L_i^{c\prime}}\{\phi_3(\alpha_1, \alpha_2)\}$ is the spatial part of the electric excitation; ω is the excitation frequency. The equation for the modal participation factor is

$$\ddot{\eta}_k + 2\xi_k\omega_k\dot{\eta}_k + \omega_k^2\eta_k = \hat{F}_k' e^{j\omega t}, \tag{3.5.15}$$

Where

$$\hat{F}_k' = \frac{1}{\rho h N_k}\int_{\alpha_1}\int_{\alpha_2}\left\{\sum_{i=1}^{3}\left(F_i' + L_i^{c\prime}\right)U_{ik}\right\}A_1A_2\,d\alpha_1\,d\alpha_2. \tag{3.5.16}$$

For a harmonic excitation, the steady-state response $\eta_k(t)$ is also harmonic, i.e.,

$$\eta_k(t) = \Lambda_k e^{j(\omega t - \psi_k)}. \tag{3.5.17}$$

Substituting $\eta_k(t)$ into the modal equation and simplifying the equation accordingly, one can derive the response:

$$\Lambda_k e^{-j\psi_k} = \frac{\hat{F}_k'}{\left(\omega_k^2 - \omega^2\right) + 2j\zeta_k\omega_k\omega}, \tag{3.5.18}$$

where $j = \sqrt{-1}$. The magnitude Λ_k and phase lag ψ_k are defined as

$$\Lambda_k = \frac{\hat{F}_k'}{\omega_k^2\sqrt{\left[1 - \left(\frac{\omega}{\omega_k}\right)^2\right]^2 + 4\zeta_k^2\left(\frac{\omega}{\omega_k}\right)^2}}, \tag{3.5.19a}$$

$$\psi_k = \tan^{-1}\left[\frac{2\zeta_k\left(\frac{\omega}{\omega_k}\right)}{1 - \left(\frac{\omega}{\omega_k}\right)^2}\right]. \tag{3.5.19b}$$

The magnitude transfer function $H(\omega)$ in frequency domain is defined as

$$H(\omega) = \frac{1}{\omega_k^2 \sqrt{\left[1 - \left(\frac{\omega}{\omega_k}\right)^2\right]^2 + 4\zeta_k^2 \left(\frac{\omega}{\omega_k}\right)^2}}. \tag{3.5.20}$$

Note that the steady-state response can be excited by either the mechanical or electric inputs. As a common practice in vibration control of active piezoelectric structures, electric forces/moments are used to counteract the (elastic) mechanical forces/moments such that the vibration can be suppressed and controlled. Detailed control techniques and feedback algorithms are discussed next.

3.5.3 Feedback Algorithms and Distributed Controls

As discussed previously, there are two kinds of excitation forces for active structures composed of piezoelectric materials. One is the mechanical force; the other is the electric force introduced by the converse piezoelectric effect. If Love's operator of the converse piezoelectric effect $L_i^c(\phi_1, \phi_2, \phi_3)$ is designed as a function of mechanical motions, such as displacement, velocity, or acceleration, a closed-loop feedback control system can be established. In these cases, the voltages fed back to the piezoelectric actuators induce electric forces/moments which can be used to counteract and control the structural vibrations. In the following derivations, three generic feedback algorithms are discussed: 1) a displacement feedback, 2) a velocity feedback, and 3) an acceleration feedback. It is assumed that the i-th feedback voltage is only a function of the state in the i-th direction. Feedback forces in spatial domain are defined first, followed by modal forces and equations.

3.5.3.1 Spatial Feedback Forces: Love's Operators

1) **Displacement Feedback**

In the displacement feedback, it is assumed that Love's operator $L_i^c\{\phi_i\}$ is a function of displacements, i.e.,

$$L_i^c\{\phi_i\} = \overset{*}{\mathsf{F}}_1\{u_i(\alpha_1, \alpha_2, t)\}, \quad i = 1, 2, 3, \tag{3.5.21a}$$

where $\overset{*}{\mathsf{F}}_1$ denotes a generic spatial/time function and i denotes the i-th direction. (In practical applications of active piezoelectric structures, usually only the transverse direction, i.e., i = 3, is considered.) The feedback can be further expressed in a modal coordinate expression:

$$L_i^c\{\phi_i\} = \sum_{m=1}^{\infty} \mathcal{G}_{im}^{df}(\alpha_1, \alpha_2)\ \eta_m(t), \tag{3.5.21b}$$

where $\mathcal{G}_{im}^{df}(\alpha_1, \alpha_2)$ is the ***displacement modal feedback function***, i.e., a spatial function. It is assumed that the displacement is composed of all participating modes, i.e., m = 1,2,3,…,∞. The velocity and acceleration feedbacks can be defined accordingly.

2) **Velocity Feedback**

$$L_i^c\{\phi_i\} = \overset{*}{F}_2\{\dot{u}_i(\alpha_1, \alpha_2, t)\}, \tag{3.5.22a}$$

$$L_i^c\{\phi_i\} = \sum_{m=1}^{\infty} \mathcal{G}_{i\ m}^{v\ f}(\alpha_1, \alpha_2)\ \dot{\eta}_m(t). \tag{3.5.22b}$$

3) **Acceleration Feedback**

$$L_i^c\{\phi_i\} = \overset{*}{F}_3\{\ddot{u}_i(\alpha_1, \alpha_2, t)\}, \tag{3.5.23a}$$

$$L_i^c\{\phi_i\} = \sum_{m=1}^{\infty} \mathcal{G}_{i\ m}^{a\ f}(\alpha_1, \alpha_2)\ \ddot{\eta}_m(t). \tag{3.5.23b}$$

Note that $\mathcal{G}_{im}^{df}(\alpha_1, \alpha_2)$, $\mathcal{G}_{im}^{vf}(\alpha_1, \alpha_2)$, and $\mathcal{G}_{im}^{af}(\alpha_1, \alpha_2)$ are defined as the spatially distributed ***modal feedback functions*** with respect to the displacement, velocity, and acceleration feedback, respectively. In general, the feedback signal can be defined by either points, lines, or finite distributed areas. (Note that the feedback signal can also be expressed in a modal expansion form, e.g., a point signal can be represented by two delta functions multiplied by the modal coordinate.) The feedback forces corresponding to the three feedback algorithms were defined above. Modal equations with the feedback forces are defined next. In later derivations, it is assumed that all external mechanical forces are neglected.

3.5.3.2 Modal Feedback Forces

Substituting Eqs. (3.5.21b), (3.5.22b) and (3.5.23b) into Eq. (3.5.7a) respectively gives the k-th modal control force expressions as:

1) **Displacement Feedback**

$$\hat{F}_k = \frac{1}{\rho h N_k}\int_{\alpha_1}\int_{\alpha_2}\sum_{j=1}^{3}\left\{\sum_{m=1}^{\infty}\mathcal{G}_{jm}^{df}(\alpha_1,\alpha_2)\eta_m(t)U_{jk}(\alpha_1,\alpha_2)\right\}A_1A_2d\alpha_1d\alpha_2, \tag{3.5.24}$$

2) **Velocity Feedback**

$$\hat{F}_k = \frac{1}{\rho h N_k}\int_{\alpha_1}\int_{\alpha_2}\sum_{j=1}^{3}\left\{\sum_{m=1}^{\infty}\mathcal{G}_{jm}^{vf}(\alpha_1,\alpha_2)\dot{\eta}_m(t)U_{jk}(\alpha_1,\alpha_2)\right\}A_1A_2d\alpha_1d\alpha_2, \tag{3.5.25}$$

3) **Acceleration Feedback**

$$\hat{F}_k = \frac{1}{\rho h N_k}\int_{\alpha_1}\int_{\alpha_2}\sum_{j=1}^{3}\left\{\sum_{m=1}^{\infty}\mathcal{G}_{jm}^{af}(\alpha_1,\alpha_2)\ddot{\eta}_m(t)U_{jk}(\alpha_1,\alpha_2)\right\}A_1A_2d\alpha_1d\alpha_2, \tag{3.5.26}$$

where the external mechanical forces are not considered and

$$N_k = \int_{\alpha_1}\int_{\alpha_2}\sum_{j=1}^{3}U_{jk}^2A_1A_2d\alpha_1d\alpha_2. \tag{3.5.27}$$

3.5.3.3 Modal Equations

Substituting the expressions of modal control forces into Eq. (3.5.8) yields the modal equations with feedback control forces for three feedback algorithms.

1) Displacement Feedback:

$$\begin{aligned}&\ddot{\eta}_k + \frac{c}{\rho h}\dot{\eta}_k + \omega_k^2\eta_k\\&-\frac{1}{\rho h N_k}\sum_{j=1}^{3}\sum_{m=1}^{\infty}\int_{\alpha_1}\int_{\alpha_2}\{\mathcal{G}_{jm}^{df}(\alpha_1,\alpha_2)\eta_m(t)U_{jk}(\alpha_1,\alpha_2)\}A_1A_2d\alpha_1d\alpha_2\end{aligned}. \tag{3.5.28}$$

2) Velocity Feedback:

$$\begin{aligned}&\ddot{\eta}_k + \frac{c}{\rho h}\dot{\eta}_k + \omega_k^2\eta_k \\ &= \frac{1}{\rho h N_k}\sum_{j=1}^{3}\sum_{m=1}^{\infty}\int_{\alpha_1}\int_{\alpha_2}\left(\mathcal{G}_{jm}^{vf}(\alpha_1,\alpha_2)\dot{\eta}_m(t)U_{jk}(\alpha_1,\alpha_2)\right)A_1A_2 d\alpha_1 d\alpha_2\end{aligned} \quad (3.5.29)$$

3) Acceleration Feedback:

$$\begin{aligned}&\ddot{\eta}_k + \frac{c}{\rho h}\dot{\eta}_k + \omega_k^2\eta_k \\ &= \frac{1}{\rho h N_k}\sum_{j=1}^{3}\sum_{m=1}^{\infty}\int_{\alpha_1}\int_{\alpha_2}\left(\mathcal{G}_{jm}^{af}(\alpha_1,\alpha_2)\ddot{\eta}_m(t)U_{jk}(\alpha_1,\alpha_2)\right)A_1A_2 d\alpha_1 d\alpha_2\end{aligned} \quad (3.5.30)$$

Note that generic modal equations with the displacement, velocity, and acceleration feedback were established. A generic shell control system can be described by an infinite number of second order modal equations. Summing all participating modes with their respective modal participating coordinates establishes the total response of the shell continuum. In practical applications, however, there are usually a finite number of modes contributing the total dynamic responses. Note that spillovers could occur when cross coupling terms introduced by the residual modes come into feedback in an active structural control system. That is, feedback control forces for controlled modes appear in the governing equations of other uncontrolled modes due to modal interactions among all participating modes; this is due to the integration of control forces on the modal shapes in these equations. This integration implies that control forces appear in all other modal coordinates. Detailed discussion on observation and control spillovers will be presented in Chaps. 6, 7, and 8. Generic distributed control concepts based on distributed piezoelectric modal (shaped and convolved) actuators are presented next. For simplicity, only the **transverse** mode is considered in the following derivations.

3.6 Distributed Convolving Actuators and Controls

In the three generic modal control equations, there is no observation spillover; but control spillover problems still exist. One method to eliminate the control spillover is to design spatially distributed modal gain functions, i.e., $\mathcal{G}_{3n}^{df}(\alpha_1,\alpha_2)$, $\mathcal{G}_{3n}^{vf}(\alpha_1,\alpha_2)$, and $\mathcal{G}_{3n}^{af}(\alpha_1,\alpha_2)$, with spatially distributed characteristics such that they are orthogonal to other natural modes, i.e. $k \neq p$. According to the orthogonality of natural mode shapes, these spatially distributed gains can be designed corresponding to the mode shapes of controlled modes, e.g.,

$$\mathcal{G}^{\mathrm{f}}_{3n}(\alpha_1, \alpha_2) = \mathcal{G}^{f}_{3n} U_{3n}(\alpha_1, \alpha_2). \tag{3.6.1}$$

where $\mathcal{G}^{\mathrm{f}}_{3n}$ is a constant and $U_{3n}(\alpha_1, \alpha_2)$ is the n-th mode shape function in the transverse direction. (Note that only the transverse oscillation is considered in the formulation, although the procedures can be extended to encompass all three coordinate directions.) In this case, the control forces appear only in the equations of controlled modes such that control spillovers prevented. (Note that only the transverse electric voltage ϕ_3 is considered in the later derivations, although the procedures can be used for all three directions.)

3.6.1 Modal Gain Function

As discussed above, the transverse modal feedback force is a spatial function. Accordingly, for simplicity, the ***modal feedback function*** $\mathcal{G}^{\mathrm{df}}_{im}(\alpha_1, \alpha_2)$ is assumed to be the mode shape function multiplied by a constant gain. Thus, the n-th ***modal gain function*** in the operator can be expressed as:

1) **Displacement Feedback**

$$L^c_i\{\phi_3\} = \mathcal{G}^{\mathrm{df}}_i U_{3n}(\alpha_1, \alpha_2)\eta_n(t), \tag{3.6.2}$$

2) **Velocity Feedback**

$$L^c_i\{\phi_3\} = \mathcal{G}^{\mathrm{vf}}_i U_{3n}(\alpha_1, \alpha_2)\dot{\eta}_n(t), \tag{3.6.3}$$

3) **Acceleration Feedback**

$$L^c_i\{\phi_3\} = \mathcal{G}^{\mathrm{af}}_i U_{3n}(\alpha_1, \alpha_2)\ddot{\eta}_n(t), \tag{3.6.4}$$

where $\mathcal{G}^{\mathrm{df}}_i$, $\mathcal{G}^{\mathrm{vf}}_i$, and $\mathcal{G}^{\mathrm{af}}_i$ are weighting factors (constants) which are independent of spatial coordinates and time.

3.6.2 Modal Feedback Control Force

With the expressions above, one can define the ***modal feedback control forces*** $\hat{F}_k$ for the n-th mode as follows.

1) Displacement Feedback

$$\hat{F}_k = \frac{\mathcal{G}_3^{df}}{\rho h N_k}\eta_n(t)\int_{\alpha_1}\int_{\alpha_2} U_{3n}(\alpha_1,\alpha_2)U_{3k}(\alpha_1,\alpha_2)A_1A_2d\alpha_1 d\alpha_2, \tag{3.6.5a}$$

where

$$\text{i)}\quad \hat{F}_k = \frac{\mathcal{G}_3^{df}}{\rho h N_n}\eta_n(t)\int_{\alpha_1}\int_{\alpha_2} U_{3n}^2 A_1A_2d\alpha_1 d\alpha_2, \quad (\text{for } k = n), \tag{3.6.5b}$$

$$\text{ii)}\quad \hat{F}_k = 0\,, \quad (\text{for } k \neq n). \tag{3.6.5c}$$

2) Velocity Feedback

$$\hat{F}_k = \frac{\mathcal{G}_3^{vf}}{\rho h N_k}\dot{\eta}_n(t)\int_{\alpha_1}\int_{\alpha_2} U_{3n}(\alpha_1,\alpha_2)U_{3k}(\alpha_1,\alpha_2)A_1A_2d\alpha_1 d\alpha_2, \tag{3.6.6a}$$

where

$$\text{i)}\quad \hat{F}_k = \frac{\mathcal{G}_3^{vf}}{\rho h N_n}\dot{\eta}_n(t)\int_{\alpha_1}\int_{\alpha_2} U_{3n}^2 A_1A_2d\alpha_1 d\alpha_2 \quad (\text{for } k = n), \tag{3.6.6b}$$

$$\text{ii)}\quad \hat{F}_k = 0\,, \quad (\text{for } k \neq n). \tag{3.6.6c}$$

3) Acceleration Feedback

$$\hat{F}_k = \frac{\mathcal{G}_3^{af}}{\rho h N_k}\ddot{\eta}_n(t)\int_{\alpha_1}\int_{\alpha_2} U_{3n}(\alpha_1,\alpha_2)U_{3k}(\alpha_1,\alpha_2)A_1A_2d\alpha_1 d\alpha_2, \tag{3.6.7a}$$

where

$$\text{i)}\quad \hat{F}_k = \frac{\mathcal{G}_3^{af}}{\rho h N_n}\ddot{\eta}_n(t)\int_{\alpha_1}\int_{\alpha_2} U_{3n}^2 A_1A_2d\alpha_1 d\alpha_2\,, \quad (\text{for } k = n), \tag{3.6.7b}$$

$$\text{ii)}\quad \hat{F}_k = 0 \quad (\text{for } k \neq n). \tag{3.6.7c}$$

Specific designs and effectiveness of the distributed shaped convolving actuators will be studied in Chaps. 8 and 9.

In the following case studies, the derived active piezoelectric shell control theory is simplified to a piezoelectric bimorph beam and a piezoelectric cylindrical shell; their control effectivenesses are studied. It is assumed that the bimorph beam is cantilevered; the cylindrical shell is simply supported on all four edges. Frequency control of the cantilever bimorph beam is studied (Tzou and Zhong, 1991b). (Note that static actuation of the bimorph beam is evaluated and its theoretical, finite element and experimental data are compared in the Finite Element chapter, Chap. 13.) Damping control of the cylindrical shell panel via distributed in-plane membrane forces is investigated (Tzou and Zhong, 1991b).

3.7 Frequency Control of a Piezoelectric Bimorph Beam

A piezoelectric bimorph beam is made of two piezoelectric layers with opposite polarity, Fig. 3.7.

Due to the reversed polarity in these two layers, the resultant stresses (due to the converse piezoelectric effect) in these two layers are of opposite sign when an external voltage is applied across the beam thickness. These two stresses cause a coupling effect leading to a bending effect of the bimorph beam. Frequency control of the bimorph beam (a simple active piezoelectric structure) via a displacement feedback is studied. Analytical solutions will be compared with laboratory results. From Fig. 3.7, the resultant normal stresses introduce a moment with respect to the neutral axis. Thus, the beam bends due to the induced bending. The moment M_{11} can be simply calculated as

$$M_{11} = \frac{bh^2}{4} e_{31} E_3, \tag{3.7.1}$$

where b is the width; h is the thickness; e_{31} is a piezoelectric constant; E_3 is a transverse electric field, $E_3 = \phi_3/h$ in a conventional expression. In order to achieve the structural adaptivity and to control structural dynamics, a closed-loop control system is designed, Fig. 3.8. Note that the feedback electrical field is a function of the tip displacement based on a measurement from a displacement transducer—a proximeter.

The feedback voltage ϕ applied to the beam is proportional to the displacement at the free end. That is,

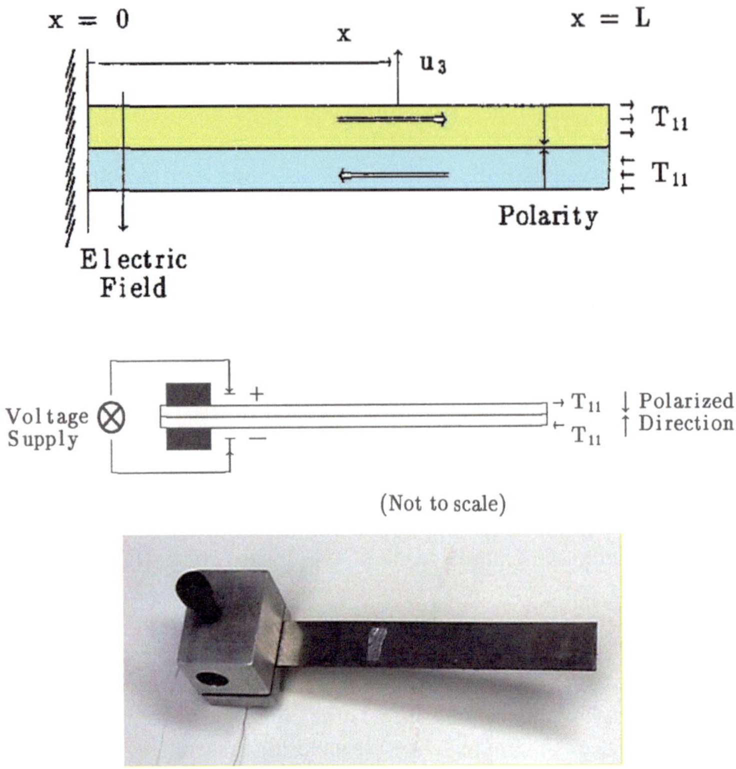

Fig. 3.7 A cantilever piezoelectric bimorph beam

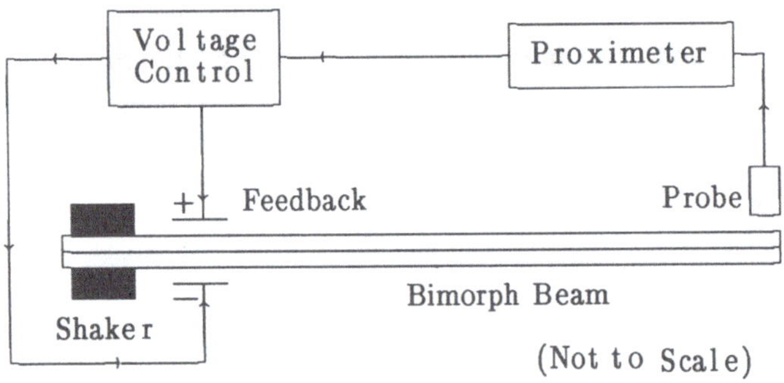

Fig. 3.8 A laboratory feedback control system for the active piezoelectric bimorph beam

$$\phi = \mathcal{G} S u_3(x = L, t), \tag{3.7.2}$$

where $\mathcal{G}$ is the control gain and S is the transducer sensitivity. Substituting Eq. (3.7.2) into Eq. (3.7.1) gives

$$M_{11} = \frac{bhe_{31}\mathcal{G}\,S}{4} u_3(L,t). \tag{3.7.3}$$

For a transversely vibrating beam, the Lamé parameters are $A_1 = 1$ and $A_2 = 1$ and radii of curvature are $R_1 = R_2 = \infty$. Simplifying the piezoelectric shell equation in the transverse direction to a transversely vibrating beam with a uniform electrical field (the electrode resistance is neglected), one can derive the equation of motion:

$$YI\frac{\partial^4 u_3}{\partial x^4} + \rho\tilde{A}\frac{\partial^2 u_3}{\partial t^2} = 0, \tag{3.7.4}$$

where Y is Young's modulus; I is the area moment of inertia; $\tilde{A}$ is the cross-section area. For a cantilever piezoelectric bimorph beam subjected to an external electrical field E_3, the resultant moment at the free end is equal to zero, i.e., the summation of electric and mechanical moments is zero or the mechanical moment equals the electric (control) moment. Thus, the control effect is introduced via a ***boundary control*** at the free end. This boundary difference results in a set of different characteristic equations and a new set of controlled natural frequencies accordingly.

Thus, boundary conditions at the free-end of the bimorph beam are defined as: shear force $Q_{13} = 0$ and moment $M_{11} = \tilde{K}u_3(x = L, t)$.

$$\frac{\partial^3 u_3}{\partial x^3} = 0, \tag{3.7.5a}$$

$$\frac{\partial^2 u_3}{\partial x^2} = -\frac{\tilde{K}}{YI}u_3; \tag{3.7.5b}$$

where

$$\tilde{K} = \frac{bhe_{31}\mathcal{G}\,S}{4}. \tag{3.7.6}$$

Note that the boundary conditions at the clamped end are the same as any conventional cantilever beam, i.e., zero displacement and zero slope at x = 0:

$$u_3(0,t) = 0, \tag{3.7.7a}$$

$$\frac{\partial u_3(x = 0, t)}{\partial x} = 0. \tag{3.7.7b}$$

Since the frequency control is evaluated, free vibration analysis is carried out next.

3.7.1 Free Vibration Analysis

It is assumed the beam is oscillating harmonically at one of the natural frequencies, i.e., $u_3(x,t) = U_3(x)e^{j\omega t}$. Substituting the harmonical motion into the equation of motion gives

$$\frac{\partial^4 U_3}{\partial x^4} - \lambda^4 U_3 = 0, \tag{3.7.8}$$

where the eigenvalue λ is defined by

$$\lambda^4 = \frac{\omega^2 \rho'}{YI}, \tag{3.7.9}$$

where $\rho' = \rho\tilde{A}$ and ω is the circular natural frequency in radian/s. Using the Laplace transform method, one can obtain a solution as:

$$\begin{aligned} U_3(x) = U_3(0)A(\lambda x) + \frac{1}{\lambda}\frac{dU_3(0)}{dx}B(\lambda x) + \frac{1}{\lambda^2}\frac{d^2U_3(0)}{dx^2}C(\lambda x) \\ + \frac{1}{\lambda^3}\frac{d^3U_3(0)}{dx^3}D(\lambda x), \end{aligned} \tag{3.7.10}$$

where

$$A(\lambda x) = \frac{1}{2}[\cosh(\lambda x) + \cos(\lambda x)], \tag{3.7.11a}$$

$$B(\lambda x) = \frac{1}{2}[\sinh(\lambda x) + \sin(\lambda x)], \tag{3.7.11b}$$

$$C(\lambda x) = \frac{1}{2}[\cosh(\lambda x) - \cos(\lambda x)], \tag{3.7.11c}$$

$$D(\lambda x) = \frac{1}{2}[\sinh(\lambda x) - \sin(\lambda x)]. \tag{3.7.11d}$$

The solution also depends on boundary conditions. Substituting the boundary conditions into Eq. (3.7.10) yields a *characteristic equation*:

$$\cos h(\lambda L)\cos(\lambda L) + 1 = \frac{2\tilde{K}}{YI\lambda^2}\sin h(\lambda L)\sin(\lambda L). \tag{3.7.12}$$

Considering a three-interval difference, one can derive a finite difference characteristic equation as

$$\lambda^3 - \left(2\tilde{K}' + 14\right)\lambda^2 + \left(33 + 20\tilde{K}'\right)\lambda - \left(18\tilde{K}' + 4\right) = 0, \qquad (3.7.13)$$

where $\tilde{K}' = \frac{\tilde{K}\Delta^2}{YI} = \frac{bhe_{31}\mathcal{G}S\Delta^2}{4YI}$ and Δ is the difference. Note that eigen-solutions depend on $\tilde{K}'$ when material and geometry are specified. More specifically, eigen solutions are functions of feedback gains $\mathcal{G}$ and the transducer sensitivity S. Thus, natural frequencies and mode shapes can be manipulated via the feedback control systems, i.e., controlling the gain can vary the natural frequency of the beam. Thus, the piezoelectric bimorph beam system is adaptive and active control can be achieved.

3.7.2 *Finite Difference Results*

For a piezoelectric bimorph beam with b = 10 mm, h = 1 mm, L = 100 mm, $Y = 4.704 \times 10^9 N/m^2$, $d_{31} = \frac{e_{31}}{Y} = 23 \times 10^{-12} C/N$, the first three natural frequencies of the (original) uncontrolled system are 23.9 Hz, 112.4 Hz, and 222.2 Hz. Control of natural frequency with different resultant feedback gains (i.e., $\mathcal{G}S$) are investigated. These calculations are plotted in Figs. 3.9, 3.10 and 3.11 for the first, the second, and the third natural frequencies, respectively. Note that each natural frequency is evaluated from negative to positive feedback.

It is observed that all three natural frequencies increase when the resultant control gain increases. However, the control effectiveness to each mode is different.

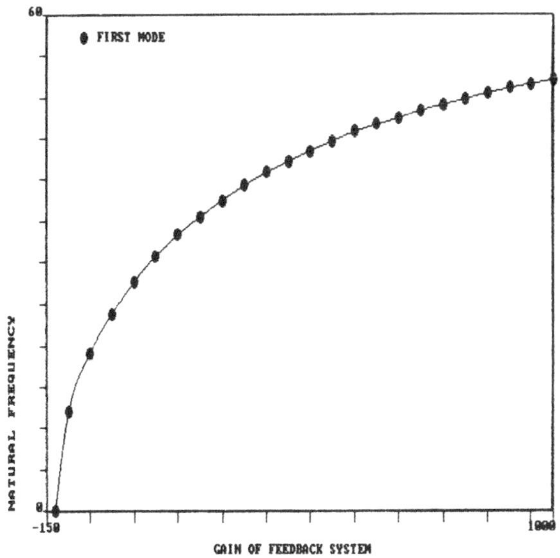

Fig. 3.9 Control of the first natural frequency

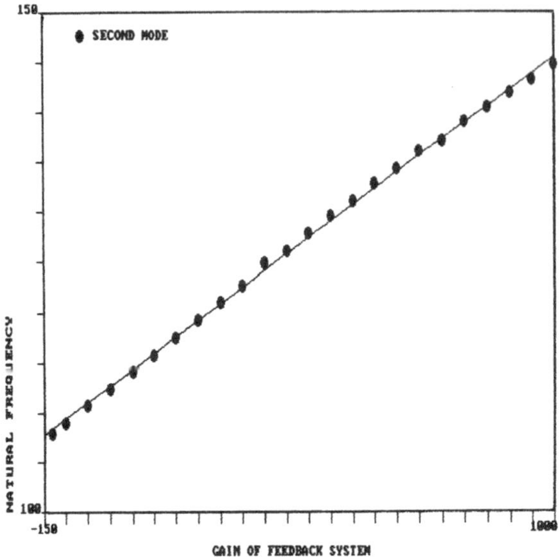

Fig. 3.10 Control of the second natural frequency

Note that the ***displacement feedback control*** (free-end displacement) is considered in this case, which leads to controlling the natural frequencies.

3.7.3 Laboratory Validation

In order to experimentally prove the active structure theory derived earlier, a piezoelectric bimorph cantilever beam was design and tested in the StrucTronics & Design Lab at the University of Kentucky. The bimorph beam was mounted on a vibration shaker and white noise signals were input into the shaker which directly excites the beam, Figs. 3.7 and 3.8. The displacement of free-end was picked up by a displacement sensor—a proximeter. This displacement signal was amplified and then fed back into the piezoelectric beam. Note that this model was only tested for its first mode with gains from −100 to +100, due to equipment limitations. The frequency spectrum of the model was analyzed by a signal processing system.

It should be pointed out that the experiment was carried out in a slightly different way due to physical characteristics of the piezoelectric beam and the transducer used for measuring the tip displacement. Since a proximeter, an eddy—current sensor, was used to measure the free-end displacement, a metal mass (tip mass) was attached to the beam tip to introduce the eddy—current interference. Thus, this

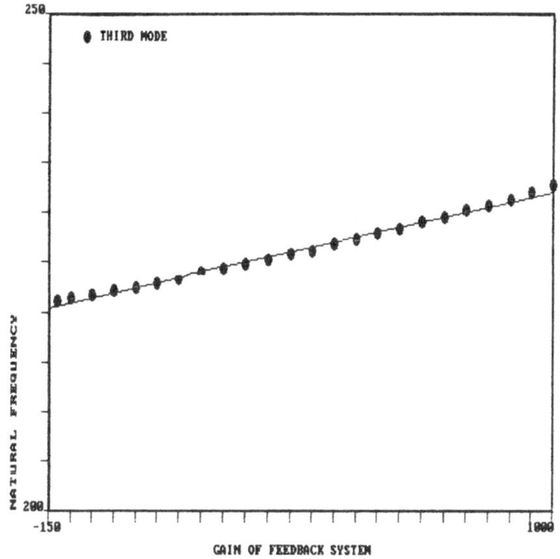

Fig. 3.11 Control of the third natural frequency

model is different from the above model which does not have a tip concentrated mass. Consequently, a new mathematical model was formulated and finite difference (five points) solutions were derived. Analytical solutions and experimental data of the first mode are summarized in Table 3.1.

The difference between the analytical solutions and experimental data is introduced primarily by the tip-mass location. In the finite difference model, the tip-mass was considered at the free-end. However, the mass is about 5% inside in the experimental model, which introduces a higher natural frequency. Besides, the electrode stiffness and the bonding material were not considered in the theoretical model and the finite difference calculation, which could lead to the numerical solutions being lower than the experimental data. (Again, note that the finite

Table 3.1 Comparison between theory and experiment

Gain	−100	−50	0	50	100
Feedback Voltage (V)	−200	−100	0	100	200
Analytical Result (Hz)	8.53	10.48	12.39	14.25	16.04
Experiment Data (Hz)	9	11	13	15	17
Error	5.2%	4.7%	4.7%	5.0%	5.6%

difference results presented in Figs. 3.9, 3.10 and 3.11 are for a cantilever piezoelectric bimorph beam without tip mass.) (Furthermore, static actuation of the bimorph beam is evaluated and its theoretical, finite element and experimental data are compared in the Finite Element chapter, Chap. 13.)

3.8 Damping Control of Piezoelectric Cylindrical Shells

Application of the generic active piezoelectric shell theory to distributed control of a piezoelectric cylindrical thin shell structure is demonstrated in this case. Damping control of the cylindrical shell using distributed in plane membrane forces is studied. It is assumed that the thin piezoelectric shell is made of a symmetrical hexagonal piezoelectric material, class $C_{6v} = 6\,\text{mm}$, which has ten material constants (Tiersten, 1969; Tzou & Zhong, 1990). The top and bottom surfaces are covered by thin electrodes whose material properties are negligible. A transverse electric field E_3 is applied across the electroded surfaces. Figure 3.12 shows the cylindrical shell panel, with thickness h, and its coordinate system.

3.8.1 Modeling and Analysis

For a cylindrical shell panel, the fundamental form is defined as

$$(ds)^2 = (1)(dx)^2 + (R)^2(d\theta)^2, \tag{3.8.1}$$

where ds is the infinitesimal distance; x defines the longitudinal direction; θ defines the circumferential direction. R is the radius of curvature of the θ-axis. Thus,

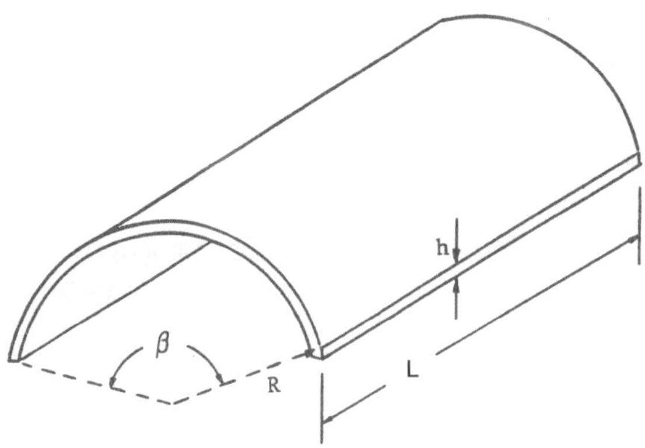

Fig. 3.12 A piezoelectric cylindrical shell panel

$A_1 = 1$, $A_2 = R$, $R_1 = \infty$, and $R_2 = R$. Substituting these four parameters into the active piezoelectric shell equation (transverse vibration only) and simplifying, one can derive the piezoelectric shell equation in the transverse direction u_3 as:

$$D'\left(-\frac{\partial^4 u_3}{\partial x^4} + \frac{1}{R^2}\frac{\partial^3 u_3}{\partial x^2 \partial \theta} - \frac{2}{R^2}\frac{\partial^4 u_3}{\partial x^2 \partial \theta^2} + \frac{1}{R^4}\frac{\partial^3 u_3}{\partial \theta^3} - \frac{1}{R^4}\frac{\partial^4 u_3}{\partial \theta^4}\right) - K'\left(\frac{\mu}{R}\frac{\partial u_x}{\partial x} + \frac{1}{R^2}\frac{\partial u_\theta}{\partial \theta} + \frac{u_3}{R^2}\right) - \rho h \ddot{u}_3 = \frac{N^e_{\theta\theta}}{R}, \tag{3.8.2}$$

where D' and K' are respectively the bending and membrane stiffness defined in Sect. 2.4 of Chap. 2. Note that the terms inside the first parenthesis are contributed by bending effects and those inside the second parenthesis are related to in-plane membrane effects. Because the radius of curvature R of the θ axis (the circumferential direction) is finite, an in-plane electric membrane force $N^e_{\theta\theta}$ is left in the system equation; this force can be used to control the shell. Besides open-loop controls, the input electric field E_3 can be made as a function of either displacement or velocity in closed-loop control systems. In this study, a ***velocity feedback control*** is considered and its effect on damping control is evaluated. Note that the feedback voltage is normalized with respect to modal oscillation amplitudes in the later analysis. A closed-loop control schematic is illustrated in Fig. 3.13.

It is assumed that the cylindrical shell panel is simply supported on all four edges, and the shell is made of a polymeric piezoelectric polyvinylidene fluoride (PVDF) material with $e_{31} = e_{32} = 4.6 \times 10^{-2}\,C/m^2$, $\varepsilon_{33} = 1.026 \times 10^{-6}\,F/m$, and capacitance $C = 3.8 \times 10^{-6}\,F/m$. The shell dimension is 10 cm long, 1 mm thick, and the radius of curvature is 5 cm. Three different sizes: 60°, 90°, and 120° curvature angles are studied and effects of shell deepness compared. Note that since the shell is made of a PVDF polymeric material, $D' \cong D$ and $K' \cong K$.

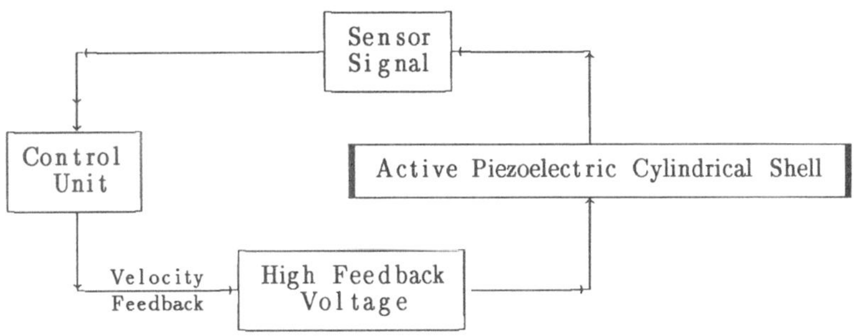

Fig. 3.13 A feedback control system for active cylindrical shells

The natural modes of the simply supported cylindrical shell in three principal directions are

$$U_x(x,\theta) = A\cos\frac{m\pi x}{L}\sin\frac{n\pi\theta}{\beta}, \tag{3.8.3a}$$

$$U_\theta(x,\theta) = B\sin\frac{m\pi x}{L}\cos\frac{n\pi\theta}{\beta}, \tag{3.8.3b}$$

$$U_3(x,\theta) = C\sin\frac{m\pi x}{L}\sin\frac{n\pi\theta}{\beta}; \tag{3.8.3c}$$

where *A, B,* and *C* are constants, i.e., amplitudes of modal oscillations. (Detailed free vibration analysis and mode shapes are presented in Chap. 9.) Using the modal expansion technique and considering the velocity feedback, one can derive the mn-th modal equation as

$$\ddot{\eta}_{mn} + \frac{1}{\rho h}\left(c_{mn} + \frac{16 e_{31}\phi_3}{3R\pi^2}\delta'(m,n)\right)\dot{\eta}_{mn} + \omega_{mn}^2\eta_{mn} = 0, \tag{3.8.4}$$

where η_{mn} is the modal coordinate; c_{mn} is the inherent damping factor; ϕ_3 is the feedback voltage; $\delta'(m,n)$ is defined as

$$\delta'(m,n) = (-\cos m\pi + 1)(-\cos n\pi + 1). \tag{3.8.5}$$

Thus, the controlled damping ratio ζ'_{mn} of the piezoelectric shell after the velocity feedback becomes

$$\zeta'_{mm} = \frac{1}{2\rho h\omega_{mn}}\left(c_{mn} + \frac{16 e_{31} V_3}{3R\pi^2}\delta'(m,n)\right). \tag{3.8.6}$$

The initial mn-th modal damping ratio ζ_{mn} is defined as $[c_{mn}/(2\rho h\omega_{mn})]$. It is observed that the enhanced damping, the second term: $\frac{16 e_{31}\phi_3}{3R\pi^2}\delta'(m,n)$, vanishes for all even m or n because $\delta'(m,n) = 0$. Thus, this distributed in-plane membrane control forces are effective only for odd natural modes (both m and n are odd numbers) of the cylindrical shell. Substituting all material and geometry parameters into the above equation, one can estimate the damping change due to feedback voltages ϕ_3.

3.8.2 Results and Discussion

Damping variations of the first three odd modes for all three curvature angles, i.e., 60°, 90°, and 120°, are plotted in Figs. 3.14, 3.15 and 3.16.

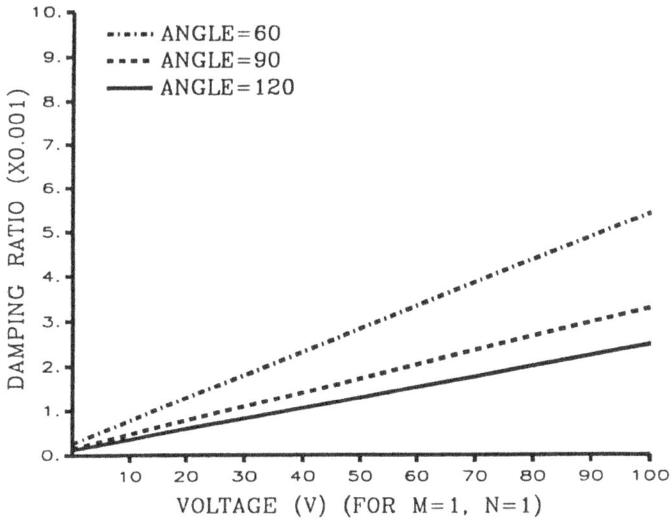

Fig. 3.14 Damping control for the first odd mode, (m = 1, n = 1)

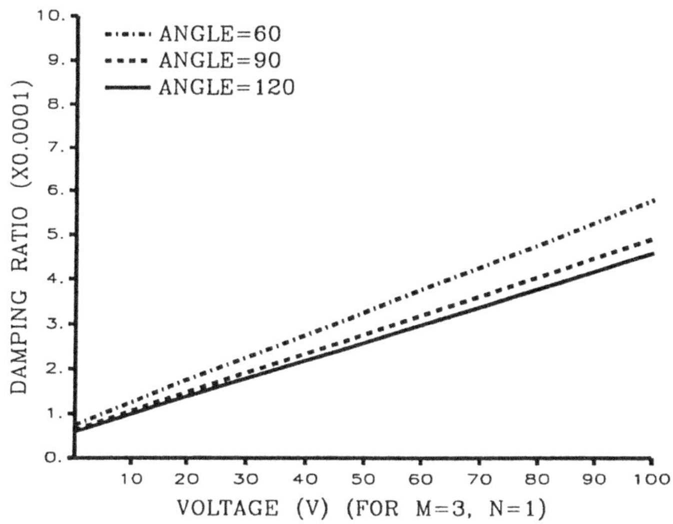

Fig. 3.15 Damping control for the second odd mode, (m = 3, n = 1)

Recall that m denotes the x-direction and n the θ-direction. The transverse natural mode is $U_3(x,\theta) = C \sin\frac{m\pi x}{L} \sin\frac{n\pi\theta}{\beta}$. (Detailed mode shapes of the cylindrical shell are presented in Chap. 9.) Damping variations for the first three odd modes are controlled by the feedback voltages. In general, the control effects increase as the feedback voltage increases. However, in practice, this voltage should

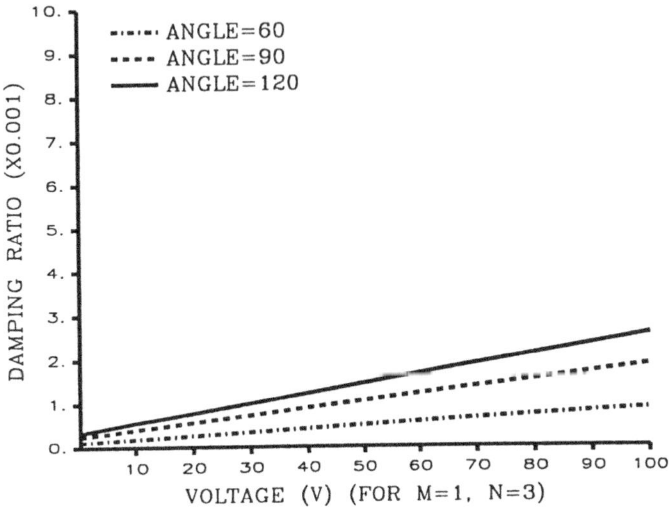

Fig. 3.16 Damping control for the third odd mode, (m = 1, n = 3)

not exceed the breakdown voltage of the piezoelectric material. Figure 3.15 shows the control effect to the (m = 3, n = 1) mode, i.e., $\sin(3\pi x/L)\sin(\pi\theta/\beta)$. Since the control forces are circumferential, i.e., in the θ-direction, the control effect to the (3,1) mode is very small. The other two modes, (m = 1, n = 1) and (m = 1, n = 3), Figs. 3.14 and 3.16, have relatively significant circumferential oscillations. Thus, circumferential control effects to these two modes are relatively significant.

Note that the modal damping ratio is inversely proportional to the modal frequency as indicated in the derived equation. For the first two modes, (m = 1, n = 1) and (m = 3, n = 1), the natural frequencies for $\beta = 120°$ shell are higher than those of the $\beta = 90°$ and $\beta = 60°$ shells. Thus, the damping ratio lines are below the other two in Figs. 3.14 and 3.15. However, it is reversed for the (m = 1, n = 3) mode, as shown in Fig. 3.16. (Detailed discussions on the natural frequencies and mode shapes of a circular cylindrical shell are presented in Chap. 9.)

In summary, a distributed in-plane electric force term, in the θ-direction, was left in the governing equation and it was used as a control force in distributed vibration controls. The velocity feedback was used in a closed-loop control system. In general, system damping ratios were enhanced when feedback voltages increased. Analytical solutions showed that this control force is only effective for odd modes and ineffective for even modes, due to voltage cancellations of antisymmetrical modes. Since this control force is circumferential in the θ-direction, control effects for natural modes with dominating circumferential oscillations were more significant.

3.9 Summary

In this chapter, applications of the generic piezoelectric shell theories to a number of common piezoelectric continua were presented. A four-step reduction procedure was introduced and it was demonstrated in two geometries. The first case was a piezoelectric plate which includes 1) a thick plate and 2) a thin plate. The derived system equations of the thick piezoelectric plate were completely identical to published results (Tiersten, 1969). The second case was a piezoelectric shell of revolution which represents another class of shell continua e.g., piezoelectric spheres, cylinders, cones, etc., which were discussed in detail. Applications of the generic shell vibration theory to other piezoelectric continua can be further explored. Note that the theory was derived based on a symmetrical hexagonal piezoelectric structure—class $C_{6v} = 6mm$.

A concept on **active** piezoelectric shell structures was proposed; a generic theory of an active piezoelectric shell continuum was derived from the piezoelectric thin shell theory in Sect. 2.4. Mathematical models of the active shell structures for three feedback algorithms, i.e., 1) a displacement feedback, 2) a velocity feedback, and 3) an acceleration feedback, were formulated. The electric force and moment components in the piezoelastic equations can be used to control the active shell continuum. A set of revised piezoelastic equations with distributed membrane control forces was then formulated and control applications discussed.

Frequency control of a piezoelectric bimorph beam via the displacement feedback was studied first. Since the beam was cantilevered, control effect was introduced through a boundary control at the free end. Natural frequencies can be actively controlled, however, control effectiveness to each mode varies. Analytical solutions were compared favorably with experimental data.

Damping control (introduced by the velocity feedback) of a cylindrical shell with simply supported boundary conditions was studied. The system equation was derived by simplifying the generic shell equation using four system parameters, i.e., two Lamé parameters and two radii of curvature. A distributed in-plane electric force term, in the θ-direction, was left in the equation and it was used as a circumferential control force in distributed shell controls. The velocity feedback was used in a closed-loop control system. In general, system damping ratios were enhanced when feedback voltages increased. Analytical solutions showed that this control force is effective only for odd natural modes and ineffective for even modes, due to voltage cancellations of antisymmetrical modes. Since this control force is in the θ-direction, control effects for natural modes with significant circumferential oscillations were more significant. In practical applications, however, the maximum voltage is restricted by the breakdown voltage of piezoelectric materials. In addition, high feedback voltage could also heat up the piezoelectric shell. This temperature effect was not considered in this study.

Appendix: Thick Piezoelectric Plate Equations

In *Linear Piezoelectric Plate Vibration* (Tiersten, 1969; p. 58), the electromechanical equations of a piezoelectric thick plate are written in a tensor expression:

$$\begin{aligned} &c_{11}u_{1,11} + (c_{11} + c_{66})u_{2,12} + (c_{13} + c_{44})u_{3,13} + c_{66}u_{1,22} \\ &\quad + c_{44}u_{1,33} + (e_{31} + e_{15})\varphi_{,13} = \rho\ddot{u}_1 \end{aligned}, \tag{A.1}$$

$$\begin{aligned} &c_{66}u_{2,11} + (c_{66} + c_{12})u_{1,12} + c_{11}u_{2,22} + (c_{13} + c_{44})u_{3,23} \\ &+ c_{44}u_{2,33} + (e_{31} + e_{15})\varphi_{,23} = \rho\ddot{u}_2 \end{aligned}, \tag{A.2}$$

$$\begin{aligned} &c_{44}u_{3,11} + (c_{44} + c_{13})u_{1,31} + c_{44}u_{3,22} + (c_{44} + c_{13})u_{2,23} \\ &+ c_{33}u_{3,33} + e_{15}\varphi_{,11} + e_{15}\varphi_{,22} + e_{33}\varphi_{,33} = \rho\ddot{u}_3 \end{aligned}. \tag{A.3}$$

References

Soedel, W., 1981, *Vibrations of Shells and Plates*, Dekker, New York.

Thomson, W.T., 1981, *Theory of Vibration with Applications*, Prentice-Hall, Englewood Cliffs, N.J.

Tiersten, H.F., 1969, *Linear Piezoelectric Plate Vibrations*, Plenum Press, New York.

Tzou, H.S. and Pandita, S., 1987, “A Multi-Purpose Dynamic and Tactile Sensor for Robot Manipulators,” (with Pandita, S.), *Journal of Robotic Systems*, Vol. (4.6), pp. 719–741, 1987.

Tzou, H.S., 1989, “Integrated Distributed Sensing and Active Vibration Suppression of Flexible Manipulators using Distributed Piezoelectrics,” *Journal of Robotic Systems*, Vol.(6), No.6, pp.745–767, December 1989.

Tzou, H.S., 1992, “A New Distributed Sensation and Control Theory for “Intelligent” Shells,” *Journal of Sound and Vibration*, Vol.152, No.3, pp.335–350, March 1992.

Tzou, H.S. and Zhong, J.P., 1990, “Electromechanical Dynamics of Piezoelectric Shell Distributed Systems, Part-1&2,” *Robotics Research-1990*, ASME-DSC-Vol.26, pp. 207–211, 1990 ASME Winter Annual Meetings, Dallas, Texas, Nov. 25–30, 1990; “Electromechanics and Vibrations of Piezoelectric Shell Distributed Systems,” *ASME Journal of Dynamic Systems, Measurements, and Control*, 1993.

Tzou, H.S. and Zhong, J.P., 1991a, “Adaptive Piezoelectric Shell Structures: Theory and Experiments,” *AIAA/ASME/ASCE/AHA/ASC 32nd Structures, Structural Dynamics and Materials Conference*, pp. 2290–2296, Paper No. AIAA-91-1238-CP, Baltimore, Maryland, April 8–10, 1991. *Mechanical Systems and Signal Processing*, Vol. (7), No. (3), May 1993.

Tzou, H.S. and Zhong, J.P., 1991b, “Control of Piezoelectric Cylindrical Shells via Distributed In-Plane Membrane Forces,” *Controls for Aerospace Systems*, DSC-Vol.35, pp. 15–20, Distributed Control of Flexible Structures, Aerospace Panel, Dynamic Systems and Control Division, 1991 ASME WAM, Atlanta, GA, December 1–6, 1991.

Tzou, H.S. and Zhong, J.P., 1991c, “Theory on Hexagonal Symmetrical Piezoelectric Thick Shells Applied to Smart Structures,” *Structural Vibration and Acoustics*, Edrs. Huang, Tzou, et al., ASME-DE-Vol.34, pp. 7–15, Symposium on Intelligent Structural Systems, 1991 ASME 13th Biennial Conference on Mechanical Vibration and Noise, Miami, Florida, September 22–25, 1991.

Tzou, H.S., 1992, "Active Piezoelectric Shell Continua," *Intelligent Structural Systems*, H.S. Tzou & G.L. Anderson, (Editors), pp. 9–74, Kluwer Academic Publishers, Aug. 1992.

Tzou, H.S., 2000, "Integrated Piezoelectric Sensor Actuator Design for Distributed Identification and Control of "Smart" Machines and Flexible Robots: Part 1: Theory and Experiments," *Intelligent Systems and Robotics*, G.W. Zobrist and C.Y Ho, (Editors), pp. 92–122, Gordon and Breach Science Publishers (2000OPA), Netherlands, 2000.

Tzou, H.S., Zhu, Y.T. and Hagiwara, I., 2002, "Distributed Precision Control of Structronic Shells and Common Shapes – A New Approach," *Recent Research Developments in Sound & Vibration*, ISBN: 81-7895-031-6, Vol.1, Part-2, pp. 613–646, TransWorld Research, 2002.

Zhong, J.P., 1991, *A Study of Piezoelectric Shell Dynamics Applied to DIstributed Structural Identification and Control*, Ph.D. Thesis, Department of Mechanical Engineering, University of Kentucky, Lexington, KY. (3/PzShl.BkPzSm1).

Chapter 4
Distributed Sensing and Control of Elastic Shells

The inherent natural damping of continua is usually not sufficient enough to effectively and quickly suppress the oscillations of high-performance flexible structural systems, e.g., aerospace structures. Thus, effective vibration control techniques become necessary. Vibration control techniques are generally classified into two major categories: ***passive*** and ***active*** (Reinhorn & Manolis, 1985). The passive vibration control techniques are based on energy absorption or dissipation principles (Tzou, 1988a), e.g., viscoelastic dampers, dynamic absorbers, shock absorbers, friction dampers, etc. Active vibration control techniques usually rely on counteracting mechanisms which generate opposing forces or moments to actively counteract the undesirable oscillations. One of the major advantages of active devices over passive devices is the "self-adaptivity" which offers variable control for various operational environments. However, because of this active actuation capability, external power sources (e.g., electrical, hydraulic, pneumatic, etc.) and a decision-making "brain" (such as a central processing unit) are usually required (Tzou & Gadre, 1988, 1990). In addition, active controls also depend on accurate measurements of current dynamic states. Thus, sensors or transducers are also essential to monitor the structural states in active controls. Both sensors and actuators used here are made of distributed piezoelectric layers in this chapter, as well as in the book.

Structures are **distributed** in nature, i.e., structural behaviors are functions of ***spatial*** and ***time*** variables; they are classified as ***continua*** or ***distributed-parameter systems*** (DPS's). Vibration suppression and control of distributed parameter systems always represents a challenge, both in theory and practice. Theoretical development for control of distributed parameter systems has been constantly advancing during the past three decades (Butkovskii, 1962; Wang, 1966; Sakawa, 1966; Lions, 1968; Tzafestas, 1970; Robinson, 1971; Brichkin et al. 1973; Meirovitch, 1983; Vidyasagar, 1988; Balas, 1988; Tzou, 1987, 1988b, 1989a; Zimmerman, Inman & Juang, 1988). However, due to practical limitations, e.g., availability of distributed sensors and actuators, real world applications are relatively limited. Note that conventional sensors and actuators are mostly **discrete** types which measure or

H. (HS) Tzou, *Piezoelectric Shells*, Solid Mechanics and Its Applications 247,
https://doi.org/10.1007/978-94-024-1258-1_4

control spatially ***discrete*** locations. These devices are ineffective if they are placed at modal nodes or lines. One solution to this problem is to use *distributed* sensors and actuators which are spatially distributed so that they are sensitive to spatially distributed structural behaviors. (Problems and solutions related to spatially distributed sensors and actuators will be discussed in Chaps. 7–9.) In this chapter, a generic deep elastic shell continuum laminated with distributed piezoelectric layers is regarded as a generic DPS or a generic smart structronic shell system from which distributed sensing and control theories can be derived. One of the piezoelectric layers is used as a distributed sensor for structural measurements based on the ***direct piezoelectric effect***. The other is used as a distributed actuator for structural controls via the ***converse piezoelectric effect***. (Recall that the generic shell can be easily simplified to many other common continua or structures, e.g., shells of revolutions, paraboloidal shells, toroidal shells, spherical shells, cylindrical shells, conical shells, plates, rings, etc. (Tzou, 1992).

Distributed vibration control of flexible structures using synthetic piezoelectric ceramics and/or polymeric polyvinylidene fluoride (PVDF) has been investigated by many researchers. Usov and Surygin (1984) studied damping changes of a semiconductor structure using piezofilm. Sirlin (1987) used a PVDF as an active isolator for spacecraft applications. Plump, Hubbard, and Baily (1987) used a piezoelectric film to enhance the damping ratio of a cantilever beam. Tzou also applied a PVDF film as an active damper in a flexible structure (1987) and an active vibration isolator/exciter (Tzou & Gadre, 1988 & 1990). Crawley and de Luis (1987) investigated the use of piezoceramic actuators as elements of intelligent structures. Lee and Moon (1988) proposed a piezoelectric modal sensor/actuator design for a plate and applied the design to a two-dimensional thin beam. Hanagud and Obal (1988) identified the dynamic coupling coefficients in a structure with piezoceramic sensors and actuators. Fanson and Garba (1988) presented an experimental study of active members made of piezoelectric ceramics. Baz and Poh (1988) evaluated the performance of an active control system with piezoceramic actuators. Cudney et al. (1989) proposed a distributed multi-layered actuator using the deep beam theory. Crawley and Anderson (1990) developed detailed piezoceramic actuator modeling for beams. A theory on multi-layered shells coupled with distributed piezoelectric shell actuators was also proposed and evaluated (Tzou and Gadre, 1989; Tzou, 1989b). A generic distributed piezoelectric identification and vibration control theory was proposed for a shell continuum with surface coupled distributed sensor and actuator layers (Tzou, 1990). Hagood et al. (1990) proposed a constitutive dynamic modeling of piezoelectric actuator and control. Distributed active vibration control of a cylindrical shell and a plate coupled with the PVDF was also investigated (Tzou and Tseng, 1988a&b). Distributed piezoelectric sensor and actuator have been applied to flexible robot manipulators (Tzou, 1989c; Tzou, Tseng, & Wan, 1990). In addition to theoretical and experimental studies, distributed vibration control of a beam modeled by piezoelectric beam finite elements was investigated by Obal (1986). Distributed sensing and active vibration control of a triple layered plate with top and bottom piezoelectric layers was studied by Tzou and Tseng (1990). A blade shallow shell was studied by Tzou and Tseng (1988a). This chapter is devoted to the development of distributed sensing and control theories for a generic deep elastic shell continuum laminated with distributed piezoelectric sensor and actuator layers, so called a generic smart structronic shell system.

As discussed in Chap. 1, there are many synthetic and natural piezoelectric materials. In general, most of them are crystalline solids which are dense, brittle and difficult to fabricate into complex shapes. In the sensing and control of flexible continua, a flexible piezoelectric material is desirable for two reasons. First, the piezoelectric material needs to be closely coupled with the flexible continua, but not to change the dynamic characteristics, e.g., natural frequencies and modes. Second, the material must be flexible, not brittle, so that it will not break during strenuous structural vibrations. Thus, a tough and pliant polymeric piezoelectric polyvinylidene fluoride (PVDF) is selected in this study. (Fundamental properties of the PVDF are discussed in Appendix.)

In this chapter, a generic deep elastic shell continuum laminated with distributed piezoelectric layers is regarded as a generic DPS or a generic smart structronic shell system from which distributed sensing and control theories are derived. As discussed previously, one of the piezoelectric layers serves as a distributed sensor and the other as a distributed actuator. Detailed electromechanics of the distributed piezoelectric sensor and actuator are analyzed and governing system equations are derived accordingly. The modal expansion method is incorporated with the theories to describe the general sensing and control equations in modal coordinates. Applications of the generic theories to other common geometries are demonstrated in case studies. Generic sensor/actuator design principles, i.e., ***the segmentation technique*** **and** ***the shaping technique***, of the distributed sensors and actuators are also discussed. Distributed energy harvesting of piezoelectric laminated linear and nonlinear shell structures is presented in Chaps. 11 and 12.

4.1 System Definitions

In this study, a generic distributed parameter system is an elastic double-curvature shell continuum sandwiched between two flexible piezoelectric shell layers, Fig. 4.1 where α_1 and α_2 axes define the neutral surface. The top piezoelectric layer serves as a distributed sensor and the bottom a distributed actuator. It is assumed that the piezoelectric layer is much thinner than the elastic shell thickness and the layers fully cover the shell continuum. Localization and discretization of the sensor/actuator can be achieved by step-functions and/or Dirac delta functions. (Segmentation and shaping of distributed sensors/actuators will be discussed in later chapters.) Note that multi-layers of piezoelectric sheets could also be coupled with the flexible shell. One group can serve as a distributed sensor (or sensors) and the other serves as a distributed shell actuator (or actuators) for active vibration suppression and control.

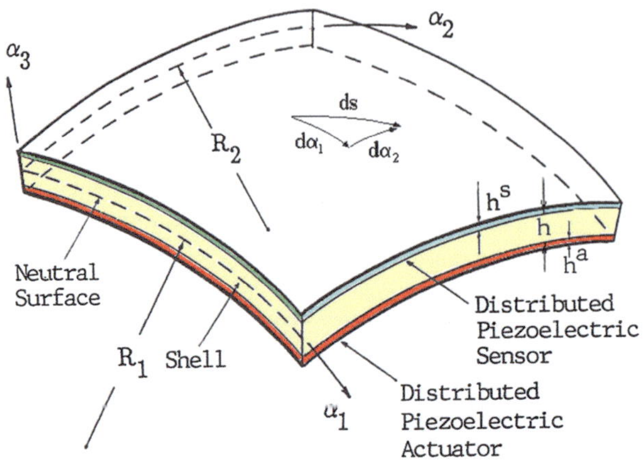

Fig. 4.1 A deep elastic shell with distributed piezoelectric shell sensor/actuator layers

The infinitesimal distance ds (Fig. 4.1) of a shell element in a curvilinear coordinate system, (α_1, α_2, and α_3) is defined by a fundamental form:

$$(ds)^2 = A_1^2(d\alpha_1)^2 + A_2^2(d\alpha_2)^2, \tag{4.1.1}$$

where A_1 and A_2 are the Lamé parameters. Equations of motion of the elastic shell can be derived from either Hamilton's principle or simplification of generic piezoelectric shell equations, Sect. 2.5 of Chap. 2.

$$\begin{aligned}&-\frac{\partial(N_{11}A_2)}{\partial\alpha_1}-\frac{\partial(N_{21}A_1)}{\partial\alpha_2}-N_{12}\frac{\partial A_1}{\partial\alpha_2}+N_{22}\frac{\partial A_2}{\partial\alpha_1}-\frac{1}{R_1}\left[\frac{\partial(M_{11}A_2)}{\partial\alpha_1}\right.\\&\left.+\frac{\partial(M_{21}A_1)}{\partial\alpha_2}+M_{12}\frac{\partial A_1}{\partial\alpha_2}-M_{22}\frac{\partial A_2}{\partial\alpha_1}\right]+A_1A_2\rho h\ddot{u}_1=A_1A_2F_1\end{aligned}, \tag{4.1.2}$$

$$\begin{aligned}&-\frac{\partial(N_{12}A_2)}{\partial\alpha_1}-\frac{\partial(N_{22}A_1)}{\partial\alpha_2}-N_{21}\frac{\partial A_2}{\partial\alpha_1}+N_{11}\frac{\partial A_1}{\partial\alpha_2}-\frac{1}{R_2}\left[\frac{\partial(M_{12}A_2)}{\partial\alpha_1}\right.\\&\left.+\frac{\partial(M_{22}A_1)}{\partial\alpha_2}+M_{21}\frac{\partial A_2}{\partial\alpha_1}-M_{11}\frac{\partial A_1}{\partial\alpha_2}\right]+A_1A_2\,\rho\,h\,\ddot{u}_2=A_1A_2F_2\end{aligned}, \tag{4.1.3}$$

$$\begin{aligned}&-\frac{\partial}{\partial\alpha_1}\left\{\frac{1}{A_1}\left[\frac{\partial(M_{11}A_2)}{\partial\alpha_1}+\frac{\partial(M_{21}A_1)}{\partial\alpha_2}+M_{12}\frac{\partial A_1}{\partial\alpha_2}-M_{22}\frac{\partial A_2}{\partial\alpha_1}\right]\right\}\\&-\frac{\partial}{\partial\alpha_2}\left\{\frac{1}{A_2}\left[\frac{\partial(M_{12}A_2)}{\partial\alpha_1}+\frac{\partial(M_{22}A_1)}{\partial\alpha_2}+M_{21}\frac{\partial A_2}{\partial\alpha_1}-M_{11}\frac{\partial A_1}{\partial\alpha_2}\right]\right\},\\&+A_1A_2\left(\frac{N_{11}}{R_1}+\frac{N_{22}}{R_2}\right)+A_1A_2\rho h\ddot{u}_3=A_1A_2F_3\end{aligned} \tag{4.1.4}$$

where $\ddot{u}_i$ is the acceleration in the i-th direction; F_i is the externally applied mechanical load; ρ is the mass density; h is the shell thickness; R_1 and R_2 are the curvature radii of α_1 and α_2 axes, respectively. Based on the derivations in Sect. 2.4

of Chapter 2, the resultant forces N_{ij} and moments M_{ij} can be redefined in terms of generic displacements u_1, u_2, and u_3 on the neutral surface.

4.1.1 Membrane Forces

$$N_{11} = K\left\{\left[\frac{1}{A_1}\frac{\partial u_1}{\partial \alpha_1} + \frac{u_2}{A_1 A_2}\frac{\partial A_1}{\partial \alpha_2} + \frac{u_3}{R_1}\right] + \mu\left[\frac{1}{A_2}\frac{\partial u_2}{\partial \alpha_2} + \frac{u_1}{A_1 A_2}\frac{\partial A_2}{\partial \alpha_1} + \frac{u_3}{R_2}\right]\right\}, \tag{4.1.5}$$

$$N_{22} = K\left\{\left[\frac{1}{A_2}\frac{\partial u_2}{\partial \alpha_2} + \frac{u_1}{A_1 A_2}\frac{\partial A_2}{\partial \alpha_1} + \frac{u_3}{R_2}\right] + \mu\left[\frac{1}{A_1}\frac{\partial u_1}{\partial \alpha_1} + \frac{u_2}{A_1 A_2}\frac{\partial A_1}{\partial \alpha_2} + \frac{u_3}{R_1}\right]\right\}, \tag{4.1.6}$$

$$N_{12} = \frac{K(1-\mu)}{2}\left\{\frac{A_2}{A_1}\frac{\partial}{\partial \alpha_1}\left(\frac{u_2}{A_2}\right) + \frac{A_1}{A_2}\frac{\partial}{\partial \alpha_2}\left(\frac{u_1}{A_1}\right)\right\}, \tag{4.1.7}$$

where μ is Poisson's ratio, and K is the membrane stiffness, $K = Yh/(1-\mu^2)$ where Y is Young's modulus of the shell.

4.1.2 Bending Moments

$$\begin{aligned} M_{11} = D\Bigg\{&\left[\frac{1}{A_1}\frac{\partial}{\partial \alpha_1}\left(\frac{u_1}{R_1} - \frac{1}{A_1}\frac{\partial u_3}{\partial \alpha_1}\right) + \frac{1}{A_1 A_2}\left(\frac{u_2}{R_2} - \frac{1}{A_2}\frac{\partial u_3}{\partial \alpha_2}\right)\frac{\partial A_1}{\partial \alpha_2}\right] + \mu\left[\frac{1}{A_2}\frac{\partial}{\partial \alpha_2}\left(\frac{u_2}{R_2} - \frac{1}{A_2}\frac{\partial u_3}{\partial \alpha_2}\right),\right. \\ &\left.+ \frac{1}{A_1 A_2}\left(\frac{u_1}{R_1} - \frac{1}{A_1}\frac{\partial u_3}{\partial \alpha_1}\right)\frac{\partial A_2}{\partial \alpha_1}\right]\Bigg\} \end{aligned} \tag{4.1.8}$$

$$\begin{aligned} M_{22} = D\Bigg\{&\left[\frac{1}{A_2}\frac{\partial}{\partial \alpha_2}\left(\frac{u_2}{R_2} - \frac{1}{A_2}\frac{\partial u_3}{\partial \alpha_2}\right) + \frac{1}{A_1 A_2}\left(\frac{u_1}{R_1} - \frac{1}{A_1}\frac{\partial u_3}{\partial \alpha_1}\right)\frac{\partial A_2}{\partial \alpha_1}\right] + \mu\left[\frac{1}{A_1}\frac{\partial}{\partial \alpha_1}\left(\frac{u_1}{R_1} - \frac{1}{A_1}\frac{\partial u_3}{\partial \alpha_1}\right),\right. \\ &\left.+ \frac{1}{A_1 A_2}\left(\frac{u_2}{R_2} - \frac{1}{A_2}\frac{\partial u_3}{\partial \alpha_2}\right)\frac{\partial A_1}{\partial \alpha_2}\right]\Bigg\} \end{aligned} \tag{4.1.9}$$

$$\begin{aligned} M_{12} = \frac{D(1-\mu)}{2}\Bigg\{&\frac{A_2}{A_1}\frac{\partial}{\partial \alpha_1}\left[\frac{1}{A_2}\left(\frac{u_2}{R_2} - \frac{1}{A_2}\frac{\partial u_3}{\partial \alpha_2}\right)\right] \\ &+ \frac{A_1}{A_2}\frac{\partial}{\partial \alpha_2}\left[\frac{1}{A_1}\left(\frac{u_1}{R_1} - \frac{1}{A_1}\frac{\partial u_3}{\partial \alpha_1}\right)\right]\Bigg\}; \end{aligned} \tag{4.1.10}$$

where D is the bending stiffness and $D = Yh^3/[12(1-\mu^2)]$ (Note that the resultant forces and moments could also include piezoelectricity induced feedback forces and moments which are to be derived later.)

4.1.3 Modal Expansion Method

Dynamic response of a shell structure can be determined by ***the modal expansion method*** in which the total response is a sum of all participating modes U_{ik} weighted with their corresponding modal participation factors η_k's:

$$u_i(\alpha_1, \alpha_2, t) = \sum_{k=1}^{\infty} \eta_k(t) U_{ik}(\alpha_1, \alpha_2) \quad i = 1, 2, 3, \tag{4.1.11}$$

where i denotes the principal direction, Sect. 3.5 of Chap. 3. Since for a distributed system, the number of modes is infinite, i.e., k = 1,2,3,...,∞. This modal expression is used to estimate the modal contribution of distributed piezoelectric sensors. Simplification of the theory to commonly occurring geometries is demonstrated in case studies.

4.2 Distributed Sensing of Elastic Shells

Recall that the top piezoelectric layer on the generic shell distributed system serves as a distributed sensor. In this section, a distributed sensing theory based on the direct piezoelectric effect and the shell strains/deformations is developed. It is assumed that the distributed piezoelectric sensor layer is much thinner than that of the shell structure. Thus, the piezoelectric sensor strains are assumed constant and equal to the outer surface strains of the shell. (Note that in the distributed sensing application, only the direct piezoelectric effect is considered.)

4.2.1 Sensor Output Signal

Assuming the electric charge is balanced in the piezoelectric sensor layer (an insulator), one can derive an electrical charge equation by using the Gauss theorem (Tzou, 1988b):

$$\nabla\{D_i\} = 0, \tag{4.2.1}$$

where ∇ is a differential operator. (Governing equations from the linear piezoelectricity theory are summarized in APPENDIX.) From the thin shell configuration defined earlier, only the transverse electric field E_3 is considered so that the strains and dielectric displacement D_3 are independent of the α_3. According to Maxwell's equation, the electric field E_i can be related to the electric potential ϕ by

$$\{E_i\} = -\nabla\phi. \tag{4.2.2}$$

From Eqs. (4.9.1) and (4.9.2) in APPENDIX, the voltage across the electrodes can be obtained by integrating the electric field over the thickness of the piezoelectric shell sensor layer, i.e.,

$$\phi = -\int^{h^S} E_3 d\alpha_3 = h^S\left(h_{31}S_{11}^S + h_{32}S_{22}^S - \beta_{33}D_3\right), \tag{4.2.3}$$

where h^s denotes the piezoelectric sensor layer thickness; S_{11}^S and S_{22}^S are the normal strains in the α_1 and α_2 directions, respectively; the superscript "s" denotes the distributed sensor layer. It is assumed that the piezoelectric material is not sensitive to in-plane twisting shear strain S_{12}. In addition, since the shell is thin, the transverse shear strains S_{13} and S_{23} are neglected. The piezoelectric sensor layer is coupled with the elastic shell; thus, the normal strains in the sensor layer can be estimated by (Sect. 2.4):

$$\begin{aligned} S_{11}^S = &\left\{\left[\frac{1}{A_1}\frac{\partial u_1}{\partial \alpha_1} + \frac{u_2}{A_1A_2}\frac{\partial A_1}{\partial \alpha_2} + \frac{u_3}{R_1}\right]\right. \\ &\left. + r_1^S\left[\frac{1}{A_1}\frac{\partial}{\partial \alpha_1}\left(\frac{u_1}{R_1} - \frac{1}{A_1}\frac{\partial u_3}{\partial \alpha_1}\right) + \frac{1}{A_1A_2}\frac{\partial A_1}{\partial \alpha_2}\left(\frac{u_2}{R_2} - \frac{1}{A_2}\frac{\partial u_3}{\partial \alpha_2}\right)\right]\right\}, \end{aligned} \tag{4.2.4}$$

$$\begin{aligned} S_{22}^S = &\left\{\left[\frac{1}{A_2}\frac{\partial u_2}{\partial \alpha_2} + \frac{u_1}{A_1A_2}\frac{\partial A_2}{\partial \alpha_1} + \frac{u_3}{R_2}\right]\right. \\ &\left. + r_2^S\left[\frac{1}{A_2}\frac{\partial}{\partial \alpha_2}\left(\frac{u_2}{R_2} - \frac{1}{A_2}\frac{\partial u_3}{\partial \alpha_2}\right) + \frac{1}{A_1A_2}\frac{\partial A_2}{\partial \alpha_1}\left(\frac{u_1}{R_1} - \frac{1}{A_1}\frac{\partial u_3}{\partial \alpha_1}\right)\right]\right\}, \end{aligned} \tag{4.2.5}$$

where r_1^s and r_2^s denote the distances measured from the neutral surface to the mid-plane of the sensor layer; u_1, u_2 and u_3 are the displacements in three principal directions. Note that the total strains are contributed by two components: a membrane component and a bending component. All terms inside the first bracket are the membrane strain terms; all terms leading by r_i^s are bending strain terms. Bending strains are functions of distances r_i^s measured from the neutral surface. Rearranging Eq. (4.2.3), one can write the electric displacement D_3^S as

$$D_3^S = \frac{1}{\beta_{33}}\left(h_{31}S_{11}^S + h_{32}S_{22}^S - \frac{\phi}{h^S}\right). \tag{4.2.6}$$

Since D_3^S is defined as the charge per unit area, one can integrate Eq. (4.2.6) over the electroded surface S^e to estimate a total surface charge. An open-circuit voltage ϕ^S condition can be obtained by setting the charge zero, i.e.,

$$\begin{aligned}\phi^S &= \frac{h^S}{S^e}\int_{S^e}\left(h_{31}S_{11}^S + h_{32}S_{22}^S\right)dS^e \\ &= \frac{h^S}{S^e}\int_{S^e}\left(h_{31}S_{11}^S + h_{32}S_{22}^S\right)A_1A_2\,d\alpha_1\,d\alpha_2\end{aligned}. \tag{4.2.7}$$

Substituting the strains into Eq. (4.2.7) yields the distributed sensor output $\phi^S(\alpha_1,\alpha_2)$ in terms of displacements and other system parameters.

$$\begin{aligned}\phi^S(\alpha_1,\alpha_2) = \frac{h^S}{S^e}\int_{S^e}\Bigg[h_{31}\Bigg\{&\left[\frac{1}{A_1}\frac{\partial u_1}{\partial\alpha_1} + \frac{u_2}{A_1A_2}\frac{\partial A_1}{\partial\alpha_2} + \frac{u_3}{R_1}\right] \\ &+ r_1^S\left[\frac{1}{A_1}\frac{\partial}{\partial\alpha_1}\left(\frac{u_1}{R_1} - \frac{1}{A_1}\frac{\partial u_3}{\partial\alpha_1}\right) + \frac{1}{A_1A_2}\frac{\partial A_1}{\partial\alpha_2}\left(\frac{u_2}{R_2} - \frac{1}{A_2}\frac{\partial u_3}{\partial\alpha_2}\right)\right]\Bigg\} \\ + h_{32}\Bigg\{&\left[\frac{1}{A_2}\frac{\partial u_2}{\partial\alpha_2} + \frac{u_1}{A_1A_2}\frac{\partial A_2}{\partial\alpha_1} + \frac{u_3}{R_2}\right] \\ &+ r_2^S\left[\frac{1}{A_2}\frac{\partial}{\partial\alpha_2}\left(\frac{u_2}{R_2} - \frac{1}{A_2}\frac{\partial u_3}{\partial\alpha_2}\right) + \frac{1}{A_1A_2}\frac{\partial A_2}{\partial\alpha_1}\left(\frac{u_1}{R_1} - \frac{1}{A_1}\frac{\partial u_3}{\partial\alpha_1}\right)\right]\Bigg\}\Bigg]dS^e\end{aligned}. \tag{4.2.8}$$

Note that this equation is valid for both surface coupled and internally embedded distributed piezoelectric sensors. The total signal is contributed by two strain components, i.e., membrane and bending, as discussed previously. Thus, this sensor equation can be used in 1) *membrane sensor* applications, such as for in-plane contraction and expansion, and 2) *bending sensor* applications, such as for bending vibrations. Furthermore, if the shear strain S_{12}^S reserved and normal strains eliminated, the sensor signal equation can be rewritten in terms of the in-plane shear effect and used as *shear* sensors.

To examine the modal contribution to the sensor output, one can further substitute the modal expression into Eq. (4.2.8).

$$\begin{aligned}
\phi^S(\alpha_1,\alpha_2) = \frac{h^S}{S^e}\int_{S^e}\Bigg[h_{31}\Bigg\{\Bigg[\frac{1}{A_1}\frac{\partial}{\partial\alpha_1}\sum_{k=1}^{\infty}\eta_k(t)U_{1k}(\alpha_1,\alpha_2)\\
+\frac{1}{A_1A_2}\sum_{k=1}^{\infty}\eta_k(t)U_{2k}(\alpha_1,\alpha_2)\frac{\partial A_1}{\partial\alpha_2}+\frac{1}{R_1}\sum_{k=1}^{\infty}\eta_k(t)U_{3k}(\alpha_1,\alpha_2)\Bigg]\\
+r_1^S\Bigg[\frac{1}{A_1}\frac{\partial}{\partial\alpha_1}\left(\frac{1}{R_1}\sum_{k=1}^{\infty}\eta_k(t)U_{1k}(\alpha_1,\alpha_2)-\frac{1}{A_1}\frac{\partial}{\partial\alpha_1}\sum_{k=1}^{\infty}\eta_k(t)U_{3k}(\alpha_1,\alpha_2)\right)\\
+\frac{1}{A_1A_2}\frac{\partial A_1}{\partial\alpha_2}\left(\frac{1}{R_2}\sum_{k=1}^{\infty}\eta_k(t)U_{2k}(\alpha_1,\alpha_2)-\frac{1}{A_2}\frac{\partial}{\partial\alpha_2}\sum_{k=1}^{\infty}\eta_k(t)U_{3k}(\alpha_1,\alpha_2)\right)\Bigg]\Bigg\}\\
+h_{32}\Bigg\{\Bigg[\frac{1}{A_2}\frac{\partial}{\partial\alpha_2}\sum_{k=1}^{\infty}\eta_k(t)U_{2k}(\alpha_1,\alpha_2)\\
+\frac{1}{A_1A_2}\sum_{k=1}^{\infty}\eta_k(t)U_{1k}(\alpha_1,\alpha_2)\frac{\partial A_2}{\partial\alpha_1}+\frac{1}{R_2}\sum_{k=1}^{\infty}\eta_k(t)U_{3k}(\alpha_1,\alpha_2)\Bigg]\\
+r_2^S\Bigg[\frac{1}{A_2}\frac{\partial}{\partial\alpha_2}\left(\frac{1}{R_2}\sum_{k=1}^{\infty}\eta_k(t)U_{2k}(\alpha_1,\alpha_2)-\frac{1}{A_2}\frac{\partial}{\partial\alpha_2}\sum_{k=1}^{\infty}\eta_k(t)U_{3k}(\alpha_1,\alpha_2)\right)\\
+\frac{1}{A_1A_2}\frac{\partial A_2}{\partial\alpha_1}\Bigg(\frac{1}{R_1}\sum_{k=1}^{\infty}\eta_k(t)U_{1k}(\alpha_1,\alpha_2)\\
-\frac{1}{A_1}\frac{\partial}{\partial\alpha_1}\sum_{k=1}^{\infty}\eta_k(t)U_{3k}(\alpha_1,\alpha_2)\Bigg)\Bigg]\Bigg\}\Bigg]A_1A_2d\alpha_1d\alpha_2
\end{aligned} \tag{4.2.9}$$

Equation (4.2.9) shows that the total output of the distributed sensor is a summation of all participating modes and the specific contribution of each mode is determined by the modal participation factor. The modal participation factors η_k's are zero for those modes not participating in the shell oscillation.

4.2.2 Local Point Voltage

If the surface average and integration is removed from Eqs. (4.2.8) and (4.2.9), a discrete local voltage amplitude can be estimated by specifying the location (α_1^*, α_2^*) interested. It is assumed that there is a finite electrode (or point electrode) centering around the point (α_1^*, α_2^*) of interest.

$$
\begin{aligned}
\phi^S(\alpha_1^*,\alpha_2^*) = h^S\Bigg[h_{31}\Bigg\{ & \Bigg[\frac{1}{A_1}\frac{\partial}{\partial\alpha_1}\sum_{k=1}^{\infty}\eta_k(t)U_{1k}(\alpha_1^*,\alpha_2^*) \\
& + \frac{1}{A_1A_2}\sum_{k=1}^{\infty}\eta_k(t)U_{2k}(\alpha_1^*,\alpha_2^*)\frac{\partial A_1}{\partial\alpha_2} + \frac{1}{R_1}\sum_{k=1}^{\infty}\eta_k(t)U_{3k}(\alpha_1^*,\alpha_2^*)\Bigg] \\
& + r_1^S\Bigg[\frac{1}{A_1}\frac{\partial}{\partial\alpha_1}\Bigg(\frac{1}{R_1}\sum_{k=1}^{\infty}\eta_k(t)U_{1k}(\alpha_1^*,\alpha_2^*) \\
& - \frac{1}{A_1}\frac{\partial}{\partial\alpha_1}\sum_{k=1}^{\infty}\eta_k(t)U_{3k}(\alpha_1^*,\alpha_2^*)\Bigg) \\
& + \frac{1}{A_1A_2}\frac{\partial A_1}{\partial\alpha_2}\Bigg(\frac{1}{R_2}\sum_{k=1}^{\infty}\eta_k(t)U_{2k}(\alpha_1^*,\alpha_2^*) \\
& - \frac{1}{A_2}\frac{\partial}{\partial\alpha_2}\sum_{k=1}^{\infty}\eta_k(t)U_{3k}(\alpha_1^*,\alpha_2^*)\Bigg)\Bigg]\Bigg\} \\
+ h_{32}\Bigg\{ & \Bigg[\frac{1}{A_2}\frac{\partial}{\partial\alpha_2}\sum_{k=1}^{\infty}\eta_k(t)U_{2k}(\alpha_1^*,\alpha_2^*) \\
& + \frac{1}{A_1A_2}\sum_{k=1}^{\infty}\eta_k(t)U_{1k}(\alpha_1^*,\alpha_2^*)\frac{\partial A_2}{\partial\alpha_1} + \frac{1}{R_2}\sum_{k=1}^{\infty}\eta_k(t)U_{3k}(\alpha_1,\alpha_2)\Bigg] \\
& + r_2^S\Bigg[\frac{1}{A_2}\frac{\partial}{\partial\alpha_2}\Bigg(\frac{1}{R_2}\sum_{k=1}^{\infty}\eta_k(t)U_{2k}(\alpha_1^*,\alpha_2^*) \\
& - \frac{1}{A_2}\frac{\partial}{\partial\alpha_2}\sum_{k=1}^{\infty}\eta_k(t)U_{3k}(\alpha_1^*,\alpha_2^*)\Bigg) \\
& + \frac{1}{A_1A_2}\frac{\partial A_2}{\partial\alpha_1}\Bigg(\frac{1}{R_1}\sum_{k=1}^{\infty}\eta_k(t)U_{1k}(\alpha_1^*,\alpha_2^*) \\
& - \frac{1}{A_1}\frac{\partial}{\partial\alpha_1}\sum_{k=1}^{\infty}\eta_k(t)U_{3k}(\alpha_1^*,\alpha_2^*)\Bigg)\Bigg]\Bigg\}\Bigg]
\end{aligned}
\tag{4.2.10}
$$

It is assumed that the voltage is developed on a finite electrode area. In this case, dynamic states, i.e., a *potential map* or *contour*, of the whole shell continuum can be defined.

4.2.3 *Modal Identification—Modal Voltage*

It is desirable to identify the mode shapes of the shell continuum via the distributed piezoelectric sensor. Since mode shapes are distinct for the natural modes, the induced potential (spatial) distributions—***modal voltages***—should also be unique. To construct a *modal voltage* of a natural mode, one needs to calculate a number of local point signals and graphically connect the local amplitudes together for the natural mode. The ***potential map*** or ***contour*** represent the modal voltage of the natural mode. Note that a finite separation of surface electrode is required to prevent the global voltage average since electric charges are freely moving on the electrodes. The local output signal $\phi_k^S(\alpha_1^*, \alpha_2^*)$ of the k-th natural mode can be expressed as

$$
\begin{aligned}
\phi_k^S\left(\alpha_1^*, \alpha_2^*\right) = h^S\Bigg[h_{31}\Bigg\{\Bigg[&\frac{1}{A_1}\frac{\partial}{\partial\alpha_1}\eta_k(t)U_{1k}\left(\alpha_1^*, \alpha_2^*\right) + \frac{1}{A_1 A_2}\eta_k(t) \\
&\cdot U_{2k}\left(\alpha_1^*, \alpha_2^*\right)\frac{\partial A_1}{\partial\alpha_2} + \frac{1}{R_1}\eta_k(t)U_{3k}\left(\alpha_1^*, \alpha_2^*\right)\Bigg] \\
+ r_1^S\Bigg[&\frac{1}{A_1}\frac{\partial}{\partial\alpha_1}\left(\frac{1}{R_1}\eta_k(t)U_{1k}\left(\alpha_1^*, \alpha_2^*\right) - \frac{1}{A_1}\frac{\partial}{\partial\alpha_1}\eta_k(t)\right. \\
&\left.\cdot U_{3k}\left(\alpha_1^*, \alpha_2^*\right)\right) + \frac{1}{A_1 A_2}\frac{\partial A_1}{\partial\alpha_2}\left(\frac{1}{R_2}\eta_k(t)U_{2k}\left(\alpha_1^*, \alpha_2^*\right)\right. \\
&\left.- \frac{1}{A_2}\frac{\partial}{\partial\alpha\alpha_2}\eta_k(t)U_{3k}\left(\alpha_1^*, \alpha_2^*\right)\right)\Bigg]\Bigg\} \\
+ h_{32}\Bigg\{\Bigg[&\frac{1}{A_2}\frac{\partial}{\partial\alpha_2}\eta_k(t)U_{2k}\left(\alpha_1^*, \alpha_2^*\right) + \frac{1}{A_1 A_2}\eta_k(t) \\
&\cdot U_{1k}\left(\alpha_1^*, \alpha_2^*\right)\frac{\partial A_2}{\partial\alpha_1} + \frac{1}{R_2}\eta_k(t)U_{3k}(\alpha_1, \alpha_2)\Bigg] \\
+ r_2^S\Bigg[&\frac{1}{A_2}\frac{\partial}{\partial\alpha_2}\left(\frac{1}{R_2}\eta_k(t)U_{2k}\left(\alpha_1^*, \alpha_2^*\right) - \frac{1}{A_2}\frac{\partial}{\partial\alpha\alpha_2}\eta_k(t)\right. \\
&\left.\cdot U_{3k}\left(\alpha_1^*, \alpha_2^*\right)\right) + \frac{1}{A_1 A_2}\frac{\partial A_2}{\partial\alpha_1}\left(\frac{1}{R_1}\eta_k(t)U_{1k}\left(\alpha_1^*, \alpha_2^*\right)\right. \\
&\left.- \frac{1}{A_1}\frac{\partial}{\partial\alpha_1}\eta_k(t)U_{3k}\left(\alpha_1^*, \alpha_2^*\right)\right)\Bigg]\Bigg\}\Bigg]
\end{aligned}
\tag{4.2.11}
$$

In general, the output signal of a distributed piezoelectric sensor is contributed by strains which can be further classified into two components: 1) the in-plane membrane strain and 2) the out-of-plane bending strain. Depending on the placement of the sensor, the contribution of each strain component could be different. For example, only the bending strain contributes to the output signal if the shell experiences bending oscillations only. On the other hand, the membrane strain contributes to the output if the shell only experiences membrane oscillations. A case study on ***modal voltages*** of a cantilever plate is presented in Chap. 13. Note that the piezoelectric sensor layer has a uniform thickness h^s. Detailed electromechanics of ***distributed shell modal sensors*** (**shaped** and/or **convolved**) is presented in Chaps. 8 and 9. Observation spillover problems (Meirovitch & Baruh, 1981; Balas, 1978) and solutions of distributed sensors are discussed in Chaps. 7–9. Optimal control of parabolic shells is presented in Chap. 10. Energy harvesting of piezoelectric laminated shells is presented in Chap. 11.

4.3 Distributed Vibration Control of Shells

It is assumed that the piezoelectric actuator layer is unconstrained and is free from external in-plane normal forces. Thus, the stress effects can be neglected in the derivation, i.e., a stress-free condition. (This stress-free condition implies that the boundaries of the actuator layer can not be fully constrained.) The induced strains due to imposed control voltages, the converse piezoelectric effect, are used to estimate actuation authorities counteracting the shell oscillation. It is also assumed that the applied control voltage $\phi^a(\alpha_1, \alpha_2)$ is much more significant than the self-generated voltage ϕ, from the direct effect, in the distributed actuator. Thus, the self-generated voltage ϕ is neglected in the active vibration control system. ϕ^a can be either a reference voltage in an open-loop control, or the distributed sensor signal ϕ^s in a closed-loop control. In this section, this distributed vibration control mechanism is analyzed and three feedback algorithms, namely 1) the direct proportional feedback, 2) the negative velocity feedback, and 3) a Lyapunov feedback, are proposed. System dynamic equations including all control effects are also derived.

In the shell definition, the bottom piezoelectric layer is designated as a distributed actuator for active vibration suppression and control. It is assumed that the distributed piezoelectric layer is made of a bi-axially polarized piezoelectric material. Thus, a voltage ϕ^a applied to the distributed actuator layer introduces two in-plane normal strains (in both α_1 and α_2 directions) in the actuator layer due to the converse piezoelectric effect.

$$S^a_{11} = (d_{31}\phi^a)/h^a, \tag{4.3.1}$$

$$S^a_{22} = (d_{32}\phi^a)/h^a, \tag{4.3.2}$$

where d_{31} is the piezoelectric strain constant, h^a is the actuator thickness, and the superscript "a" denotes the distributed piezoelectric actuator. (The induced strains and the resultant effect are illustrated in Fig. 4.2.) Note that these strains are generated in the distributed actuator layer which is located a distance away from the shell neural surface. Thus, these strains introduce counteracting control actions for the shell structure. The sign of the feedback voltage should be carefully controlled so that the induced control actions counteract the shell oscillations.

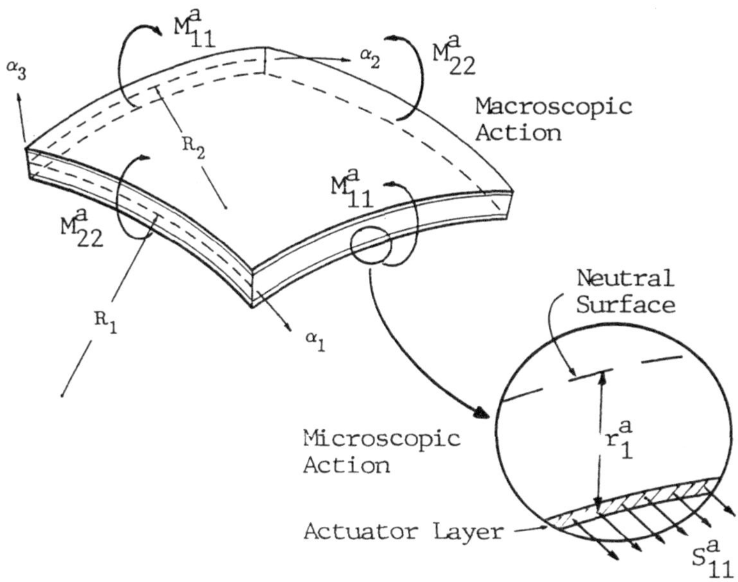

Fig. 4.2 Distributed control by a distributed piezoelectric actuator

4.3.1 Control Forces/Moments and System Equations

The in-plane effective control forces N^a_{ii} and moments M^a_{ii} induced by the imposed actuator voltage ϕ^a can be expressed as

$$N^a_{11} = d_{31} Y_p \phi^a, \tag{4.3.3}$$

$$N^a_{22} = d_{32} Y_p \phi^a, \tag{4.3.4}$$

$$M^a_{11} = r^a_1 d_{31} Y_p \phi^a, \tag{4.3.5}$$

$$M^a_{22} = r^a_2 d_{32} Y_p \phi^a; \tag{4.3.6}$$

where Y_p is Young's modulus of the piezoelectric actuator and r^a_i is the effective moment arm (distance measured from the neutral surface to the mid-plane of the piezoelectric actuator). The in-plane twisting (shear) effect is not considered. Note that the imposed actuator voltage ϕ^a is determined by control algorithms (i.e., open or closed loop control) which will be discussed later. Substituting these induced normal forces and counteracting moments into the equation of motions of the laminated shell yields

$$
\begin{aligned}
&-\frac{\partial(\tilde{N}_{11}A_2)}{\partial\alpha_1}-\frac{\partial(\tilde{N}_{21}A_1)}{\partial\alpha_2}-\tilde{N}_{12}\frac{\partial A_1}{\partial\alpha_2}+\tilde{N}_{22}\frac{\partial A_2}{\partial\alpha_1}\\
&-\frac{1}{R_1}\left[\frac{\partial(\tilde{M}_{11}A_2)}{\partial\alpha_1}+\frac{\partial(\tilde{M}_{21}A_1)}{\partial\alpha_2}+\tilde{M}_{12}\frac{\partial A_1}{\partial\alpha_2}-\tilde{M}_{22}\frac{\partial A_2}{\partial\alpha_1}\right],\\
&+A_1A_2\rho h\ddot{u}_1=A_1A_2F_1
\end{aligned}
\tag{4.3.7}
$$

$$
\begin{aligned}
&-\frac{\partial(\tilde{N}_{12}A_2)}{\partial\alpha_1}-\frac{\partial(\tilde{N}_{22}A_1)}{\partial\alpha_2}-\tilde{N}_{21}\frac{\partial A_2}{\partial\alpha_1}+\tilde{N}_{11}\frac{\partial A_1}{\partial\alpha_2}\\
&-\frac{1}{R_2}\left[\frac{\partial(\tilde{M}_{12}A_2)}{\partial\alpha_1}+\frac{\partial(\tilde{M}_{22}A_1)}{\partial\alpha_2}+\tilde{M}_{21}\frac{\partial A_2}{\partial\alpha_1}-\tilde{M}_{11}\frac{\partial A_1}{\partial\alpha_2}\right],\\
&+A_1A_2\rho h\ddot{u}_2=A_1A_2F_2
\end{aligned}
\tag{4.3.8}
$$

$$
\begin{aligned}
&-\frac{\partial}{\partial\alpha_1}\left\{\frac{1}{A_1}\left[\frac{\partial(\tilde{M}_{11}A_2)}{\partial\alpha_1}+\frac{\partial(\tilde{M}_{21}A_1)}{\partial\alpha_2}+\tilde{M}_{12}\frac{\partial A_1}{\partial\alpha_2}-\tilde{M}_{22}\frac{\partial A_2}{\partial\alpha_1}\right]\right\}\\
&-\frac{\partial}{\partial\alpha_2}\left\{\frac{1}{A_2}\left[\frac{\partial(\tilde{M}_{12}A_2)}{\partial\alpha_1}+\frac{\partial(\tilde{M}_{22}A_1)}{\partial\alpha_2}+\tilde{M}_{21}\frac{\partial A_2}{\partial\alpha_1}-\tilde{M}_{11}\frac{\partial A_1}{\partial\alpha_2}\right]\right\},\\
&+A_1A_2\left[\frac{\tilde{N}_{11}}{R_1}+\frac{\tilde{N}_{22}}{R_2}\right]+A_1A_2\rho h\ddot{u}_3=A_1A_2F_3
\end{aligned}
\tag{4.3.9}
$$

where the "$\sim$" terms include the both original elastic and induced feedback control effects induced by the converse piezoelectric effect. These resultant forces and moments are modified to include the induced normal forces and counteracting moments, i.e., $\tilde{N}_{ij}=N_{ij}-d_{ij}h^aY_p\phi^a$ and $\tilde{M}_{ij}=M_{ij}-M^a_{ij}$. N_{ij} and M_{ij} are defined in Sect. 4.1. Note that the in-plane twisting effect is assumed negligible, i.e., $\tilde{M}_{12}=M_{12}$ and $\tilde{N}_{12}=N_{12}$. In addition, material and stiffness properties of the piezoelectric actuator layer are not considered. The feedback voltage ϕ^a is determined by the control algorithms, i.e., the direct proportional feedback, negative velocity feedback, and Lyapunov feedback, which are discussed next.

4.3.2 Direct Proportional Feedback Control

In the direct proportional feedback control, the feedback voltage ϕ^a is generated by amplifying the sensor output ϕ^S directly, i.e.,

$$
\phi^a=\mathcal{G}\phi^S, \tag{4.3.10}
$$

where $\mathcal{G}$ denotes the voltage amplified ratio—a *feedback control gain*—which can be adjusted depending on the performance requirement of the system. (Since the sensor signal is a function of strains ultimately contributed by displacements (u_1, u_2, u_3, t), this control scheme usually controls system frequencies of the laminated shell system.) ϕ^S (u_1, u_2, u_3, t) is defined in Eqs. (4.2.8) and (4.2.9). (Note that the indices inside the parentheses are redefined in terms of displacements and time. They were previously defined in spatial coordinates α_1 and α_2.) Thus, the feedback induced effective forces and moments are

$$N_{11}^a = \mathcal{G} d_{31} Y_p \phi^S, \tag{4.3.11}$$

$$N_{22}^a = \mathcal{G} d_{32} Y_p \phi^S, \tag{4.3.12}$$

$$M_{11}^a = \mathcal{G} r_1^a d_{31} Y_p \phi^S, \tag{4.3.13}$$

$$M_{22}^a = \mathcal{G} r_2^a d_{32} Y_p \phi^S. \tag{4.3.14}$$

These forces and moments can be substituted into the laminated shell system equations to give the closed-loop system equations. Note that this control algorithm leads to a change of system frequencies. The amplitude could also change due to the frequency change. Detailed discussions are presented in Sect. 6.2 of Chap. 6.

4.3.3 Negative Velocity Feedback Control

The feedback signal can be differentiated so that a strain rate (related to the velocity) information is obtained. The velocity feedback can enhance the system damping and therefore effectively control the oscillation amplitude. Note that for the velocity feedback, the sensor signal ϕ^S used in Eqs. (4.3.11)-(4.3.14) needs to be replaced by $\dot{\phi}^S$.

$$N_{11}^a = -\mathcal{G} d_{31} Y_p \dot{\phi}^S, \tag{4.3.15}$$

$$N_{22}^a = -\mathcal{G} d_{32} Y_p \dot{\phi}^S, \tag{4.3.16}$$

$$M_{11}^a = -\mathcal{G} r_1^a d_{31} Y_p \dot{\phi}^S, \tag{4.3.17}$$

$$M_{22}^a = -\mathcal{G} r_2^a d_{32} Y_p \dot{\phi}^S, \tag{4.3.18}$$

where

$$\dot{\phi}^{S} = \frac{\partial}{\partial t}\left[\phi^{S}(u_1, u_2, u_3, t)\right] = \phi^{S}(\dot{u}_1, \dot{u}_2, \dot{u}_3, t). \tag{4.3.19}$$

4.3.4 Lyapunov Feedback Control

In Lyapunov feedback control, the feedback voltage amplitude is constant and the sign is opposite to the velocity (Tzou, 1988b). The sensor information $\phi^S(u_1, u_2, u_3, t)$ is basically contributed by shell oscillations in all three directions. Thus, the amplitude of feedback signal can be expressed as

$$\phi^{a} = -\mathcal{G}\,\mathrm{sgn}\frac{\partial}{\partial t}\left[\phi^{S}(u_1, u_2, u_3, t)\right], \tag{4.3.20}$$

where $\mathcal{G}$ is the feedback gain and "sgn" is a signum function, i.e., sgn $[z] = -1$ if $z < 0$, 0 if $z = 0$, and +1 if $z > 0$. The forces and moments can be written as

$$N_{11}^{a} = -\mathcal{G}\, d_{31} Y_p \mathrm{sgn}\left(\dot{\phi}^{S}\right), \tag{4.3.21}$$

$$N_{22}^{a} = -\mathcal{G}\, d_{32} Y_p \mathrm{sgn}\left(\dot{\phi}^{S}\right), \tag{4.3.22}$$

$$M_{11}^{a} = -\mathcal{G}\, r_1^a d_{31} Y_p \mathrm{sgn}\left(\dot{\phi}^{S}\right), \tag{4.3.23}$$

$$M_{22}^{a} = -\mathcal{G}\, r_2^a d_{32} Y_p \mathrm{sgn}\left(\dot{\phi}^{S}\right). \tag{4.3.24}$$

Note that when ϕ^S is used in the feedback control, an averaged dynamic state of the shell is considered. In practical applications, a single point transverse velocity $\dot{u}_3\left(\alpha_1^*, \alpha_2^*, t\right)$ (i.e., a discrete location) can also be used in the feedback control. (Note the assumptions discussed previously.) Thus, the feedback amplitude becomes

$$\phi^{a} = -\ \mathcal{G}\ \mathrm{sgn}\left[\dot{u}_3\left(\alpha_1^*, \alpha_2^*, t\right)\right]. \tag{4.3.25}$$

where (α_1^*, α_2^*) denotes the specific location. Thus, the effective forces and moments are defined as

$$N_{11}^{a} = -\mathcal{G}\, d_{31} Y_p \mathrm{sgn}\left[\dot{u}_3\left(\alpha_1^*, \alpha_2^*, t\right)\right], \tag{4.3.26}$$

$$N_{22}^{a} = -\mathcal{G}\, d_{32} Y_p \mathrm{sgn}\left[\dot{u}_3\left(\alpha_1^*, \alpha_2^*, t\right)\right], \tag{4.3.27}$$

$$M_{11}^{a} = -\mathcal{G}\, r_1^{a} d_{31} Y_p \operatorname{sgn}\left[\dot{u}_3\left(\alpha_1^*, \alpha_2^*, t\right)\right], \tag{4.3.28}$$

$$M_{22}^{a} = -\mathcal{G}\, r_2^{a} d_{31} Y_p \operatorname{sgn}\left[\dot{u}_3\left(\alpha_1^*, \alpha_2^*, t\right)\right]. \tag{4.3.29}$$

4.3.5 Modal Feedback Gains

In addition, since the sensor signal can be represented by the modal expansion, the system feedback gain $\mathcal{G}$ can be extended to a ***modal feedback gain*** $\hat{\mathcal{G}}_k$ from which control of each individual mode can be achieved. Using the proportional feedback as an example, one can write the control forces and moments with modal feedback gains $\hat{\mathcal{G}}_k$ and mode shapes as

$$\begin{aligned}
N_{ii}^{a} = d_{3i} Y_p \frac{h^S}{S^e} \int_{S^e} \Bigg[& h_{31} \Bigg\{ \Bigg[\frac{1}{A_1} \frac{\partial}{\partial \alpha_1} \sum_{k=1}^{\infty} \hat{\mathcal{G}}_k \eta_k U_{1k} \\
& + \frac{1}{A_1 A_2} \sum_{k=1}^{\infty} \hat{\mathcal{G}}_k \eta_k U_{2k} \frac{\partial A_1}{\partial \alpha_2} + \frac{1}{R_1} \sum_{k=1}^{\infty} \hat{\mathcal{G}}_k \eta_k U_{3k} \Bigg] \\
& + r_1^S \Bigg[\frac{1}{A_1} \frac{\partial}{\partial \alpha_1} \left(\frac{1}{R_1} \sum_{k=1}^{\infty} \hat{\mathcal{G}}_k \eta_k U_{1k} - \frac{1}{A_1} \frac{\partial}{\partial \alpha_1} \sum_{k=1}^{\infty} \hat{\mathcal{G}}_k \eta_k U_{3k} \right) \\
& + \frac{1}{A_1 A_2} \frac{\partial A_1}{\partial \alpha_2} \left(\frac{1}{R_2} \sum_{k=1}^{\infty} \hat{\mathcal{G}}_k \eta_k U_{2k} - \frac{1}{A_2} \frac{\partial}{\partial \alpha_2} \sum_{k=1}^{\infty} \hat{\mathcal{G}}_k \eta_k U_{3k} \right) \Bigg] \\
& + h_{32} \Bigg\{ \Bigg[\frac{1}{A_2} \frac{\partial}{\partial \alpha_2} \sum_{k=1}^{\infty} \hat{\mathcal{G}}_k \eta_k U_{2k} + \frac{1}{A_1 A_2} \sum_{k=1}^{\infty} \hat{\mathcal{G}}_k \eta_k U_{1k} \frac{\partial A_2}{\partial \alpha_1} + \frac{1}{R_2} \sum_{k=1}^{\infty} \hat{\mathcal{G}}_k \eta_k U_{3k} \Bigg] \\
& + r_2^S \Bigg[\frac{1}{A_2} \frac{\partial}{\partial \alpha_2} \left(\frac{1}{R_2} \sum_{k=1}^{\infty} \hat{\mathcal{G}}_k \eta_k U_{2k} - \frac{1}{A_2} \frac{\partial}{\partial \alpha_2} \sum_{k=1}^{\infty} \hat{\mathcal{G}}_k \eta_k U_{3k} \right) \\
& + \frac{1}{A_1 A_2} \frac{\partial A_2}{\partial \alpha_1} \left(\frac{1}{R_1} \sum_{k=1}^{\infty} \hat{\mathcal{G}}_k \eta_k U_{1k} - \frac{1}{A_1} \frac{\partial}{\partial \alpha_1} \sum_{k=1}^{\infty} \hat{\mathcal{G}}_k \eta_k U_{3k} \right) \Bigg] \Bigg\} \Bigg] A_1 A_2 \, d\alpha_1 \, d\alpha_2;
\end{aligned} \tag{4.3.30}$$

$$
\begin{aligned}
M_{ii}^{a} = r_i^a d_{3i} Y_p \frac{h^S}{S^e} \int_{S^e} \Bigg[h_{31} \Bigg\{ \Bigg[& \frac{1}{A_1} \frac{\partial}{\partial \alpha_1} \sum_{k=1}^{\infty} \hat{\mathcal{G}}_k \eta_k U_{1k} \\
& + \frac{1}{A_1 A_2} \sum_{k=1}^{\infty} \hat{\mathcal{G}}_k \eta_k U_{2k} \frac{\partial A_1}{\partial \alpha_2} + \frac{1}{R_1} \sum_{k=1}^{\infty} \hat{\mathcal{G}}_k \eta_k U_{3k} \Bigg] \\
& + r_1^S \Bigg[\frac{1}{A_1} \frac{\partial}{\partial \alpha_1} \left(\frac{1}{R_1} \sum_{k=1}^{\infty} \hat{\mathcal{G}}_k \eta_k U_{1k} - \frac{1}{A_1} \frac{\partial}{\partial \alpha_1} \sum_{k=1}^{\infty} \hat{\mathcal{G}}_k \eta_k U_{3k} \right) \\
& + \frac{1}{A_1 A_2} \frac{\partial A_1}{\partial \alpha_2} \left(\frac{1}{R_2} \sum_{k=1}^{\infty} \hat{\mathcal{G}}_k \eta_k U_{2k} - \frac{1}{A_2} \frac{\partial}{\partial \alpha_2} \sum_{k=1}^{\infty} \hat{\mathcal{G}}_k \eta_k U_{3k} \right) \Bigg] \Bigg\} \\
+ h_{32} \Bigg\{ \Bigg[& \frac{1}{A_2} \frac{\partial}{\partial \alpha_2} \sum_{k=1}^{\infty} \hat{\mathcal{G}}_k \eta_k U_{2k} + \frac{1}{A_1 A_2} \sum_{k=1}^{\infty} \hat{\mathcal{G}}_k \eta_k U_{1k} \frac{\partial A_2}{\partial \alpha_1} + \frac{1}{R_2} \sum_{k=1}^{\infty} \hat{\mathcal{G}}_k \eta_k U_{3k} \Bigg] \\
& + r_2^S \Bigg[\frac{1}{A_2} \frac{\partial}{\partial \alpha_2} \left(\frac{1}{R_2} \sum_{k=1}^{\infty} \hat{\mathcal{G}}_k \eta_k U_{2k} - \frac{1}{A_2} \frac{\partial}{\partial \alpha_2} \sum_{k=1}^{\infty} \hat{\mathcal{G}}_k \eta_k U_{3k} \right) \\
& + \frac{1}{A_1 A_2} \frac{\partial A_2}{\partial \alpha_1} \left(\frac{1}{R_1} \sum_{k=1}^{\infty} \hat{\mathcal{G}}_k \eta_k U_{1k} - \frac{1}{A_1} \frac{\partial}{\partial \alpha_1} \sum_{k=1}^{\infty} \hat{\mathcal{G}}_k \eta_k U_{3k} \right) \Bigg] \Bigg\} \Bigg] A_1 A_2 d\alpha_1 d\alpha_2.
\end{aligned}
\tag{4.3.31}
$$

Similarly, a new set of system dynamic equations can be derived. It can be observed that all three actuator parameters (i.e., piezoelectric constant d_{ij}, modulus of elasticity Y_p, and the moment arm r_j^a) are of importance to the overall control effects. In general, a higher piezoelectric constant, stiffer piezoelectric actuator, and longer moment arm contribute better control effects. These individual effects will be studied in Chap. 7.

It should be noted that control spillover could occur in the above formulations which include cross coupling terms. This control spillover (Meirovitch & Baruh, 1981; Balas, 1978) of distributed actuators is presented in Chaps. 7–9. Designs of distributed shaped and convolved actuators are discussed in Chaps. 8 and 9, optimal control of parabolic shells is presented in Chap. 10, and energy harvesting in Chapts. 11 and 12.

4.4 State Equation

System dynamic equations of the distributed shell/sensor/actuator system can be transferred into the state equation. Define the elastic terms associated with the elastic shell by generic Love's operator L_i's (i = 1,2,3) and the feedback control terms as H_i's (i = 1,2,3).

$$\begin{aligned}L_1 = & \frac{1}{\rho h A_1 A_2}\left[\frac{\partial(N_{11}A_2)}{\partial\alpha_1} + \frac{\partial(N_{21}A_1)}{\partial\alpha_2} + N_{12}\frac{\partial A_1}{\partial\alpha_2} - N_{22}\frac{\partial A_2}{\partial\alpha_1}\right. \\ & \left. + \frac{1}{R_1}\left(\frac{\partial(M_{11}A_2)}{\partial\alpha_1} + \frac{\partial(M_{21}A_1)}{\partial\alpha_2} + M_{12}\frac{\partial A_1}{\partial\alpha_2} - M_{22}\frac{\partial A_2}{\partial\alpha_1}\right)\right],\end{aligned} \quad (4.4.1)$$

$$\begin{aligned}L_2 = & \frac{1}{\rho h A_1 A_2}\left[\frac{\partial(N_{12}A_2)}{\partial\alpha_1} + \frac{\partial(N_{22}A_1)}{\partial\alpha_2} + N_{21}\frac{\partial A_2}{\partial\alpha_1} - N_{11}\frac{\partial A_1}{\partial\alpha_2}\right. \\ & \left. + \frac{1}{R_2}\left(\frac{\partial(M_{12}A_2)}{\partial\alpha_1} + \frac{\partial(M_{22}A_1)}{\partial\alpha_2} + M_{21}\frac{\partial A_2}{\partial\alpha_1} - M_{11}\frac{\partial A_1}{\partial\alpha_2}\right)\right],\end{aligned} \quad (4.4.2)$$

$$\begin{aligned}L_3 = & \frac{1}{\rho h A_1 A_2}\left\{\frac{\partial}{\partial\alpha_1}\left(\frac{1}{A_1}\right)\left[\frac{\partial(M_{11}A_2)}{\partial\alpha_1} + \frac{\partial(M_{21}A_1)}{\partial\alpha_2}\right.\right. \\ & \left. + M_{12}\frac{\partial A_1}{\partial\alpha_2} - M_{22}\frac{\partial A_2}{\partial\alpha_1}\right] + \frac{\partial}{\partial\alpha_2}\left(\frac{1}{A_2}\right)\left[\frac{\partial(M_{12}A_2)}{\partial\alpha_1} + \frac{\partial(M_{22}A_1)}{\partial\alpha_2}\right. \\ & \left.\left. + M_{21}\frac{\partial A_2}{\partial\alpha_1} - M_{11}\frac{\partial A_1}{\partial\alpha_2}\right] - A_1 A_2\left(\frac{N_{11}}{R_1} + \frac{N_{22}}{R_2}\right)\right\}.\end{aligned} \quad (4.4.3)$$

$$\begin{aligned}H_1 = & \frac{1}{\rho h A_1 A_2}\left[\frac{\partial(N^a_{11}A_2)}{\partial\alpha_1} + \frac{\partial(N^a_{21}A_1)}{\partial\alpha_2} + N^a_{12}\frac{\partial A_1}{\partial\alpha_2} - N^a_{22}\frac{\partial A_2}{\partial\alpha_1}\right. \\ & \left. + \frac{1}{R_1}\left(\frac{\partial(M^a_{11}A_2)}{\partial\alpha_1} + \frac{\partial(M^a_{21}A_1)}{\partial\alpha_2} + M^a_{12}\frac{\partial A_1}{\partial\alpha_2} - M^a_{22}\frac{\partial A_2}{\partial\alpha_1}\right)\right],\end{aligned} \quad (4.4.4)$$

$$\begin{aligned}H_2 = & \frac{1}{\rho h A_1 A_2}\left[\frac{\partial(N^a_{12}A_2)}{\partial\alpha_1} + \frac{\partial(N^a_{22}A_1)}{\partial\alpha_2} + N^a_{21}\frac{\partial A_2}{\partial\alpha_1} - N^a_{11}\frac{\partial A_1}{\partial\alpha_2}\right. \\ & \left. + \frac{1}{R_2}\left(\frac{\partial(M^a_{12}A_2)}{\partial\alpha_1} + \frac{\partial(M^a_{22}A_1)}{\partial\alpha_2} + M^a_{21}\frac{\partial A_2}{\partial\alpha_1} - M^a_{11}\frac{\partial A_1}{\partial\alpha_2}\right)\right],\end{aligned} \quad (4.4.5)$$

$$\begin{aligned}H_3 = & \frac{1}{\rho h A_1 A_2}\left\{\frac{\partial}{\partial\alpha_1}\left(\frac{1}{A_1}\right)\left[\frac{\partial(M^a_{11}A_2)}{\partial\alpha_1} + \frac{\partial(M^a_{21}A_1)}{\partial\alpha_2} + M^a_{12}\frac{\partial A_1}{\partial\alpha_2}\right.\right. \\ & \left. - M^a_{22}\frac{\partial A_2}{\partial\alpha_1}\right] + \frac{\partial}{\partial\alpha_2}\left(\frac{1}{A_2}\right)\left[\frac{\partial(M^a_{12}A_2)}{\partial\alpha_1} + \frac{\partial(M^a_{22}A_1)}{\partial\alpha_2}\right. \\ & \left.\left. + M^a_{21}\frac{\partial A_2}{\partial\alpha_1} - M^a_{11}\frac{\partial A_1}{\partial\alpha_2}\right] - A_1 A_2\left(\frac{N^a_{11}}{R_1} + \frac{N^a_{22}}{R_2}\right)\right\}.\end{aligned} \quad (4.4.6)$$

Note that the in-plane control shear forces and moments can be neglected, i.e., $N^a_{12} = N^a_{21} = 0$ and $M^a_{12} = M^a_{21} = 0$, as discussed previously. Substituting L_i's and H_i's into the original system equation and transferring them into state space gives

$$\frac{\partial}{\partial t}\begin{pmatrix} u_1 \\ u_2 \\ u_3 \\ \dot{u}_1 \\ \dot{u}_2 \\ \dot{u}_3 \end{pmatrix} = \begin{pmatrix} 0 & 0 & 0 & 1 & 0 & 0 \\ 0 & 0 & 0 & 0 & 1 & 0 \\ 0 & 0 & 0 & 0 & 0 & 1 \\ L_1 & 0 & 0 & 0 & 0 & 0 \\ 0 & L_2 & 0 & 0 & 0 & 0 \\ 0 & 0 & L_3 & 0 & 0 & 0 \end{pmatrix}\begin{pmatrix} u_1 \\ u_2 \\ u_3 \\ \dot{u}_1 \\ \dot{u}_2 \\ \dot{u}_3 \end{pmatrix} + [I]\begin{pmatrix} 0 \\ 0 \\ 0 \\ F_1/\rho h + H_1 \\ F_2/\rho h + H_2 \\ F_3/\rho h + H_3 \end{pmatrix}, \quad (4.4.7)$$

where [I] denotes an identity matrix. Symbolically,

$$\frac{\partial \mathbf{U}}{\partial t} = \mathbf{AU} + \mathbf{Bm}, \quad (4.4.8)$$

where **U**, **A**, and **B** are defined in the above state equation, and

$$\mathbf{m} = (0 \quad 0 \quad 0 \quad F_1/\rho h + H_1 \quad F_2/\rho h + H_2 \quad F_3/\rho h + H_3)^t, \quad (4.4.9)$$

where "t" denotes the vector or matrix transpose. The state equation, Eqs. (4.4.7) or (4.4.8), is for a generic shell continuum (distributed parameter system) with distributed piezoelectric sensors/actuators. The state equation is defined in a generic form which includes the in-plane control shear force and moments. Depending on the four system parameters, i.e., two Lamé parameters and two radii, this equation can be simplified to apply to many other common geometries. Simplification procedures, similar to those in Chap. 3, are presented in case studies.

4.5 Examples

Similar to the generic piezoelectric shell equations, the generic theories for the distributed sensor and actuator laminated on an elastic deep shell, i.e., the generic smart structronic shell system, can be simplified to apply to a large number of common shell and non-shell structures with laminated distributed sensors and actuators. In this section, three examples, 1) a hemispheric shell, 2) a flat plate, and 3) an Euler-Bernoulli beam laminated with distributed piezoelectric sensor and actuator layers, are demonstrated using the reduction procedure discussed in Section §3.1 of Chap. 3. Governing open- or closed-control system equations for the smart structures are derived. It is assumed that the bending effect dominates the shell oscillations in distributed sensor/actuator applications.

4.5.1 Case 1: A Hemispheric Shell

A hemispheric shell laminated with fully covered distributed piezoelectric sensor and actuator layers is shown in Fig. 4.3. Distributed sensation and control of the

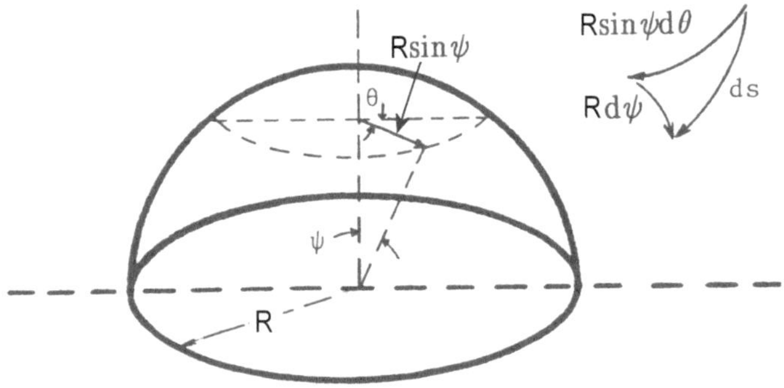

Fig. 4.3 A hemispheric shell with distributed piezoelectric sensor/actuator layers

hemispheric shell are discussed (Tzou, 1992). The fundamental form defining the hemispheric shell is

$$(ds)^2 = R^2(d\psi)^2 + R^2 \sin^2\psi(d\theta)^2, \tag{4.5.1}$$

where R is the radius of the shell; ψ and θ are defined in **Fig.** 4.3. Thus, the Lamé parameters are $A_1 = R$ and $A_2 = R \sin @$; two neutral surface coordinates are $\alpha_1 = \psi$ and $\alpha_2 = \theta$, and radii are $R_1 = R_2 = R$.

Substituting the system parameters into Eq. (4.2.8) yields

$$\phi^S = \frac{h^S}{2\pi}\int_0^{\pi}\int_0^{\pi}\left\{\frac{h_{31}r_\psi^S}{R^2}\left[\frac{\partial}{\partial\psi}\left(u_\psi - \frac{\partial u_3}{\partial\psi}\right)\right]\right.$$
$$\left. + \frac{h_{32}r_\theta^S}{R^2}\left[\frac{1}{\sin\psi}\frac{\partial}{\partial\theta}\left(u_\theta - \frac{1}{\sin\psi}\frac{\partial u_3}{\partial\theta}\right) + \cot\psi\left(u_\psi - \frac{\partial u_3}{\partial\psi}\right)\right]\right\}\sin\psi\, d\psi\, d\theta \tag{4.5.2}$$

Note that the moment arms $r_\varphi^S = r_\theta^S$ for a piezoelectric sensor layer with a uniform thickness. All terms inside brace { } multiplied by h^s represent a local voltage amplitude if the location (ψ^*, θ^*) is specified. Using Eqs. (4.3.11)–(4.3.14), Eqs. (4.3.15)–(4.3.18), and Eqs. (4.3.21)–(4.3.24), one can derive the effective control forces and moments for each control algorithm respectively. The closed-loop system equation can also be derived using the geometric parameters defined above. For demonstration purpose, only the system equation for the transverse oscillation is derived.

$$
\begin{aligned}
&-\frac{\partial}{\partial\psi}\left[\frac{\partial\left(\tilde{M}_{\psi\psi}\sin\psi\right)}{\partial\psi}+\frac{\partial M_{\psi\theta}}{\partial\theta}-\cos\psi\,\tilde{M}_{\theta\theta}\right]\\
&-\frac{\partial}{\partial\theta}\left\{\frac{1}{\sin\psi}\left[\frac{\partial\left(M_{\psi\theta}\sin\psi\right)}{\partial\psi}+\frac{\partial\tilde{M}_{\theta\theta}}{\partial\theta}+\cos\psi\,M_{\psi\theta}\right]\right\}\\
&+R\sin\psi\left[\tilde{N}_{\psi\psi}+\tilde{N}_{\theta\theta}\right]+R^2\sin\psi\,\rho\,h\,\ddot{u}_3=R^2\sin\psi\,F_3
\end{aligned}
\qquad (4.5.3)
$$

$\tilde{N}_{ij}$ and $\tilde{M}_{ij}$ are the resultant forces and moments, i.e., $\tilde{N}_{ii} = N_{ii} - N^a_{ii}$ and $\tilde{M}_{ii} = M_{ii} - M^a_{ii}$. N_{ii} and M_{ii} can be derived by substituting the geometric parameters into the general force and moment equations in Sect. 4.1. In-plane shear effects in the distributed actuator are neglected. The derived mechanical force and moment expressions are

$$
N_{\psi\psi}=\frac{K}{R}\left[\left(\frac{\partial u_\psi}{\partial\psi}+u_3\right)+\mu\left(\frac{1}{\sin\psi}\frac{\partial u_\theta}{\partial\theta}+u_\psi\cot\psi+u_3\right)\right], \qquad (4.5.4)
$$

$$
N_{\theta\theta}=\frac{K}{R}\left[\left(\frac{1}{\sin\psi}\frac{\partial u_\theta}{\partial\theta}+u_\psi\cot\psi+u_3\right)+\mu\left(\frac{\partial u_\psi}{\partial\psi}+u_3\right)\right], \qquad (4.5.5)
$$

$$
N_{\psi\theta}=\frac{K(1-\mu)}{2R}\left[\sin\psi\frac{\partial}{\partial\psi}\left(\frac{u_\theta}{\sin\psi}\right)+\frac{1}{\sin\psi}\frac{\partial u_\psi}{\partial\theta}\right]; \qquad (4.5.6)
$$

$$
\begin{aligned}
M_{\psi\psi}=\frac{D}{R^2}&\left\{\left[\frac{\partial}{\partial\psi}\left(u_\psi-\frac{\partial u_3}{\partial\psi}\right)\right]+\mu\left[\frac{1}{\sin\psi}\frac{\partial}{\partial\theta}\left(u_\theta-\frac{1}{\sin\psi}\frac{\partial u_3}{\partial\theta}\right)\right.\right.\\
&\left.\left.+\cot\psi\left(u_\psi-\frac{\partial u_3}{\partial\psi}\right)\right]\right\}
\end{aligned}, \qquad (4.5.7)
$$

$$
\begin{aligned}
M_{\theta\theta}=\frac{D}{R^2}&\left\{\left[\frac{1}{\sin\psi}\frac{\partial}{\partial\theta}\left(u_\theta-\frac{1}{\sin\psi}\frac{\partial u_3}{\partial\theta}\right)+\cot\psi\left(u_\psi-\frac{\partial u_3}{\partial\psi}\right)\right]\right.\\
&\left.+\mu\left[\frac{\partial}{\partial\psi}\left(u_\psi-\frac{\partial u_3}{\partial\psi}\right)\right]\right\}
\end{aligned}, \qquad (4.5.8)
$$

$$
\begin{aligned}
M_{\psi\theta}=\frac{D(1-\mu)}{2R^2}&\left\{-\cot\psi\left(u_\theta-\frac{1}{\sin\psi}\frac{\partial u_3}{\partial\theta}\right)\right.\\
&\left.+\frac{\partial}{\partial\psi}\left(u_\theta-\frac{1}{\sin\psi}\frac{\partial u_3}{\partial\theta}\right)+\frac{1}{\sin\psi}\frac{\partial}{\partial\theta}\left(u_\psi-\frac{\partial u_3}{\partial\psi}\right)\right\}
\end{aligned}. \qquad (4.5.9)
$$

4.5.2 Case 2: A Rectangular Plate

The second case is a zero-curvature shell—a plate—with infinite radii of curvatures for the two neutral surface coordinate axes x and y, i.e., $R_1 = R_2 = \infty$, **Fig.** 4.4 (Tzou, 1990). It is assumed that the rectangular plate has a width "b" and a length "a". The fundamental form is defined as

$$(ds)^2 = (dx)^2 + (dy)^2. \tag{4.5.10}$$

Thus, the Lamé parameters are $A_1 = 1$ and $A_2 = 1$. It is assumed that bi-axially oriented piezoelectric sensor and actuator layers are aligned with the coordinate axes x and y. (Otherwise, an orientation matrix needs to be defined.) Distributed sensation and control, i.e., the direct feedback control and Lyapunov control, of the rectangular plate are derived using the generic shell governing equations.

It is assumed that the rectangular plate experiences only a transverse vibration. For a distributed sensor stretching from x_1 to x_2 in the x direction and from y_1 to y_2 in the y direction, the distributed sensor output can be derived from Eqs. (4.2.8) and (4.2.9):

$$\phi^S = -\frac{h^S}{S^e}\int_{S^e}\left(h_{31}r_x^S\frac{\partial^2 u_3}{\partial x^2} + h_{32}r_y^S\frac{\partial^2 u_3}{\partial y^2}\right)dS^e. \tag{4.5.11}$$

The output signal can be further expressed in a modal expression form:

$$\begin{aligned}
\phi^S &= -\frac{h^S}{S^e}\int_x\int_y\left\{h_{31}r_x^S\frac{\partial^2}{\partial x^2}\left[\sum_{k=1}^{\infty}\eta_k U_{3k}\right] + h_{32}r_y^S\frac{\partial^2}{\partial y^2}\left[\sum_{k=1}^{\infty}\eta_k U_{3k}\right]\right\}dx\,dy \\
&= -\frac{h^S}{S^e}\int_{x_1}^{x_2}\int_{y_1}^{y_2}\left\{h_{31}r_x^S\frac{\partial^2}{\partial x^2}\left[\sum_{k=1}^{\infty}\eta_k U_{3k}\right] + h_{32}r_y^S\frac{\partial^2}{\partial y^2}\left[\sum_{k=1}^{\infty}\eta_k U_{3k}\right]\right\}dx\,dy
\end{aligned}, \tag{4.5.12}$$

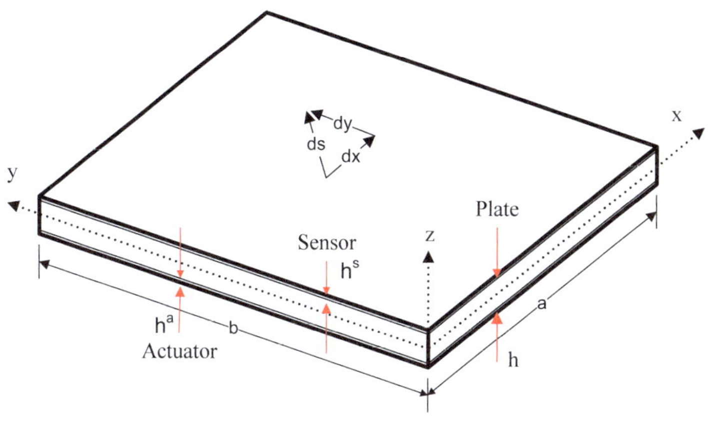

Fig. 4.4 A rectangular plate laminated with piezoelectric sensor/actuator layers

where $r_x^s = r_y^s$ for a uniform thickness piezoelectric layer. For a fully covered distributed sensor, the integration covers the whole plate area, i.e., $\mathbf{S}^e = ab$; thus, the integration is $\int_0^a \int_0^b [\ldots] d_x d_y$. For a distributed patch sensor covering an area defined by a_1 and a_2 in the x direction and b_1 and b_2 in the y direction, the integration is defined as $\int_{a_1}^{a_2} \int_{b_1}^{b_2} [\ldots] d_x d_y$.

Based on the control forces and moments defined for the control algorithms, one can derive the corresponding control forces and moments and then the closed-loop governing equation accordingly. Note that the in-plane control forces N_{xx}^a and N_{yy}^a are canceled out because of the infinite radii of curvatures, i.e., $R_1 = R_2 = \infty$. Substituting the moments and geometric parameters into the generic shell equation, one can also derive the plate system equation in the transverse direction.

$$\rho h \frac{\partial^2 u_3}{\partial t^2} + D\left[\frac{\partial^4 u_3}{\partial x^4} + 2\frac{\partial^4 u_3}{\partial x^2 \partial y^2} + \frac{\partial^4 u_3}{\partial y^4}\right] - \frac{\partial^2 M_{xx}^a}{\partial x^2} - \frac{\partial^2 M_{yy}^a}{\partial y^2} = F_3, \qquad (4.5.13)$$

where the effective control moments are determined by the control algorithms. Note that the control moments are spatially distributed. In the case when the control moments are not spatial functions, the control effects can be introduced from boundaries via the ***boundary controls*** presented in later chapters.

4.5.3 *Case 3: An Euler-Bernoulli Beam*

Note that the plate can be reduced to an Euler-Bernoulli beam, **Fig.** 4.5, by considering only one effective axis, e.g., the x direction.

Thus, the sensing equation, Eq. (4.5.11), can be reduced to

$$\phi^S = -\frac{bh^S}{S^e} \int_x \left(h_{31} r_x^S \frac{\partial^2 u_3}{\partial x^2} \right) dx, \qquad (4.5.14)$$

where b is the beam width and x defines the sensor length in the x direction. The integration depends on the boundary conditions and the effective area of the distributed sensor layer. (For example, the sensing signal is a function of the free-end slope for a cantilever beam with a fully covered sensing layer, Chap. 6.) This sensing equation can also be expressed in modal coordinates by assuming $u_3 = \sum_{K=1}^{\infty} \eta_k u_{3k}$. The control moment M_{xx}^a of the beam case can also be derived from the control algorithms. The closed-loop system equation, Eq. (4.5.13), of the rectangular plate can be simplified to apply to the beam case:

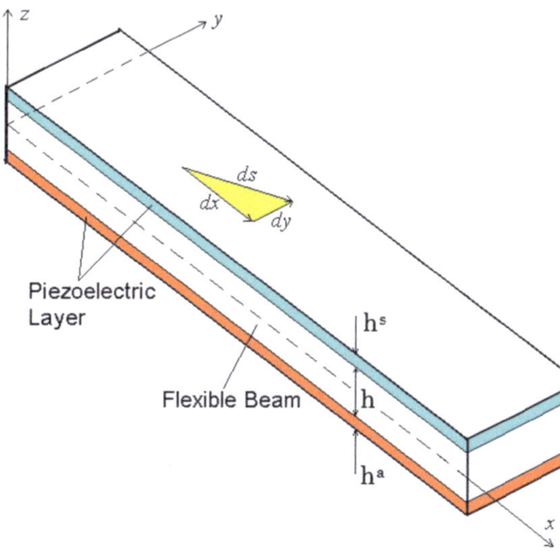

Fig. 4.5 An Euler-Bernoulli beam laminated with piezoelectric sensor/actuator layers

$$\rho\tilde{A}\frac{\partial^2 u_3}{\partial t^2} + YI\frac{\partial^4 u_3}{\partial x^4} - b\frac{\partial^2 M_{xx}^a}{\partial x^2} = \tilde{F}_3, \tag{4.5.15}$$

where $\tilde{A} = bh$, I is the area moment of inertia, and $\tilde{F}_3 = bF_3$.

Applications to other geometrical configurations can be achieved by following similar procedures.

4.6 Design Principles of Distributed Sensors/Actuators

Similar to the generic piezoelectric shell equations in Chaps. 2 and 3, the generic distributed sensor and actuator theories for the laminated elastic deep shell, i.e., the generic smart structronic shell system, can be simplified to apply to a large number of common shell and non-shell structures with laminated distributed sensors and actuators. Three examples, 1) a hemispheric shell, 2) a flat plate, and 3) an Euler-Bernoulli beam laminated with distributed piezoelectric sensor and actuator layers, were demonstrated above. Generic design principles, i.e., ***the segmentation and shaping techniques***, of the distributed sensors and actuators are discussed next.

4.6.1 The Segmentation Technique

Figure 4.6a illustrates the segmentation technique, in which a distributed sensor/actuator layer is segmented into a number of regular-shape patches with equal or

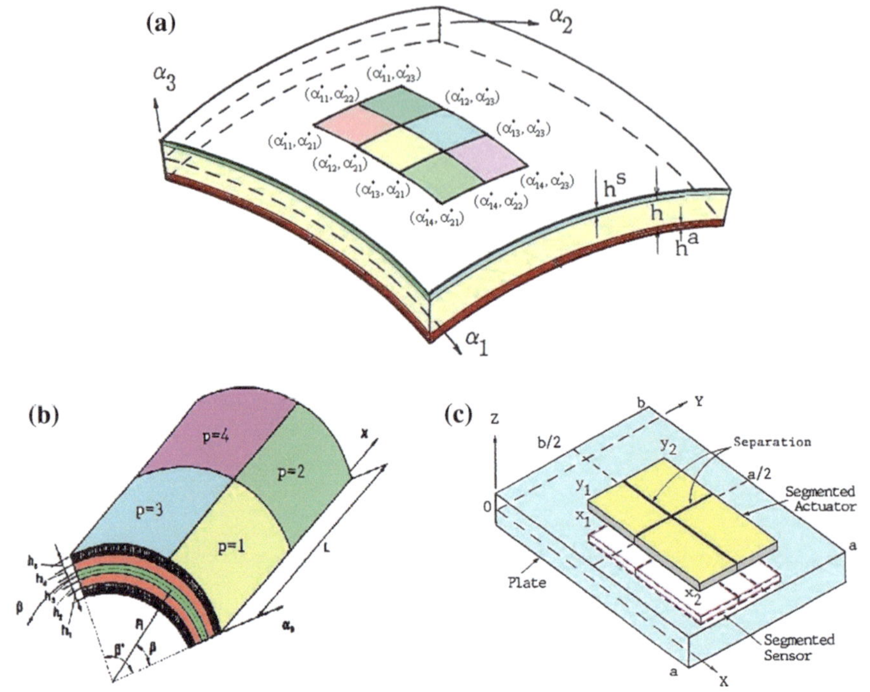

Fig. 4.6 The segmentation technique of distributed sensor/actuator

unequal sizes whose locations are clearly denoted by their corner coordinates. Each segmented patch or a group of patches can be individually monitored or controlled. When the size of parches approaches to infinitesimal, these elements become distributed **neurons** or "**muscle**" elements, either monitoring microscopic behaviors or leading to microscopic control actions. Figure 4.6b shows four-segmented patches on a composite cylindrical shell panel; Fig. 4.6c shows four-segmented patches on a zero-curvature shell—a plate and its detailed sensing and control behaviors and effectiveness are presented in Chap. 7.

4.6.2 The Shaping Technique

On the other hand, the shaping technique involves either the planner shape variation, Fig. 4.7a, or the thickness variation, Fig. 4.7b. These shapes usually are modal sensitive and they are designed based on specific mode shape functions or modal strain functions. Accordingly, these shaped sensors/actuators are only effective to selected modal oscillations of the DPS. For example, Fig. 4.7c illustrates designs of

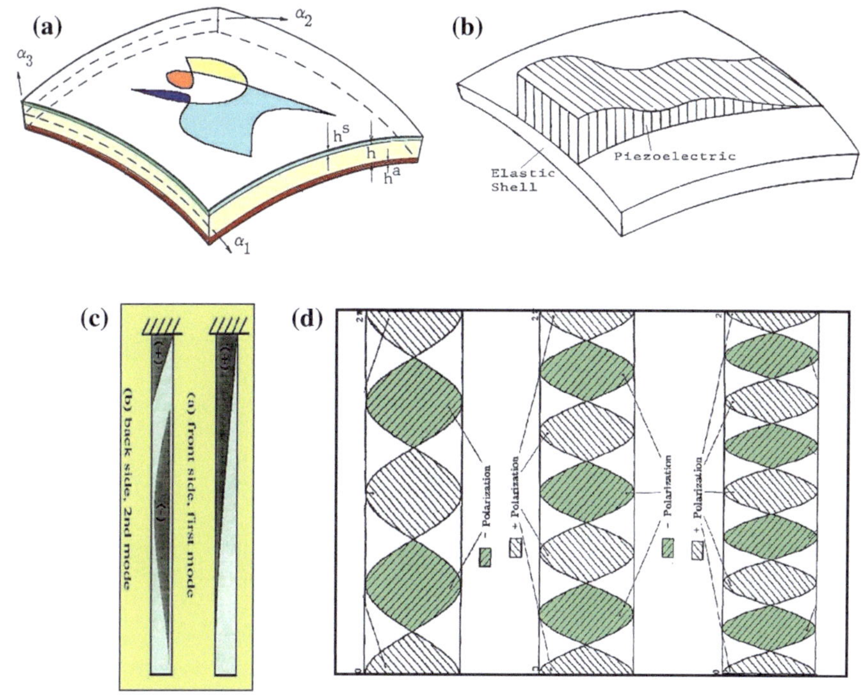

Fig. 4.7 The shaping technique of distributed sensor/actuator

the first and second orthogonal modal sensors/actuators for a cantilever beam; Fig. 4.7d illustrates those of the second to fourth modes of a free-floating ring. Application of the shaping technique is demonstrated in a ring shell case presented in Chap. 8.

Applications of the generic structronic shell equations and sensor/actuator equations to other standard DPSs or standard shell/non-shell structures (i.e., common geometries) are illustrated in Fig. 4.8. Again, Fig. 4.7a shows the surface shaping; Fig. 4.7b shows the thickness shaping; Fig. 4.7c shows the surface shaping, i.e., modal sensor/actuator, on a cantilever beam; Fig. 4.7d shows the surface shaping, i.e., convolving orthogonal sensor/actuator shaping, of a ring, presented in Chapter 8. These techniques can be applied to all shell and plate structures and systems.

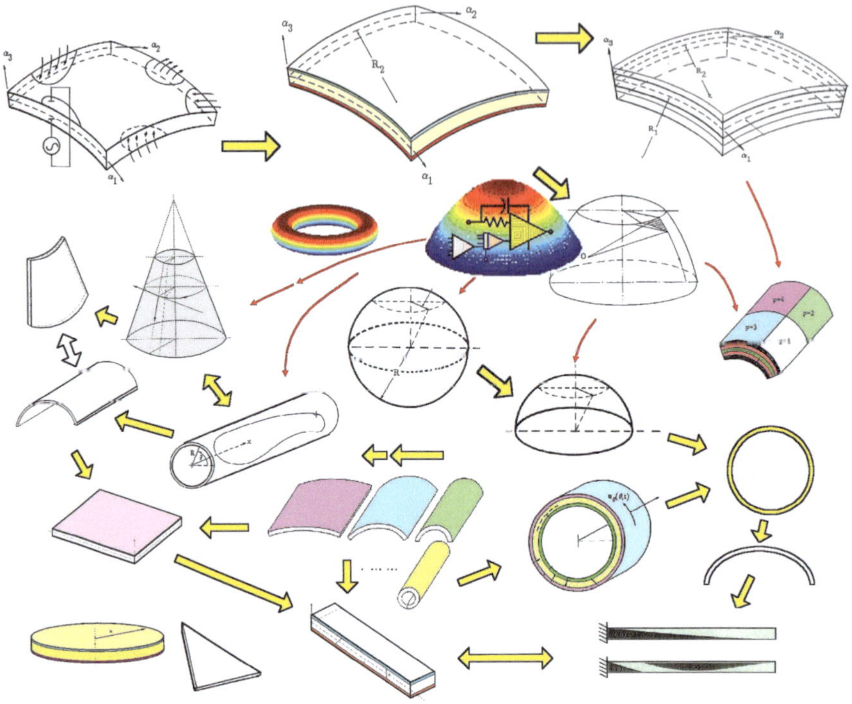

Fig. 4.8 Applications of generic shell theories

4.7 Summary

Distributed sensing and control of a generic distributed parameter system (DPS) or a generic smart structronic shell system, i.e., a deep elastic shell laminated with distributed piezoelectric sensor and actuator layers, was proposed and corresponding generic theories derived. Based on the direct piezoelectric effect, the distributed sensor can be used to monitor shell oscillations; the converse effect enables the distributed actuators to manipulate structural behaviors and to suppress structural vibrations. Two generic sensor/actuator design principles, i.e., **the segmentation technique** and **the shaping technique**, were also presented.

A new distributed sensing theory and a distributed vibration control theory for the generic shell DPS were developed. The sensor theory shows that the distributed sensor output is contributed by all vibration modes with their corresponding modal participation factors. That is, the distributed sensor can **theoretically** measure all vibration modes of the shell. A voltage distribution contour (***modal voltage***) or ***potential map*** can be constructed by graphically connecting all local voltage amplitudes, calculated from local strains, on the laminated shell. (It is assumed that each sensor spot has a finite or point electrode area.) Similarly, the distributed actuator can **theoretically** control all vibration modes using either **1) the direct**

proportional feedback, 2) the negative velocity feedback, or 3) the Lyapunov feedback via **the segmentation design or the shaping design**. The derived theories are very generic, which can be simplified to apply to other common geometries, e.g., beams, plates, rings, cylinder shells, spherical shells, conical shells, etc. The derived governing equations were then transformed into the state space, and a state equation of the generic structronic shell system, in matrix form, was derived.

Note that there is a potential problem when calculating averaged distributed sensor voltages. A zero (or minimal) output could occur for anti-symmetrical modes of a symmetrical structure due to the surface charge cancellations. In this case, a point or finite area (segmented, patched or shaped) voltage signal could be used in the feedback control system as discussed in the Lyapunov control case. The control effectiveness of piezoelectric actuators depends on the moment arm, Young's modulus, and piezoelectric constants of the actuator material, in addition to the feedback voltage and control gains. Note that a feedback voltage higher than a ***breakdown voltage*** could destroy the dipole molecular structure, and consequently make the actuator become ineffective.

Three common geometries, a hemispheric shell, a rectangular plate, and an Euler-Bernoulli beam, were used to demonstrate the usefulness of the generic structronic shell theories. Based on four geometric parameters, i.e., two Lamé parameters and two radii of curvatures, the original sensation, control, and closed-loop system equations can be easily simplified to apply to these geometries. Similar procedures can be applied to other common geometries, such as shells of revolutions, paraboloidal shells, toroidal shells, spherical shells, cylindrical shells, conical shells, plates, rings, etc., Figure 4.8. The derived closed-form plate equation shows that the in-plane control forces are canceled out due to the infinite radii of curvature. However, both control forces and moments are preserved due to non-zero radii in the hemispheric shell case.

Note that the derived theory is limited to stress-free conditions, i.e., boundaries of the sensor/actuator layers can not be fully constrained (fixed) and the induced strains are used to estimate induced actuation effects, e.g., membrane control forces and bending control moments. In the feedback controls, both distributed and single point (discrete) voltage can be used in both the direct proportional feedback control and Lyapunov control. Since the modal voltage of the distributed piezoelectric sensor can be calculated in the time-domain, vibration control can be extended to a multi-input and multi-output (MIMO) control with a variable gain matrix. This distributed MIMO control algorithm needs to be further studied. Note that distributed sensors/actuators could also introduce "spillover" problems. Solutions to observation and control spillovers are to be discussed in Chaps. 7, 8, and 9. Distributed energy harvesting of piezoelectric laminated linear and nonlinear shell structures is presented in Chaps. 11 and 12.

Appendix

Piezoelectricity Theory

Two fundamental equations are used in the derivation of distributed sensor theory:

$$\{T\} = [c^D]\{S\} - [h]^t\{D\}, \tag{4.9.1}$$

$$\{E\} = [\beta^S]\{D\} - [h]\{S\}, \tag{4.9.2}$$

where $\{T\}$ is the stress vector (i.e., $\{T\} = \{T_{11}\ T_{22}\ T_{33}\ T_{23}\ T_{31}\ T_{12}\}^t$); $[c^D]$ is the elasticity matrix evaluated at constant dielectric displacement; $\{S\}$ is the strain vector (i.e., $\{S\} = \{S_{11}\ S_{22}\ S_{33}\ S_{23}\ S_{31}\ S_{12}\}^t$); [h] is the piezoelectric constant matrix; $\{D\}$ is the electric displacement vector; $[.]^t$ indicates the matrix transpose; $\{E\}$ is the electric field vector; $[\beta^S]$ is the dielectric impermeability matrix evaluated at constant strain.

Piezoelectric Matrix of Polyvinylidene Fluoride (PVDF)

Polymeric polyvinylidene fluoride (PVDF) has a mm2 structure (Kawai, 1969). The piezoelectric matrix [d] of a PVDF polymer can be expressed as (Sessler, 1981)

$$[d_{ij}] = \begin{bmatrix} 0 & 0 & 0 & 0 & d_{15} & 0 \\ 0 & 0 & 0 & d_{24} & 0 & 0 \\ d_{31} & d_{32} & d_{33} & 0 & 0 & 0 \end{bmatrix}. \tag{4.9.3}$$

Note that the piezoelectric coefficient d_{24} is equal to d_{15} for a PVDF electrically polarized and not mechanical stretched.

References

Balas, M.J., 1988, "Nonlinear Finite-Dimensional Control of a Class of Nonlinear Distributed Parameter Systems Using Residual Mode Filters," *Recent Development in Control of Nonlinear and Distributed Parameter Systems*, ASME-DSC-Vol.10, pp.19-22, December 1988.

Balas, M.J., 1978, "Active Control of Flexible Systems," *Journal of Optimization Theory and Applications*, Vol.25, No.3, July, pp.415-436.

Baz, A. and Poh, S., 1988, "Performance of an Active Control System with Piezoelectric Actuators," *Journal of Sound _ Vibration*, Vol.126, No.8, pp.327-343.

Brichkin, L.A., Butkovskii, A.G., and Pustyl'nikou, L.M., 1973, "Application of Finite Integral Transformation to Optimal Control Problems", *Automatika Telemekhanika* **7**, pp.13-24.

Butkovskii, A.G., 1962, "The Maximum Principle for Optimum Systems with Distributed Parameters", *Automation and Remote Control*, Vol.22, pp.1429-1438.

Crawley, E.F. and deLuis, J., 1987, "Use of Piezoelectric Actuators as Elements of Intelligent Structures," *AIAA Journal*, Vol.25, No.10, pp.1373-1385.

Crawley, E.F. and Anderson, E.H., 1990, "Detailed Models of Piezoceramic Actuation of Beams," *Journal of Intelligent Material Systems*, Vol.1.1, pp.4-25.

Cudney, H.H., Inman, D.J., & Y. Oshman, 1989, "Distributed Parameter Actuators for Structural Control," Proceedings of 1989 American Control Conference, pp.1189-1194.

Fanson, J.L. and Garba, J.A., 1988, "Experimental Studies of Active Members in Control of Large Space Structures," AIAA Paper 88-2207, *Proc. of AIAA/ASME/AHS 29th Structures, Structural Dynamics, and Materials Conference*, pp.9-17.

Hagood, N., Chung, W., and A. von Flotow, 1990, "Modeling of Piezoelectric Actuator Dynamics for Active Structural Control," AIAA Paper No.90-1087, 31st Structures, Structural Dynamics and Materials Conference, Long Beach, CA, April 2-4, 1990.

Hanagud, S. and Obal, M.W., 1988, "Identification of Dynamic Coupling Coefficients in a Structure with Piezoelectric Sensors and Actuators," *Proc. of AIAA/ASME/AHS 29th Structures, Structural Dynamics, and Materials Conference*, (Paper No.88-2418), Part-3, pp.1611-1620.

Kawai, H., 1969, "The Piezoelectricity of Poly(vinylidene Fluoride)," Jpn. J. Appl. Phys., Vol.8, pp. 975-976.

Lee, C. K. & Moon, F.C., 1988, "Modal Sensors/Actuators," IBM Research Report, RJ 6306 (61975), June.

Lions, J.L., 1968, "*Optimal Control of Systems Governed by Partial Differential Equations*", Dunod and Gauthier-Villars, Paris.

Meirovitch, L. and Silverberg, L.M., 1983, "Globally Optimal Control of Self-Adjoint Distributed Systems," *Optimal Applications _ Methods*, Vol.4, pp.365-386.

Meirovitch, L. and Baruh, 1981, "Effect of Damping on Observation Spillover Instability, *Journal of Optimization Theory and Applications*, Vol.35, No.1, Sept. pp.31-44.

Obal, M.W., 1986, *Vibration Control of Flexible Structures Using Piezoelectric Devices as Sensors and Actuators*, Ph.D. Thesis, Georgia Institute of Technology.

Plump, J.M., Hubbard, J. E., and Baily, T., 1987, "Nonlinear Control of a Distributed System: Simulation and Experimental Results," ASME *J. Dynamic Systems, Measurement, and Control*, pp.133-139.

Reinhorn A.M. and Manolis, G.D., 1985, "Current state of knowledge on structural control," *Sound and Vibration Digest*, Vol.17, pp.7-16.

Robinson, A.C., 1971, "A Survey of Optimal Control of Distributed Parameter Systems", *Automatica* ***7***, pp.371-388.

Sakawa, Y., 1966, "Optimal Control of Certain Type of Linear Distributed-Parameter Systems", *IEEE Trans. on Automatic Control*, Vol. AC-11, pp.35-41.

Sessler, G.M., 1981, "Piezoelectricity in Polyvinylidene Fluoride," *J. Acoust. Soc. Am.*, 70(6), pp. 1596-1608.

Sirlin, S.W, 1987, "Vibration Isolation for Spacecraft Using the Piezoelectric Polymer PVF_2," *Proc. of the 114th Meeting of the Acoustics Society of America*, Nov.

Tzafestas, S.G., 1970, "Optimal Distributed-Parameter Control Using Classical Variational Theory", *Int. J. Control* ***12***, pp.593-608.

Tzou, H. S., 1987, "Active Vibration Control of Flexible Structures Via Converse Piezoelectricity", *Developments in Mechanics*, Vol.14-C, pp.201-206.

Tzou, H.S., 1988a, "Dynamic analysis and passive control of viscoelastically damped nonlinear dynamic contacts," *J. Finite Elements in Analysis and Design*, Vol.(4), No.3, pp.232-238.

Tzou, H.S., 1988b, "Integrated Sensing and Adaptive Vibration Suppression of Distributed Systems," *Recent Development in Control of Nonlinear and Distributed Parameter Systems*, ASME-DSC-Vol.10, pp.51-58, December.

Tzou, H.S., 1989a, "Distributed Sensing and Feedback Controls of Distributed Parameter Systems," *High-Performance Computing*, Edited by J.-L. Delhaye and E. Gelenbe, North-Holland, ELSEVIER Science Publishers B.V., Amsterdam, Netherlands, pp.95-107.

Tzou, H.S., 1989b, "Theoretical Development of a Layered Thin Shell with Internal Distributed Controllers," *Failure Prevention and Reliability-1989*, pp.17-20; 1989 ASME Technical Design Conference, Montreal, Canada, Sept.17-20.

Tzou, H.S., 1989c, "Integrated Distributed Sensing and Active Vibration Suppression of Flexible Manipulators using Distributed Piezoelectrics," *Journal of Robotic Systems*, Vol.6, No.6, pp. 745-767, December.

Tzou, H.S., 1990, "Distributed Modal Identification and Vibration Control of Continua: Theory and Applications," *Proceedings of 1990 American Control Conference*, pp. 1237-1243, San Diego, CA, May 23-25, 1990; *ASME Journal of Dynamic Systems, Measurements, and Control*, Vol.(113), No.(3), pp.494-499, September 1991.

Tzou, H.S., 1992, "A New Distributed Sensor and Actuator Theory for "Intelligent" Shells," *Journal of Sound _ Vibration*, Vol.(153), No.(2), pp. 335-350, March 1992.

Tzou, H. S. & Gadre, M., 1988, "Active Vibration Suppression by Piezoelectric Polymer with Variable Feedback Gain," *AIAA Journal*, Vol.26, No.8, pp.1014-1017.

Tzou, H. S. & Gadre, M., 1989, "Theoretical Analysis of a Muli-Layered Thin Shell Coupled with Piezoelectric Shell Actuators for Distributed Vibration Controls," *Journal of Sound and Vibration*, Vol.132, No.3, pp.433-450, August.

Tzou H.S. & Gadre, M., 1990, "Active Vibration Isolation and Excitation by a Piezoelectric Slab with Constant Feedback Gains," *Journal of Sound _ Vibration* Vol.136, No.3, pp.477-490, February.

Tzou, H.S. and Tseng, C.I., 1988a, "Active Vibration Control of Distributed Parameter Systems by Finite Element Method," *Computers in Engineering 1988*, Vol.3, pp.599-604.

Tzou, H.S. and Tseng, C.I., 1988b, "Development of a Thin Piezoelectric Finite Element Applied to Distributed Sensing and Active Vibration Controls," 88-ASME/CIE-2, ASME Winter Annual Meeting, Chicago, Ill, 1988. (Invited Paper).

Tzou, H.S. and Tseng, C.I., 1990, "Distributed Piezoelectric Sensor/Actuator Design for Dynamic Measurement/Control of Distributed Parameter Systems: A Piezoelectric Finite Element Approach," *Journal of Sound & Vibration*, Vol.138, No.1, pp.17-34, April.

Tzou, H.S., Tseng, C.I. & Wan, G.C., 1990, "Distributed Structural Dynamics Control of Flexible Manipulators, Part 2: Distributed Sensor and Active Electromechanical Actuator," *Journal of Computers and Structures*, Vol.35, No.6, pp.679-687.

Usov, V.S. and Surygin, A.I., 1984, "Variation of SAW Velocity and Damping in a Piezofilm-Semiconductor Structure under the Effect of a Constant Transverse Electric Field," *Radioelektronika*, (ISSN 0021-3470), Vol.27. (Nov. 1984).

Vidyasagar, M., 1988, "Control of Distributed Parameter System Using the Coprime Factorization Approach," *Recent Development in Control of Nonlinear and Distributed Parameter Systems*, ASME-DSC-Vol.10, pp.1-10, December.

Wang, P.K.C., 1966, "On the Feedback Control of Distributed Parameter Systems", *Int. J. on Control*, Vol. 3, No. 3, pp. 255-273.

Zimmerman, D.C., Inman, D.J. and Juang, J.N., 1988, "Low Authority- Threshold Control for Large Flexible Structures," AIAA Paper 88-2270, *Proc. of AIAA/ASME/AHS 29th Structures, Structural Dynamics, and Materials Conference*, pp.459-469. (4/ShlScrAct-New.BkPzSm2)

Chapter 5
Multi-layered Shell Actuators

Due to the rapid development of adaptive structures and large flexible systems, "***intelligent***" or "***smart***" structures with built in sensors, actuators, and control electronics are increasingly in demand (Tzou & Anderson, 1920; Wada, Fanson, & Crawley, 1989). Conventional composite materials have been feasible for decades, and they are constantly being advanced in many features, such as strength/weight ratios, directional properties, high-temperature tolerance, etc. A step beyond conventional incremental improvements is an addition of "*intelligence*", i.e., sensors, actuators, control electronics, central processing unit (CPU), etc., to the composites. This chapter presents a theoretical development of a multi-layered thin shell with internal distributed actuators for distributed vibration controls of active shell structures.

The internal distributed actuator can be composed of either electromechanical or electromagnetic sensitive materials, e.g., piezoelectrics, shape memory alloys, electrostrictive materials, magnetostrictive materials, electrorheological materials, etc. (Wada, et al., 1989; Tzou & Anderson, 1920; Tzou, Lee and Arnold, 2004). Shell oscillations induced by external excitations or disturbances can be measured by sensors, either discrete or distributed, then sensor signals can be processed and fed back to the internal distributed actuator, actively suppressing the undesirable oscillation (Baily & Hubbard, 1987; Tzou, 1987; Tzou & Tseng, 1988a). Studies on a simple geometry-beam-with various damping treatments or actuators have been presented (Baily & Hubbard, 1987; Hanagud & Obal, 1988; Crawley & de Luis, 1987; Tzou, 1987). A finite element formulation of a "layered" piezoelectric thin plate has been proposed, and the effectiveness of the distributed piezoelectric element to distributed sensing and control evaluated (Tzou & Tseng, 1988b & 1988c). A theory on distributed sensing and control of a generic thin shell coupled with distributed sensor and actuator was proposed (Tzou, 1988). In this chapter, a generic theory for a multi-layered deep shell with arbitrary internal distributed actuators is developed. Thin shell equations are based on Kirchhoff-Love's theory and Hamilton's principle (Love, 1888; Soedel, 1981). The generic shell equations can be simplified to apply to many other common geometries and structures, e.g.,

H. (HS) Tzou, *Piezoelectric Shells*, Solid Mechanics and Its Applications 247,
https://doi.org/10.1007/978-94-024-1258-1_5

cylindrical panel, plate (rectangular or circular), beam, etc., by a reduction procedure, presented in Sect. 3.1 of Chap. 3. Demonstration examples are also presented.

5.1 A Multi-layered Shell with Distributed Actuators

In this section, general assumptions associated with the theoretical development are discussed first. A generic layered shell with induced local strain/stress resulting from the feedback control loop is evaluated next. Later, Love's equations of motion for a single-layer thin deep shell is extended to apply to a deep shell with multiple layers, some of which are internal distributed actuator layers subjected to localized electromechanical control actions. In later derivations, emphasis is placed on incorporating the local induced control actions (e.g., due to a distributed control force) with a multi-layered deep shell element. Detailed procedures for deriving the shell equations were presented in Chap. 2.

5.1.1 Assumptions

It is assumed that the layered thin shell is composed of several thin shell layers (with the same dimensions in the α_1 and α_2 directions) perfectly bonded together. The combined shell thickness of the layered shell is still thin, so that Kirchhoff-Love's assumptions apply. A multi-layered shell with its curvilinear coordinate system $(\alpha_1, \alpha_2, \& \alpha_3)$ is illustrated in Fig. 5.1. Each shell layer may have different thickness h_i. The radii of curvature (R_1 and R_2) are much greater than the shell thickness variation, i.e., $R_j \gg h_j, j = 1,2$. Thus, all shell layers are defined with the same Lamé parameters A_1 and A_2. The neutral surfaces of the shell layers are all parallel to each other, and all transverse distances α_3 and displacements u_3 are measured from a reference surface—the "modulus-weighted" mid-surface. Note that all deformations resulting from the transverse shears and rotatory inertias are neglected.

If a forcing function is locally applied to one or more layers, the induced additional deformations and strains appear in these layers. (This forcing function could be a voltage applied to the actuator layers made of electromechanical sensitive materials, such as piezoelectrics, electro-/magneto/photos-strictive, shape-memory or other controllable materials.) Note that this action is assumed **internal** and relatively **small**, so that displacement discontinuity between layers is prevented. Only the induced forces and moments affect the whole multi-layered shell structure. Note that the control action is induced via an *equivalent formulation* arising from the electromechanical phenomena; this generic formulation is then applied to the layered shell actuators. (If the induced control action is directly derived from the piezoelectric effect, the material symmetry needs to be considered, see Chap. 2.)

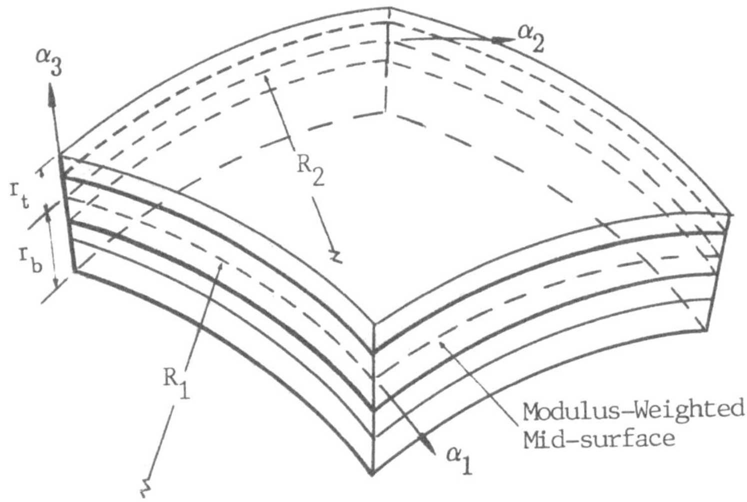

Fig. 5.1 A multi-layered thin shell with internal distributed actuators

It is assumed that only the in-plane forces and moments are considered. The resultant forces and bending moments for the entire layered shell continuum are derived by summing all membrane forces N_{jk} and bending moments M_{jk} of each shell layer, i.e.,

$$N_{jk} = \int_{r_b}^{r_t} \sigma_{jk} d\alpha_3 = \sum_{i=1}^{n} \int_{r_i}^{r_{i+1}} \sigma^{\tau}_{jk_i} d\alpha_3 = \sum_{i=1}^{n} N_{jk_i} \quad \text{for} \quad j,k = 1,2. \qquad (5.1.1)$$

$$M_{jk} = \int_{r_b}^{r_t} \alpha_3 \sigma_{jk} d\alpha_3 = \sum_{i=1}^{n} \int_{r_i}^{r_{i+1}} \alpha_3 \sigma^{\tau}_{jk_i} d\alpha_3 = \sum_{i=1}^{n} M_{jk_i} \quad \text{for} \quad j,k = 1,2. \quad (5.1.2)$$

r_t denotes the distance measured from the weighted mid-surface to the top surface of the layered shell; r_b defines the distance from the mid-surface to the shell bottom surface; r_i and r_{i+1} denote the distances measured from the mid-surface to the bottom of the i-th and (i + 1)-th shell layers, respectively. Note that $r_{i+1} - r_i = h_i$, the i-th layer thickness. α_3 defines the moment arm—a distance measured from the reference surface to the mid-surface of the shell layer; σ_{jk} is the resultant mechanical stress of the layered shell; $\sigma^{\tau}_{jk_i}$ denotes the total stress of the i-th shell layer; the superscript τ denotes the total effect. $\sigma^{\tau}_{jk_i}$ is contributed by two components: 1) an inherent elastic component and 2) an induced actuation component (i.e., for control purpose).

Dynamic equations for the i-th shell layer are derived by using Hamilton's principle. Dynamic equations for the entire multi-layered shell are formulated by incorporating all component membrane forces and bending moments of all shell layers.

5.1.2 Analysis of the i-th Distributed Actuator Layer

In this section, stresses and strains consisting of the elastic component and the induced component are defined; they are used to define forces and moments of the i-th shell layer in the next section.

A **three dimensional** (3D) infinitesimal distance ds_i in the i-th shell layer is defined by the 3D ***fundamental form*** (illustrated as ds_2 in Fig. 2.2):

$$(ds_i)^2 = A_1^2(1+\frac{\alpha_3}{R_1})^2(d\alpha_1)^2 + A_2^2(1+\frac{\alpha_3}{R_2})^2(d\alpha_2)^2 + (d\alpha_3)^2, \tag{5.1.3}$$

where A_1 and A_2 are the Lamé parameters and the subscript "i" denotes the i-th layer. α_3 is the distance measured from the "modulus weighted" mid-surface; R_j is the radius of curvature of the α_j axis; $d\alpha_j$ is an infinitesimal distance in the α_j direction. Note that the infinitesimal distance defined on the *neutral surface*, $(ds_i)^2 = A_1^2(d\alpha_1)^2 + A_2^2(d\alpha_2)^2$, was discussed in Chaps. 2 and 3. Equation (5.1.3) is reduced to the conventional expression by eliminating the α_3 component.

For convenience and simplicity, one can define

$$g_{jj}(\alpha_1,\alpha_2,\alpha_3) = A_j^2(1+\frac{\alpha_3}{R_j})^2, \quad j = 1,2, \tag{5.1.4a}$$

$$g_{33}(\alpha_1,\alpha_2,\alpha_3) = 1. \tag{5.1.4b}$$

The basic mechanical stress-strain (σ and S) relations are defined as

$$S_{11_i} = \frac{1}{Y_i}[\sigma_{11_i} - \mu_i(\sigma_{22_i} + \sigma_{33_i})], \tag{5.1.5a}$$

$$S_{22_i} = \frac{1}{Y_i}[\sigma_{22_i} - \mu_i(\sigma_{11_i} + \sigma_{33_i})], \tag{5.1.5b}$$

where Y_i is Young's modulus; μ_i is Poisson's ratio; subscripts "$_{ii}$" indicate stress/strain directions in the α_1, α_2, and α_3 directions, respectively. The sub-subscript "i" denotes the i-th shell layer. Displacement U_j and infinitesimal coordinate changes ξ_j' arising from the finite deformation are related by

$$U_j = \xi_j'\sqrt{g_{jj}(\alpha_1,\alpha_2,\alpha_3)}, \quad j = 1,2,3. \tag{5.1.6}$$

For the i-th distributed shell actuator layer, it is assumed that a local control force can be introduced by a feedback signal. The induced strains for the i-th layer are denoted by S_{11_i}' and S_{22_i}' (superscript $'$ denotes the induced component). The induced displacements for the i-th layer are denoted by Ψ_{1_i}, Ψ_{2_i} and Ψ_{3_i}. (Note that the induced displacement components are internal to the i-th layer such that displacement continuity is still warranted. Only the resulting force and moment affect the whole layered shell.) For a thin shell, the displacements in the α_1 and α_2 directions are assumed to be linear variation in the α_3 direction, and the U_3 is independent of α_3. Thus, the displacements are

$$U_{1_i}(\alpha_1, \alpha_2, \alpha_3) = u_1(\alpha_1, \alpha_2) + \alpha_3 \beta_{1_i}^{\tau}(\alpha_1, \alpha_2) + \Psi_{1_i}(\alpha_1, \alpha_2), \tag{5.1.7a}$$

$$U_{2_i}(\alpha_1, \alpha_2, \alpha_3) = u_2(\alpha_1, \alpha_2) + \alpha_3 \beta_{2_i}^{\tau}(\alpha_1, \alpha_2) + \Psi_{2_i}(\alpha_1, \alpha_2), \tag{5.1.7b}$$

$$U_{3_i}(\alpha_1, \alpha_2, \alpha_3) = u_3(\alpha_1, \alpha_2) + \Psi_{3_i}(\alpha_1, \alpha_2, \alpha_3), \tag{5.1.7c}$$

where $\beta_{1_i}^{\tau}$ and $\beta_{2_i}^{\tau}$ are the total deformation angles which need to be solved using Kirchhoff-Love's assumptions. The superscript τ denotes the total effect. The last term Ψ_{3_i} in Eq. (5.1.7c) can be neglected if 1) the layer is thin and 2) only the in-plane (α_1 and α_2) actuator motion is considered. Again, Ψ_{3_i} is considered an internal action within the i-th layer itself. Only the resultant force is transmitted to the adjacent layers; otherwise, separation between layers could occur. For generality, this term is still preserved. For a thin shell, the shear strains are negligible, i.e., $S_{13_i} = 0$ and $S_{23_i} = 0$. Thus,

$$S_{13_i} = A_1\left(1 + \frac{\alpha_3}{R_1}\right) \frac{\partial}{\partial \alpha_3}\left[\frac{U_{1_i}}{A_1\left(1 + \frac{\alpha_3}{R_1}\right)}\right] + \frac{1}{A_1\left(1 + \frac{\alpha_3}{R_1}\right)} \frac{\partial U_{3_i}}{\partial \alpha_1} = 0. \tag{5.1.8}$$

Using Love's assumptions $S_{13_i} = 0$ and substituting U_{1_i} and g_{ii} into the above equation, one can derive

$$\sqrt{g_{11}} \frac{\partial}{\partial \alpha_3}\left[\frac{u_1 + \Psi_{1_i} + \alpha_3 \beta_{1_i}^{\tau}}{\sqrt{g_{11}}}\right] + \frac{1}{\sqrt{g_{11}}} \frac{\partial}{\partial \alpha_1}(u_3 + \Psi_{3_i}) = 0. \tag{5.1.9}$$

Thus, the total deflection angle $\beta_{1_i}^{\tau}$ in the α_1 direction of the i-th layer is

$$\beta_{1_i}^{\tau} = \frac{u_1}{R_{1_i}} - \frac{1}{A_1} \frac{\partial u_3}{\partial \alpha_1} + \left[\frac{\Psi_{1_i}}{R_{1_i}} - \left(\frac{1}{A_1} \frac{\Psi_{3_i}}{\partial \alpha_1}\right)\right]. \tag{5.1.10}$$

Note that the terms inside the bracket are introduced by the induced deflection on the i-th shell actuator layer. Thus, the total angle can be divided into two components: one is the original deflection angle due to the shell oscillation itself and the other is due to the active control action, i.e.,

$$\beta_{1_i}^{\tau} = \beta_{1_i} + \beta_{1_i}', \tag{5.1.11}$$

where

$$\beta_{1_i} = \frac{u_1}{R_1} - \frac{1}{A_1}\frac{\partial u_3}{\partial \alpha_1}, \tag{5.1.12a}$$

$$\beta'_{1_i} = \frac{\Psi_{1_i}}{R_1} - \frac{1}{A_1}\frac{\partial \Psi_{3_i}}{\partial \alpha_1}. \tag{5.1.12b}$$

Note that the induced term is denoted by a superscript ‘. Similarly,

$$\beta^{\tau}_{2_i} = \beta_{2_i} + \beta'_{2_i}, \tag{5.1.13}$$

where

$$\beta_{2_i} = \frac{u_2}{R_2} - \frac{1}{A_2}\frac{\partial u_3}{\partial \alpha_2}, \tag{5.1.14a}$$

$$\beta'_{2_i} = \frac{\Psi_{2_i}}{R_2} - \frac{1}{A_2}\frac{\partial \Psi_{3_i}}{\partial \alpha_2}. \tag{5.1.14b}$$

The total normal strains $S^{\tau}_{11_i}$ and $S^{\tau}_{22_i}$ also need to be defined in terms of two components mentioned above.

$$S^{\tau}_{11_i} = \frac{1}{\sqrt{g_{11}}}\left(\frac{\partial U_{1_i}}{\partial \alpha_1} + \frac{U_{2_i}}{A_2}\frac{\partial A_1}{\partial \alpha_2} + U_{3_i}\frac{A_1}{R_1}\right). \tag{5.1.15a}$$

$$S^{\tau}_{22_i} = \frac{1}{\sqrt{g_{22}}}\left(\frac{\partial U_{2_i}}{\partial \alpha_2} + \frac{U_{1_i}}{A_1}\frac{\partial A_2}{\partial \alpha_1} + U_{3_i}\frac{A_2}{R_2}\right). \tag{5.1.15b}$$

Substituting Eqs. (5.1.7a, 5.1.7b, & 5.1.7c) into Eq. (5.1.15a) and assuming that

$$\begin{cases} \alpha_3/R_1 \ll 1, \\ \alpha_3/R_2 \ll 1, \end{cases} \tag{5.1.16}$$

one can derive the total normal strains:

$$S^{\tau}_{11_i} = \frac{1}{A_1}\frac{\partial}{\partial \alpha_1}\left(u_1 + \Psi_{1_i} + \alpha_3\beta^{\tau}_{1_i}\right) + \frac{1}{A_1A_2}\left(u_2 + \Psi_{2_i} + \alpha_3\beta^{\tau}_{2_i}\right)\frac{\partial A_1}{\partial \alpha_2} + \frac{u_3 + \Psi_{3_i}}{R_1}, \tag{5.1.17}$$

$$S^{\tau}_{22_i} = \frac{1}{A_2}\frac{\partial}{\partial \alpha_2}\left(u_2 + \Psi_{2_i} + \alpha_3\beta^{\tau}_{2_i}\right) + \frac{1}{A_1A_2}\left(u_1 + \Psi_{1_i} + \alpha_3\beta^{\tau}_{1_i}\right)\frac{\partial A_2}{\partial \alpha_1} + \frac{u_3 + \Psi_{3_i}}{R_2}. \tag{5.1.18}$$

The transverse normal strain $S^{\tau}_{33_i}$ is defined by

$$S_{33_i}^{\tau} = \frac{\partial U_{3_i}}{\partial \alpha_3} = \frac{\partial u_{3_i}}{\partial \alpha_3} + \frac{\partial \Psi_{3_i}}{\partial \alpha_3}. \tag{5.1.19}$$

Note that $\frac{\partial u_{3_i}}{\partial \alpha_3} = 0$, since u_3 is a function of α_1 and α_2 only, Eq. (5.1.7c). As discussed previously, $\frac{\partial \Psi_{3_i}}{\partial \alpha_3}$ is an internal action. Only the resulting force will be transmitted to the adjacent layers; otherwise, a separation between adjacent layers could occur. In addition, the control action is primarily an in-plane action. The total strain can be further classified into two components: 1) a membrane strain $\hat{S}_{nn_i}^{\tau}$ (independent of thickness) and 2) a bending strain $\hat{k}_{nn_i}^{\tau}$. Separating the membrane and bending strains in the α_1 and α_2 directions gives the detailed expressions.

1) Total Membrane Strains:

$$\hat{S}_{11_i}^{\tau} = \frac{1}{A_1}\frac{\partial}{\partial \alpha_1}(u_1 + \Psi_{1_i}) + \frac{u_2 + \Psi_{2_i}}{A_1 A_2}\frac{\partial A_1}{\partial \alpha_2} + \frac{u_3 + \Psi_{3_i}}{R_1}, \tag{5.1.20a}$$

$$\hat{S}_{22_i}^{\tau} = \frac{1}{A_2}\frac{\partial}{\partial \alpha_2}(u_2 + \Psi_{2_i}) + \frac{u_1 + \Psi_{1_i}}{A_1 A_2}\frac{\partial A_2}{\partial \alpha_1} + \frac{u_3 + \Psi_{3_i}}{R_2}, \tag{5.1.20b}$$

$$\hat{S}_{12_i}^{\tau} = \frac{A_2}{A_1}\frac{\partial}{\partial \alpha_1}\left(\frac{u_2 + \Psi_{2_i}}{A_2}\right) + \frac{A_1}{A_2}\frac{\partial}{\partial \alpha_2}\left(\frac{u_1 + \Psi_{1_i}}{A_1}\right); \tag{5.1.20c}$$

2) Total Bending Strains:

$$\hat{k}_{11_i}^{\tau} = \frac{1}{A_1}\frac{\partial \beta_{1_i}^{\tau}}{\partial \alpha_1} + \frac{\beta_{2_i}^{\tau}}{A_1 A_2}\frac{\partial A_1}{\partial \alpha_2}, \tag{5.1.21a}$$

$$\hat{k}_{22_i}^{\tau} = \frac{1}{A_2}\frac{\partial \beta_{2_i}^{\tau}}{\partial \alpha_2} + \frac{\beta_{1_i}^{\tau}}{A_1 A_2}\frac{\partial A_2}{\partial \alpha_1}, \tag{5.1.21b}$$

$$\hat{k}_{12_i}^{\tau} = \frac{A_2}{A_1}\frac{\partial}{\partial \alpha_1}\left(\frac{\beta_{2_i}^{\tau}}{A_2}\right) + \frac{A_1}{A_2}\frac{\partial}{\partial \alpha_2}\left(\frac{\beta_{1_i}^{\tau}}{A_1}\right). \tag{5.1.21c}$$

Separate and rewrite the induced membrane strains $\hat{S}_{nn_i}^{\prime}$'s and bending strains $\hat{k}_{nn_i}^{\prime}$'s due to the localized deformation induced by the control action on the i-th layer. The induced membrane strains and bending strains can be further derived.

1) Induced Membrane Strains:

$$\hat{S}'_{11_i} = \frac{1}{A_1}\frac{\partial \Psi_{1_i}}{\partial \alpha_1} + \frac{\Psi_{2_i}}{A_1 A_2}\frac{\partial A_1}{\partial \alpha_2} + \frac{\Psi_{3_i}}{R_1}, \tag{5.1.22a}$$

$$\hat{S}'_{22_i} = \frac{1}{A_2}\frac{\partial \Psi_{2_i}}{\partial \alpha_2} + \frac{\Psi_{1_i}}{A_1 A_2}\frac{\partial A_2}{\partial \alpha_1} + \frac{\Psi_{3_i}}{R_2}, \tag{5.1.22b}$$

$$\hat{S}'_{12_i} = \frac{A_2}{A_1}\frac{\partial}{\partial \alpha_1}\left(\frac{\Psi_{2_i}}{A_2}\right) + \frac{A_1}{A_2}\frac{\partial}{\partial \alpha_2}\left(\frac{\Psi_{1_i}}{A_1}\right); \tag{5.1.22c}$$

2) Induced Bending Strains:

$$\begin{aligned}\hat{k}'_{11_i} &= \frac{1}{A_1}\frac{\partial \beta'_{1_i}}{\partial \alpha_1} + \frac{\beta'_{2_i}}{A_1 A_2}\frac{\partial A_1}{\partial \alpha_2} \\ &= \frac{1}{A_1}\frac{\partial}{\partial \alpha_1}\left(\frac{\Psi_{1_i}}{R_1} - \frac{1}{A_1}\frac{\partial \Psi_{3_i}}{\partial \alpha_1}\right) + \frac{1}{A_1 A_2}\left(\frac{\Psi_{2_i}}{R_2} - \frac{1}{A_2}\frac{\partial \Psi_{3_i}}{\partial \alpha_2}\right)\frac{\partial A_1}{\partial \alpha_2},\end{aligned} \tag{5.1.23a}$$

$$\begin{aligned}\hat{k}'_{22_i} &= \frac{1}{A_2}\frac{\partial \beta'_{2_i}}{\partial \alpha_2} + \frac{\beta}{A_1 A_2}\frac{\partial A_2}{\partial \alpha_1} \\ &= \frac{1}{A_2}\frac{\partial}{\partial \alpha_2}\left(\frac{\Psi_{2_i}}{R_2} - \frac{1}{A_2}\frac{\partial \Psi_{3_i}}{\partial \alpha_2}\right) + \frac{1}{A_1 A_2}\left(\frac{\Psi_{1_i}}{R_1} - \frac{1}{A_1}\frac{\partial \Psi_{3_i}}{\partial \alpha_1}\right)\frac{\partial A_2}{\partial \alpha_1},\end{aligned} \tag{5.1.23b}$$

$$\begin{aligned}\hat{k}'_{12_i} &= \frac{A_2}{A_1}\frac{\partial}{\partial \alpha_1}\left(\frac{\beta'_{2_i}}{A_2}\right) + \frac{A_1}{A_2}\frac{\partial}{\partial \alpha_2}\left(\frac{\beta'_{1_i}}{A_1}\right) \\ &= \frac{A_2}{A_1}\frac{\partial}{\partial \alpha_1}\left[\frac{1}{A_2}\left(\frac{\Psi_{2_i}}{R_2} - \frac{1}{A_2}\frac{\partial \Psi_{3_i}}{\partial \alpha_2}\right)\right] + \frac{A_1}{A_2}\frac{\partial}{\partial \alpha_2}\left[\frac{1}{A_1}\left(\frac{\Psi_{1_i}}{R_1} - \frac{1}{A_1}\frac{\partial \Psi_{3_i}}{\partial \alpha_1}\right)\right].\end{aligned} \tag{5.1.23c}$$

Note that Eq. (5.1.22a) and Eq. (5.1.23a) can also be used to represent the original (elastic) membrane and bending strains if the variable Ψ_j is replaced by u_j. The total strains can be represented as a summation of the original strain and the induced strain.

$$\hat{S}^{\tau}_{11_i} = \hat{S}_{11_i} + \hat{S}'_{11_i}, \tag{5.1.24a}$$

$$\hat{k}^{\tau}_{11_i} = \hat{k}_{11_i} + \hat{k}'_{11_i}, \tag{5.1.24b}$$

$$\hat{S}^{\tau}_{22_i} = \hat{S}_{22_i} + \hat{S}'_{22_i}, \tag{5.1.24c}$$

$$\hat{k}^{\tau}_{22_i} = \hat{k}_{22_i} + \hat{k}'_{22_i}, \tag{5.124d}$$

$$\hat{S}^{\tau}_{12_i} = \hat{S}_{12_i} + \hat{S}'_{12_i}, \tag{5.1.24e}$$

$$\hat{k}^{\tau}_{12_i} = \hat{k}_{12_i} + \hat{k}'_{12_i}. \tag{5.1.24f}$$

Thus,

$$\begin{aligned} S^{\tau}_{11_i} &= \left(\hat{S}_{11_i} + \alpha_3 \hat{k}_{11_i}\right) + \left(\hat{S}'_{11_i} + \alpha_3 \hat{k}'_{11_i}\right) \\ &= S_{11_i} + S'_{11_i} \end{aligned}, \tag{5.1.25a}$$

$$\begin{aligned} S^{\tau}_{22_i} &= \left(\hat{S}_{22_i} + \alpha_3 \hat{k}_{22_i}\right) + \left(\hat{S}'_{22_i} + \alpha_3 \hat{k}'_{22_i}\right) \\ &= S_{22_i} + S'_{22_i} \end{aligned}, \tag{5.1.25b}$$

$$\begin{aligned} S^{\tau}_{12_i} &= \left(\hat{S}_{12_i} + \alpha_3 \hat{k}_{12_i}\right) + \left(\hat{S}'_{12_i} + \alpha_3 \hat{k}'_{12_i}\right) \\ &= S_{12_i} + S'_{12_i} \end{aligned}. \tag{5.1.25c}$$

The mechanical **stress-strain relations** must be developed in terms of the original (elastic) and the induced components,

$$\begin{aligned} \sigma^{\tau}_{11_i} &= \frac{Y_i}{1-\mu_i^2}\left(S^{\tau}_{11_i} + \mu_i S^{\tau}_{22_i} + \mu_i S'_{33_i}\right) \\ &= \frac{Y_i}{1-\mu_i^2}\left(S_{11_i} + \mu_i S_{22_i}\right) + \frac{Y_i}{1-\mu_i^2}\left(S'_{11_i} + \mu_i S'_{22_i} + \mu_i S'_{33_i}\right), \\ &= \sigma_{11_i} + \sigma'_{11_i} \end{aligned} \tag{5.1.26}$$

where

$$\sigma_{11_i} = \frac{Y_i}{1-\mu_i^2}\left(S_{11_i} + \mu_i S_{22_i}\right), \tag{5.1.27a}$$

$$\sigma'_{11_i} = \frac{Y_i}{1-\mu_i^2}\left(S'_{11_i} + \mu_i S'_{22_i} + \mu_i S'_{33_i}\right). \tag{5.1.27b}$$

Similarly,

$$\sigma^{\tau}_{22_i} = \sigma_{22_i} + \sigma'_{22_i}, \tag{5.1.28a}$$

$$\sigma^{\tau}_{12_i} = G_i\left(S_{12_i} + S'_{12_i}\right) = \sigma_{12_i} + \sigma'_{12_i}, \tag{5.1.28b}$$

where G_i is the shear modulus and

$$G_i = \frac{Y_i}{2(1+\mu_i)}. \tag{5.1.29}$$

5.1.3 Resultant Forces and Moments of the Layered Shell

The stress and strain equations of the i-th shell actuator layer have been derived. Next, resultant forces and bending moments for the entire multi-layered shell will be formulated using the stresses derived previously. (Note that only the induced stresses affect the entire layered shell.) Figure 5.2 shows the generalized forces and moments for the i-th shell layer.

It is assumed that the resultant forces and moments can be calculated as a sum of all components induced by all layers. Thus, the resultant forces N_{jk}'s and moments M_{jk}'s for the entire multi-layered shell, from r_b (bottom surface) to r_t (top surface), can be evaluated as:

$$N_{jk} = \int_{r_b}^{r_t} \sigma_{jk} d\alpha_3 = \sum_{i=1}^{n} \int_{r_i}^{r_{i+1}} \sigma_{jk_i}^{\tau} d\alpha_3 = \sum_{i=1}^{n} N_{jk_i} \quad \text{for} \quad j,k = 1, 2. \tag{5.1.30}$$

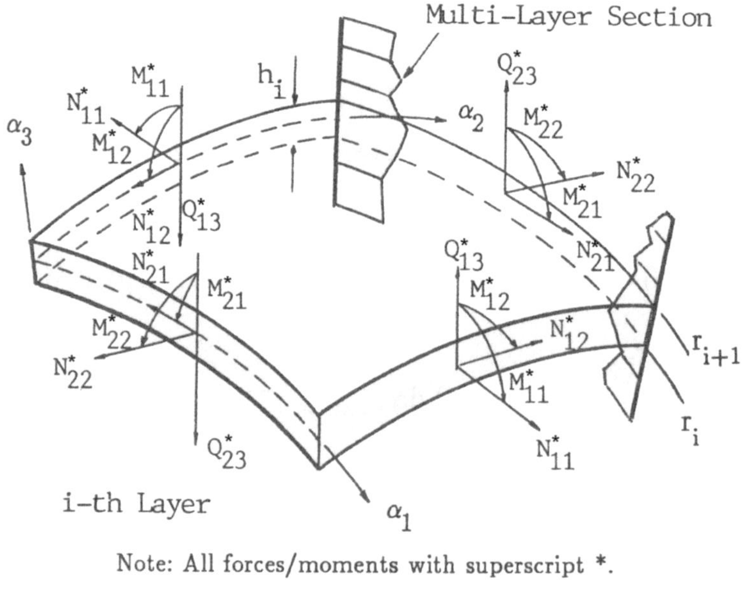

Fig. 5.2 Generalized forces and moments for the i-th layer shell

$$M_{jk} = \int_{r_b}^{r_t} \alpha_3 \sigma_{jk} d\alpha_3 = \sum_{i=1}^{n} \int_{r_i}^{r_{i+1}} \alpha_3 \sigma^{\tau}_{jk_i} d\alpha_3 = \sum_{i=1}^{n} M_{jk_i} \quad \text{for} \quad j,k = 1,2. \tag{5.1.31}$$

The distances r_x's were defined earlier; $(r_{i+1} - r_i = h_i)$ is the thickness of the i-th shell layer. Thus,

$$N_{11} = \int_{r_b}^{r_t} \sigma_{11} d\alpha_3 = \sum_{i=1}^{n} \int_{r_i}^{r_{i+1}} \sigma^{\tau}_{11_i} d\alpha_3 = \sum_{i=1}^{n} \int_{r_i}^{r_{i+1}} \left(\sigma_{11_i} + \sigma'_{11_i}\right) d\alpha_3. \tag{5.1.32}$$

Substituting Eq. (5.1.26) into Eq. (5.1.32) yields

$$\begin{aligned} N_{11} &= \sum_{i=1}^{n} \int_{r_i}^{r_{i+1}} \frac{Y_i}{1-\mu_i^2} \left[\left(S_{11_i} + \mu_i S_{22_i}\right) + S'_{11_i} + \mu_i \left(S'_{22_i} + S'_{33_i}\right)\right] d\alpha_3 \\ &= \sum_{i=1}^{n} \frac{Y_i}{1-\mu_i^2} \left\{ (r_{i+1} - r_i) \left[\hat{S}_{11_i} + \hat{S}'_{11_i} + \mu_i \left(\hat{S}_{22_i} + \hat{S}'_{22_i} \right) \right] \right. \\ &\quad + \frac{(r_{i+1}^2 - r_i^2)}{2} \left[\hat{k}_{11_i} + \hat{k}'_{11_i} + \mu_i \left(\hat{k}_{22_i} + \hat{k}'_{22_i} \right) \right] \\ &\quad \left. + \int_{r_i}^{r_{i+1}} \mu_i S'_{33_i} d\alpha_3 \right\} \end{aligned} \tag{5.1.33}$$

Define a membrane stiffness K_i for the i-th shell layer:

$$K_i = \frac{Y_i h_i}{1-\mu_i^2}. \tag{5.1.34}$$

Note that $h_i = (r_{i+1} - r_i)$, the thickness of the i-th layer. Applying Eq. (5.1.25a) and Eq. (5.1.33) gives

$$\begin{aligned} N_{11} = \sum_{i=1}^{n} &\left[K_i \left(\hat{S}^{\tau}_{11_i} + \mu_i \hat{S}^{\tau}_{22_i} \right) \right. \\ &\left. + \frac{Y_i (r_{i+1}^2 - r_i^2)}{2\left(1-\mu_i^2\right)} \left(\hat{k}^{\tau}_{11_i} + \mu_i \hat{k}^{\tau}_{22_i} \right) + \mathbb{C}_N \right], \end{aligned} \tag{5.1.35}$$

$$\begin{aligned} N_{22} = \sum_{i=1}^{n} &\left[K_i \left(\hat{S}^{\tau}_{22_i} + \mu_i \hat{S}^{\tau}_{11_i} \right) \right. \\ &\left. + \frac{Y_i (r_{i+1}^2 - r_i^2)}{2\left(1-\mu_i^2\right)} \left(\hat{k}^{\tau}_{22_i} + \mu_i \hat{k}^{\tau}_{11_i} \right) + \mathbb{C}_N \right], \end{aligned} \tag{5.1.36}$$

$$N_{12} = \sum_{i=1}^{n} \left(G_i \hat{S}_{12_i}^{\tau} + G_i \frac{r_{i+1}^2 - r_i^2}{2} \hat{k}_{12_i}^{\tau} \right), \tag{5.1.37}$$

where

$$\mathbb{C}_N = \frac{Y_i \mu_i}{1 - \mu_i^2} \int_{r_i}^{r_{i+1}} S'_{33_i} d\alpha_3. \tag{5.1.38}$$

The resultant moments of the entire multi-layered shell can be evaluated in a similar way.

$$M_{11} = \int_{r_b}^{r_t} \alpha_3 \sigma_{11} d\alpha_3 = \sum_{i=1}^{n} \int_{r_i}^{r_{i+1}} \alpha_3 \sigma_{11_i}^{\tau} d\alpha_3. \tag{5.1.39}$$

Substituting Eq. (5.1.26) into Eq. (5.1.39) yields

$$M_{11} = \sum_{i=1}^{n} \left\{ \frac{Y_i}{1 - \mu_i^2} \int_{r_i}^{r_{i+1}} \alpha_3 \left[S_{11_i}^{\tau} + \mu_i \left(S_{22_i}^{\tau} + S'_{33_i} \right) \right] d\alpha_3 \right\}. \tag{5.1.40}$$

Integrating over the thickness and separating the membrane and bending components, one can derive

$$M_{11} = \sum_{i=1}^{n} \left[\frac{Y_i (r_{i+1}^2 - r_i^2)}{2(1 - \mu_i^2)} \left(\hat{S}_{11_i}^{\tau} + \mu_i \hat{S}_{22_i}^{\tau} \right) + \mathbb{C}_M + \frac{Y_i (r_{i+1}^3 - r_i^3)}{3(1 - \mu_i^2)} \left(\hat{k}_{11_i}^{\tau} + \mu_i \hat{k}_{22_i}^{\tau} \right) \right]. \tag{5.1.41}$$

Similarly,

$$M_{22} = \sum_{i=1}^{n} \left[\frac{Y_i (r_{i+1}^2 - r_i^2)}{2(1 - \mu_i^2)} \left(\hat{S}_{22_i}^{\tau} + \mu_i \hat{S}_{11_i}^{\tau} \right) + \mathbb{C}_M + \frac{Y_i (r_{i+1}^3 - r_i^3)}{3(1 - \mu_i^2)} \left(\hat{k}_{22_i}^{\tau} + \mu_i \hat{k}_{11_i}^{\tau} \right) \right], \tag{5.1.42}$$

$$M_{12} = \sum_{i=1}^{n} \left[G_i \frac{r_{i+1}^2 - r_i^2}{2} \hat{S}_{12_i}^{\tau} + G_i \frac{r_{i+1}^3 - r_i^3}{3} \hat{k}_{12_i}^{\tau} \right], \tag{5.1.43}$$

where

$$\mathbb{C}_M = \frac{Y_i \mu_i}{1-\mu_i^2} \int_{r_i}^{r_{i+1}} \alpha_3 S'_{33_i} d\alpha_3. \tag{5.1.44}$$

Note that $\mathbb{C}_N$ and $\mathbb{C}_M$ are relatively insignificant since 1) the control actions are primarily in-plane actions and 2) the shell is thin. G_i is the shear modulus as defined earlier. Now the resultant forces and moments have been evaluated. Transverse shear forces Q_{13} and Q_{23} will be derived later. The dynamic equations of the multi-layered shell actuator will be derived next.

5.1.4 Dynamic Equations of the Multi-layered Shell

In this section, dynamic equations for the i-th distributed shell actuator layer are developed first. For the system equations of the entire multi-layered shell, the corresponding equations are integrated over all shell layers to obtain the equations in terms of the resultant membrane forces and bending moments.

Hamilton's principle states that all energy variations of the i-th shell layer are zero, except at $t = t_1$ and $t = t_0$ where t_1 and t_0 are arbitrary, i.e.,

$$\delta \int_{t_0}^{t_1} [\mathcal{K}_i - \mathcal{U}_i] dt = 0, \tag{5.1.45}$$

where $\mathcal{K}_i$ is the kinetic energy; $\mathcal{U}_i$ is the total potential energy of the strain energy, the work done by the boundary forces and moments, and the energy associated with distributed loads. The subscript i denotes the i-th shell layer; δ denotes a variation symbol. Substituting all these energy terms into the equation and taking all variations zero, one can derive the equations of motion of the i-th shell layer:

$$\begin{aligned} &-\frac{\partial}{\partial \alpha_1}(N_{11_i} A_2) - \frac{\partial}{\partial \alpha_2}(N_{21_i} A_1) - N_{12_i} \frac{\partial A_1}{\partial \alpha_2} + N_{22_i} \frac{\partial A_2}{\partial \alpha_1} \\ &- A_1 A_2 \frac{Q_{13_i}}{R_1} + A_1 A_2 \rho_i h_i \ddot{u}_1 = A_1 A_2 F_{1_i} \end{aligned}, \tag{5.1.46}$$

$$\begin{aligned} &-\frac{\partial}{\partial \alpha_1}(N_{12_i} A_2) - \frac{\partial}{\partial \alpha_2}(N_{22_i} A_1) - N_{21_i} \frac{\partial A_2}{\partial \alpha_1} + N_{11_i} \frac{\partial A_1}{\partial \alpha_2} \\ &- A_1 A_2 \frac{Q_{23_i}}{R_2} + A_1 A_2 \rho_i h_i \ddot{u}_2 = A_1 A_2 F_{2_i} \end{aligned}, \tag{5.1.47}$$

$$\begin{aligned} &-\frac{\partial}{\partial \alpha_1}(Q_{13_i} A_2) - \frac{\partial}{\partial \alpha_2}(Q_{23_i} A_1) + A_1 A_2 \left[\frac{N_{11_i}}{R_1} + \frac{N_{22_i}}{R_2} \right] \\ &+ A_1 A_2 \rho_i h_i \ddot{u}_3 = A_1 A_2 F_{3_i} \end{aligned}, \tag{5.1.48}$$

where Q_{13_i} and Q_{23_i} are defined by

$$Q_{13_i}A_1A_2 = \frac{\partial}{\partial\alpha_1}(M_{11_i}A_2) + \frac{\partial}{\partial\alpha_2}(M_{21_i}A_1) + M_{12_i}\frac{\partial A_1}{\partial\alpha_2} - M_{22_i}\frac{\partial A_2}{\partial\alpha_1}, \quad (5.1.49)$$

$$Q_{23_i}A_1A_2 = \frac{\partial}{\partial\alpha_1}(M_{12_i}A_2) + \frac{\partial}{\partial\alpha_2}(M_{22_i}A_1) + M_{21_i}\frac{\partial A_2}{\partial\alpha_1} - M_{11_i}\frac{\partial A_1}{\partial\alpha_2}. \quad (5.1.50)$$

Note that F_{1_i}, F_{2_i} and F_{3_i} are the external mechanical forces in the principal directions of the i-th layer. To obtain the overall Love's equations for the multi-layered shell, the corresponding equations are integrated over all the layers so that the resultant equations are in terms of the resultant forces and moments for the entire shell. The method of obtaining the dynamic equations for the entire multi-layered shell is demonstrated in the first equation:

$$\sum_{i=1}^{n}\left[-\frac{\partial}{\partial\alpha_1}(N_{11_i}A_2) - \frac{\partial}{\partial\alpha_2}(N_{21_i}A_1) - N_{12_i}\frac{\partial A_1}{\partial\alpha_2} + N_{22_i}\frac{\partial A_2}{\partial\alpha_1} \right. \\ \left. -A_1A_2\frac{Q_{13_i}}{R_1} + A_1A_2\rho_i h_i \ddot{u}_1\right] = \sum_{i=1}^{n}[A_1A_2F_{1_i}] \quad (5.1.51)$$

Applying Eq. (5.1.30) and Eqs. (5.1.35)–(5.1.37) yields

$$-\frac{\partial}{\partial\alpha_1}(N_{11}A_2) - \frac{\partial}{\partial\alpha_2}(N_{21}A_1) - N_{12}\frac{\partial A_1}{\partial\alpha_2} + N_{22}\frac{\partial A_2}{\partial\alpha_1} - A_1A_2\frac{Q_{13}}{R_1}, \\ + A_1A_2\rho h\ddot{u}_1 = A_1A_2F_1 \quad (5.1.52)$$

where $Q_{13} = \sum_{i=1}^{n} Q_{13_i}$ and is obtained from

$$\sum_{i=1}^{n}\left[\frac{\partial}{\partial\alpha_1}(N_{11_i}A_2) + \frac{\partial}{\partial\alpha_2}(N_{21_i}A_1) + M_{12_i}\frac{\partial A_1}{\partial\alpha_2} - M_{22_i}\frac{\partial A_2}{\partial\alpha_1}. \right. \\ \left. - Q_{13_i}A_1A_2\right] = 0 \quad (5.1.53)$$

In this manner, the modified Love's equations of motion for the entire multi-layered shell with internal distributed actuators are

$$-\frac{\partial}{\partial\alpha_1}(N_{11}A_2) - \frac{\partial}{\partial\alpha_2}(N_{21}A_1) - N_{12}\frac{\partial A_1}{\partial\alpha_2} + N_{22}\frac{\partial A_2}{\partial\alpha_1} - A_1A_2\frac{Q_{13}}{R_1}, \\ + A_1A_2\rho h\ddot{u}_1 = A_1A_2F_1 \quad (5.1.54)$$

$$-\frac{\partial}{\partial\alpha_1}(N_{12}A_2) - \frac{\partial}{\partial\alpha_2}(N_{22}A_1) - N_{21}\frac{\partial A_2}{\partial\alpha_1} + N_{11}\frac{\partial A_1}{\partial\alpha_2} - A_1A_2\frac{Q_{23}}{R_2}, \\ + A_1A_2\rho h\ddot{u}_2 = A_1A_2F_2 \quad (5.1.55)$$

$$
\begin{aligned}
&-\frac{\partial}{\partial\alpha_1}(Q_{13}A_2)-\frac{\partial}{\partial\alpha_2}(Q_{23}A_1)+A_1A_2\left[\frac{N_{11}}{R_1}+\frac{N_{22}}{R_2}\right],\\
&+A_1A_2\rho h\ddot{u}_3=A_1A_2F_3
\end{aligned}
\tag{5.1.56}
$$

where Q_{13} and Q_{23} are defined by

$$
Q_{13}A_1A_2=\frac{\partial}{\partial\alpha_1}(M_{11}A_2)+\frac{\partial}{\partial\alpha_2}(M_{21}A_1)+M_{12}\frac{\partial A_1}{\partial\alpha_2}-M_{22}\frac{\partial A_2}{\partial\alpha_1},
\tag{5.1.57}
$$

$$
Q_{23}A_1A_2=\frac{\partial}{\partial\alpha_1}(M_{12}A_2)+\frac{\partial}{\partial\alpha_2}(M_{22}A_1)+M_{21}\frac{\partial A_2}{\partial\alpha_1}-M_{11}\frac{\partial A_1}{\partial\alpha_2}.
\tag{5.1.58}
$$

The resultant forces and moments of the multi-layered thin shell are defined in Eqs. (5.1.35)–(5.1.37) and Eqs. (5.1.41)–(5.1.43), i.e.,

$$
M_{jk}=\sum_{i=1}^{n}M_{jk_i}\quad j,k=1,2;
\tag{5.1.59a}
$$

$$
N_{jk}=\sum_{i=1}^{n}N_{jk_i}\quad j,k=1,2;
\tag{5.1.59b}
$$

$$
F_j=\sum_{i=1}^{n}F_{j_i}\quad j=1,2,3;
\tag{5.1.59c}
$$

$$
Q_{k3}=\sum_{i=1}^{n}Q_{k3_i}\quad k=1,2;
\tag{5.1.59d}
$$

$$
\rho h=\sum_{i=1}^{n}\rho_i h_i.
\tag{5.1.59e}
$$

The derived generic equations of motion for the multi-layered thin shell can then be simplified to various common geometries, such as cylinder, sphere, plate, beam, etc. Boundary conditions should be defined according to practical situations. It can be easily shown that these equations reduce to those of a single layer elastic shell - Love's elastic shell equations (Soedel, 1981).

Two examples are provided to demonstrate the usefulness of the generic multi-layered shell equations derived above. The first case is a multi-layered plate actuator (rectangular or circular) with internal actuator layers. This design configuration is similar to the multi-shell structure, however, the plate has infinite radii of curvatures, $R_1 = R_2 = \infty$. The second case is a beam with a piezoelectric actuator layer on the top. These system equations will be derived by simplifying the generic multi-layered shell equations.

5.2 A Multi-layered Plate Actuator

A plate is a special case of shell with zero curvatures, i.e., the radii of curvatures are infinite, Fig. 5.3. That is, $R_1 = \infty$ and $R_2 = \infty$. Thus,

$$\frac{1}{R_1} = \frac{1}{R_2} = 0. \tag{5.2.1}$$

The total strains consisting of the original strains and the induced strains can be simplified and further organized in membrane strains and bending strains:

$$\begin{aligned} S^{\tau}_{11_i} &= S_{11_i} + S'_{11_i} \\ &= \left(\hat{S}_{11_i} + \alpha_3 \hat{k}_{11_i}\right) + \left(\hat{S}'_{11_i} + \alpha_3 \hat{k}'_{11_i}\right) \\ &= \left(\hat{S}_{11_i} + \hat{S}'_{11_i}\right) + \alpha_3 \left(\hat{k}_{11_i} + \hat{k}'_{11_i}\right) \\ &= \left\{\frac{1}{A_1}\frac{\partial}{\partial \alpha_1}(u_1 + \Psi_{1_i}) + \frac{u_2 + \Psi_{2_i}}{A_1 A_2}\frac{\partial A_1}{\partial \alpha_2}\right\} \\ &\quad + \alpha_3 \left\{\frac{1}{A_1}\frac{\partial \beta^{\tau}_{1_i}}{\partial \alpha_1} + \frac{\beta^{\tau}_{2_i}}{A_1 A_2}\frac{\partial A_1}{\partial \alpha_2}\right\} \end{aligned} . \tag{5.2.2a}$$

Similarly,

$$\begin{aligned} S^{\tau}_{22_i} &= \left\{\frac{1}{A_2}\frac{\partial}{\partial \alpha_2}(u_2 + \Psi_{2_i}) + \frac{u_1 + \Psi_{1_i}}{A_1 A_2}\frac{\partial A_2}{\partial \alpha_1}\right\} \\ &\quad + \alpha_3 \left\{\frac{1}{A_2}\frac{\partial \beta^{\tau}_{2_i}}{\partial \alpha_2} + \frac{\beta^{\tau}_{1_i}}{A_1 A_2}\frac{\partial A_2}{\partial \alpha_1}\right\} \end{aligned} , \tag{5.2.2b}$$

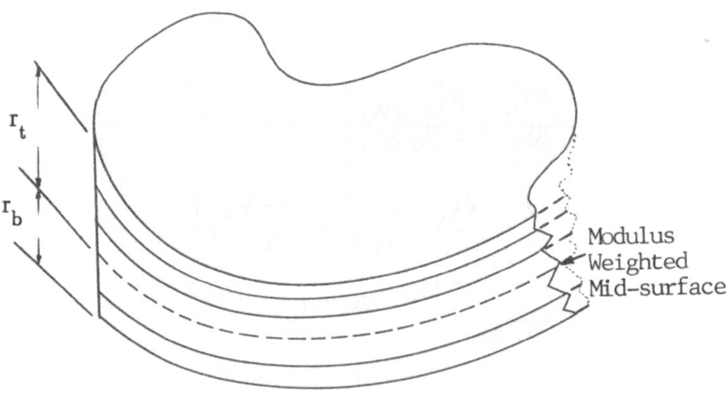

Fig. 5.3 A multi-layered plate actuator

$$S^{\tau}_{12_i} = \left\{ \frac{A_2}{A_1}\frac{\partial}{\partial \alpha_1}\left(\frac{u_2 + \Psi_{2_i}}{A_2}\right) + \frac{A_1}{A_2}\frac{\partial}{\partial \alpha_2}\left(\frac{u_1 + \Psi_{1_i}}{A_1}\right)\right\} + \alpha_3 \left\{ \frac{A_2}{A_1}\frac{\partial}{\partial \alpha_1}\left(\frac{\beta^{\tau}_{2_i}}{A_2}\right) + \frac{A_1}{A_2}\frac{\partial}{\partial \alpha_2}\left(\frac{\beta^{\tau}_{1_i}}{A_1}\right)\right\}, \tag{5.2.2c}$$

$$S^{\tau}_{33_i} = \left\{ \frac{\partial \Psi_{3_i}}{\partial \alpha_3} \right\}. \tag{5.2.2d}$$

And the total rotation angles, Eqs. (5.1.11)–(5.1.14a), are redefined as

$$\beta^{\tau}_{1_i} = \beta_{1_i} + \beta'_{1_i} = -\frac{1}{A_1}\frac{\partial u_3}{\partial \alpha_1} - \frac{1}{A_1}\frac{\partial \Psi_{3_i}}{\partial \alpha_1}, \tag{5.2.3a}$$

$$\beta^{\tau}_{2_i} = \beta_{2_i} + \beta'_{2_i} = -\frac{1}{A_2}\frac{\partial u_3}{\partial \alpha_2} - \frac{1}{A_2}\frac{\partial \Psi_{3_i}}{\partial \alpha_2}. \tag{5.2.3b}$$

The **stress-strain relations** are written as a sum of two terms: the original and the induced. The resulting dynamic equations for the multi-layered plate can be written as

$$-\frac{\partial(N_{11}A_2)}{\partial \alpha_1} - \frac{\partial(N_{21}A_1)}{\partial \alpha_2} - N_{12}\frac{\partial A_1}{\partial \alpha_2} + N_{22}\frac{\partial A_2}{\partial \alpha_1} + A_1A_2\rho h\ddot{u}_1 = A_1A_2F_1, \tag{5.2.4}$$

$$-\frac{\partial(N_{12}A_2)}{\partial \alpha_1} - \frac{\partial(N_{22}A_1)}{\partial \alpha_2} - N_{21}\frac{\partial A_2}{\partial \alpha_1} + N_{11}\frac{\partial A_1}{\partial \alpha_2} + A_1A_2\rho h\ddot{u}_2 = A_1A_2F_2 \tag{5.2.5}$$

$$-\frac{\partial(Q_{13}A_2)}{\partial \alpha_1} - \frac{\partial(Q_{23}A_1)}{\partial \alpha_2} + A_1A_2\rho h\ddot{u}_3 = A_1A_2F_3, \tag{5.2.6}$$

where Q_{13} and Q_{23} are defined in Eqs. (5.1.57)–(5.1.58). The resultant forces N_{ij} 's and moments M_{ij} 's are defined in Eqs. (5.1.35)–(5.1.37) and Eqs. (5.1.41)–(5.1.43), respectively. If only the transverse vibration is considered, substituting Eqs. (5.1.57)–(5.1.58) into Eq. (5.2.6) yields

$$\begin{aligned} &-\frac{\partial}{\partial \alpha_1}\left[\frac{1}{A_1}\left(\frac{\partial}{\partial \alpha_1}(M_{11}A_2) + \frac{\partial}{\partial \alpha_2}(M_{21}A_1) + M_{12}\frac{\partial A_1}{\partial \alpha_2} - M_{22}\frac{\partial A_2}{\partial \alpha_1}\right)\right] \\ &-\frac{\partial}{\partial \alpha_2}\left[\frac{1}{A_2}\left(\frac{\partial}{\partial \alpha_1}(M_{12}A_2) + \frac{\partial}{\partial \alpha_2}(M_{22}A_1) + M_{21}\frac{\partial A_2}{\partial \alpha_1} - M_{11}\frac{\partial A_1}{\partial \alpha_2}\right)\right] \\ &+ A_1A_2\rho h\ddot{u}_3 = A_1A_2F_3 \end{aligned} \tag{5.2.7}$$

Substituting resultant forces and moments into the above equation yields an expression in terms of displacements $u_i'S$. The final equation is for a generic plate, which can be further simplified to other plates using two Lamé constants.

5.2.1 A Rectangular Plate

$$A_1 = 1, \quad A_2 = 1, \quad d\alpha_1 = dx, \quad \text{and} \quad d\alpha_2 = dy. \tag{5.2.8}$$

5.2.2 A Circular Plate

$$A_1 = 1, \quad A_2 = \mathbb{R}, \quad d\alpha_1 = d\mathbb{R}, \quad \text{and} \quad d\alpha_2 = d\theta, \tag{5.2.9}$$

where $\mathbb{R}$ denotes the distance measured from the origin to the location where the infinitesimal distance "ds" is defined. Note that $\mathbb{R} = R$ (radius of the circular plate) on the outer boundary.

5.2.3 An Elliptical Plate

$$\begin{aligned} &A_1 = A_2 = \left(a^2 - b^2\right)\left(\sin^2 v + \sinh^2 u\right), \\ &d\alpha_1 = du, \quad \text{and} \quad d\alpha_2 = dv \end{aligned}, \tag{5.2.10}$$

where a and b are the major and the minor radii of the ellipse, respectively.

5.3 A Piezoelectric/Elastic Beam

Consider a two-layered composite beam made of a distributed PVDF actuator layer perfectly bounded on a fiberglass beam, Fig. 5.4. These two layers have the same dimension with α_1 in the longitudinal direction, α_2 the width direction, and α_3 the transverse direction. The subscripts "1" and "2" refer to the fiberglass and the PVDF, respectively. The radii of curvatures are: $R_x = \infty$ and $R_y = \infty$, thus, $1/R_x = 0$ and $1/R_y = 0$.

The distributed actuator layer is made of an electromechanical sensitive material —a piezoelectric PVDF material, and it is subjected to an electric potential $\phi(\alpha_1, t)$ applied across its thickness. Due to the converse piezoelectric effect, an electromechanical induced strain will be generated in the layer. A longitudinal normal strain S_{11_2} is generated if the actuator layer is a mono-axially oriented piezoelectric

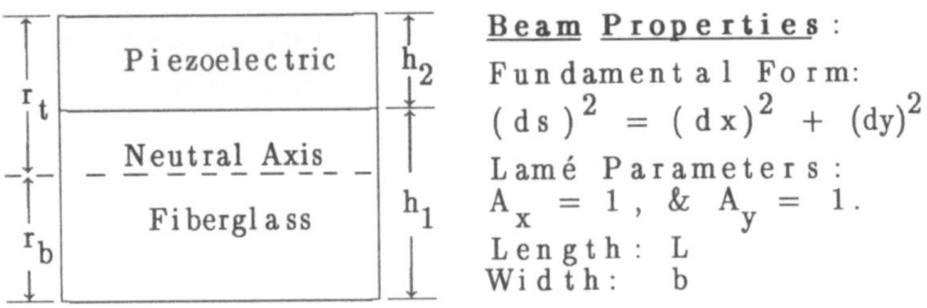

Fig. 5.4 A fiberglass beam with a distributed piezoelectric actuator

material. Variations in the α_2 direction, the width direction, are not considered, i.e. $\frac{\partial(\cdot)}{\partial\alpha_2} = 0$. The dynamic equations of this composite beam can be derived by using the multi-layered shell equations developed earlier. The neutral axis is determined by a standard moment analysis.

$$r_b = \frac{Y_1 h_1^2 + Y_2 h_2^2 + 2Y_2 h_1 h_2}{2(Y_1 h_1 + Y_2 h_2)}. \tag{5.3.1}$$

r_b is the distance measured from the neutral surface to the bottom surface. In the later analysis, each of the beam layer will be analyzed respectively, and then these two set of equations will be summed together to a layered system equation by following the procedures presented earlier. Note that only transverse vibration is considered in this case.

5.3.1 *Fiberglass Layer (i = 1)*

This layer is not subjected to any localized forcing functions. Note that the terms associated with the plexiglas layer is denoted by a subscript or sub-subscript 1. Hence, the original and induced components are defined as

$$\begin{aligned} &\Psi_{1_1} = \Psi_{2_1} = \Psi_{3_1} = 0; \beta'_{1_1} = \beta^{\tau}_{2_1} = 0; \\ &\hat{S}'_{11_1} = S^{\tau}_{22_1} = S^{\tau}_{12_1} = S'_{33_1} = 0; \\ &\beta_{1_1} = -\frac{\partial u_3}{\partial \alpha_1}; \quad \hat{S}_{11_1} = \frac{\partial u_1}{\partial \alpha_1}; \quad \hat{k}_{11_1} = -\frac{\partial^2 u_3}{\partial \alpha_1^2} \end{aligned}. \tag{5.3.2}$$

From Eqs. (5.1.35)–(5.1.37), normal forces are defined as

$$N_{11_1} = \int_{-r_b}^{h_1 - r_b} \frac{Y_1}{1-\mu_1^2}\left(\hat{S}_{11_1} + \alpha_3 \hat{k}_{11_1}\right) d\alpha_3, \tag{5.3.3a}$$

$$N_{22_1} = N_{12_1} = 0. \tag{5.3.3b}$$

From Eqs. (5.1.41)–(5.1.43), moments are defined as

$$M_{11_1} = \int_{-r_b}^{h_1 - r_b} \frac{Y_1}{1-\mu_1^2}\left(\hat{S}_{11_1} + \alpha_3 \hat{k}_{11_1}\right) \alpha_3 d\alpha_3, \tag{5.3.4a}$$

$$M_{22_1} = M_{12_1} = 0. \tag{5.3.4b}$$

The transverse dynamic equation of the thin shell actuator is then reduced to

$$-\frac{\partial^2 M_{11_1}}{\partial \alpha_1^2} + \rho_1 h_1 \ddot{u}_3 = F_{3_1}. \tag{5.3.5}$$

5.3.2 *Piezoelectric Control Layer (i = 2)*

This layer is subjected to a localized forcing function—the control feedback voltage $\phi(\alpha_1, t)$—which induces strains in the piezoelectric layer only. Note that the terms associated with the PVDF layer is denoted by a subscript or sub-subscript 2.

$$\begin{aligned} &\Psi_{2_2} = \Psi_{3_2} = 0; \beta_{2_2} = 0; \\ &\beta_{1_2} = -\frac{\partial u_3}{\partial \alpha_1}; \hat{S}_{11_2} = \frac{\partial u_1}{\partial \alpha_1}; \hat{S}'_{11_2} = \frac{\partial \Psi_{1_2}}{\partial \alpha_1}. \end{aligned} \tag{5.3.6}$$

It is recognized that $\hat{S}'_{11_2}$ is the induced longitudinal strain induced in the piezoelectric layer due to the applied voltage. This is given by

$$\hat{S}'_{11_2} = d_{31}\phi(\alpha_1, t), \tag{5.3.7a}$$

$$\hat{k}_{11_2} = -\frac{\partial^2 u_3}{\partial \alpha_1^2}, \tag{5.3.7b}$$

$$S^{\tau}_{22_2} = S^{\tau}_{12_2} = 0, \tag{5.3.7c}$$

where d_{31} is the piezoelectric constant. Normal forces and moments are defined as

$$N_{11_2} = \int_{h_1 - r_b}^{h_1 + h_2 - r_b} \frac{Y_2}{1 - \mu_2^2} \left(\hat{S}_{11_2} + \hat{S}'_{11_2} + \alpha_3 \hat{k}_{11_2} \right) d\alpha_3, \tag{5.3.8a}$$

$$N_{22_2} = N_{12_2} = 0, \tag{5.3.8b}$$

$$M_{11_2} = \int_{h_1 - r_b}^{h_1 + h_2 - r_b} \frac{Y_2}{1 - \mu_2^2} \left(\hat{S}_{11_2} + \hat{S}'_{11_2} + \alpha_3 \hat{k}_{11_2} \right) \alpha_3 d\alpha_3, \tag{5.3.8c}$$

$$M_{22_2} = M_{12_2} = 0. \tag{5.3.8d}$$

The transverse dynamic equation of the PVDF layer is reduced to

$$-\frac{\partial^2 M_{11_2}}{\partial \alpha_1^2} + \rho_2 h_2 \ddot{u}_3 = F_{3_2}. \tag{5.3.9}$$

5.3.3 Governing Equation of Motion for the Entire Composite Beam

Summing all forces and moments of each layer yields

$$-\frac{\partial^2}{\partial\alpha_1^2}(M_{11_1}+M_{11_2})+(\rho_1 h_1+\rho_2 h_2)\ddot{u}_3 = F_{3_1}+F_{3_2}. \tag{5.3.10}$$

Using the simplified notation described earlier and considering the transverse motion only, one can derive a transverse vibration equation for the entire two-layered composite beam.

$$-\frac{\partial^2 M_{11}}{\partial\alpha_1^2}+\rho h\ddot{u}_3 = F_3. \tag{5.3.11}$$

Defining the moments of inertia about the "modulus weighted" neutral axis by using the parallel axis theorem, one can derive:

$$I_1 = \frac{bh_1^3}{12}+bh_1\left(\frac{h_1}{2}-r_b\right)^2, \tag{5.3.12a}$$

$$I_2 = \frac{bh_2^3}{12}+bh_2\left(h_1+\frac{h_2}{2}-r_b\right)^2. \tag{5.3.12b}$$

Assume $(1-\mu_1^2)\approx(1-\mu_2^2)\approx 1$. Substituting Eq. (5.3.7b) into Eq. (5.3.8c) and carrying out the integration, one can obtain

$$bM_{11} = -YI\frac{\partial^2 u_3}{\partial\alpha_1^2}+\tilde{\lambda}\phi(\alpha_1,t), \tag{5.3.13}$$

where

$$YI = Y_1 I_1 + Y_2 I_2 \tag{5.3.14}$$

$$\hat{\lambda} = \frac{d_{31}Y_1 Y_2 h_1(h_1+h_2)}{2(Y_1 I_1+Y_2 I_2)} \tag{5.3.15}$$

In this manner, the equation of motion for the transverse vibration is derived as

$$\frac{\partial^2}{\partial\alpha_1^2}\left(YI\frac{\partial^2 u_3}{\partial\alpha_1^2}-\hat{\lambda}\phi(\alpha_1,t)\right)+\rho\tilde{A}\ddot{u}_3 = \tilde{F}_3 \tag{5.3.16}$$

where $\tilde{A} = bh$ and $\tilde{F}_3 = bF_3$. The second term in the parenthesis is interpreted as a control moment induced by the feedback voltage, relative to the weighted mid-axis of the two layered beam. It is observed that the control moment is proportional to the feedback voltage. This equation is identical to that derived in a reference (Baily & Hubbard, 1987) via a conventional approach.

5.4 Summary

In this chapter, a theoretical development of a multi-layered thin shell distributed actuator was presented. The distributed actuator layers can be made of electromechanical sensitive materials which respond to externally supplied voltages and generate local control forces for active distributed vibration controls. Based on the assumptions, dynamic equations for the generic multi-layered thin shell actuator (with distributed control layers) were developed using Kirchhoff-Love's theory and Hamilton's principle. The system equations are generic and can be simplified to apply to many other common geometries and structures, such as plates (e.g., circular or rectangular), other conventional shells (e.g., cylindrical shell, spheres), beams, etc. The common geometries can be defined by the fundamental form, Lamé parameters, radii of curvatures, etc. It should be noted that the deformations resulting from transverse shears and rotatory inertias were neglected in the derivations.

Two demonstration examples, 1) a multi-layered thin plate actuator and 2) a PVDF/fiberglass composite beam, were presented. The system equations of the above examples were derived using the developed multi-layered thin shell equations via a direct reduction. The first case was a generic plate actuator which can be either rectangular, circular, or elliptical depending on the geometrical parameters, e.g., Lamé parameters, fundamental form, etc. The derived system equation of the second case was identical to the equation presented by other researchers.

Furthermore, following the simplification procedures presented in Chaps. 3 and 4, one can simplify the multi-layer double-curvature actuator governing equations to account for many other shell and non-shell structures and systems (Tzou, Zhu and Hagiwara, 2002; Howard, et al, 2001; Tzou and Zhou, 1997). Detailed multi-layer micro-thermo-electromechanics and sensing/control effectiveness of various 1D, 2D plate DPSs, and advanced shell DPSs, such as ring shells (Hu, Li and Tzou, 2013a; Smithmaitrie and Tzou, 2005; Tzou, Zhong and Natori, 1993), cylindrical shells (Li, et al., 2012, 2015; DeHaven, et al., 2007; Tzou, Bao and Venkayya, 1996a, b; Tzou and Zhong, 1996), spherical shells (Zhou and Tzou, 2000; Smithmaitrie and Tzou, 2004; Tzou, Smithmaitrie and Ding., 2002), conical shells (Tzou, Chai and Wang, 2003, Wang, et al, 2006; Chai, Tzou and Higuchi, 2006; Chai, DeHaven and Tzou, 2005), paraboloidal shells (Hu, Li and Tzou, 2013b; Tzou, Ding, 2004; Tzou, Ding and Hagiwara, 2002; Yue, Deng and Tzou, 2008), toroidal shells (Tzou, Chai and Wang, 2004; Tzou and Wang, 2002, 2003), etc., can be evaluated.

References

Baily, T. and Hubbard, J.E., 1987, "Distributed Piezoelectric Polymer Active Vibration Control of a Cantilever Beam," *J. of Guidance, Control, and Dynamics,* Vol.8 No.5, pp.605-611.

Chai, W.K., DeHaven, J.G. and Tzou, H.S., 2005, "Distributed Micro-control Characteristics of Shallow Conical Shell Sections," *Journal of Vibration and Control*, Vol.11, No.11, pp.1397–1411.

Chai, W.K., Tzou, H.S. and Higuchi, K., 2006, "Micro-actuation Characteristics of Conical Shell Sections," *Journal of Sound & Vibration*. Vol.293, No.1–2, pp.286–298.

Crawley, E.F. and de Luis, J., 1987, "Use of Piezoelectric Actuators as Elements of Intelligent Structures," *AIAA Journal*, Vol.25, No.10, pp.1373-1385.

DeHaven, J.G., Han, Y. and Tzou, 2007, "Transition of Membrane/bending Neural Signals on Transforming Adaptive Shells," *Journal of Vibration and Control*, Vol.13, No.7, pp.1007–1029.

Hanagud, S. and Obal, M.W., 1988, "Identification of Dynamic Coupling Coefficients in a Structure with Piezoelectric Sensors and Actuators," *Proc. of AIAA/ASME/AHS 29th Structures, Structural Dynamics, and Materials Conference*, (Paper No.88-2418), Part-3, pp.1611-1620.

Howard, R.V., Chai, W.K. and Tzou, H.S., 2001, "Modal Voltages of Linear and Non-Linear Structures Using Distributed Artificial Neurons," *Mechanical Systems and Signal Processing* 15, 629–640.

Hu, S.D., Li, H. and Tzou, H.S., 2013a, "Flexoelectric responses of circular rings," *Journal of Vibration and Acoustics* -Transactions of the ASME, 2013, 135(2): 021003.

Hu, S.D., Li, H. and Tzou, H.S., 2013b, "Distributed Neural Signals on Parabolic Cylindrical Shells," *Journal of Sound and Vibration* 332, 2984–3001.

Li, H.Y., Li, H. and Tzou, H.S., 2015, "Frequency Control of Beams and Cylinrical Shells with Light-Activated Shape Memory Polymers," ASME *Journal of Vibration and Acoustics*, 37, p.011006.

Li, H., Hu, S. D., Tzou, H. S., and Chen, Z. B., 2012, "Optimal Vibration Control of Conical Shells with Collocated Helical Sensor/Actuator Pairs," *Journal of Theoretical and Applied Mechanics*, 50(3), pp.769–784.

Love, A.E.N., 1888, "On the Small Free Vibrations and Deformations of Thin Elastic Shells," *Phil. Trans. Royal Society (London)*, Vol.179A, pp.491-546.

Smithmaitrie, P. and Tzou, H.S., 2004, Micro-control actions of actuator patches laminated on hemispherical shells, *Journal of Sound and Vibration*, 277, 691–710.

Smithmaitrie, P. and Tzou, H.S., 2005, "Electro-dynamics, Micro-actuation and Design of Ultrasonic Curvilinear Arc Stators," *Journal of Sound & Vibration*, Vol.284, pp.635–650.

Soedel, W., 1981, *Vibrations of Shells and Plates*, Marcel Dekker Inc., New York.

Tzou, H.S., 1987, "Active Vibration Controls of Flexible Structures Via Converse Piezoelectricity," *Developments in Mechanics*, Vol.14(c), pp.1201-1206. The 20th Midwest Mechanics Conference, West Lafayette, IN, August 1987.

Tzou, H.S., 1988, "Integrated Sensing and Adaptive Vibration Suppression of Distributed Systems," *Recent Advances in Control of Nonlinear and Distributed Parameter Systems*, ASME-DSC-Vol.(10), pp.51-58, 1988 ASME WAM, Chicago, Illinois, Nov.27-Dec.2, 1988.

Tzou, H.S. and Anderson, G.L. (Editors), *Intelligent Structural Systems*, ISBN No.0-7923-1920-6, 488 pages, Book, Kluwer Academic Publishers, August 1992.

Tzou, H.S., Bao, Y., Venkayya, V.B., 1996a, "Study of Segmented Transducers Laminated on Cylindrical Shells, Part 1: Sensor Patches," *Journal of Sound & Vibration*, Vol.197, No.2, pp.207–224.

Tzou, H.S., Bao, Y., Venkayya, V.B., 1996b, "Study of Segmented Transducers Laminated on Cylindrical Shells, Part 2: Actuator Patches," *Journal of Sound & Vibration*, Vol.197, No.2, pp.225–249, October 1996.

Tzou, H. S., Ding, J. H., and Hagiwara, I., 2002, "Micro-Control Actions of Segmented Actuator Patches Laminated on Deep Paraboloidal Shells," *JSME International Journal Series C-Mechanical Systems Machine Elements and Manufacturing*, 45(1), pp.8–15.

Tzou, H.S., Chai, W.K. and Wang, D.W., 2003, "Modal Voltages and Micro-Signal Analysis of Conical Shells of Revolution," *Journal of Sound and Vibration* 260, 589–609.

Tzou, H. S., Chai, W. K., and Wang, D. W., 2004, "Micro-Control Actions and Location Sensitivity of Actuator Patches Laminated on Toroidal Shells," *Journal of Vibration and Acoustics-Transactions of the ASME*, 126(2), pp.284–297.

Tzou, H. S. and Ding, J. H., 2004, "Optimal Control of Precision Paraboloidal Shell Structronic Systems," *Journal of Sound and Vibration*, 276(1–2), pp.273–291.

Tzou, H.S. Lee, H.-J. and Arnold, S.M., "Smart Materials, Precision Sensors/Actuators, Smart Structures and Structronic Systems," *Mechanics of Advanced Materials and Structures,* Vol.11, pp.367–393, 2004.

Tzou, H. S., Smithmaitrie, P. and Ding, J. H., 2002, "Sensor Electromechanics and Distributed Signal Analysis of Piezo(Electric)-Elastic Spherical Shells," *Mechanical Systems and Signal Processing* (*Journal of*), Vol.16(2–3), pp.185–199.

Tzou, H.S. and Tseng, C.I., 1988a, "Sensing and Adaptive Vibration Control of Flexible Distributed Mechanical Systems," *Machine Dynamics and Engr. Applications,* Xian Jiaotong University Press, China, Vol.1, pp.G1-G6, August 1988.

Tzou, H.S. and Tseng, C.I., 1988b, "Active Vibration Controls of Distributed Parameter Systems by Finite Element Method," *ASME Computers in Engineering 1988*, Vol.3, pp.599-604.

Tzou, H.S. and Tseng, C.I., 1988c, "Development of a Thin Piezoelectric Finite Element Applied to Distributed Sensing and Vibration Controls," ASME Paper No. 88-WA/CIE-2, 1988 WAM, Chicago, Illinois, Nov.27-Dec.2, 1988.

Tzou, H.S. and Wang, D.W., 2002, "Micro-Sensing Characteristics and Modal Voltages Of Linear/Non-Linear Toroidal Shells," *Journal of Sound and Vibration*, 2002, 254(2): 203–218.

Tzou, H.S. and Wang, D.W., 2003, "Vibration Control of Toroidal Shells with Parallel Aand Diagonal Piezoelectric Actuators," *Journal of Pressure Vessel Technology* -Transactions of the ASME, 2003, 125(2): 171–176.

Tzou, H.S. and Zhong, J., 1996, "Spatially Filtered Vibration Control of Cylindrical Shells," *Shock and Vibration Journal*, Vol.3, No.4, pp.269–278.

Tzou, H.S., and Zhou, Y., 1997, "Nonlinear Piezothermoelasticity and Multi field Actuations, Part 2: Control of Nonlinear Buckling and Dynamics," ASME Transactions, *Journal of Vibration & Acoustics*, Vol.119, pp.382–389.

Tzou, H.S., Zhong, J.P., and Natori, M.C., 1993, "Sensor Mechanics of Distributed Shell Convolving Sensors Applied to Flexible Rings," ASME Transactions, *Journal of Vibration & Acoustics*, Vol.115, No.1, pp.40–46.

Tzou, H.S., Zhu, Y.T. and Hagiwara, I, 2002, "Distributed Precision Control of Structronic Shells and Common Shapes– A New Approach," *Recent Research Developments in Sound & Vibration*, ISBN: 81-7895-031-6, Vol.1, Part-2, pp.613–646, TransWorld Research, 2002.

Wada, B.K, Fanson, J.L. & Crawley, E.F., 1989, "Adaptive Structures," *Adaptive Struatures*, ASME AD-Vol.15, pp.1-8, 1989 ASME Winter Annual Meeting, San Francisco, CA, December. (5/MultiShlAct.BkPzSm2).

Wang, D.W., Tzou, H.S., Arnold, S.M. and Lee, H.J., 2006, "Control of Static Shape, Dynamic Oscillation and Thermally Induced Vibration of Nozzles," ASME Transactions, *Journal of Pressure Vessel Technology*, Vol.128, pp.357–363.

Yue, H. H., Deng, Z. Q., and Tzou, H.S., 2008, "Optimal Actuator Locations and Precision Micro-Control Actions on Free Paraboloidal Membrane Shells," *Communications in Nonlinear Science and Numerical Simulation*, 13(10), pp.2298–2307.

Zhou, Y.H., and Tzou, H.S., 2000, "Active Control of Nonlinear Piezoelectric Circular Shallow Spherical Shells," *International Journal of Solids and Structures*, 37(12), pp.1663–1677.

Chapter 6
Boundary Control of Beams

Distributed vibration control of a laminated elastic beam is studied in this chapter. It is assumed that two thin layers of a mono-axially oriented piezoelectric material, i.e., d_{31} only and $d_{32} = 0$, are respectively bonded on the top and bottom surfaces of the elastic beam. One layer serves as a distributed sensor and the other as a distributed actuator. The effective axis of the piezoelectric layers is aligned with the x-axis to ensure the maximum piezoelectric effects in sensor and actuator applications. It is intended to use the distributed sensor signal as a feedback reference in a closed-loop feedback control system. Two control algorithms, namely a ***displacement feedback*** and a ***velocity feedback***, are implemented and their control effectiveness evaluated. In the displacement feedback, the distributed sensor signal is amplified and fed back to the distributed piezoelectric actuator. (Note that the sensor signal is proportional to strains which can be ultimately expressed in terms of displacements as presented in Chap. 4. Thus, the conventional "displacement" feedback is used. In general, the dominating vibration component contributes to a higher strain level and consequently to a higher percentage of the total output signal.) In the velocity feedback, this sensor output is differentiated, amplified, and then fed back into the distributed actuator. (Note that the signal used in the velocity feedback is actually a strain/unit-time or strain rate. Since the strain is ultimately expressed in terms of displacements, the time derivative of the displacement is the velocity. Thus, a conventional term—"velocity feedback"—is used.) Control effectivenesses, natural frequency and damping, of these two control algorithms are evaluated. Note that the control action to the laminated cantilever beam is introduced via ***boundary controls***. The ***Laplace transform method*** is used to obtain the solutions of the laminated beam system equation.

H. (HS) Tzou, *Piezoelectric Shells*, Solid Mechanics and Its Applications 247,
https://doi.org/10.1007/978-94-024-1258-1_6

6.1 System Definition: Beam with Sensor and Actuator

A cantilever beam coupled with a distributed piezoelectric sensor and a distributed piezoelectric actuator is illustrated in Fig. 6.1.

It is assumed that the piezoelectric layers are perfectly bonded on the top and bottom surfaces of the elastic beam and the layers are much thinner than the elastic beam. Physical properties of the bonding material and the piezoelectric layers are not considered in the analyses. (Note that physical properties of the distributed PVDF actuator were considered in Sect. 5.3, Chap. 5.) As discussed in Chap. 4, the beam system equation, in the transverse direction α_3, can be derived as

$$\rho h \ddot{u}_3 + D\frac{\partial^4 u_3}{\partial x^4} - \frac{\partial^2 (M_{11}^a)}{\partial x^2} = F_3, \tag{6.1.1a}$$

where ρ is the mass density; h is the beam thickness; D is the bending stiffness $D = (Yh^3)/[12(1 - \mu^2)$; Y is Young's modulus; μ is Poisson's ratio; M_{11}^a is the control moment induced by the distributed actuator; F_3 is the mechanical excitation. Note that $\alpha_1 = x$, and Poisson's ratio is usually neglected in beam applications.

For a beam with a rectangular cross section, this equation is reduced to

$$\rho\tilde{A}\ddot{u}_3 + YI\frac{\partial^4 u_3}{\partial x^4} - \frac{b\partial^2 (M_{11}^a)}{\partial x^2} = bF_3, \tag{6.1.1b}$$

where $\tilde{A} = bh$; $I = bh^3/12$ and b is the beam width and h is the beam thickness. Note that Poisson's ratio is neglected.

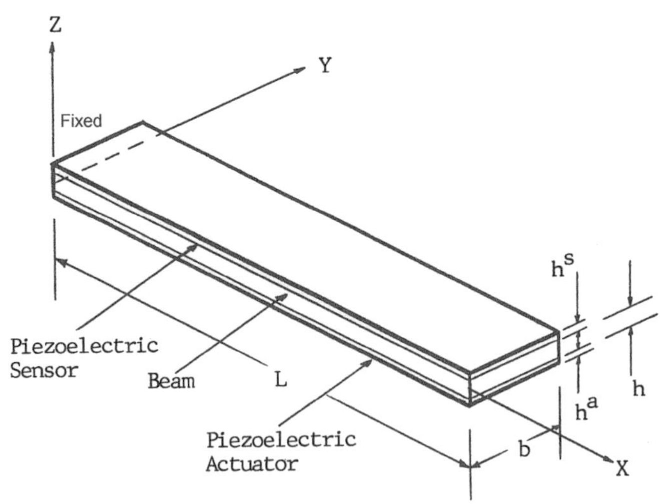

Fig. 6.1 A cantilever beam with distributed sensor and actuator layers

The output signal ϕ^S of a distributed sensor, from $x = x_1$ to $x = x_2$ and $x_2 > x_1$, contributed by the bending effect can be estimated by the normal strain in the x direction:

$$\begin{aligned}
\phi^S &= \frac{h^S}{x_2 - x_1} \int_{\alpha_1} \left[h_{31} S_{11}^S\right] d\alpha_1 \\
&= -\frac{h^S}{x_2 - x_1} \int_{x_1}^{x_2} \left[h_{31} r_1^S \left(\frac{\partial^2 u_3}{\partial x^2}\right)\right] dx\,, \qquad (6.1.2a) \\
&= -\frac{h^S}{x_2 - x_1} h_{31} r_1^S \left(\frac{\partial u_3}{\partial x}\right)\Bigg|_{x_1}^{x_2}
\end{aligned}$$

where h^s is the sensor thickness; $(x_2 - x_1)$ is the sensor length; h_{31} is the piezoelectric constant; S_{11}^s is the in-plane normal strain in the x direction; r_1^s is the distance measured from the neutral axis to the mid-plane of the sensor layer. The sensor equation shows that the output signal is proportional to the slopes at both ends of the sensor layer. Note that the sensor signal is zero if the slopes at $x = x_1$ is equal to that at $x = x_2$, e.g., anti-symmetrical modes of a simply supported beam laminated with a symmetrically distributed sensor layer. (Other restrictions of this setup will be discussed in later chapters.) For a fully distributed sensor in Fig. 6.1, i.e., $x_1 = 0$ and $x_2 = L$, the sensor signal becomes

$$\phi^S = -\frac{h^S}{L} h_{31} r_1^S \left(\frac{\partial u_3}{\partial x}\right)\Bigg|_0^L. \qquad (6.1.2b)$$

Next, two closed-loop feedback algorithms: 1) a displacement feedback and 2) a velocity feedback will be presented.

6.2 Displacement Feedback Control

In the displacement feedback control, the sensor signal is used directly in a closed-loop feedback system. Note that the sensor output is a function of strains and the strains are contributed by deflections. Amplifying the sensor signal by a feedback gain $\mathcal{G}$ gives the feedback voltage ϕ^a:

$$\phi^a = \mathcal{G}\phi^S = -\frac{\mathcal{G} h^S h_{31} r_1^S}{L} \left(\frac{\partial u_3}{\partial x}\right)\Bigg|_0^L. \qquad (6.2.1)$$

Thus, the distributed control moment M_{11}^a, Sect. 4.3 in Chap. 4, becomes

$$\begin{aligned} M_{11}^{a} &= r_1^a d_{31} Y_p \phi^a \\ &= -\frac{\mathcal{G} h^S r_1^a d_{31} Y_p r_1^S h_{31}}{L} \left(\frac{\partial u_3}{\partial x}\right)\Big|_0^L, \\ &= \tilde{\xi} \left(\frac{\partial u_3}{\partial x}\right)\Big|_0^L \end{aligned} \tag{6.2.2}$$

where r_1^a is the effective moment arm; d_{31} is the piezoelectric constant; Y_p is Young's modulus of the piezoelectric actuator layer; $\tilde{\xi}$ is defined as

$$\tilde{\xi} = \frac{\mathcal{G} h^S r_1^a d_{31} Y_p r_1^S h_{31}}{L}. \tag{6.2.3}$$

Substituting Eq. (6.2.2) into Eq. (6.1.1) and setting $F_3 = 0$, one can derive the homogeneous equation of motion:

$$\rho h \ddot{u}_3 + D \frac{\partial^4 u_3}{\partial x^4} - \frac{\partial^2}{\partial x^2} \left[\tilde{\xi} \left(\frac{\partial u_3}{\partial x} \right) \Big|_0^L \right] = 0. \tag{6.2.4}$$

Note that the distributed moment M_{11}^a, Eq. (6.2.2), is not a function of spatial coordinates. (It is assumed that the resistance of the surface electrode is neglected so that the voltage is uniformly distributed.) Neglecting Poisson's ratio in the bending stiffness D and multiplying the equation by a beam width b, one can derive

$$\rho \tilde{A} \ddot{u}_3 + YI \frac{\partial^4 u_3}{\partial x^4} = 0, \tag{6.2.5}$$

where $\tilde{A}(=bh)$ is the cross-section area and $I = bh^3/12$ is the area moment of inertia. Boundary conditions of a cantilever beam are zero displacement and zero slope at x = 0, and zero resultant moment and zero shear force at x = L. (Note that the resultant moment at the free end is composed of a mechanical component and an electric component.) Thus, boundary conditions of the laminated beam can be written accordingly.

1) At clamped end (x = 0):

a) Displacement:

$$u_3(0) = 0, \tag{6.2.6}$$

b) Slope:

$$\frac{du_3(0)}{dx} = 0; \tag{6.2.7}$$

2) At free end (x = L):

a) Moment:

$$M_{11}(L) = LbM^{a}_{11}(L)$$
$$\Rightarrow -YI\frac{d^2u_3(L)}{dx^2} = Lb\tilde{\xi}\frac{du_3(L)}{dx}, \tag{6.2.8}$$

b) Shear force: Q(L) = 0

$$\Rightarrow \frac{d^3u_3(L)}{dx^3} = 0. \tag{6.2.9}$$

Note that the resultant moment includes both mechanical and electric effects. This boundary condition contributed by feedback voltages will be used to control the cantilever beam—***boundary controls***. Thus, for a laminated beam with fully distributed sensor and actuator layers, the control effect is introduced via the controlled electric boundary moments at the free end. In later derivations, control effectiveness due to ***boundary moment control*** will be analyzed and results presented.

6.2.1 *Free Vibration Analysis*

For the free vibration analysis, it is assumed that every point in the elastic beam system oscillates harmonically at a natural frequency, i.e.,

$$u_3(x,t) = U_3(x)e^{j\omega t}. \tag{6.2.10}$$

where U_3 is the oscillation amplitude and ω is the natural frequency. Note that $j = \sqrt{-1}$. Substituting this equation into the beam system equation gives

$$YI\frac{d^4U_3}{dx^4} - \rho\tilde{A}\omega^2U_3 = 0$$
$$\Rightarrow \frac{d^4U_3}{dx^4} - \lambda^4_{di}U_3 = 0. \tag{6.2.11}$$

where the i-th eigenvalue λ_{di} is defined as

$$\lambda^4_{di} = \frac{\rho\tilde{A}\omega^2}{YI}. \tag{6.2.12}$$

Note that the subscript d is for the displacement feedback and i denotes the i-th beam mode. ω can be expressed as ω_i—the i-th natural frequency corresponding to the eigenvalue λ_{di}. Following the assumption of harmonic motion at natural frequencies, one can rewrite the boundary conditions in the displacement coordinate U_3:

1) At clamped end (x = 0):

a) Displacement:

$$U_3(0) = 0. \tag{6.2.13}$$

b) Slope:

$$\frac{dU_3(0)}{dx} = 0. \tag{6.2.14}$$

2) At free end (x = L):

a) Moment:

$$\begin{aligned} M_{11}(L) &= LbM_{11}^{a}(L) \\ \Rightarrow -YI\frac{d^2U_3(L)}{dx^2} &= Lb\tilde{\xi}\frac{dU_3(L)}{dx}. \end{aligned} \tag{6.2.15}$$

b) Shear force:

$$\begin{aligned} Q(L) &= 0 \\ \Rightarrow \frac{d^3U_3(L)}{dx^3} &= 0. \end{aligned} \tag{6.2.16}$$

A free-vibration equation and boundary conditions of the cantilever beam laminated with distributed piezoelectric layers are defined. Next, the Laplace transform method is used to solve the system equation. Taking Laplace transformation of Eq. (6.2.11) gives

$$U_3(s) = \frac{1}{s^4 - \lambda_{di}^4}\left[s^3U_3(0) + s^2\frac{dU_3(0)}{dx} + s\frac{d^2U_3(0)}{dx^2} + \frac{d^3U_3(0)}{dx^3}\right]. \tag{6.2.17}$$

Substituting all boundary conditions into the above equation yields

$$\begin{aligned} U_3(s) &= \frac{1}{s^4 - \lambda_{di}^4}\left[s\frac{d^2U_3(0)}{dx^2} + \frac{d^3U_3(0)}{dx^3}\right] \\ &= \frac{s}{s^4 - \lambda_{di}^4}U_3''(0) + \frac{1}{s^4 - \lambda_{di}^4}U_3'''(0), \end{aligned} \tag{6.2.18}$$

where $U_3''(0) = \frac{d^2U_3(0)}{dx^2}$, and $U_3'''(0) = \frac{d^3U_3(0)}{dx^3}$ are constants. Taking the inverse transformation yields

$$\mathcal{L}^{-1}[U_3(s)] = U_3''(0)\mathcal{L}^{-1}\left(\frac{s}{s^4 - \lambda_{di}^4}\right) + U_3'''(0)\mathcal{L}^{-1}\left(\frac{1}{s^4 - \lambda_{di}^4}\right). \tag{6.2.19}$$

Using the partial fraction method, one can derive

$$\begin{aligned} U_3(x) &= U_3''(0)\mathcal{L}^{-1}\left(\frac{b_1}{s - a_1} + \frac{b_2}{s - a_2} + \frac{b_3}{s - a_3} + \frac{b_4}{s - a_4}\right) \\ &\quad + U_3'''(0)\mathcal{L}^{-1}\left(\frac{b_5}{s - a_1} + \frac{b_6}{s - a_2} + \frac{b_7}{s - a_3} + \frac{b_8}{s - a_4}\right). \end{aligned} \tag{6.2.20}$$

Thus, the solution becomes

$$\begin{aligned} U_3(x) &= U_3''(0)(b_1 e^{a_1 x} + b_2 e^{a_2 x} + b_3 e^{a_3 x} + b_4 e^{a_4 x}) \\ &\quad + U_3'''(0)(b_5 e^{a_1 x} + b_6 e^{a_2 x} + b_7 e^{a_3 x} + b_8 e^{a_4 x}), \end{aligned} \tag{6.2.21}$$

where $a_1,\ldots,a_4$, and $b_1,\ldots,b_8$ can be determined by substituting λ_{di} into Eq. (6.2.19) and employing the partial fraction expansion method. The first, second, and third derivatives of U_3 with respect to x can be derived as

$$U_3(x) = U_3''(0)d_1(x) + U_3'''(0)d_2(x), \tag{6.2.22}$$

$$\frac{dU_3(x)}{dx} = U_3''(0)d_3(x) + U_3'''(0)d_4(x), \tag{6.2.23}$$

$$\frac{d^2U_3(x)}{dx^2} = U_3''(0)d_5(x) + U_3'''(0)d_6(x), \tag{6.2.24}$$

$$\frac{d^3U_3(x)}{dx^3} = U_3''(0)d_7(x) + U_3'''(0)d_8(x), \tag{6.2.25}$$

where

$$d_1(x) = \sum_{i=1}^{4} b_i e^{a_i x}, \qquad d_2(x) = \sum_{i=1}^{4} b_{i+4} e^{a_i x}, \tag{6.2.26a, b}$$

$$d_3(x) = \sum_{i=1}^{4} a_i b_i e^{a_i x}, \qquad d_4(x) = \sum_{i=1}^{4} a_i b_{i+4} e^{a_i x}, \tag{6.2.26c, d}$$

$$d_5(x) = \sum_{i=1}^{4} a_i^2 b_i e^{a_i x}, \qquad d_6(x) = \sum_{i=1}^{4} a_i^2 b_{i+4} e^{a_i x}, \tag{6.2.26e, f}$$

$$d_7(x) = \sum_{i=1}^{4} a_i^3 b_i e^{a_i x}, \qquad d_8(x) = \sum_{i=1}^{4} a_i^3 b_{i+4} e^{a_i x}. \tag{6.2.26g, h}$$

Taking the other two boundary conditions (at x = L) into consideration and substituting Eqs. (6.2.22)–(6.2.25) into Eqs. (6.2.15) and (6.2.16), one can derive

a)
$$\begin{aligned} & -YI\frac{d^2U_3(L)}{dx^2} = Lb\tilde{\xi}\frac{dU_3(L)}{dx} \\ & \Rightarrow -YI\left[U_3''(0)d_5(L) + U_3'''(0)d_6(L)\right] = Lb\tilde{\xi}\left[U_3''(0)d_3(L) + U_3'''(0)d_4(L)\right], \end{aligned} \tag{6.2.27}$$

b)
$$\begin{aligned} & \frac{d^3U_3(L)}{dx^3} = 0 \\ & \Rightarrow U_3''(0)d_7(L) + U_3'''(0)d_8(L) = 0. \end{aligned} \tag{6.2.28}$$

Rearranging the above two equations, Eqs. (6.2.27) and (6.2.28), into a matrix form yields

$$\left[\begin{array}{c|c} YId_5(L)+Lb\tilde{\xi}d_3(L) & YId_6(L)+Lb\tilde{\xi}d_4(L) \\ \hline d_7(L) & d_8(L) \end{array}\right]\left\{\begin{array}{c} U_3''(0) \\ U_3'''(0) \end{array}\right\} = \{0\}. \tag{6.2.29}$$

For non-trivial solutions,

$$\left\{\begin{array}{c} U_3''(0) \\ U_3'''(0) \end{array}\right\} \neq 0, \tag{6.2.30}$$

and the determinant of the coefficient matrix equals zero. Thus,

$$\det\left[\begin{array}{c|c} YId_5(L)+Lb\tilde{\xi}d_3(L) & YId_6(L)+Lb\tilde{\xi}d_4(L) \\ \hline d_7(L) & d_8(L) \end{array}\right] = 0. \tag{6.2.31}$$

Expanding the determinant gives a ***characteristic equation*** of the laminated beam system with the displacement feedback. Using a numerical method, one can estimate its roots (eigenvalues) λ_{di}'s satisfying the characteristic equation and further calculate natural frequencies ω_{di}'s for the system. Note that the constant $\tilde{\xi}$, Eq. (6.2.3), is determined by a number of parameters: feedback gains, piezoelectric constants, Young's modulus, sensor thickness, and sensor/actuator distances.

6.2.2 Control Effectiveness: Displacement Feedback

In this section, control effectiveness of the displacement feedback is evaluated. Variations of natural frequencies and damping ratios due to feedback gains are studied and results discussed. Material properties and physical dimensions of a beam laminated with distributed polyvinylidene fluoride (PVDF) sensor and actuator layers are summarized in Table 6.1.

Using the physical properties listed in Table 6.1, one can determine the natural frequency variations due to the displacement feedback. The numerical results are summarized in Table 6.2.

It is observed that the frequency variations due to displacement feedbacks are not significant in all three fundamental modes. Damping ζ_i changes resulting from the change of natural frequency ω_i can also be evaluated by (Sects. 3.5 and 3.8 in Chap. 3)

$$\zeta_i = \frac{c}{2\rho h \omega_i}, \tag{6.2.32}$$

where c is the equivalent viscous damping factor; ω_i is the natural frequency of the i-th mode; ρ and h are the density and thickness of the beam. As the natural

Table 6.1 Properties of the laminated beam model

1) Plexiglas beam:	
Y (Young's modulus)	3.1028×10^9 (N/m^2)
ρ (Mass density)	1190.0 (Kg/m^3)
h (Thickness)	1.6×10^{-3} (m)
b (Width)	0.01 (m)
L (Length)	0.1 (m)
μ (Poisson's ratio)	0.3
D (Bending stiffness)	$Yh^3/[12(1-\mu^2)] = 1.1638$ (N − m)
I (Area moment of inertia)	$bh^3/12 = 3.4133 \times 10^{-12}$ (m^4)
2) Polyvinylidene Fluoride (PVDF) sensor/actuator layer:	
Y_p (Young's modulus)	2.00×10^9 (N/m^2)
ρ_p (Mass density)	1800.0 (Kg/m^3)
h^s (Thickness of sensor)	40 (μm)
h^a (Thickness of actuator)	40 (μm)
r (Sensor distance)	$(h^s + h)/2 = 8.2 \times 10^{-4}$ (m)
r (Actuator distance)	$(h^a + h)/2 = 8.2 \times 10^{-4}$ (m)
d_{31} (Piezoelectric constant)	$2.3 \times 10^{-11} (\frac{m/m}{V/m})$
h_{31} (Piezoelectric constant)	$g_{31} \times Y_p = 2.16 \times 10^{-1} \times 2.0 \times 10^9$ $4.32 \times 10^8 (\frac{V/m}{m/m})$

Table 6.2 Relationship between natural frequencies and displacement feedback gains

Displacement feedback gain	Natural frequency f (Hz)		
	1st mode	2nd mode	3rd mode
0.00	41.735	261.550	732.350
100.00	41.670	261.426	732.207
200.00	41.505	261.302	732.110
300.00	41.539	261.176	731.990
400.00	41.474	261.051	731.869
500.00	41.380	260.925	731.750
600.00	41.342	260.799	731.629
700.00	41.275	260.674	731.518
800.00	41.207	260.546	731.387
900.00	41.140	260.420	731.266
1000.00	41.072	260.293	731.145

Table 6.3 Relationship between damping ratios and displacement feedback gains

Displacement feedback gain	Damping ratio ($\times 10^{-3}$)		
	1st mode	2nd mode	3rd mode
0.00	20.000	3.19136	1.13976
100.00	20.031	3.19287	1.13994
200.00	20.062	3.19439	1.14013
300.00	20.094	3.19593	1.14032
400.00	20.126	3.19746	1.14050
500.00	20.172	3.19901	1.14069
600.00	20.190	3.20054	1.14087
700.00	20.223	3.20208	1.14105
800.00	20.256	3.20365	1.14125
900.00	20.289	3.20521	1.14145
1000.00	20.323	3.20677	1.14163

frequency ω_i becomes smaller, the inferred equivalent damping ratio ζ_i increases accordingly. (Note that c is assumed constant.) The resulting damping changes are summarized in Table 6.3. Note that initial damping ratios are assumed.

Again, the variations are very small. Note that this damping changes are introduced by the change of controlled natural frequencies, Eq. (6.2.32), in the displacement feedback as discussed previously.

6.3 Velocity Feedback Control

In this control algorithm, the sensor signal is differentiated, amplified, and then fed back into the distributed piezoelectric actuator. In this section, system equation and its characteristic equation will be derived. Note that the Laplace transform method is used to derive analytical solutions. The velocity feedback signal ϕ^a can be defined as

$$\begin{aligned} \phi^a &= \mathcal{G}\frac{\partial \phi^S}{\partial t} \\ &= -\frac{\mathcal{G}h^S h_{31} r_1^S}{L}\frac{\partial}{\partial t}\left[\left(\frac{\partial u_3}{\partial x}\right)\Big|_0^L\right]. \end{aligned} \tag{6.3.1}$$

Using the velocity signal ϕ^a and following the definitions in Sect. 4.3 of Chap. 4, one can derive a distributed moment M_{11}^a in the following form:

$$\begin{aligned} M_{11}^a &= r_1^a d_{31} Y_p \phi^a \\ &= -\frac{\mathcal{G}h^S r_1^a d_{31} Y_p r_1^S h_{31}}{L}\frac{\partial}{\partial t}\left[\left(\frac{\partial u_3}{\partial x}\right)\Big|_0^L\right] \\ &= \tilde{\xi}\frac{\partial}{\partial t}\left[\left(\frac{\partial u_3}{\partial x}\right)\Big|_0^L\right], \end{aligned} \tag{6.3.2}$$

where

$$\tilde{\xi} = -\frac{\mathcal{G}h^S r_1^a d_{31} Y_p r_1^S h_{31}}{L}. \tag{6.3.3}$$

Substituting Eq. (6.3.2) into the system equation and setting external excitation zero, i.e., $F_3 = 0$, yields

$$\rho h\ddot{u}_3 + D\frac{\partial^4 u_3}{\partial x^4} - \frac{\partial^2}{\partial x^2}\left\{\tilde{\xi}\frac{\partial}{\partial t}\left[\left(\frac{\partial u_3}{\partial x}\right)\Big|_0^L\right]\right\} = 0. \tag{6.3.4}$$

Again, note that the distributed moment is not a function of coordinate; the Poisson's ratio effect can be neglected. Thus, the beam system equation is simplified to

$$\rho\tilde{A}\ddot{u}_3 + YI\frac{\partial^4 u_3}{\partial x^4} = 0. \tag{6.3.5}$$

Since the distributed moment is not explicitly contained in the system equation, control actions are introduced through the electric boundary conditions—***boundary controls*** as discussed previously.

6.3.1 Free Vibration Analysis

Assume the laminated beam system oscillates harmonically at natural frequencies, i.e.,

$$u_3(x,t) = U_3(x)e^{j\hat{\omega}t}. \tag{6.3.6}$$

Then, the system equation becomes

$$\begin{aligned} &YI\frac{d^4U_3}{dx^4} - \rho\tilde{A}\hat{\omega}^2U_3 = 0 \\ &\Rightarrow \frac{d^4U_3}{dx^4} - \lambda_{vk}^4U_3 = 0, \end{aligned} \tag{6.3.7}$$

where λ_{vk} is the k-th eigenvalue of the system with the velocity feedback, denoted by the subscript v; λ_{vk} is defined as

$$\lambda_{vk}^4 = \frac{\rho\tilde{A}\hat{\omega}^2}{YI}. \tag{6.3.8}$$

Boundary conditions at the free end are modified to reflect the velocity signal used in feedback controls. The fixed end boundary conditions are the same as before.

1) At clamped end (x = 0):

a) Displacement:

$$U_3(0) = 0, \tag{6.3.9}$$

b) Slope:

$$\frac{dU_3(0)}{dx} = 0. \tag{6.3.10}$$

2) At free end (x = L):

a) Moment: $M_{11}(L) = LbM_{11}^a(L)$

$$\begin{aligned} &\Rightarrow -YI\frac{d^2u_3(L)}{dx^2} = Lb\tilde{\xi}\frac{\partial}{\partial t}\left[\left(\frac{du_3}{dx}\right)\Big|_0^L\right] \\ &\Rightarrow -YI\frac{d^2U_3(L)}{dx^2} = j\hat{\omega}Lb\tilde{\xi}\frac{dU_3(L)}{dx}, \end{aligned} \tag{6.3.11}$$

b) Shear force: Q(L) = 0

$$\Rightarrow \frac{d^3U_3(L)}{dx^3} = 0. \tag{6.3.12}$$

Note that the boundary control moment has a component of $j\hat{\omega}$ introduced by the velocity feedback and $j = \sqrt{-1}$. In order to satisfy the boundary condition, a complex frequency $\hat{\omega}$ and a complex transverse displacement U_3 need to be used in the later analysis. Assume both the displacement U_3 and the natural frequency $\hat{\omega}$ have a real component and an imaginary component:

$$U_3(x) = U_r(x) + jU_i(x), \tag{6.3.13}$$

where the subscript r denotes the real component and i the imaginary component, and

$$\hat{\omega} = \omega_1 + j\omega_2. \tag{6.3.14}$$

Substituting Eq. (6.3.14) into Eq. (6.3.6) yields

$$u_3(x,t) = U_3(x)e^{j(\omega_1 + j\omega_2)t}. \tag{6.3.15}$$

Since $\hat{\omega}$ is complex, the real part of the exponential function, i.e., $j(j\omega_2)$, represents the decay rate (a convergence envelop or oscillation envelop); the imaginary part, i.e., $j\omega_1$, represents the damped natural frequency (Ewins, 1986; Zaveri, 1985). It is assumed that the beam system is lightly damped. Concerning the complex transverse displacement $U_3(x)$, the magnitude of U_3 represents the displacement; the phase angle of U_3 represents the phase differences among various parts of a distributed structure (Zaveri, 1985).

Substituting U_3 and $\hat{\omega}$ into the system equation, Eq. (6.3.7), and separating the real and imaginary parts of the equation, one can derive

$$\frac{d^4U_r(x)}{dx^4} - pU_r(x) + qU_i(x) = 0, \tag{6.3.16a}$$

$$\frac{d^4U_i(x)}{dx^4} - pU_i(x) - qU_r(x) = 0, \tag{6.3.16b}$$

where p and q are the real and imaginary parts of the k-th eigenvalue λ_{vk}^4; they are defined as:

$$\lambda_{vk}^4 = \frac{\rho\tilde{A}\hat{\omega}^2}{YI} = p + jq, \tag{6.3.17}$$

i.e.,

$$p = \frac{\rho\tilde{A}\left(\omega_1^2 - \omega_2^2\right)}{YI} \quad \text{and} \quad q = \frac{2\rho\tilde{A}\omega_1\omega_2}{YI}.$$

Taking the Laplace transforms of Eqs. (6.3.16a) and (6.3.16b) and applying the fixed end boundary conditions, Eqs. (6.3.9) and (6.3.10), one can derive

$$\left(s^4 - p\right)U_r(s) + qU_i(s) = U_r'''(0) + sU_r''(0), \tag{6.3.18}$$

$$\left(s^4 - p\right)U_i(s) - qU_r(s) = U_i'''(0) + sU_i''(0), \tag{6.3.19}$$

where $U_r'''(0) = \frac{d^3U_{r(0)}}{dx^3}$, $U_r''(0) = \frac{d^2U_{r(0)}}{dx^2}$, $U_i'''(0) = \frac{d^3U_{i(0)}}{dx^3}$, and $U_i''(0) = \frac{d^2U_{i(0)}}{dx^2}$ are constants determined by the boundary conditions. Solving Eqs. (6.3.18)–(6.3.19) simultaneously gives

$$\begin{aligned} U_r(s) = \frac{1}{\left(s^4 - p\right)^2 + q^2}&\left[\left(s^4 - p\right)U_r'''(0) + \left(s^5 - ps\right)U_r''(0) - qU_i'''(0)\right. \\ &\left.-qsU_i''(0)\right], \end{aligned} \tag{6.3.20}$$

$$\begin{aligned} U_i(s) = \frac{1}{\left(s^4 - p\right)^2 + q^2}&\left[qU_r'''(0) + qsU_r''(0) + \left(s^4 - p\right)U_i'''(0)\right. \\ &\left.+\left(s^5 - ps\right)U_i''(0)\right]. \end{aligned} \tag{6.3.21}$$

Taking the inverse Laplace transformation and using the partial fractions, one can derive $U_r(x)$ and $U_i(x)$:

$$\begin{aligned} U_r(x) = U_r'''(0)\mathcal{L}^{-1}&\left[\frac{m_1}{s - z_1} + \frac{m_2}{s - z_2} + \frac{m_3}{s - z_3} + \frac{m_4}{s - z_4} + \frac{m_5}{s - z_5} + \frac{m_6}{s - z_6} + \frac{m_7}{s - z_7} + \frac{m_8}{s - z_8}\right] \\ + U_r''(0)\mathcal{L}^{-1}&\left[\frac{m_9}{s - z_1} + \frac{m_{10}}{s - z_2} + \frac{m_{11}}{s - z_3} + \frac{m_{12}}{s - z_4} + \frac{m_{13}}{s - z_5} + \frac{m_{14}}{s - z_6} + \frac{m_{15}}{s - z_7} + \frac{m_{16}}{s - z_8}\right] \\ + U_i'''(0)\mathcal{L}^{-1}&\left[\frac{m_{17}}{s - z_1} + \frac{m_{18}}{s - z_2} + \frac{m_{19}}{s - z_3} + \frac{m_{20}}{s - z_4} + \frac{m_{21}}{s - z_5} + \frac{m_{22}}{s - z_6} + \frac{m_{23}}{s - z_7} + \frac{m_{24}}{s - z_8}\right] \\ + U_i''(0)\mathcal{L}^{-1}&\left[\frac{m_{25}}{s - z_1} + \frac{m_{26}}{s - z_2} + \frac{m_{27}}{s - z_3} + \frac{m_{28}}{s - z_4} + \frac{m_{29}}{s - z_5} + \frac{m_{30}}{s - z_6} + \frac{m_{31}}{s - z_7} + \frac{m_{32}}{s - z_8}\right], \end{aligned} \tag{6.3.22}$$

$$\begin{aligned}U_i(x) &= U_r'''(0)\mathcal{L}^{-1}\left[\frac{n_1}{s-z_1}+\frac{n_2}{s-z_2}+\frac{n_3}{s-z_3}+\frac{n_4}{s-z_4}+\frac{n_5}{s-z_5}+\frac{n_6}{s-z_6}+\frac{n_7}{s-z_7}+\frac{n_8}{s-z_8}\right]\\
&+U_r''(0)\mathcal{L}^{-1}\left[\frac{n_9}{s-z_1}+\frac{n_{10}}{s-z_2}+\frac{n_{11}}{s-z_3}+\frac{n_{12}}{s-z_4}+\frac{n_{13}}{s-z_5}+\frac{n_{14}}{s-z_6}+\frac{n_{15}}{s-z_7}+\frac{n_{16}}{s-z_8}\right]\\
&+U_i'''(0)\mathcal{L}^{-1}\left[\frac{n_{17}}{s-z_1}+\frac{n_{18}}{s-z_2}+\frac{n_{19}}{s-z_3}+\frac{n_{20}}{s-z_4}+\frac{n_{21}}{s-z_5}+\frac{n_{22}}{s-z_6}+\frac{n_{23}}{s-z_7}+\frac{n_{24}}{s-z_8}\right]\\
&+U_i''(0)\mathcal{L}^{-1}\left[\frac{n_{25}}{s-z_1}+\frac{n_{26}}{s-z_2}+\frac{n_{27}}{s-z_3}+\frac{n_{28}}{s-z_4}+\frac{n_{29}}{s-z_5}+\frac{n_{30}}{s-z_6}+\frac{n_{31}}{s-z_7}+\frac{n_{32}}{s-z_8}\right].\end{aligned} \tag{6.3.23}$$

The real and imaginary components of the complex displacement solution in Eq. (6.3.13) are

$$\begin{aligned}U_r(x) = {} & U_r'''(0)[m_1e^{z_1x}+m_2e^{z_2x}+m_3e^{z_3x}+m_4e^{z_4x}\\
&+m_5e^{z_5x}+m_6e^{z_6x}+m_7e^{z_7x}+m_8e^{z_8x}]\\
&+U_r''(0)[m_9e^{z_1x}+m_{10}e^{z_2x}+m_{11}e^{z_3x}+m_{12}e^{z_4x}\\
&+m_{13}e^{z_5x}+m_{14}e^{z_6x}+m_{15}e^{z_7x}+m_{16}e^{z_8x}]\\
&+U_i'''(0)[m_{17}e^{z_1x}+m_{18}e^{z_2x}+m_{19}e^{z_3x}+m_{20}e^{z_4x}\\
&+m_{21}e^{z_5x}+m_{22}e^{z_6x}+m_{23}e^{z_7x}+m_{24}e^{z_8x}]\\
&+U_i''(0)[m_{25}e^{z_1x}+m_{26}e^{z_2x}+m_{27}e^{z_3x}+m_{28}e^{z_4x}\\
&+m_{29}e^{z_5x}+m_{30}e^{z_6x}+m_{31}e^{z_7x}+m_{32}e^{z_8x}],\end{aligned} \tag{6.3.24}$$

$$\begin{aligned}U_i(x) = {} & U_r'''(0)[n_1e^{z_1x}+n_2e^{z_2x}+n_3e^{z_3x}+n_4e^{z_4x}\\
&+n_5e^{z_5x}+n_6e^{z_6x}+n_7e^{z_7x}+n_8e^{z_8x}]\\
&+U_r''(0)[n_9e^{z_1x}+n_{10}e^{z_2x}+n_{11}e^{z_3x}+n_{12}e^{z_4x}\\
&+n_{13}e^{z_5x}+n_{14}e^{z_6x}+n_{15}e^{z_7x}+n_{16}e^{z_8x}]\\
&+U_i'''(0)[n_{17}e^{z_1x}+n_{18}e^{z_2x}+n_{19}e^{z_3x}+n_{20}e^{z_4x}\\
&+n_{21}e^{z_5x}+n_{22}e^{z_6x}+n_{23}e^{z_7x}+n_{24}e^{z_8x}]\\
&+U_i''(0)[n_{25}e^{z_1x}+n_{26}e^{z_2x}+n_{27}e^{z_3x}+n_{28}e^{z_4x}\\
&+n_{29}e^{z_5x}+n_{30}e^{z_6x}+n_{31}e^{z_7x}+n_{32}e^{z_8x}],\end{aligned} \tag{6.3.25}$$

where $z_1,.., z_8$, $m_1,.., m_{32}$, and $n_1,.., n_{32}$ can be determined by the partial fraction expansions from Eqs. (6.3.22)–(6.3.23) for a given λ_{vk}. Condensing Eqs. (6.3.24) and (6.3.25) in terms of $e_n(x)$ and $f_n(x)$ yields

$$U_r(x) = e_1(x)U_r'''(0)+e_2(x)U_r''(0)+e_3(x)U_i'''(0)+e_4(x)U_i''(0), \tag{6.3.26}$$

$$U_i(x) = f_1(x)U_r'''(0)+f_2(x)U_r''(0)+f_3(x)U_i'''(0)+f_4(x)U_i''(0). \tag{6.3.27}$$

Note that $e_n(x)$ and $f_n(x)$ are corresponding to the terms inside the parentheses in Eqs. (6.3.24) and (6.3.25).

$$\begin{aligned} e_1(x) = [&m_1e^{z_1x} + m_2e^{z_2x} + m_3e^{z_3x} + m_4e^{z_4x} \\ &+ m_5e^{z_5x} + m_6e^{z_6x} + m_7e^{z_7x} + m_8e^{z_8x}], \end{aligned} \tag{6.3.28a}$$

$$\begin{aligned} e_2(x) = [&m_9e^{z_1x} + m_{10}e^{z_2x} + m_{11}e^{z_3x} + m_{12}e^{z_4x} \\ &+ m_{13}e^{z_5x} + m_{14}e^{z_6x} + m_{15}e^{z_7x} + m_{16}e^{z_8x}], \end{aligned} \tag{6.3.28b}$$

$$\begin{aligned} e_3(x) = [&m_{17}e^{z_1x} + m_{18}e^{z_2x} + m_{19}e^{z_3x} + m_{20}e^{z_4x} \\ &+ m_{21}e^{z_5x} + m_{22}e^{z_6x} + m_{23}e^{z_7x} + m_{24}e^{z_8x}], \end{aligned} \tag{6.3.29a}$$

$$\begin{aligned} e_4(x) = [&m_{25}e^{z_1x} + m_{26}e^{z_2x} + m_{27}e^{z_3x} + m_{28}e^{z_4x} \\ &+ m_{29}e^{z_5x} + m_{30}e^{z_6x} + m_{31}e^{z_7x} + m_{32}e^{z_8x}], \end{aligned} \tag{6.3.29b}$$

$$\begin{aligned} f_1(x) = [&n_1e^{z_1x} + n_2e^{z_2x} + n_3e^{z_3x} + n_4e^{z_4x} \\ &+ n_5e^{z_5x} + n_6e^{z_6x} + n_7e^{z_7x} + n_8e^{z_8x}], \end{aligned} \tag{6.3.30a}$$

$$\begin{aligned} f_2(x) = [&n_9e^{z_1x} + n_{10}e^{z_2x} + n_{11}e^{z_3x} + n_{12}e^{z_4x} \\ &+ n_{13}e^{z_5x} + n_{14}e^{z_6x} + n_{15}e^{z_7x} + n_{16}e^{z_8x}], \end{aligned} \tag{6.3.30b}$$

$$\begin{aligned} f_3(x) = [&n_{17}e^{z_1x} + n_{18}e^{z_2x} + n_{19}e^{z_3x} + n_{20}e^{z_4x} \\ &+ n_{21}e^{z_5x} + n_{22}e^{z_6x} + n_{23}e^{z_7x} + n_{24}e^{z_8x}], \end{aligned} \tag{6.3.31a}$$

$$\begin{aligned} f_4(x) = [&n_{25}e^{z_1x} + n_{26}e^{z_2x} + n_{27}e^{z_3x} + n_{28}e^{z_4x} \\ &+ n_{29}e^{z_5x} + n_{30}e^{z_6x} + n_{31}e^{z_7x} + n_{32}e^{z_8x}]. \end{aligned} \tag{6.3.31b}$$

The first, second, and third derivatives of U_r and U_i with respect to x can be derived as

$$\frac{dU_r(x)}{dx} = e_5(x)U_r'''(0) + e_6(x)U_r''(0) + e_7(x)U_i''''(0) + e_8(x)U_i''(0), \tag{6.3.32}$$

$$\frac{dU_r^2(x)}{dx^2} = e_9(x)U_r''''(0) + e_{10}(x)U_r''(0) + e_{11}(x)U_i''''(0) + e_{12}(x)U_i''(0), \tag{6.3.33}$$

$$\frac{dU_r^3(x)}{dx^3} = e_{13}(x)U_r'''(0) + e_{14}(x)U_r''(0) + e_{15}(x)U_i'''(0) + e_{16}(x)U_i''(0); \tag{6.3.34}$$

and

$$\frac{dU_i(x)}{dx} = f_5(x)U_r'''(0) + f_6(x)U_r''(0) + f_7(x)U_i'''(0) + f_8(x)U_i''(0), \tag{6.3.35}$$

$$\frac{dU_i^2(x)}{dx^2} = f_9(x)U_r'''(0) + f_{10}(x)U_r''(0) + f_{11}(x)U_i'''(0) + f_{12}(x)U_i''(0), \tag{6.3.36}$$

$$\frac{dU_i^3(x)}{dx^3} = f_{13}(x)U_r'''(0) + f_{14}(x)U_r''(0) + f_{15}(x)U_i'''(0) + f_{16}(x)U_i''(0). \tag{6.3.37}$$

Substituting the complex displacement, Eq. (6.3.13), and the complex frequency, Eq. (6.3.14), into boundary conditions, Eqs. (6.3.11) & (6.3.12), and separating the real and imaginary parts of these equations, one can derive the following four equations.

a)

$$\frac{d^3U_3(L)}{dx^3} = 0$$
$$\Rightarrow \frac{d^3U_r(L)}{dx^3} = 0, \tag{6.3.38}$$

$$\Rightarrow \frac{d^3U_i(L)}{dx^3} = 0; \tag{6.3.39}$$

b)

$$-YI\frac{d^2U_3(L)}{dx^2} = j\hat{\omega}Lb\tilde{\xi}\frac{dU_3(L)}{dx}$$
$$\Rightarrow -YI\frac{d^2U_r(L)}{dx^2} = \tilde{\gamma}\frac{dU_r(L)}{dx} - \tilde{\eta}\frac{dU_i(L)}{dx}, \tag{6.3.40}$$

$$\Rightarrow -YI\frac{d^2U_i(L)}{dx^2} = \tilde{\eta}\frac{dU_r(L)}{dx} + \tilde{\gamma}\frac{dU_i(L)}{dx}, \tag{6.3.41}$$

where

$$j\hat{\omega}Lb\tilde{\xi} = \tilde{\gamma} + j\tilde{\eta}. \tag{6.3.42}$$

Substituting Eqs. (6.3.26) & (6.3.27) and Eqs. (6.3.31a)–(6.3.37) into Eqs. (6.3.38)–(6.3.41) gives

$$e_{13}(L)U_r'''(0) + e_{14}(L)U_r''(0) + e_{15}(L)U_i'''(0) + e_{16}(L)U_i''(0) = 0, \tag{6.3.43}$$

$$f_{13}(L)U_r'''(0) + f_{14}(L)U_r''(0) + f_{15}(L)U_i'''(0) + f_{16}(L)U_i''(0) = 0, \tag{6.3.44}$$

$$\begin{aligned}&-\mathrm{YI}\left[e_9(L)U_r'''(0)+e_{10}(L)U_r''(0)+e_{11}(L)U_i'''(0)+e_{12}(L)U_i''(0)\right]\\&=\tilde{\gamma}\left[e_5(L)U_r'''(0)+e_6(L)U_r''(0)+e_7(L)U_i'''(0)+e_8(L)U_i''(0)\right]\\&\quad-\tilde{\eta}\left[f_5(L)U_r'''(0)+f_6(L)U_r''(0)+f_7(L)U_i'''(0)+f_8(L)U_i''(0)\right],\end{aligned} \tag{6.3.45}$$

$$\begin{aligned}&-\mathrm{YI}\left[f_9(L)U_r'''(0)+f_{10}(L)U_r''(0)+f_{11}(L)U_i'''(0)+f_{12}(L)U_i''(0)\right]\\&=\tilde{\eta}\left[e_5(L)U_r'''(0)+e_6(L)U_r''(0)+e_7(L)U_i'''(0)+e_8(L)U_i''(0)\right]\\&-\tilde{\gamma}\left[f_5(L)U_r'''(0)+f_6(L)U_r''(0)+f_7(L)U_i'''(0)+f_8(L)U_i''(0)\right].\end{aligned} \tag{6.3.46}$$

Rearranging the four equations above, Eqs. (6.3.43)–(6.3.46), in a matrix expression yields

$$\begin{pmatrix}A(1,1)&A(1,2)&A(1,3)&A(1,4)\\A(2,1)&A(2,2)&A(2,3)&A(2,4)\\A(3,1)&A(3,2)&A(3,3)&A(3,4)\\A(4,1)&A(4,2)&A(4,3)&A(4,4)\end{pmatrix}\begin{Bmatrix}U_r'''(0)\\U_r''(0)\\U_i'''(0)\\U_i''(0)\end{Bmatrix}=\{0\}, \tag{6.3.47}$$

where $A(1,1),..,A(4,4)$ are the coefficients determined by Eqs. (6.3.43)–(6.3.46). For no-trivial solution

$$\begin{Bmatrix}U_r'''(0)\\U_r''(0)\\U_i'''(0)\\U_i''(0)\end{Bmatrix}\neq 0, \tag{6.3.48}$$

then, the determinant of the coefficient matrix equals zero, i.e.,

$$\det\begin{pmatrix}A(1,1)&A(1,2)&A(1,3)&A(1,4)\\A(2,1)&A(2,2)&A(2,3)&A(2,4)\\A(3,1)&A(3,2)&A(3,3)&A(3,4)\\A(4,1)&A(4,2)&A(4,3)&A(4,4)\end{pmatrix}=0. \qquad ((6.3.49))$$

Expansion of the above determinant gives a *characteristic equation* of the laminated beam system with the velocity feedback and roots λ_{vk}'s of the characteristic equation are the *eigenvalues* of the beam system. Note that λ_{vk}'s are complex, from which natural frequencies $\hat{\omega}_k$'s ($\hat{\omega}_k=\omega_1+j\omega_2$) can be derived. The complex frequency $\hat{\omega}$ provides the information about the damped natural frequency and the damping ratio of the beam system.

6.3.2 Control Effectiveness: Velocity Feedback

In this section, distributed controls of a laminated cantilever beam using velocity feedbacks are studied. The physical model is identical to that described in Sect. 6.2.2. All material and physical properties are summarized in Table 6.1.

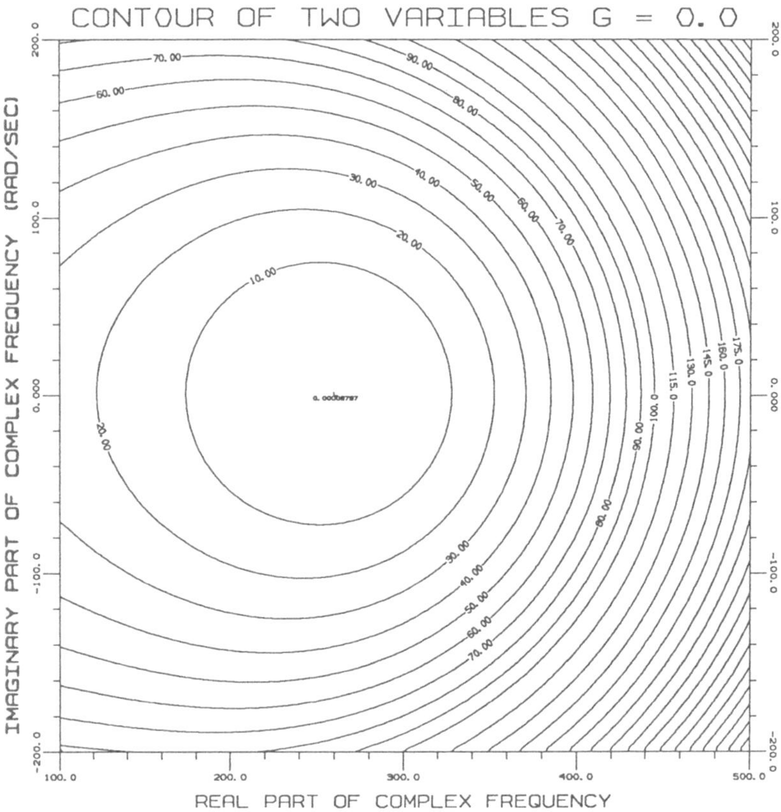

Fig. 6.2 Original complex natural frequency of the first mode

The original first three natural frequencies, in the complex plane, are presented in Figs. 6.2, 6.3 and 6.4. (Note that the imaginary part is zero.) Controlled natural frequencies for the corresponding modes are illustrated in Figs. 6.5, 6.6 and 6.7. Shifts of natural frequencies due to velocity feedbacks can be observed in these three figures.

Variations of the first three complex frequencies due to the velocity feedback are summarized in Table 6.4.

As discussed previously, a complex frequency of the k-th mode is defined as

$$\hat{\omega}_k = \omega_1 + j\omega_2, \tag{6.3.50}$$

where ω_2 defines the envelope of damped oscillation ($\omega_2 = \zeta_k \omega_n$); ω_1 is the damped natural frequency ($\omega_1 = \omega_n \sqrt{1 - \zeta_k^2}$) (Ewins, 1986); ω_n denotes the

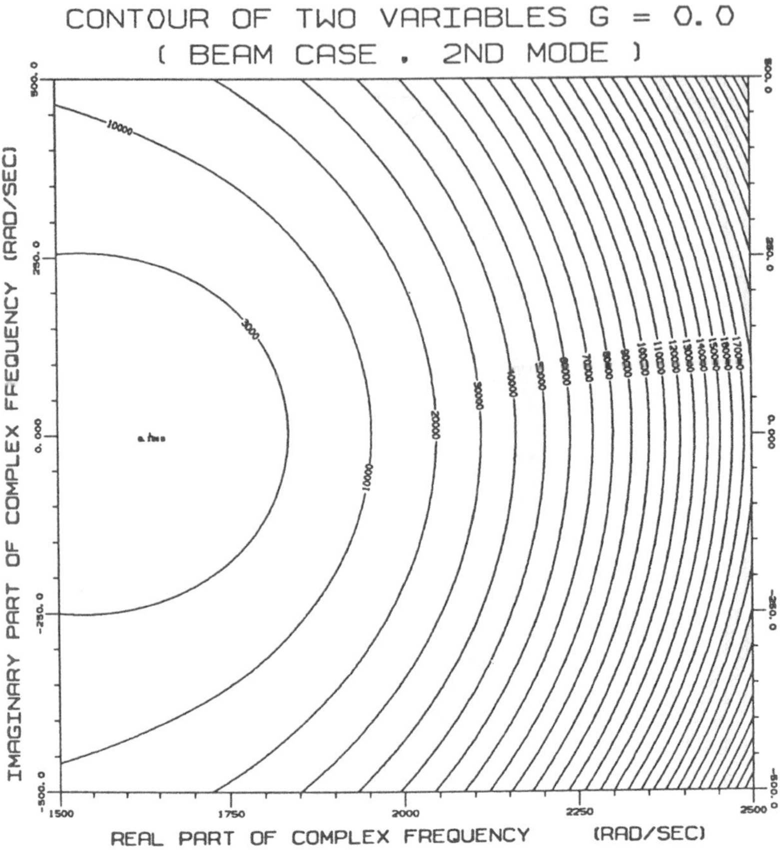

Fig. 6.3 Original complex natural frequency of the second mode

undamped natural frequency of the laminated beam without any feedback control. Thus, the damping ratio ζ_k for the k-th mode can be estimated by (Zaveri, 1985; Harris & Crede, 1976)

$$\zeta_k = \sqrt{\frac{\omega_2^2}{\omega_2^2 + \omega_1^2}} \cong \frac{\omega_2}{\omega_1}. \tag{6.3.51}$$

Damping enhancement of the beam system due to the velocity feedback can be calculated in this way and the results are summarized in Table 6.5.

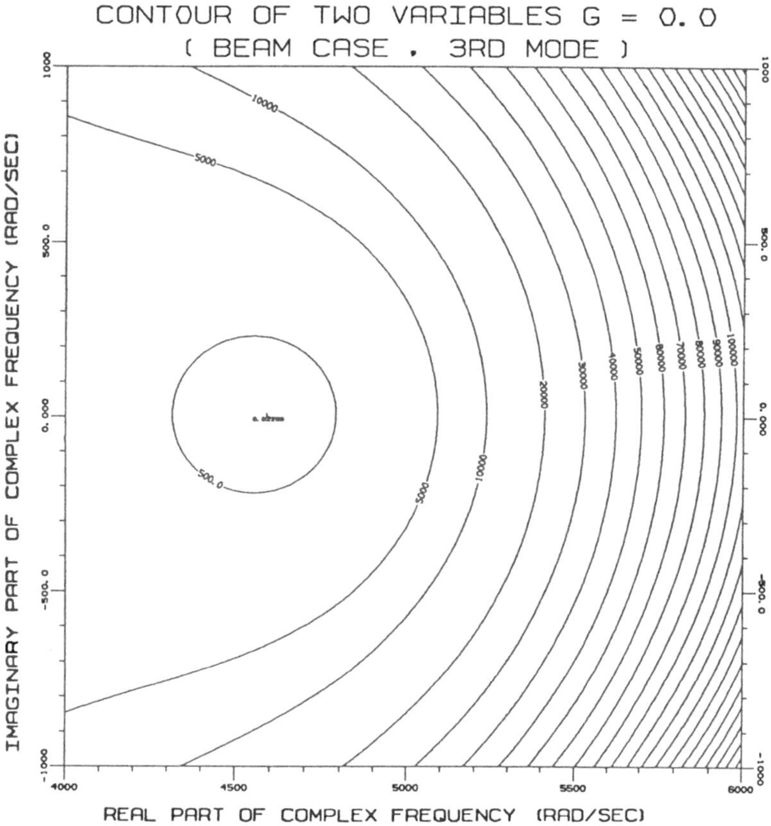

Fig. 6.4 Original complex natural frequency of the third mode

It is observed that the damping ratios increase to an optimal point and then drop down. Figures 6.8, 6.9 and 6.10 illustrate the damping variations of the first three modes. This is due to the free-end boundary condition transitioning to an equivalent "roller-slide" boundary condition, Fig. 6.11, due to extremely strong counteracting control moment at the free-end induced by the fully-distributed actuator (Tzou, Johnson, and Liu, 1994 & 1999). (This behavior is discussed next.) The phenomena were also demonstrated in laboratory experiments, Figs. 6.8 and 6.12.

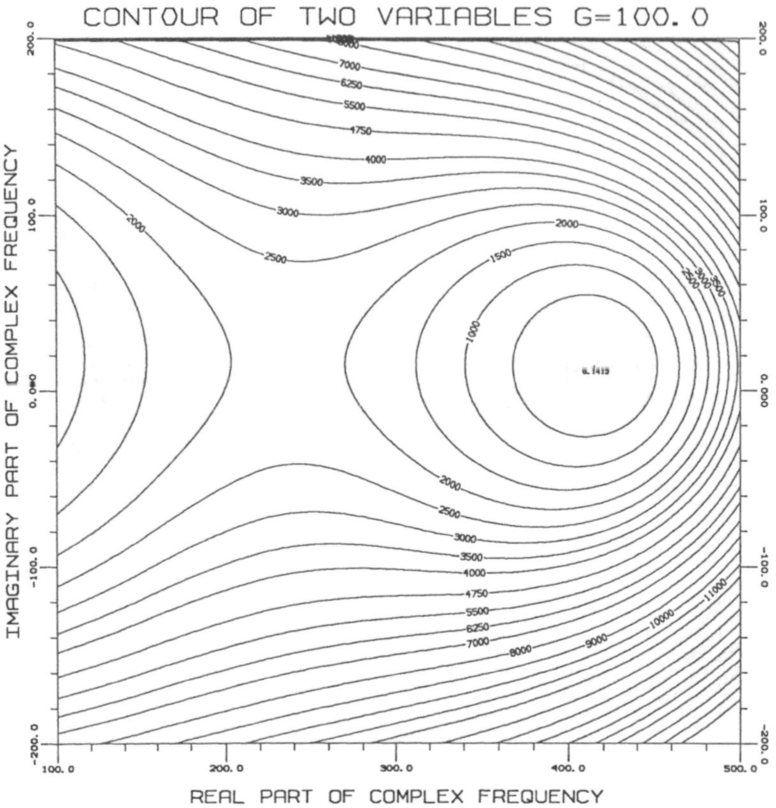

Fig. 6.5 Controlled complex natural frequency of the first mode

It is observed that the damping ratios increase to an ultimate value and drop down afterwards. (It is assumed that the piezoelectric di-pole molecular structure still sustains the high feedback voltages.) This phenomenon is caused by the high boundary feedback control effects at the free end. As discussed previously, the distributed piezoelectric actuator basically applies a counteracting moment to the cantilever beam at the free end when a high feedback voltage is injected into the

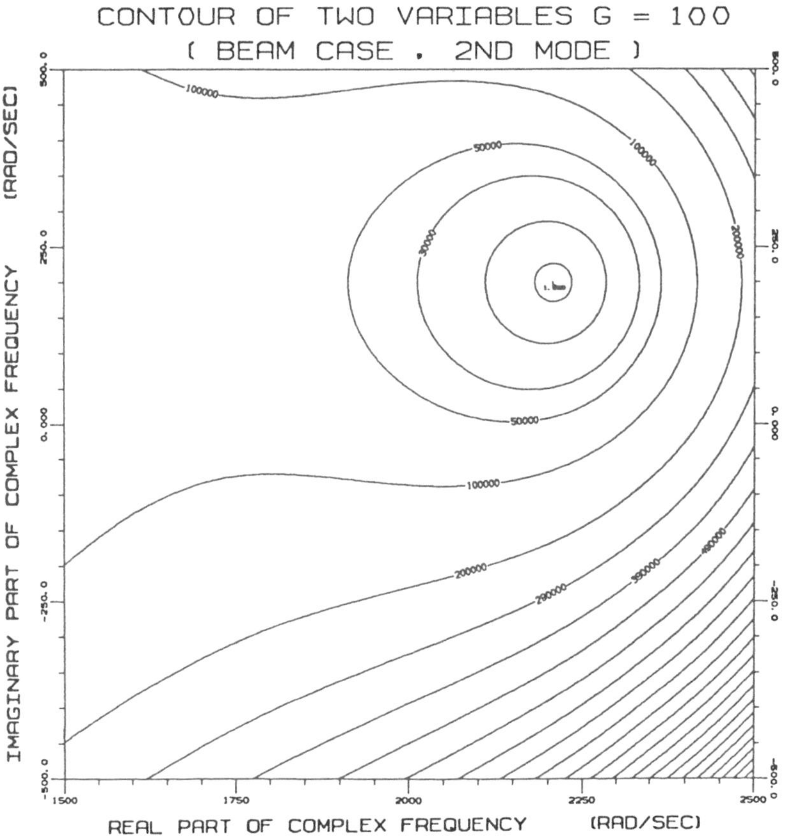

Fig. 6.6 Controlled complex natural frequency of the second mode

actuator. As the feedback voltage becomes higher and higher, the free end is further constrained and the boundary condition is gradually changing to a sliding-roller boundary condition (Tzou, Johnson, and Liu, 1994 & 1999; Liu, 1989; Johnson, 1990).

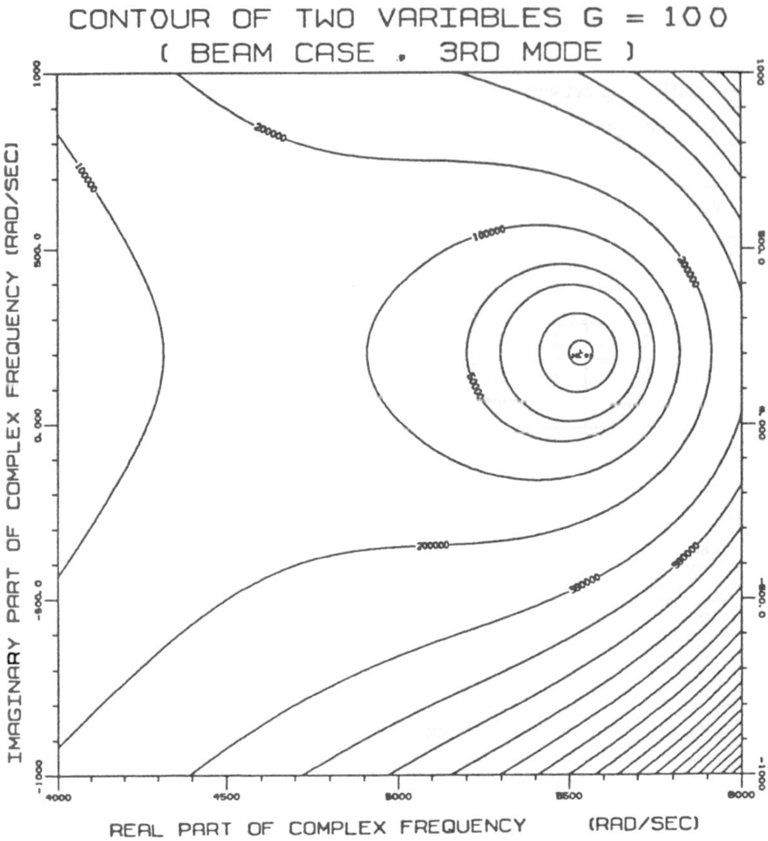

Fig. 6.7 Controlled complex natural frequency of the third mode

Table 6.4 Relationship between complex frequency and velocity feedback gain. ($\mathbf{f}_1, \mathbf{f}_2$) (Hz)

Velocity feedback gain	Natural frequency f (Hz)		
	1st mode	2nd mode	3rd mode
0.00	(41.7, 0.0)	(261.6, 0.0)	(732.4, 0.0)
100.00	(51.7, 16.2)	(351.7, 31.2)	(881.2, 31.2)
200.00	(52.7, 11.1)	(357.0, 16.0)	(884.9, 15.8)
300.00	(64.8, 7.7)	(357.9, 10.7)	(885.5, 10.5)
400.00	(65.6, 5.7)	(358.4, 8.0)	(885.7, 7.8)
500.00	(65.9, 4.6)	(358.6, 6.4)	(885.9, 6.4)
600.00	(66.1, 3.8)	(358.6, 5.3)	(885.9, 5.3)
700.00	(66.1, 3.3)	(358.6, 4.6)	(885.9, 4.5)
800.00	(66.2, 2.9)	(358.7, 3.5)	(885.9, 4.0)
900.00	(66.2, 2.6)	(358.7, 3.2)	(885.9, 3.5)
1000.00	(66.2, 2.4)	(358.7, 2.9)	(885.9, 3.2)

Table 6.5 Relationship between damping ratio and velocity feedback gain

Velocity feedback gain	Damping ratio ($\times 10^{-3}$)		
	1st mode	2nd mode	3rd mode
0.00	0.00	0.00	0.00
100.00	331.00	89.00	35.00
200.00	181.00	45.00	18.00
300.00	118.00	30.00	12.00
400.00	87.00	22.00	9.00
500.00	70.00	18.00	7.00
600.00	58.00	15.00	6.00
700.00	50.60	13.00	5.00
800.00	43.27	11.00	4.00
900.00	38.46	10.00	4.00
1000.00	36.06	9.00	4.00

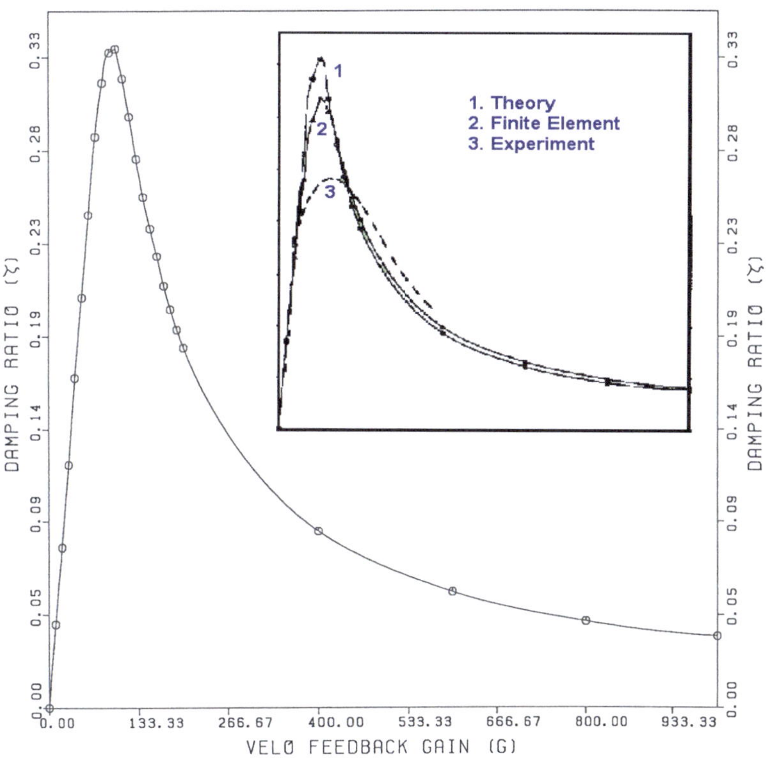

Fig. 6.8 Damping variation of the first mode with the velocity feedback

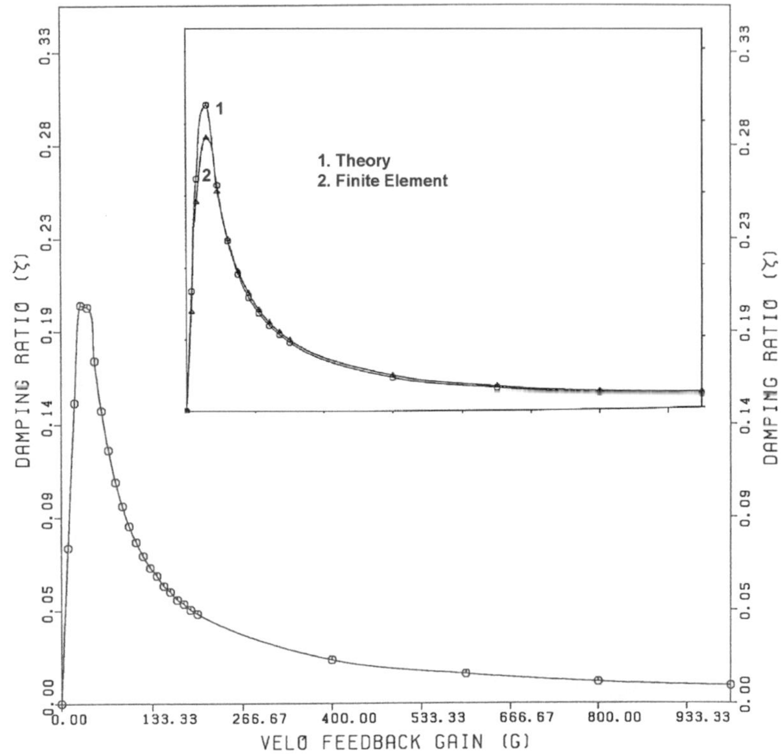

Fig. 6.9 Damping variation of the second mode with the velocity feedback

6.3.3 *Finite Element Analysis*

In order to verify this theory, the first three natural frequencies of a fixed/sliding-roller beam are calculated and compared with the controlled natural frequencies (at Gain = 1000) of the cantilever beam, Table 6.6.

In addition, the first three (undamped) natural frequencies of a double fixed-end beam are calculated using the finite element method, Fig. 6.11, and they are also listed in Table 6.6. Basically, these data support the physical explanations mentioned above.

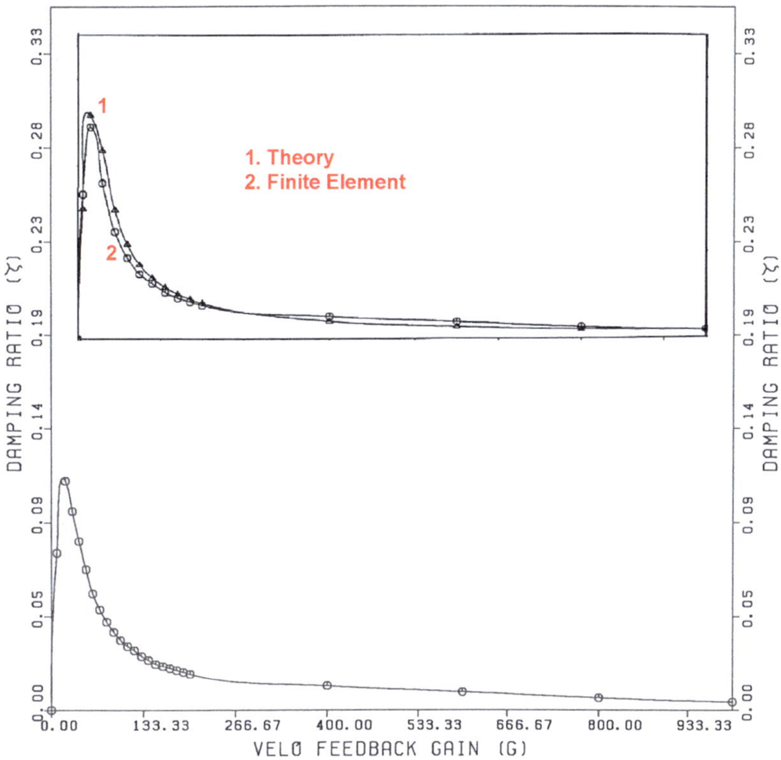

Fig. 6.10 Damping variation of the third mode with the velocity feedback

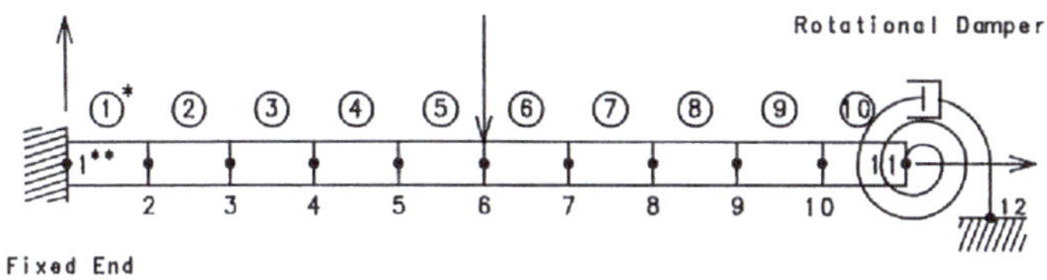

Fig. 6.11 Finite element model of a fixed-roller/slide beam with a rotational spring

6.3.4 Laboratory Experiments

Figure 6.12 shows a flowchart of the laboratory testing; Fig. 6.13 illustrates a beam model mounted on a shaker; and Fig. 6.14 shows the laboratory setup. Experimental data also suggested the same control behaviors in Fig. 6.8 (Johnson, 1990).

Table 6.6 Comparison of natural frequencies for different boundary conditions

Beams	Natural frequency f (Hz)			Solution technique
	1st	2nd	3rd	
1) Cantilever	65.8	355.9	878.8	(Theory)
Beam/actuator	65.9	357	884.3	(Finite element)
2) Fixed-roller beam	65.8	355.8	878.6	(Finite element)
3) Double fixed-end beam	66.2	358.7	885.9	(Finite element)

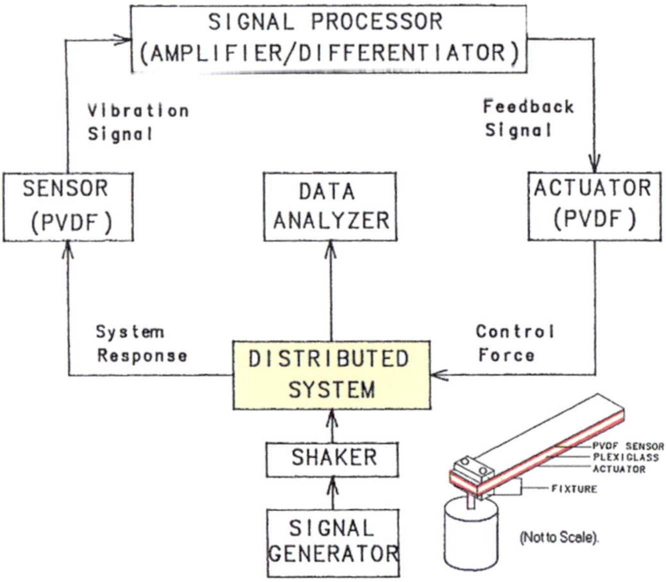

Fig. 6.12 Laboratory experiments of a PVDF laminated beam

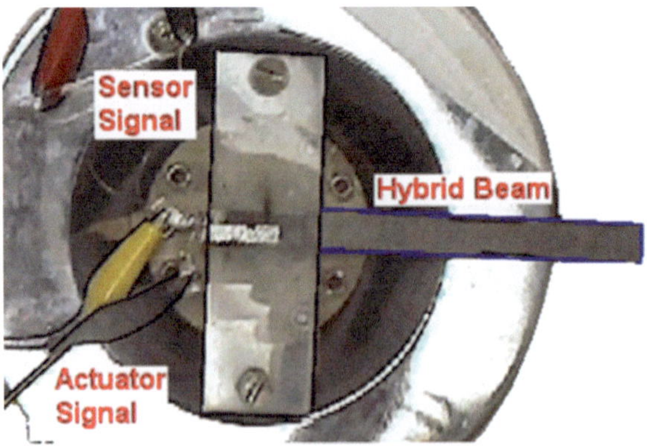

Fig. 6.13 A cantilever beam model mounted on a shaker

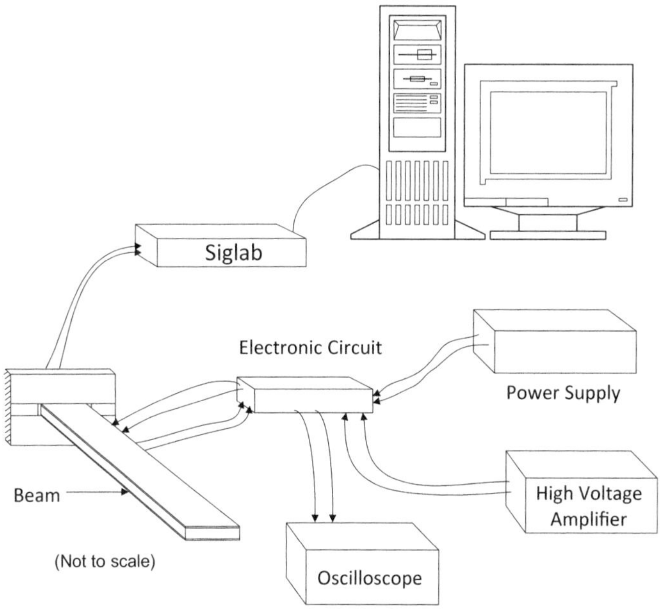

Fig. 6.14 Laboratory setup

6.4 Summary

Distributed control of a PVDF laminated cantilever beam was studied in this chapter. The laminated cantilever beam had a distributed piezoelectric sensor and a distributed actuator; both were surface bonded. Closed-loop feedback controls of the beam using the displacement and velocity signals were respectively evaluated and results compared.

The results showed that the displacement feedback controls were insignificant and the velocity feedback controls were much more effective. In the velocity feedback control, the system damping increased to an ultimate value and then gradually dropped down as the feedback gain continuously increased. This was caused by the additional constraint imposed by the boundary control moment at the free-end. The free-end boundary condition was gradually changing to a sliding-roller boundary condition as proved by finite element analyses and laboratory experiments.

References

Ewins, D.J., *Modal Testing: Theory and Practice*, 1986, Research Studies Press, England.

Harris, C.M. and Crede, C.E., 1976, *Shock and Vibration Handbook*, 2nd Ed., McGraw-Hill, 1976.

Johnson, D., 1990, *An Experimental Study on Feedback Control of a Piezoelectric Laminated Cantilever Beam*, NSF–REU Report, Department of Mechanical Engr., University of Kentucky, August 1990.

Liu, K.J, 1989, *A Theoretical Study on Distributed Active Vibration Suppression of Mechanical Systems with Integrated distributed Piezoelectric Sensor and Actuator*, MSME Thesis, Dept. of Mech. Engr., Univ. of Kentucky, 08/1989.

Tzou, H.S., Johnson, D. and Liu, K.J., "Boundary Transition and Nonlinear Control of Distributed Systems," *Stability, Vibration, and Control of Structures*, Vol.1: *Wave Motion, Intelligent Structures, and Nonlinear Mechanics*, A. Guran and D.J Inman, (Editors), pp. 163-193, World Scientific Publisher, 1994.

Tzou, H.S., Johnson, D. and Liu, K.J., "Damping Behavior of Cantilever Structronic Systems with Boundary Control," ASME Transactions, *Journal of Vibration & Acoustics*, Vol.121, pp. 402–407, July 1999.

Zaveri, K., *Modal Analysis of Large Structures – Multi–Exciter Systems*, Bruel & Kjaer Technical Review, 1985.

Chapter 7
Distributed Control of Plates with Segmented Sensors and Actuators

Integrating active materials, such as piezoelectrics, shape-memory alloys, electrostrictive materials, magnetostrictive materials, electrorheological fluids, flexoelectric materials, photostrictive materials, etc., with elastic structures transforms the structures from a completely passive system to an active adaptive system (Tzou, 1991a). With the rapid development of VLSI technologies, adding an "intelligence" to the structure could also become a reality in the near future (Tzou & Fukuda, 1991 & 1992). In the development of active piezoelectric/elastic structures, it was observed that symmetrically distributed piezoelectric sensors and actuators have observability and controllability deficiencies in monitoring and controlling elastic continua, especially with symmetrical boundary conditions, Sect. 3.8. With a single-piece symmetrically distributed sensor and actuator, anti-symmetrical structural modes may not be observable and controllable because the positive and negative sensing/control signals from different regions of the continua could cancel out each other. One method of improving the controllability and observability of distributed piezoelectric sensors/actuators is to segment them into a number of smaller pieces—sub-areas. Sensing and control effects of these segmented distributed sensors/actuators are quantitatively analyzed and studied in this chapter. Distributed vibration sensing and control of plates using segmented distributed sensors/actuators are investigated.

Theories on distributed sensing and control of shell continua using distributed piezoelectric layers were proposed recently. Applications to plate structures were also demonstrated in case studies (Tzou, 1991b; Tzou & Tseng, 1991). Ricketts studied a piezoelectric polymer flexural plate hydrophone (1981) and the frequency of completely free composite piezoelectric plates (1989). Lee (1990) proposed a theory for laminated piezoelectric plates with applications to distributed sensor/actuator designs. His formulations suggested that the distributed piezoelectric layers are capable of sensing and controlling bending, shearing, shrinking, and stretching effects of a plate. Burke and Hubbard (1990) studied distributed transducer control designs for thin plates with general boundary conditions. Modal control and observation deficiencies were also exploited. Dimitriadis, et al. (1991) investigated distributed vibration excitations of thin plates using piezoelectric actuators.

H. (HS) Tzou, *Piezoelectric Shells*, Solid Mechanics and Its Applications 247,
https://doi.org/10.1007/978-94-024-1258-1_7

As discussed in Sect. 4.6 of Chap. 4, there are two sensor/actuator design principles: one is **the segmentation technique** and the other is **the shaping technique**. In this chapter, distributed sensing and control (velocity and displacement feedback) of a simply supported plate laminated with sensor/actuator patches, i.e., the segmentation technique, is derived from the generic distributed sensing and control theories of thin shells presented in Chap. 4 (Tzou, 1991b). Modal sensing and control of the plate is derived using the modal expansion method; equivalent line control moments are also derived in the modal domain. Analytical solutions of transient and steady state responses of the simply supported plate with and without segmented distributed actuators are derived. Observation and control deficiencies of symmetrically distributed single-piece and segmented sensors/actuators are proved from analytical solutions. Advantages of segmented distributed sensors/actuators are discussed (Tzou and Fu, 1994a, b). (Emphasis is placed on the evaluation of distributed sensor/actuator segmentation.) Following the mathematical modeling and theoretical analysis, there are four case studies presented in this chapter:

1) Free Vibration Analysis: Natural frequencies and mode shapes of the plate are studied. Mode shapes are presented in Appendix.
2) Parametric Studies: Controlled damping ratios of the plate with a number of system parameters, e.g., feedback gains, mode orders, plate sizes, plate thickness, effective actuator areas, and other material properties, are investigated.
3) Time-history Analyses: Transient responses of the plate with and without feedback controls are studied.
4) Control Algorithms: Two control algorithms, namely 1) the proportional feedback and 2) the Lyapunov control, are implemented and their modal control effectiveness evaluated (Fu, 1990).

7.1 Mathematical Modeling

In general, closed-loop system equations, in three principle directions, of a generic shell continuum coupled with distributed piezoelectric sensors and actuators can be expressed in a simplified form (Tzou, 1991a & b):

$$L_i(u_1, u_2, u_3) - \rho h \ddot{u}_i = -F_i - L_i^a\left(M_{ij}^a, N_{ij}^a\right), i = 1, 2, 3, \qquad (7.1.1)$$

where i denotes the i-th direction; L_i denotes the elastic (mechanical) Love's operator (Sect. 3.4 of Chap. 3 and Sect. 4.4 of Chap. 4); u_i is the displacement; ρ is the mass density; h is the shell thickness; $\ddot{u}_i$ is the acceleration; F_i is the mechanical excitation; L_i^a denotes the Love's operator of the induced control membrane forces/bending moments; M_{ij}^a and N_{ij}^a are the actuator induced control moments and membrane forces, respectively. For an uncoupled transverse vibration of a thin elastic shell/plate, the generic system equation is reduced to

$$L_3(u_3) - \rho h\ddot{u}_3 = -F_3 - L_3^a\left(M_{ij}^a\right). \tag{7.1.2}$$

If the shell continuum has an inherent viscous damping, the system equation can be further expressed as (Soedel, 1981):

$$L_3(u_3) - \rho h\ddot{u}_3 - c\dot{u} = -F_3 - L_3^a\left(M_{ij}^a\right), \tag{7.1.3}$$

where c is the *equivalent viscous damping factor*. For piezoelectricity induced damping, the control counteracting force/moment is usually assumed to be proportional to the velocity. For a free vibration analysis (eigenvalue analysis), all external excitations (both mechanical and feedback) and damping forces are zero, i.e., $F_i = 0$, $L_i^a(M_{ij}^a, N_{ij}^a) = 0$, and $c\dot{u} = 0$. It is also assumed that all points on the continuum oscillate harmonically at one of the natural frequencies, i.e., $u_i(x, y) = U_i(x, y)e^{j\omega t}$. Thus, one can derive

$$L_i(U_{1mn}, U_{2mn}, U_{3mn}) + \rho h\omega_{mn}^2 U_{imn} = 0, \tag{7.1.4}$$

where ω_{mn} is the natural frequency of the mn-th mode, $U_{imn}(\alpha_1, \alpha_2)$ is the mode shape function—a spatial function of coordinates, and m and n are integers, m = 1,2,…,∞ and n = 1,2,…,∞. It should be noted that the controlled responses are forced responses at one of the natural frequencies when feedback controls are considered in the later analyses.

7.1.1 Plate with Distributed Sensor and Actuator

A rectangular plate with two biaxial oriented piezoelectric polymeric layers, one serving as a distributed sensor and the other a distributed actuator, is used as a physical system in this study, Fig. 7.1. Segmentation of distributed sensors and actuators will be discussed later. (Note that which piezoelectric layer serves as a sensor or actuator is not crucial. In this case, the bottom layer serves as a sensor and the top an actuator.) The piezoelectric layers are assumed to be perfectly laminated on the surfaces of the plate, and the physical properties of the bonding material are neglected. It is assumed that the transverse bending oscillation dominates the plate motion i.e., the in-plane membrane oscillations are neglected, in the later analyses.

It is assumed that the in-plane twisting piezoelectric constant is insignificant, i.e., $M_{xy}^a = 0$. The piezoelectric layers (h^s and h^a) are much thinner than the plate thickness h; thus, the effect of piezoelectric layer thickness is also neglected in the analyses. The system equation for the plate with distributed sensor and actuator layers can be written as

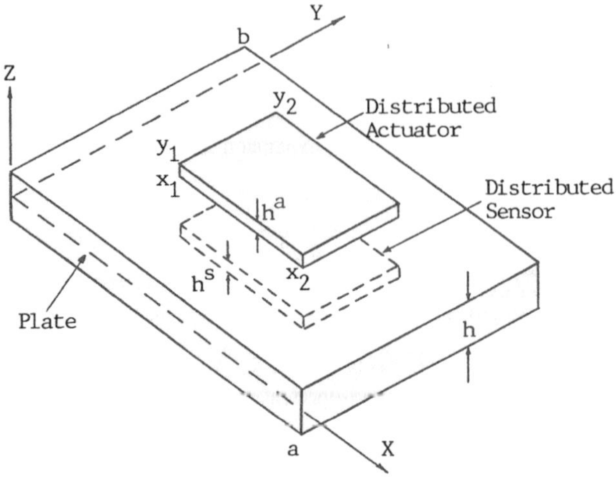

Fig. 7.1 A plate with distributed sensor and actuator layers

$$D\left(\frac{\partial^4 u_3}{\partial x^4} + 2\frac{\partial^4 u_3}{\partial x^2 \partial y^2} + \frac{\partial^4 u_3}{\partial y^4}\right) + \rho h \ddot{u}_3 + c\dot{u}_3 = F_3 + \frac{\partial^2 M^a_{xx}}{\partial x^2} + \frac{\partial^2 M^a_{yy}}{\partial y^2}, \quad (7.1.5)$$

where D is the bending stiffness, $D = (Yh^3)/12[(1-\mu^2)]$; Y is Young's modulus and μ is Poisson's ratio. M^a_{xx} and M^a_{yy} are distributed control moments (Tzou, 1991b) and they can be expressed in a modal expansion form (Tzou, 1992):

$$M^a_{xx} = \sum_{m=1}^{\infty}\sum_{n=1}^{\infty} \eta_{mn}(t) r^a_1 d_{31} Y_p \mathcal{G} \phi^S_{mn}, \quad (7.1.6)$$

$$M^a_{yy} = \sum_{m=1}^{\infty}\sum_{n=1}^{\infty} \eta_{mn}(t) r^a_2 d_{32} Y_p \mathcal{G} \phi^S_{mn}. \quad (7.1.7)$$

Note that η_{mn} is the modal participation factor; r^a_i denotes the moment arm, a distance measured from the neutral surface to the mid-plane of the actuator and $r^a_i = (h + h^a_i)/2$; d_{3_i} is the piezoelectric strain constant, Y_p is Young's modulus of the piezoelectric layer, and $\mathcal{G}$ is the feedback gain. (Note that $\mathcal{G}$ could also be defined as a ***modal feedback gain*** $\mathcal{G}_{mn}$ as discussed in Sect. 4.3 of Chap. 4.) ϕ^S_{mn} is a feedback voltage which could be a reference voltage (open-loop) or a sensor voltage (closed-loop) and it is assumed to be a spatial function. Note that the resultant control actions of the surface laminated actuator consist of two components: 1) in-plane membrane control forces and 2) out-of-plane control bending moments, Sect. 4.3 in Chap. 4. However, since the plate is flat, the radii of curvatures of the two in-plane axes are infinite. Thus, the membrane effect does not contribute any control action to the plate in the transverse oscillation.

In the case of a closed-loop feedback, ϕ^S_{mn} is the mn-th ***unit modal sensing signal*** of a distributed piezoelectric sensor expressed as a function of mode shape function. In practice, a narrow band-pass filter set at the mn-th natural frequency would be required to obtain the ***unit modal sensing signal***,

$$\phi^S_{mn}(x^*, y^*) = -h^S\left[h_{31}r^S_1\left(\frac{\partial^2 U_{3mn}(x^*, y^*)}{\partial x^2}\right) + h_{32}r^S_2\left(\frac{\partial^2 U_{3mn}(x^*, y^*)}{\partial y^2}\right)\right], \tag{7.1.8a}$$

$$\phi^S_{mn} = -\frac{h^S}{S^e}\int_{S^e}\left[h_{31}r^S_1\left(\frac{\partial^2 U_{3mn}}{\partial x^2}\right) + h_{32}r^S_2\left(\frac{\partial^2 U_{3mn}}{\partial y^2}\right)\right]dS^e, \tag{7.1.8b}$$

where h^s is the thickness of the distributed piezoelectric sensor layer; S^e is the effective electrode sensor area; h_{31} and h_{32} are the piezoelectric constants; r^s_i denotes the distance measured from the neutral surface to the mid-plane of the sensor layer and $r^s_i = (h + h^s_i)/2$. Equation (7.1.8a) denotes the point signal output; Eq. (7.1.8b) denotes the averaged signal output, see Sect. 4.2 of Chap. 4. Note that the sensing signals are primarily contributed by the bending strains of the plate in transverse oscillations. Membrane strains are not considered.

7.1.2 Modal Expansion and Vibration Controls

Based on the *modal expansion technique*, the dynamic response of a distributed system can be represented by a sum of the responses of all participating modes, i.e.,

$$u_3(x, y, t) = \sum_{m=1}^{\infty}\sum_{n=1}^{\infty}\eta_{mn}(t)U_{3mn}(x, y), \tag{7.1.9}$$

where U_{3mn} is the unit spatial part (*mode shape function*) and η_{mn} is the temporal part—an amplitude factor, i.e., the *modal participation factor*. Substituting the modal expression into the system equation leads to an equation in terms of modal participation factors. Using the modal orthogonality of natural modes, one can derive (Fu, 1990)

$$\begin{aligned}&\ddot{\eta}_{mn} + \frac{c}{\rho h}\dot{\eta}_{mn} + \omega^2_{mn}\eta_{mn}\\ &= \hat{F}_{mn} + \frac{1}{\rho h N_{mn}}\int_x\int_y\left(\frac{\partial^2 M^a_{xx}}{\partial x^2} + \frac{\partial^2 M^a_{yy}}{\partial y^2}\right)U_{3mn}dxdy\end{aligned}, \tag{7.1.10}$$

where the mn-th modal force $\hat{F}_{mn}$ induced by the mechanical excitation F_3 is defined as

$$\hat{F}_{mn} = \frac{1}{\rho h N_{mn}} \int_x \int_y F_3 U_{3mn} dxdy, \tag{7.1.11a}$$

and N_{mn} is defined by the squared mode shape functions:

$$N_{mn} = \int_x \int_y U_{3mn}^2 dxdy. \tag{7.1.11b}$$

(Note that this generic expression is only for transverse vibration modes. Generic expressions encompassing all three directions were presented in Sect. 3.5 of Chap. 3.) For a simply supported plate with a dimension of a × b, the transverse mode shape function is

$$U_{3mn} = \sin\frac{m\pi x}{a} \sin\frac{n\pi y}{b}. \tag{7.1.12}$$

Substituting U_{3mn} into N_{mn} and carrying out the surface integration, one can derive

$$N_{mn} = \frac{ab}{4}. \tag{7.1.13}$$

Using the definition of damping ratio for a single degree of freedom system, one can rewrite the modal equation as

$$\begin{aligned} &\ddot{\eta}_{mn} + 2\zeta_{mn}\omega_{mn}\dot{\eta}_{mn} + \omega_{mn}^2\eta_{mn} \\ &= \hat{F}_{mn} + \frac{1}{\rho h N_{mn}} \int_x \int_y \left(\frac{\partial^2 M_{xx}^a}{\partial x^2} + \frac{\partial^2 M_{yy}^a}{\partial y^2} \right) U_{3mn} dxdy, \end{aligned} \tag{7.1.14}$$

where the *modal damping ratio* ζ_{mn} is defined as

$$\zeta_{mn} = \frac{c}{2\rho h \omega_{mn}}. \tag{7.1.15}$$

Substituting the modal expressions of control bending moments into the modal equation yields

$$\begin{aligned} &\ddot{\eta}_{mn} + 2\zeta_{mn}\omega_{mn}\dot{\eta}_{mn} + \omega_{mn}^2\eta_{mn} \\ &= \hat{F}_{mn} + \frac{1}{\rho h N_{mn}} \int_x \int_y \left[\sum_{m=1}^{\infty} \sum_{n=1}^{\infty} \eta_{mn} \left(\frac{\partial^2 M_{xmn}}{\partial x^2} + \frac{\partial^2 M_{ymn}}{\partial y^2} \right) \right] \\ &\quad \cdot U_{3mn} dxdy, \end{aligned} \tag{7.1.16}$$

where M_{xmn} and M_{ymn} are the ***unit modal control moments*** defined by the ***unit modal sensing signal*** ϕ_{mn}^s:

$$M_{xmn} = r_1^a d_{31} Y_p \mathcal{G} \phi_{mn}^S, \tag{7.1.17}$$

$$M_{ymn} = r_2^a d_{32} Y_p \mathcal{G} \phi_{mn}^S \tag{7.1.18}$$

Since the time function of feedback control input is also a function of system modal participation factors, the mn-th term in the sum, i.e.,

$$\frac{1}{\rho h N_{mn}} \eta_{mn} \int_x \int_y \left(\frac{\partial^2 M_{xmn}}{\partial x^2} + \frac{\partial^2 M_{ymn}}{\partial y^2} \right) U_{3mn} dxdy, \tag{7.1.19}$$

represents the control effect of the mn-th mode; it can be moved to the left side and combined with the appropriate system term, e.g., η_{mn} for a *displacement feedback* or $\dot{\eta}_{mn}$ for a *velocity feedback*. (These two feedback controls will be discussed shortly.) All the other terms within the sum represent the cross coupling control effects from all other residual modes. Consequently, these two feedback related components can be separated and respectively treated in two control schemes. For convenience, define a ***modal feedback factor*** $\hat{M}_{mn}$ (Fu, 1990):

$$\hat{M}_{mn} = \frac{1}{\rho h N_{mn}} \int_x \int_y \left(\frac{\partial^2 M_{xmn}}{\partial x^2} + \frac{\partial^2 M_{ymn}}{\partial y^2} \right) U_{3mn} dxdy; \tag{7.1.20}$$

and define $\hat{T}_{mn}$ to denote the cross coupling feedback effects from the residual modes, i.e.,

$$\hat{T}_{mn} = \frac{1}{\rho h N_{mn}} \sum_{\substack{p=1 \\ p \neq m}}^{\infty} \sum_{\substack{q=1 \\ q \neq n}}^{\infty} \eta_{pq} \int_x \int_y \left(\frac{\partial^2 M_{xpq}}{\partial x^2} + \frac{\partial^2 M_{ypq}}{\partial y^2} \right) U_{3mn} dxdy. \tag{7.1.21}$$

The cross couplings could introduce spillovers resulting in unstable oscillations.

7.1.2.1 Displacement Feedback Control

For a *displacement feedback* proportional control, the modal equation can be rewritten as

$$\ddot{\eta}_{mn} + 2\zeta_{mn}\omega_{mn}\dot{\eta}_{mn} + \left(\omega_{mn}^2 - \hat{M}_{mn}\right)\eta_{mn} = \hat{F}_{mn} + \hat{T}_{mn}, \tag{7.1.22}$$

where

$$\hat{T}_{mn} = \frac{1}{\rho h N_{mn}} \sum_{\substack{p=1 \\ p \neq m}}^{\infty} \sum_{\substack{q=1 \\ q \neq n}}^{\infty} \eta_{pq} \int_x \int_y \left(\frac{\partial^2 M_{xpq}}{\partial x^2} + \frac{\partial^2 M_{ypq}}{\partial y^2} \right) U_{3mn} dxdy, \tag{7.1.23}$$

and $\hat{F}_{mn}=\frac{1}{\rho h N_{mn}}\int_x\int_y F_3 U_{3mn} dxdy$ and $N_{mn}=\int_x\int_y U_{3mn}^2 dxdy$. Both are defined for the transverse vibration modes U_{3mn} only.

7.1.2.2 Velocity Feedback Control

For a *velocity feedback* control (derivative control), the feedback time function is a first derivative of modal participation factor, and the induced moments become

$$M_{xx}^a = \sum_{m=1}^{\infty}\sum_{n=1}^{\infty} \dot{\eta}_{mn}(t) M_{xmn}, \tag{7.1.24}$$

$$M_{yy}^a = \sum_{m=1}^{\infty}\sum_{n=1}^{\infty} \dot{\eta}_{mn}(t) M_{ymn}. \tag{7.1.25}$$

Moving the common terms to the left and combining them with $\dot{\eta}_{mn}$, one obtains

$$\begin{aligned} &\ddot{\eta}_{mn} + \left(2\zeta_{mn}\omega_{mn} - \hat{M}_{mn}\right)\dot{\eta}_{mn} + \omega_{mn}^2\eta_{mn} \\ &= \hat{F}_{mn} + \hat{T}_{mn}' \end{aligned} \tag{7.1.26}$$

where

$$\hat{T}_{mn}' = \frac{1}{\rho h N_{mn}} \sum_{\substack{p=1 \\ p\neq m}}^{\infty}\sum_{\substack{q=1 \\ q\neq n}}^{\infty} \dot{\eta}_{pq} \int_x\int_y \left(\frac{\partial^2 M_{xpq}}{\partial x^2} + \frac{\partial^2 M_{ypq}}{\partial y^2}\right) U_{3mn} dxdy. \tag{7.1.27}$$

The ***modal feedback factor*** $\hat{M}_{mn}$ is treated as a system parameter in the later analyses. Once the ***modal feedback factor*** is defined, the new system parameter can be estimated without solving for the individual modal participation factor $\eta_{mn}(t)$. The *modal feedback factors* for the single sensor/actuator and quarterly segmented sensor/actuator configurations will be derived later.

Note that the above evaluation procedure is in a generic form. It can be employed for other geometric configurations whose mode shape functions, either exact or approximate, are known. These configurations include beams with most common boundary conditions, some plates, simply supported cylindrical panels or cylinders, and some other shell structures (Soedel, 1981). Again, note that the feedback is actually related to strains in the displacement feedback and "strain/unit time" in the velocity feedback.

7.2 One-Piece Symmetrically Distributed Sensor/Actuator

In a closed-loop feedback system, the induced distributed control moments are originally initiated from the distributed sensor signal. In this section, formulation of a distributed sensing signal and distributed control moments for a simply supported plate with a single-piece distributed piezoelectric sensor and an actuator is presented. Note that a simply supported plate is used in this study. (Segmentation of distributed sensor and actuator layers will be discussed in the next section.)

7.2.1 *Distributed Sensing*

The mn-th mode shape function $U_{3mn}(x,y)$, in the transverse direction, for a simply supported rectangular plate is given by $U_{3mn}(x,y) = \sin\frac{m\pi x}{a}\sin\frac{n\pi y}{b}$. The unit modal sensor signal (averaged) is $\phi^S_{mn} = -\frac{h^s}{S^e}\int_{S^e}[h_{31}r_1^s(\partial^2 U_{3mn}/\partial x^2) + h_{32}r_2^s(\partial^2 U_{3mn}/\partial y^2)]dS^e$. It is assumed that the sensor layer is fully distributed on the plate surface. Substituting the mode shape function into the sensor equation, integrating over the whole sensor surface, x = (0, a) and y = (0, b), and imposing the modal orthogonality, one can derive

$$\phi^S_{mn} = \frac{h^S}{mn}\left[h_{31}r_1^S\left(\frac{m}{a}\right)^2 + h_{32}r_2^S\left(\frac{n}{b}\right)^2\right](1-\cos m\pi)(1-\cos n\pi). \qquad (7.2.1)$$

Define a mn-th ***modal sensitivity*** S_{mn} for the mn-th mode:

$$S_{mn} = \frac{h^S}{mn}\left[h_{31}r_1^S\left(\frac{m}{a}\right)^2 + h_{32}r_2^S\left(\frac{n}{b}\right)^2\right], \qquad (7.2.2)$$

such that the sensor equation, Eq. (7.2.1), can be simplified to

$$\phi^S_{mn} = S_{mn}(1-\cos m\pi)(1-\cos n\pi). \qquad (7.2.3)$$

Note that the output signal vanishes, i.e., zero output, if the mode order either m or n is an even number. This observability problem will be further discussed later. Additional feedback controllability problem induced by the constant voltage will be answered in the next section.

7.2.2 *Distributed Vibration Control*

It is assumed that the distributed piezoelectric actuator layer covers the plate from locations x_1 to x_2 in the x direction and from y_1 to y_2 in the y direction, Fig. 7.1.

(Note that $x_1 = 0$, $x_2 = a$, $y_1 = 0$, and $y_2 = b$ for a fully distributed actuator layer.) Since the feedback voltage is constant over the whole actuator surface if the electrode resistance is ignored, the induced control moment is also uniformly distributed on the actuator covered area. As discussed previously, the moment function can be separated into a temporal function part and a spatial function part: $M^a_{xx} = \sum_{m=1}^{\infty}\sum_{n=1}^{\infty} \eta_{mn}(t)M_{xmn}$ and $M^a_{yy} = \sum_{m=1}^{\infty}\sum_{n=1}^{\infty} \eta_{mn}(t)M_{ymn}$. Thus, one can express the moment spatial distribution using a unit step function u_s:

$$u_S(x - x_i) = \begin{cases} 1, & \text{for } x > x_i; \\ 0, & x \leq x_i. \end{cases} \tag{7.2.4}$$

Then the distributed control moments M^*_{imn} 's can be expressed as

$$M^*_{xmn} = M_{xmn}[u_S(x - x_1) - u_S(x - x_2)][u_S(y - y_1) - u_S(y - y_2)], \tag{7.2.5}$$

$$M^*_{ymn} = M_{ymn}[u_S(x - x_1) - u_S(x - x_2)][u_S(y - y_1) - u_S(y - y_2)], \tag{7.2.6}$$

where the magnitudes for the proportional feedback M_{xmn} and M_{ymn} were defined previously. Substituting the sensor signal of the distributed sensor into the unit modal control moments, one obtains

$$M_{xmn} = r_1^a d_{31} Y_p \mathcal{G} S_{mn}(1 - \cos m\pi)(1 - \cos n\pi), \tag{7.2.7}$$

$$M_{ymn} = r_2^a d_{32} Y_p \mathcal{G} S_{mn}(1 - \cos m\pi)(1 - \cos n\pi). \tag{7.2.8}$$

It can be observed that the control moments for a symmetrically distributed actuator are zero if the mode order either m or n is even, i.e., anti-symmetrical modes usually.

7.2.3 Equivalent Line Moment

It is assumed that a uniform moment distribution is characterized as a resultant bending phenomenon which can be equated by a set of couples or moments acting at both ends of the distribution. Using an equivalent external distributed moment approximation and the modal expansion technique, one can derive a modal equation as

$$\ddot{\eta}_{mn} + 2\zeta_{mn}\omega_{mn}\dot{\eta}_{mn} + \omega_{mn}^2\eta_{mn} = \hat{F}_{mn} + \frac{1}{\rho h N_{mn}}\int_x\int_y\left(\frac{\partial T_{11}}{\partial x} + \frac{\partial T_{22}}{\partial y}\right)U_{3mn}\,dx\,dy, \tag{7.2.9}$$

where T_{ii} is the distributed moment acting in the i-th direction with a unit N·m/m^2 (Soedel, 1981). Comparing this equation with Eq. (7.1.14), one finds

$$\frac{\partial M_{xx}^a}{\partial x} = T_{11}, \tag{7.2.10a}$$

$$\frac{\partial M_{yy}^a}{\partial y} = T_{22}. \tag{7.2.10b}$$

Define the identity:

$$\frac{d}{dx}[u_S(x - x_i)] = \delta(x - x_i), \tag{7.2.11}$$

where δ(x) is a Dirac delta function defined as

$$\delta(x - x_i) = \begin{cases} 1, & \text{for } x = x_i; \\ 0, & x \neq x_i; \end{cases} \tag{7.2.12}$$

which has a dimension (1/m) (Soedel, 1976). Using the Dirac delta function, one can derive the equivalent line control moments in the modal domain:

$$\begin{aligned} \frac{\partial M_{xx}^a}{\partial x} &= \sum_{m=1}^{\infty}\sum_{n=1}^{\infty} \eta_{mn} M_{xmn}[\delta(x - x_1) - \delta(x - x_2)][u_S(y - y_1) - u_S(y - y_2)] \\ &= T_{11}, \end{aligned} \tag{7.2.13}$$

$$\begin{aligned} \frac{\partial M_{yy}^a}{\partial y} &= \sum_{m=1}^{\infty}\sum_{n=1}^{\infty} \eta_{mn} M_{ymn}[u_S(x - x_1) - u_S(x - x_2)][\delta(y - y_1) - \delta(y - y_2)] \\ &= T_{22}. \end{aligned} \tag{7.2.14}$$

The above two equations imply that the uniform moment is equivalent to two external ***equivalent distributed line moments*** acting at both ends of the distribution. The units of two moment definitions are consistent with each other. Thus, the equivalent line moments M_{imn}^*'s representing the control moment effects on the boundaries of actuator distribution, fully or partially distributed, are used. (Lee (1990), Burke/Hubbard (1990), and Dimitriadis, et al. (1991) also suggested the line-moment approximation in global coordinates based in their respective studies.) Modal controllability and observability of the single-piece distributed sensor and actuator, fully or partially distributed, will be emphasized in the later analyses.

7.2.4 Modal Feedback Factor and Modal Equations

Using the ***equivalent distributed line moment*** concepts, one can redefine the ***modal feedback factor*** $\hat{M}_{mn}$ as

$$\hat{M}_{mn} = \frac{1}{\rho h N_{mn}} \int_x \int_y \left(\frac{\partial^2 M^*_{xmn}}{\partial x^2} + \frac{\partial^2 M^*_{ymn}}{\partial y^2} \right) U_{3mn} dxdy, \tag{7.2.15}$$

where M^*_{xmn} and M^*_{ymn} are the *equivalent distributed line moments*. Depending on the feedback algorithms, this modal feedback factor may contribute to either damping (in the velocity feedback) or elasticity (in the displacement feedback) in the mn-th modal coordinate equation. Substituting Eqs. (7.2.5), (7.2.13) and mode shape function into the *modal feedback factor* $\hat{M}_{mn}$ and integrating the two moment terms respectively, one can obtain

$$\begin{aligned} &\int_x \int_y \frac{\partial^2 M^*_{xmn}}{\partial x^2} U_{3mn} dxdy \\ &= \int_0^a \int_0^b M_{xmn} \frac{\partial}{\partial x} [\delta(x - x_1) - \delta(x - x_2)] \cdot [u_S(y - y_1) - u_S(y - y_2)] \\ &\cdot \sin\frac{m\pi x}{a} \sin\frac{n\pi y}{b} dxdy. \end{aligned} \tag{7.2.16}$$

Integration by parts of the above equation gives

$$\begin{aligned} &\int_x \int_y \frac{\partial^2 M^*_{xmn}}{\partial x^2} U_{3mn} dxdy \\ &= -M_{xmn} \frac{mb}{na} \left(\cos\frac{m\pi x_1}{a} - \cos\frac{m\pi x_2}{a} \right) \left(\cos\frac{n\pi y_1}{b} - \cos\frac{n\pi y_2}{b} \right) \end{aligned}. \tag{7.2.17}$$

Similarly,

$$\begin{aligned} &\int_x \int_y \frac{\partial^2 M^*_{ymn}}{\partial y^2} U_{3mn} dxdy \\ &= -M_{ymn} \frac{na}{mb} \left(\cos\frac{m\pi x_1}{a} - \cos\frac{m\pi x_2}{a} \right) \left(\cos\frac{n\pi y_1}{b} - \cos\frac{n\pi y_2}{b} \right) \end{aligned}. \tag{7.2.18}$$

Then, the ***modal feedback factor*** $\hat{M}_{mn}$ for the equivalent boundary line control moments of a simply supported plate can be expressed as

$$\hat{M}_{mn} = \frac{-1}{\rho h N_{mn}} \left(M_{xmn} \frac{mb}{na} + M_{ymn} \frac{na}{mb} \right) \left(\cos\frac{m\pi x_1}{a} - \cos\frac{m\pi x_2}{a} \right) \cdot \left(\cos\frac{n\pi y_1}{b} - \cos\frac{n\pi y_2}{b} \right). \tag{7.2.19}$$

Note again that $\hat{M}_{mn}$ vanishes for all even modes if x_1 and x_2 or y_1 and y_2 are symmetrically located about the mid-span. As discussed previously, the modal feedback factor can be combined with the modal damping term in the **velocity feedback**:

$$\ddot{\eta}_{mn} + \left(2\zeta_{mn}\omega_{mn} - \hat{M}_{mn}\right)\dot{\eta}_{mn} + \omega_{mn}^2\eta_{mn} = \hat{F}_{mn} + \hat{T}'_{mn}, \tag{7.2.20}$$

where $\hat{M}_{mn}$ represents the distributed control effect and $\hat{T}'_{mn}$ denotes the cross coupling effects resulting from all other participating modes due to the closed-loop feedback. Define a controlled modal damping ratio ζ'_{mn} as

$$\zeta'_{mn} = \zeta_{mn} - \frac{\hat{M}_{mn}}{2\omega_{mn}}, \tag{7.2.21}$$

where the system inherent modal damping ratio ζ_{mn} is assumed known from laboratory experiments. Then, the modified mn-th modal equation becomes

$$\ddot{\eta}_{mn} + 2\zeta'_{mn}\omega_{mn}\dot{\eta}_{mn} + \omega_{mn}^2\eta_{mn} = \hat{F}_{mn} + \hat{T}'_{mn}. \tag{7.2.22}$$

Note that $\hat{T}'_{mn}$ is the coupling terms from the residual modes.

7.2.5 Modal Controllability and Observability

As discussed previously, the derived sensing and control equations suggest that the single-piece symmetrically distributed piezoelectric sensor and actuator have deficiencies in sensing and controlling all even modes—anti-symmetrical modes—of the simply supported plate. Thus, there are severe problems on observability and controllability for the even-order modes of the plate. According to the mode shape $\sin\frac{m\pi x}{a}\sin\frac{n\pi y}{b}$, the sensor layer is stretched at some locations and compressed at others, so that the signs of generated local electric charges are different. The (global) resultant charge is zero due to these equal and opposite instant charges.

Even if the observability problem could be solved, say by other sensing techniques, there is still modal controllability problem for the single-piece symmetrically distributed actuator. For example, when m = 1 and n = 1, the moments at both ends counteract the motion resulting in significant control effects. However, when either m or n = 2, the moment at one end suppresses the motion, while the other augments it. As a result, only very minimal, or zero, control effect is generated.

In order to ensure controllabilities for most of the modes, e.g., both even and odd modes, distributed piezoelectric sensors and actuators need to be redesigned and/or new control strategies developed. One method to improve the sensing and control performance is to segment the distributed sensor and actuator layers into a number of sub-areas so that the charge/voltage cancellation problems can be minimized or prevented. As an example, the single-piece distributed sensor and actuator layers are equally divided into four smaller pieces in which sensor signals are fed back into their collocated actuator layers. In the next section, the performances of these segmented distributed sensors and actuators are evaluated using analytical techniques.

7.3 Segmentation of Distributed Sensors and Actuators

As discussed in the previous section, theoretical derivation suggests that there are observability and controllability problems for anti-symmetrical modes if the single-piece symmetrically distributed sensor and actuator layers are used. One method of overcoming this problem is to segment both the distributed piezoelectric sensor and actuator into collocated subsections or sub-areas. That is, each pair of sensor and actuator consists of top and bottom pieces of layers at the same location of the plate respectively. Then each segmented sensor can detect the local motion state. The processed sensing signal is fed back into the collocated distributed actuator leading to a localized control effect for that sub-area only. Detailed formulation of a segmented distributed piezoelectric sensor/actuator design for the simply supported plate is presented in this section (Fu, 1990).

7.3.1 Segmented Distributed Sensors

It is assumed that the distributed piezoelectric sensor is divided into four equally sized segments, i.e., cut along the center lines. Figure 7.2 illustrates this segmentation. The four segmented sensors still cover the whole surface of the plate. A small gap is left between two adjacent sensor segments to prevent them from short circuiting, but it is ignored in the mathematical model due to its smallness. Each sensor segment responds to the local motion state and generates a signal output.

For the mn-th mode, the segment-1 sensor (with an effective electrode area S_1^e) gives an output signal $\phi_{mn}^{S_1}$ as

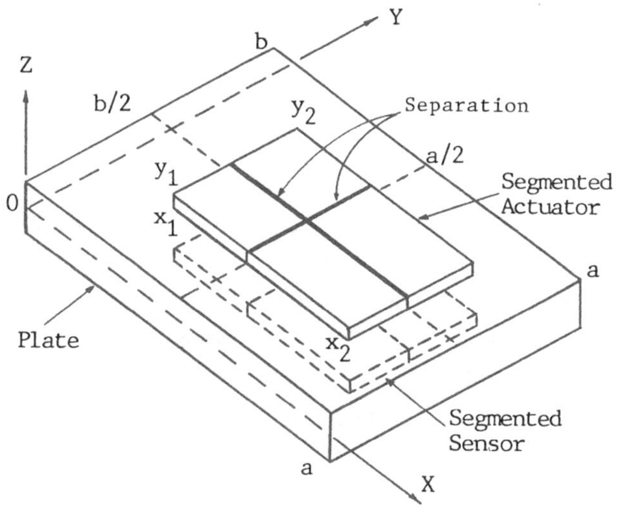

Fig. 7.2 Segmentation of distributed piezoelectric sensors and actuators

$$\begin{aligned}\phi_{mn}^{S_1} &= \frac{h^S}{S^{e_1}}\left[h_{31}r_1^S\left(\frac{m\pi}{a}\right)^2 + h_{32}r_2^S\left(\frac{n\pi}{b}\right)^2\right]\int_0^{\frac{a}{2}}\int_0^{\frac{b}{2}} \sin\frac{m\pi x}{a}\sin\frac{n\pi y}{b}dxdy \\ &= 4S_{mn}\left(1-\cos\frac{m\pi}{2}\right)\left(1-\cos\frac{n\pi}{2}\right),\end{aligned} \tag{7.3.1}$$

where S_{mn} is a ***modal sensitivity*** for the mn-th mode, i.e., $S_{mn} = \frac{h^s}{mn}\left[h_{31}r_1^s\left(\frac{m}{a}\right)^2 + h_{32}r_2^s\left(\frac{n}{b}\right)^2\right]$. And $r_1^s = r_2^s = (h+h^s)/2$. The other three segmented sensors provide output signals as

$$\phi_{mn}^{S_2} = 4S_{mn}\left(1-\cos\frac{m\pi}{2}\right)\left(\cos\frac{n\pi}{2}-\cos n\pi\right), \tag{7.3.2}$$

$$\phi_{mn}^{S_3} = 4S_{mn}\left(\cos\frac{m\pi}{2}-\cos m\pi\right)\left(\cos\frac{n\pi}{2}-\cos n\pi\right), \tag{7.3.3}$$

$$\phi_{mn}^{S_4} = 4S_{mn}\left(\cos\frac{m\pi}{2}-\cos m\pi\right)\left(1-\cos\frac{n\pi}{2}\right). \tag{7.3.4}$$

Note that the output signals will vanish only when either m or n is a multiple of four. Note that if n = m = 1, $\phi_{mn}^{S_1} = \phi_{mn}^{S_2} = \phi_{mn}^{S_3} = \phi_{mn}^{S_4}$. If n = m = 2, $\phi_{mn}^{S_1} = \phi_{mn}^{S_3}$ and $\phi_{mn}^{S_2} = \phi_{mn}^{S_4} = -\phi_{mn}^{S_1}$. In addition, the global averaged signal of the four-piece segmented sensors is equivalent to that of the single-piece sensor, Eq. (7.2.3). Thus, this segmentation technique improves the modal observability of plate natural modes.

7.3.2 *Segmented Distributed Actuators*

It is assumed that a distributed piezoelectric layer covers the center part of the plate from x_1 to x_2 in the x direction and from y_1 to y_2 in the y direction, and it is further equally divided into four segmented distributed actuators as shown in Fig. 7.2. According to the sign changes of sensor outputs at different modes, Eqs. (7.3.1)–(7.3.4), the signs of feedback voltages to each segmented actuator varies and so do distributions of induced moments. Thus, the control moment distributions can be written in the form of step function u_s's as

$$\begin{aligned} M^*_{xmn} = M_{xmn}\Big[u_S(x-x_1) - u_S\Big(x-\frac{a}{2}\Big) - (-1)^m u_S\Big(x-\frac{a}{2}\Big) \\ + (-1)^m u_S(x-x_2)\Big] \cdot \Big[u_S(y-y_1) - u_S\Big(y-\frac{b}{2}\Big) \\ -(-1)^n u_S\Big(y-\frac{b}{2}\Big) + (-1)^n u_S(y-y_2)\Big], \end{aligned} \tag{7.3.5}$$

$$\begin{aligned} M^*_{ymn} = M_{ymn}\Big[u_S(x-x_1) - u_S\Big(x-\frac{a}{2}\Big) - (-1)^m u_S\Big(x-\frac{a}{2}\Big) \\ + (-1)^m u_S(x-x_2)\Big] \cdot \Big[u_S(y-y_1) - u_S\Big(y-\frac{b}{2}\Big) \\ -(-1)^n u_S\Big(y-\frac{b}{2}\Big) + (-1)^n u_S(y-y_2)\Big]. \end{aligned} \tag{7.3.6}$$

Note that $(-1)^m$ or $(-1)^n$ are used for sign changes of different modes. The directions of control moments depending on mode-numbers m and n are consistent

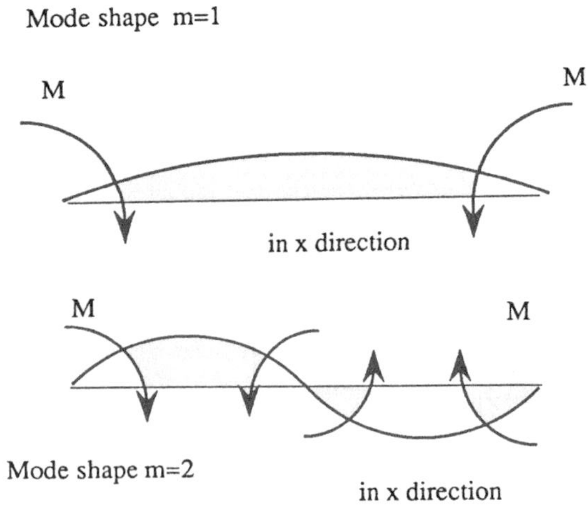

Fig. 7.3 Moment actuation of segmented actuators

with the sign changes of signals discussed previously. Figure 7.3 illustrates the control moments contributed by segmented actuators for two typical modes.

Substituting the above two generic moment expressions into Eq. (7.1.20) for the modal feedback factor and integrating each term respectively, one can derive

$$\begin{aligned}&\int_x\int_y \frac{\partial^2 M^*_{xmn}}{\partial x^2} U_{3mn}\,dxdy\\&= -M_{xmn}\frac{mb}{na}\left[\cos\frac{m\pi x_1}{a} - \cos\frac{m\pi}{2} - (-1)^m\cos\frac{m\pi}{2} + (-1)^m\cos\frac{m\pi x_2}{a}\right]\\&\cdot\left[\cos\frac{n\pi y_1}{b} - \cos\frac{n\pi}{2} - (-1)^n\cos\frac{n\pi}{2} + (-1)^n\cos\frac{n\pi y_2}{b}\right],\end{aligned} \tag{7.3.7}$$

$$\begin{aligned}&\int_x\int_y \frac{\partial^2 M^*_{ymn}}{\partial y^2} U_{3mn}\,dxdy\\&= -M_{ymn}\frac{na}{mb}\left[\cos\frac{m\pi x_1}{a} - \cos\frac{m\pi}{2} - (-1)^m\cos\frac{m\pi}{2} + (-1)^m\cos\frac{m\pi x_2}{a}\right]\\&\cdot\left[\cos\frac{n\pi y_1}{b} - \cos\frac{n\pi}{2} - (-1)^n\cos\frac{n\pi}{2} + (-1)^n\cos\frac{n\pi y_2}{b}\right].\end{aligned} \tag{7.3.8}$$

Thus, the modal feedback factor $\hat{M}_{mn}$ for the four-piece segmented actuator configuration becomes

$$\begin{aligned}\hat{M}_{mn} = &\frac{-1}{\rho h N_{mn}}\left(M_{xmn}\frac{mb}{na} + M_{ymn}\frac{na}{mb}\right)\\&\cdot\left[\cos\frac{m\pi x_1}{a} - \cos\frac{m\pi}{2} - (-1)^m\cos\frac{m\pi}{2} + (-1)^m\cos\frac{m\pi x_2}{a}\right]\\&\cdot\left[\cos\frac{n\pi y_1}{b} - \cos\frac{n\pi}{2} - (-1)^n\cos\frac{n\pi}{2} + (-1)^n\cos\frac{n\pi y_2}{b}\right].\end{aligned} \tag{7.3.9}$$

The unit modal control moments for the first colocated segmented sensor/actuator are defined as

$$M_{xmn} = r_1^a d_{31} Y_p \mathcal{G}\left[4S_{mn}\left(1-\cos\frac{m\pi}{2}\right)\left(1-\cos\frac{n\pi}{2}\right)\right], \tag{7.3.10}$$

$$M_{ymn} = r_2^a d_{32} Y_p \mathcal{G}\left[4S_{mn}\left(1-\cos\frac{m\pi}{2}\right)\left(1-\cos\frac{n\pi}{2}\right)\right]. \tag{7.3.11}$$

Note that the modal feedback factor $\hat{M}_{mn}$ would vanish only for those modes with either m or n being a multiple of four. When n = m = 1 or n = m = 3, $\hat{M}_{mn}$'s are identical to those calculated by Eq. (7.2.19) for a single piece actuator. Thus, the analysis suggests that the segmented actuator design improves the controllability for even modes without degrading the control merits for odd modes. Since lower modes are generally more important than those higher modes, only several lower modes are considered in this study, although further segmentation of actuators is possible and might provide better controllability. Note that the derived modal feedback factors for the two sensor/actuator configurations can be applied to both the displacement and velocity feedbacks as discussed previously. Detailed parametric study and time-history analyses of the plate will be presented later.

7.4 Transient Responses

It is assumed that modal excitations can be achieved such that each individual mode can be excited separately and its solution evaluated respectively. A sensor signal, due to the direct piezoelectric effect, from each segmented distributed sensor is processed (differentiated or phase shifted), amplified, and fed back to its collocated segmented distributed actuator. Signal differentiation or phase shift is used to obtain a velocity information. The difference between the above two techniques is only a constant–frequency. Note that a band-pass filter can be used to isolate the desired frequency in practice. Modal excitations can be achieved by driving the plate with input excitations and/or voltages at a given natural frequency harmonically. Steady-state responses with and without feedback can be studied in this way. Transient responses can be evaluated by turning off the excitation and observing the free decay oscillations. Time histories of these responses can be compared and damping ratios calculated. Solution techniques for the transient and steady-state responses are presented in the next two sections.

According to the modal expansion method, displacement calculation requires 1) the modal participation factor, a time function, and 2) the mode shape function—a spatial function, for each mode; the total displacement is a sum of all participating modes. The solution procedures are very involved. Thus, instead of calculating conventional displacements at specific locations, a generic mn-*th modal participation factor* is used to describe the plate motion at the mn-th natural frequency. This factor multiplied by the mn-th mode shape function (a spatial function) gives the displacement for the mn-th modal oscillation. Since only the mn-th mode is retained and all other modes are filtered out, the amplitude of the mn-th modal participation factor represents the oscillation amplitude of the mn-th mode. The modal equation of motion of the mn-th mode is written as $\ddot{\eta}_{mn} + 2\zeta_{mn}\omega_{mn}\dot{\eta}_{mn} + \omega_{mn}^2\eta_{mn} = 0$, and the natural frequency ω_{mn} for a simply supported plate is given by $\omega_{mn} = \pi^2\sqrt{\frac{D}{\rho h}}\left[\left(\frac{m}{a}\right)^2 + \left(\frac{n}{b}\right)^2\right]$ (Soedel, 1981). The initial condition $\eta_{mn}(0)$ is defined at the time instant when the plate reaches its maximum modal state, i.e., the initial modal velocity is zero, i.e., $\dot{\eta}_{mn}(0) = 0$ in its modal excitation. After that, the excitation is turned off and the plate oscillates freely. The corresponding voltage at that instant is assumed to be $\phi^S(0)$. Thus, the initial condition $\eta_{mn}(0)$ can be defined by

$$\eta_{mn}(0) = \frac{\phi^S}{\phi_{mn}^S}. \tag{7.4.1}$$

Solving the modal equation with initial conditions $\eta_{mn}(0)$ and $\dot{\eta}_{mn}(0)$ yields a transient solution:

$$\eta_{mn}(t) = e^{-\zeta_{mn}\omega_{mn}t}\left[\eta_{mn}(0)\left(\cos\gamma_{mn}t + \zeta_{mn}\omega_{mn}\frac{1}{\gamma_{mn}}\sin\gamma_{mn}t\right)\right], \qquad (7.4.2)$$

where γ_{mn} is the damped natural frequency, $\gamma_{mn} = \omega_{mn}\sqrt{1-\zeta_{mn}^2}$, and ζ_{mn} denotes the modal damping ratio in a controlled or uncontrolled system. (Note that the system is assumed **under-damped**, i.e., $\zeta_{mn} < 1$.) Time histories of the factor can be plotted, and the damping ratios are calculated and compared to evaluate the control effectiveness. Again, to infer the physical displacement of a specific location, one needs to use a spatial function, a mode shape function, multiplying the modal participation factor.

7.5 Steady-State Responses

There are two ways to provide harmonic excitations to the continua or structures: 1) a harmonic mechanical excitation and 2) a harmonic electric excitation. (The electric excitation can be achieved by injecting an electric voltage into the actuator.) The solution techniques are essentially the same. However, the electric excitations/ inputs are usually reserved for feedback controls if active controls desired. The mechanical (point) excitation force $F_3(x, y, t)$ is defined as

$$F_3(x, y, t) = p_3\delta(x - x^*)\cdot\delta(y - y^*)\sin(\omega t), \qquad (7.5.1)$$

where p_3 is the force amplitude. Note the two delta functions define the location (x^*, y^*). The modal equation with harmonic point excitation can be rewritten as

$$\ddot{\eta}_{mn} + 2\zeta_{mn}\omega_{mn}\dot{\eta}_{mn} + \omega_{mn}^2\eta_{mn} = \hat{F}_{mn}^*\sin(\omega t), \qquad (7.5.2)$$

where the (point) modal force $\hat{F}_{mn}^*$ is defined as

$$\begin{aligned}\hat{F}_{mn}^* &= \frac{p_3}{\rho h N_{mn}}\int_x\int_y \sin\frac{m\pi x}{a}\sin\frac{n\pi y}{b}\delta(x - x^*)\delta(y - y^*)dxdy \\ &= \frac{p_3}{\rho h N_{mn}}\sin\frac{m\pi x^*}{a}\sin\frac{n\pi y^*}{b},\end{aligned} \qquad (7.5.3)$$

and $N_{mn} = \int_x\int_y U_{3mn}^2 dxdy = \frac{ab}{4}$ for a simply supported plate with dimensions $a \times b$. The steady-state solution of the modal equation with a generic excitation frequency ω, i.e., $\hat{F}_{mn}^*\sin(\omega t)$, is

$$\eta_{mn} = \frac{\hat{F}^*_{mn}}{\omega^2_{mn}\sqrt{\left(1-\left(\frac{\omega}{\omega_{mn}}\right)^2\right)^2 + 4\zeta^2_{mn}\left(\frac{\omega}{\omega_{mn}}\right)^2}} \sin(\omega t - \psi), \tag{7.5.4}$$

where ω is the generic excitation frequency and ψ is a phase lag given by

$$\psi = \tan^{-1}\left[\frac{2\zeta_{mn}\left(\frac{\omega}{\omega_{mn}}\right)}{1-\left(\frac{\omega}{\omega_{mn}}\right)^2}\right]. \tag{7.5.5}$$

It is assumed that a point harmonic excitation with the mn-th natural frequency ω_{mn} is provided to the plate system, which results in an mn-th modal excitation. It is also assumed that the system damping is strong enough to constrain the oscillation amplitude from uncontrolled resonance. (Note that the purpose of the harmonic excitation at the natural frequency is to excite a particular mode so that each modal control effect can be evaluated respectively.) Changing the excitation frequency and location, one can excite various natural modes and evaluate the uncontrolled and controlled system responses. Assuming the excitation frequency is set at the mn-th natural frequency, i.e., $\omega = \omega_{mn}$, one can derive the steady-state solution as

$$\eta_{mn} = \frac{2p_3}{\rho hab\omega^2_{mn}\zeta_{mn}} \sin\frac{m\pi x^*}{a} \sin\frac{n\pi y^*}{b} \sin\left(\omega_{mn}t - \frac{\pi}{2}\right). \tag{7.5.6}$$

Note that the amplitude of modal coordinate decreases when the damping ratio increases in feedback controls.

7.6 Evaluation of Controlled Responses

Control effectivenesses of the two distributed sensor/actuator configurations, the single-piece fully distributed and the multi-piece quarterly segmented, are compared in this section. Physical interpretation of system control gains is also discussed.

7.6.1 Control Effectiveness (Proportional Velocity Feedback)

In these two sensor/actuator configurations, the modal feedback factors $\hat{M}_{mn}$'s are identical if both m and n are **odd** numbers, i.e., the four-piece segmented sensors/ actuators has the same control effect as the single-piece distributed sensor/actuator for all **odd modes**. Recall that the single-piece sensor/actuator has no control effect for all **even** modes and the four-piece segmented sensors/actuators have no effect for all **quadruple** modes (i.e., either m or n is a multiple of four). (In this study, only the four-piece segmented distributed sensors/actuators configuration is evaluated. However, further segmentation of distributed sensors and actuators is certainly feasible.) A controlled damping ratio for the mn-th mode is defined as

$$\zeta'_{mn} = \zeta_{mn} - \frac{\hat{M}_{mn}}{2\omega_{mn}}, \tag{7.6.1}$$

which is a linear function. It is intended to manipulate the second term so that the system damping ratio can be enhanced. Note that the modal control effect reduces as the modal frequency increases.

7.6.2 *Control Algorithms*

In the previous sections, the feedback control voltage was assumed to be proportional to the modal velocity of the plate system. In this way, a proportional-plus-derivative control effect takes place since the magnitude of feedback voltages varies with the vibration amplitude. It seems obvious that a stronger control effect could be obtained if the amplitude of feedback voltage is kept at a constant maximum and only the sign of the voltage is changed with respect to the direction of velocity (Lyapunov control—an on-off control). In this section, Lyapunov control of a simply supported plate, with four segmented sensors/actuators, is discussed.

7.6.2.1 Lyapunov Control

The modal equation of the system with the Lyapunov control, a constant amplitude voltage feedback control, is modified to

$$\ddot{\eta}_{mn} + 2\zeta_{mn}\omega_{mn}\dot{\eta}_{mn} + \mathrm{sgn}(\dot{\eta}_{mn})\hat{M}'_{mn} + \omega_{mn}^2\eta_{mn} = 0, \tag{7.6.2}$$

where sgn is a signum function, $\mathrm{sgn}(\cdot) = 1$ when $(\cdot) > 0$, $\mathrm{sgn}(\cdot) = 0$ when $(\cdot) = 0$, and $\mathrm{sgn}(\cdot) = -1$ when $(\cdot) < 0$. The modal feedback factor for the four-piece sensor/actuator configuration is defined as

$$\hat{M}'_{mn} = \frac{-1}{\rho h N_{mn}} \left(M_{xmn} \frac{mb}{na} + M_{ymn} \frac{na}{mb} \right) \\ \cdot \left[\cos\frac{m\pi x_1}{a} - \cos\frac{m\pi}{2} - (-1)^m \cos\frac{m\pi}{2} + (-1)^m \cos\frac{m\pi x_2}{a} \right] \\ \cdot \left[\cos\frac{n\pi y_1}{b} - \cos\frac{n\pi}{2} - (-1)^n \cos\frac{n\pi}{2} + (-1)^n \cos\frac{n\pi y_2}{b} \right], \tag{7.6.3}$$

and the control moments are defined as

$$M_{xmn} = r_1^a d_{31} Y_p |\phi^a|, \tag{7.6.4}$$

$$M_{ymn} = r_2^a d_{32} Y_p |\phi^a|, \tag{7.6.5}$$

where ϕ^a is the feedback voltage which could also be the distributed sensor output. The control effects will be studied.

7.6.2.2 Feedback Gains

Note that the distributed sensor output signal is calculated from strains which are functions of vibration amplitudes—displacements, Sect. 3.5.3 of Chap. 3. The system overall feedback gain $\mathcal{G}$ is defined as the ratio between the feedback and the sensor signals:

$$\mathcal{G} = \frac{|\phi^a|}{|\phi^S|} = \frac{\text{Amplitude of feedback voltage}}{\text{Amplitude of sensing signal voltage}}. \tag{7.6.6}$$

In the displacement feedback, this feedback gain $\mathcal{G}$ represents a "true" amplitude amplification ratio. However, the system feedback gain $\mathcal{G}$ includes a frequency component ω introduced by the signal differentiation in the velocity feedback, i.e.,

$$\mathcal{G} = \frac{\mathcal{G}' |\dot{\eta}_{mn}| \phi^S_{mn}}{|\eta_{mn}| \phi^S_{mn}} = \frac{\mathcal{G}' \omega_{mn} |\eta_{mn}|}{|\eta_{mn}|} = \mathcal{G}' \omega_{mn}. \tag{7.6.7}$$

Thus, the "true" feedback gain $\mathcal{G}'$—signal amplification ratio—in the velocity feedback is

$$\mathcal{G}' = \frac{\mathcal{G}}{\omega_{mn}}. \tag{7.6.8}$$

Note that the overall system feedback gain $\mathcal{G}$ is used as one of the comparison indices in the later analyses. However, it should be noted that a frequency ω_{mn} multiplies the "true" feedback gain.

7.7 Control Effectiveness of Segmented Actuators

A plexiglas square plate sandwiched between two thin piezoelectric polymers is used in the analyses. It is assumed that all four edges of the plate are simply supported. The top and bottom piezoelectric layers are divided into quarters, giving four equally segmented actuators and sensors. Figure 7.4 illustrates the plate model.

Note that the separation is infinitesimally small so that it is continuous elastically, but is open-circuit electrically. It is assumed that the plate has a dimension of $(0.2\,\text{m}) \times (0.2\,\text{m}) \times (1.6 \times 10^{-3}\,\text{m})$ and it is made of plexiglas. The piezoelectric sensor/actuator layers are made of a piezoelectric polyvinylidene fluoride (PVDF) polymer with a thickness of 40 μm. All material properties are summarized in Table 7.1.

To evaluate the piezoelectric actuation and vibration control effectiveness of segmented sensors/actuators, there are four analyses carried out in this study:

1) Free Vibration Analysis: This analysis serves as the fundamental and comparison basis for the other three analyses. Natural frequencies and mode shapes are presented.

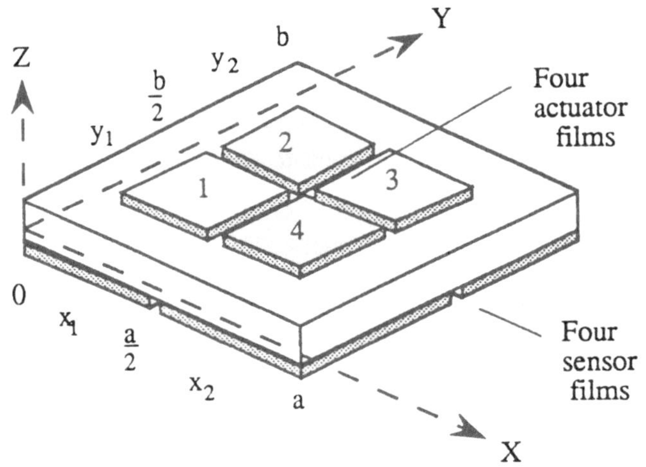

Fig. 7.4 A plate with distributed segmented sensors/actuators

Table 7.1 Material properties

Property	Plexiglas	PVDF	Units
ρ	1.19×10^3	1.80×10^3	kg/m^3
Y	3.10×10^9	2.00×10^9	N/m^2
h	1.60×10^{-3}	4.00×10^{-5}	m
μ	0.350	0.200	
d_{31}		1.0×10^{-11}	m/V
d_{32}		1.0×10^{-11}	m/V

2) Parametric Studies: This analysis focuses on the active damping ratio and its relations with respect to feedback gains, mode orders, plate sizes, actuator covering areas and other material properties.
3) Time–History Analyses: Transient responses of the plate with and without feedback controls are studied and compared.
4) Control Algorithms: Two control algorithms, namely 1) the proportional feedback and 2) Lyapunov control, are implemented and their control effectiveness compared.

7.7.1 Free Vibration Analysis

The first six natural frequencies are calculated and summarized in Table 7.2; mode shapes are presented in Appendix appended at the end of this chapter.

Note that there are symmetrical modes for modes (1,2), (1,3), (2,3), etc. Natural modes are defined by $\sin\frac{m\pi x}{a}\sin\frac{n\pi y}{b}$ where n = 1,2,…,∞ and m = 1,2,…,∞.

7.7.2 Parametric Studies

As discussed previously, control moments induced in the segmented distributed actuator counteract the motion one half and augment the motion in the other half of the displacement feedback control cycle. However, the control moments always counteract the oscillation in the velocity feedback control. Thus, only the velocity feedback control is considered in the later analyses. It was also observed that the four-piece segmented actuators and a single-piece distributed actuator provide equivalent control effects for the odd modes. Thus, odd-mode control effects of a single-piece actuator can be inferred from those of the four-piece actuator configuration. Control effectiveness with different feedback gains, mode orders, plate sizes, plate thickness, effective actuator areas and other material properties are evaluated in this section. Note that damping ratios are used as comparison indices. The initial damping ratio is assumed to be 1%, based on a laboratory experiment.

Table 7.2 Natural frequencies (Hz)

Modes	Theory
m = 1, n = 1	61.40 Hz
m = 1, n = 2	153.51 Hz
m = 2, n = 2	245.62 Hz
m = 1, n = 3	307.02 Hz
m = 2, n = 3	399.13 Hz
m = 3, n = 3	552.64 Hz

[a]Plate size: (20 cm × 20 cm); m: x modes; n: y modes

The initial condition $\eta_{mn}(0)$ for the mn-th mode is calculated based on Eq. (7.4.1) in which a constant voltage output $\phi^S(0)$ is assumed. In addition, $\dot{\eta}_{mn}(0) = 0$ at the maximum modal state when the free oscillation is initialized.

7.7.2.1 Feedback Gains

It is assumed that the total area of the four-piece segmented sensors/actuators is equal to the total surface area of the plate, i.e., it is fully covered. The inherent system damping ratio is assumed to be 1% for all modes. This assumption is not necessary; however, since the variation of damping ratios induced by the distributed control is the main concern, the inherent system damping is assumed within a reasonable range. Figures 7.5 and 7.6 show the damping variations for the first six distinct natural modes of the plate. (Note that the first six distinct modes actually represent first nine natural modes because of the modal symmetry.)

Note that the modal damping controls for all odd modes are the same for the single-piece and four-piece sensor/actuator configurations. From the calculated results, it is observed that these odd modes serve as ***breaking points*** for damping changes. The first three lower modes, (1,1), (1,2), and (2,2) modes, have the same damping variations as the first odd mode, (1,1) mode; they are considered as the first ***mode group***. For the second mode group, (1,3) and (2,2), the damping changes are the same for the second odd mode, (1,3). The same tendency could occur for the high mode groups. However, since the lower six modes already represent the first nine natural modes, further analysis was not deemed necessary.

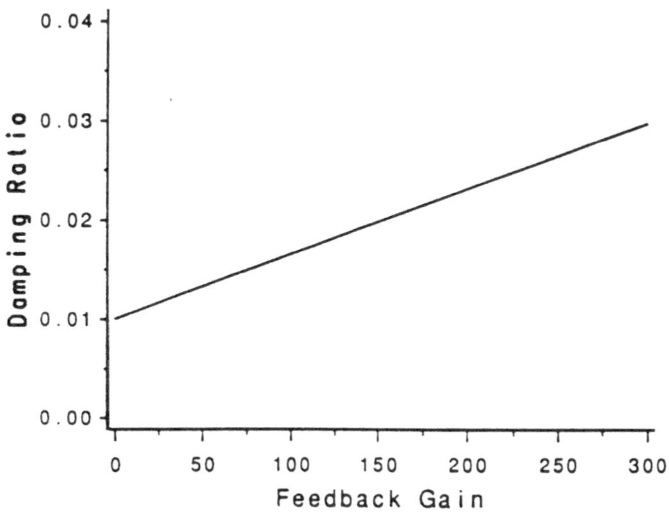

Fig. 7.5 Damping variation of the (1,1), (1,2), & (2,2) modes

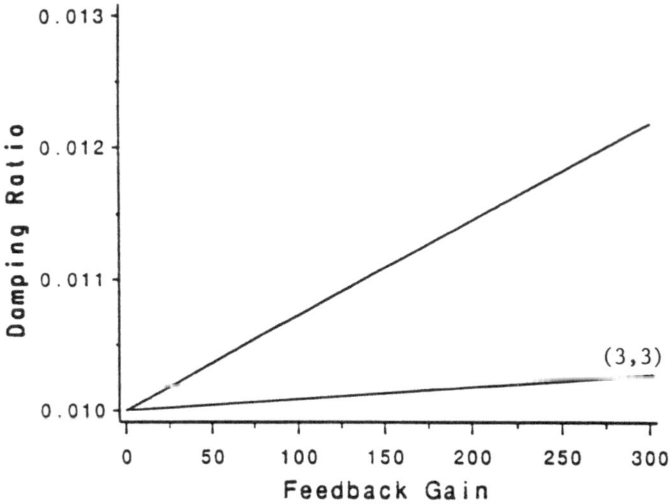

Fig. 7.6 Damping variation of the (1,3), (2,3), & (3,3) modes

Note that the feedback gain is an absolute or a total amplification ratio of signals, and the same gain is applied to all modes. The damping ratios are proportional to the feedback gains. Higher feedback gains give better control effects. However, in practice, the total feedback voltage is restricted by a breakdown voltage, usually 10–30 V/μm (d.c./a.c. voltage), for piezoelectric PVDF polymers. Besides, a sudden change of high voltages resulting from higher control gains could cause unstable oscillations. Frequency variations due to induced damping are also studied. The results suggested that the differences are insignificant.

There are a number of factors needed to be considered when examining these results. First, the free decay of a modal oscillation depends on the product of the modal damping ratio and the frequency, i.e., $e^{-\zeta_{mn}\omega_{mn}t}$. Hence, even if they have the same active damping ratio, the higher modes decay faster. Secondly, with the same magnitude of exciting force applied to a structure, the oscillation amplitude will be different for each mode; higher modes are much harder to be excited so that the sensor output signals will be relatively weak, and larger feedback gains would be needed to enhance the control effects. Thirdly, the output signals are different for different modes. Even if the same amplitude of oscillation is considered for all oscillating modes, the sensor layer deforms more severely for the higher modes. However, higher strain level does not warrant a higher output signal, because charge cancellations could occur on the surface electrodes. Fourthly, with two identical shape and different size plates, the sensor layer on the smaller plate yields higher output signals for the same level of oscillation amplitude, because the strain level is higher. Table 7.3 provides a comparison of modal output signals for two square plates, one is 20 cm × 20 cm and the other is 40 cm × 40 cm. This table confirms the third and fourth statements.

Table 7.3 Modal output signals (unit sensing output V/m) of two square plates

Size	Mode					
	m = 1 n = 1	m = 1 n = 2	m = 2 n = 2	m = 1 n = 3	m = 2 n = 3	m = 3 n = 3
$0.2 \times 0.2\,m^2$	2834 V/m	7085 V/m	11336 V/m	4723 V/m	6140 V/m	2834 V/m
$0.4 \times 0.4\,m^2$	708 V/m	1771 V/m	2833V/m	1180 V/m	1535 V/m	708 V/m

[a] $h_{31} = h_{32} = 4.32 \times 10^8$ (V/m)

7.7.2.2 Effects of Effective Actuator Area

Recall that the locations of distributed sensors and actuators were defined in a generic form, i.e., from x_1 to x_2 in the x direction and y_1 to y_2 in the y direction. However, it was assumed that the sensors and actuators symmetrically cover the plate, e.g., $(x_1 + x_2)/2$ is located at x = a/2. It is assumed that the sensor/actuator layers are equally and symmetrically divided into four pieces on a plate with a dimension of (a × b). In this section, control effects of different effective sensor/actuator areas are evaluated. Recall that the odd-mode control effects of a single-piece sensor/actuator configuration are identical to those produced by the four-piece sensor/actuator configuration.

Three cases are considered here: 1) the overall effective area is equal to the plate surface area, i.e., fully covered; 2) the effective area is equal to three-quarter of the plate area; 3) the effective area is equal to one-half of the plate area. For the first case, $x_1 = 0$, $x_2 = a$, $y_1 = 0$, and $y_2 = b$; the second case, $x_1 = a/8$, $x_2 = (7/8)a$, $y_1 = b/8$, and $y_2 = (7/8)b$; the third case, $x_1 = a/4$, $x_2 = (3/4)a$, $y_1 = b/4$, and $y_2 = (3/4)b$. Damping variations of the two natural modes: (1,1) and (2,2) are plotted versus the feedback gains, Figs. 7.7 and 7.8. It is observed that control effect

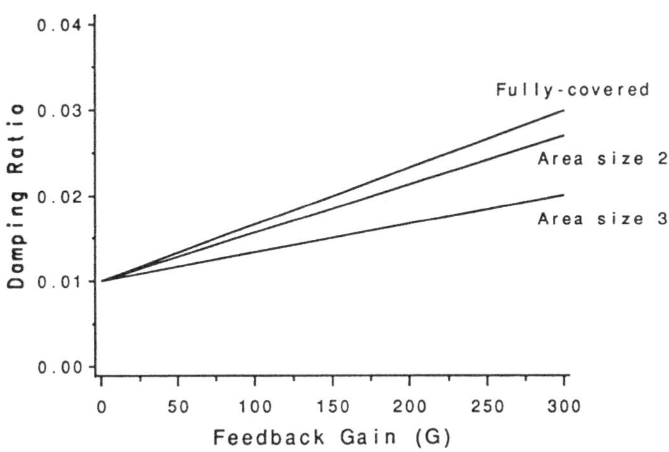

Fig. 7.7 Control effects of effective sensor/actuator area, the (1,1) mode

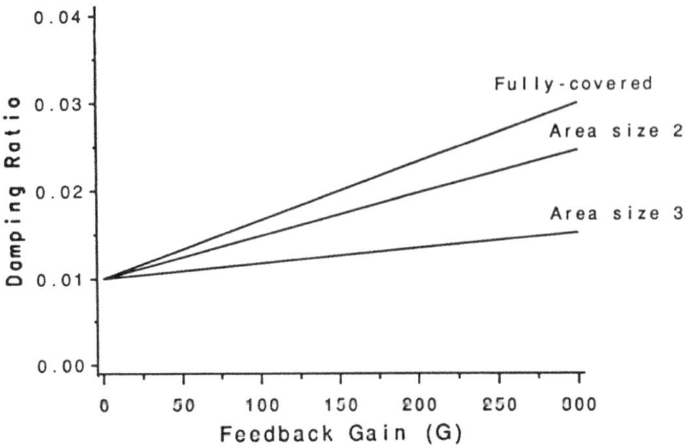

Fig. 7.8 Control effects of effective sensor/actuator area, the (2,2) mode

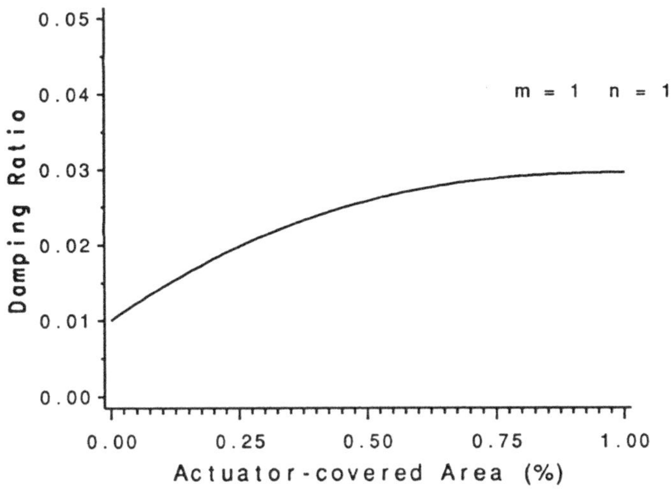

Fig. 7.9 Variations of effective actuator area, the (1,1) mode (Gain = 300)

decreases as the effective sensor/actuator shrinks due to the effective control line moments reduce as the effective actuator becomes smaller. The fully distributed sensor/actuator configuration gives the best control results. Next, damping variations versus the overall effective actuator area are calculated at a feedback gain of 300. The results are plotted in Figs. 7.9 and 7.10. It is observed that the maximum difference is about 200% for both (1,1) and (2,2) modes.

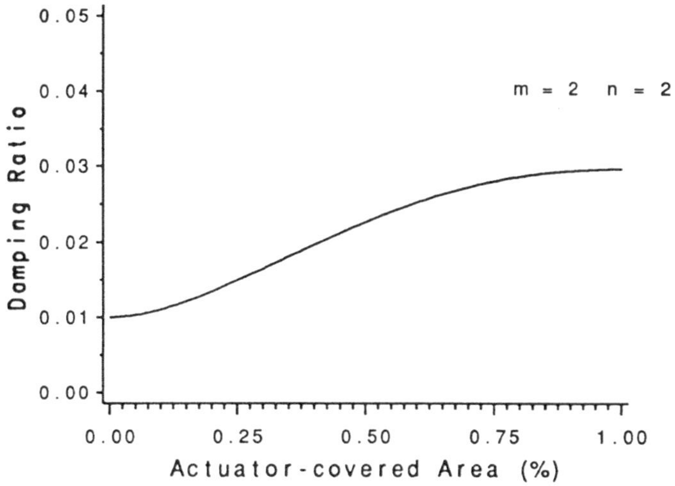

Fig. 7.10 Variations of effective actuator area, the (2,2) mode (Gain = 300)

7.7.2.3 Effects for Plate Thickness

As discussed previously, the control moments are functions of moment arms, piezoelectric constants, Young's modulus, feedback gains, and the feedback voltages. In this section, control effects of changing plate thickness, i.e., change of moment arms, are studied. It is assumed that the distributed sensors and actuators are the same as used before and they fully cover the plate surface. The results are plotted in Fig. 7.11, in which damping is plotted versus the plate thickness. The results show that the control effect decreases as the plate becomes thicker. This implies that the effect of increased bending stiffness (a cubic function of thickness) is much more significant than that of the moment arms (a linear function of thickness).

7.7.2.4 Effects of Plate Width and Length

Control effects for various plate sizes, with fully-covered segmented sensors/actuators, are also evaluated. Two cases are studied: 1) square plates with dimensions from (10 cm $\times$ 10 cm) to (50 cm $\times$ 50 cm) and 2) rectangular plates, "a" is fixed at 20 cm and "b" is changed from 5 to 50 cm. Variations of damping ratios for modes (1,1) and (2,2) of these different plates at a feedback gain of 300 are calculated and results suggest that the difference is almost negligible. (Thus, plots are not supplied.) This implies that as long as the distributed actuator changes with respect to the plate size, the damping effects remain the same. However, it should be noted that the overall control effects could be different because the natural

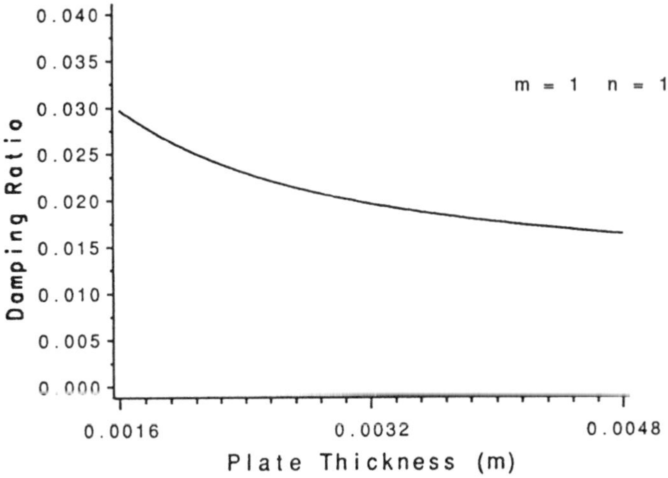

Fig. 7.11 Variations of plate thickness, the (1,1) mode (Gain = 300)

frequencies of these plates are different. (This was discussed in the first case on **Feedback Gains**.)

7.7.3 Time-History Analyses (Negative Velocity Proportional Feedback)

In the above studies, damping ratios were calculated and they were compared for a variety of different cases with the negative velocity proportional feedback. In this section, time history responses of transient oscillations are studied and compared. It was proved that basically the plate size does not alter the controlled damping ratios. Thus, a plate with a dimension of 40 cm $\times$ 40 cm is used, which has lower natural frequencies so that time histories are better distinguished. Note that only modal coordinates are plotted and compared. (For spatially distributed physical displacements of the plate, this modal coordinate needs to be multiplied by the mode shape function of that mode.) Transient responses of the plate with and without feedback controls are calculated and the envelopes are plotted in Fig. 7.12, the (1,1)-th mode, Fig. 7.13, the (1,2)-th mode, and Fig. 7.14, the (2,2)-th mode. The controlled responses are calculated based on a total feedback gain of 300. Their corresponding 10% settling times are also calculated and summarized in Table 7.4.

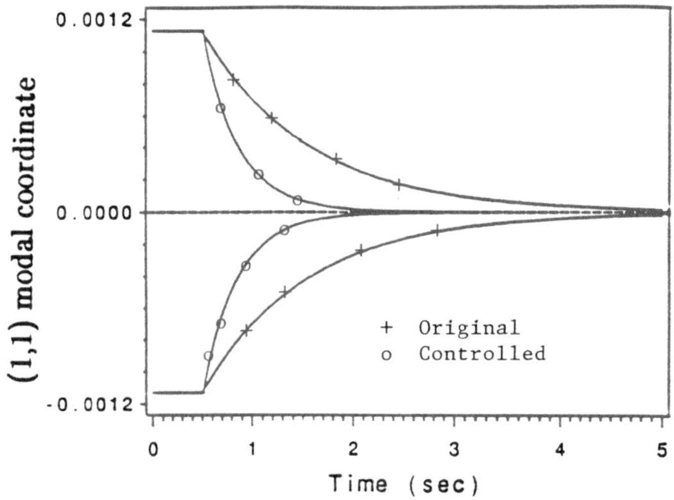

Fig. 7.12 Transient responses of the (1,1) modal coordinate (Gain = 300)

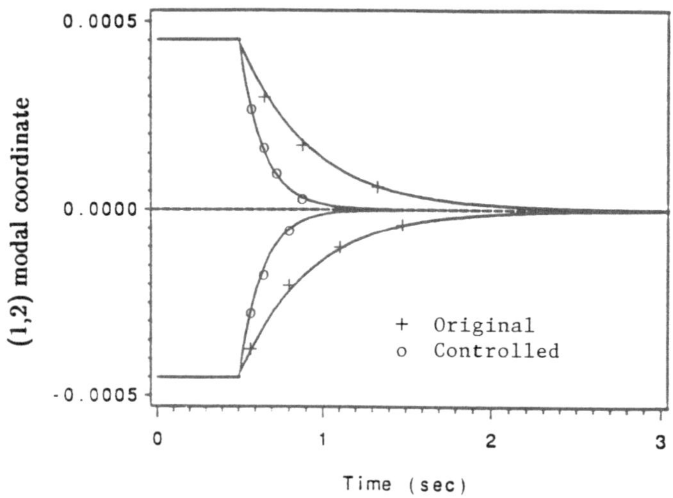

Fig. 7.13 Transient responses of the (1,2) modal coordinate

7.7.4 Control Algorithms

In the previous analyses, the feedback control voltage was assumed to be proportional to the modal velocity of plate oscillations. In this way, a proportional-plus-derivative control effect takes place since the magnitude of feedback voltage varies with vibration amplitude. It was also concluded that the control effectiveness is proportional to the feedback gain so that the maximum control effect is only

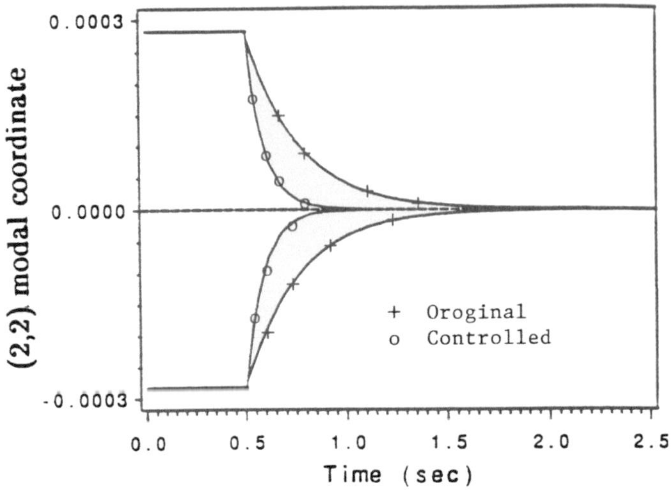

Fig. 7.14 Transient responses of the (2,2) modal coordinate

Table 7.4 Ten-percent settling times (s) for the modal transient responses

w/ or w/o control	Modes					
	m = 1 n = 1 15.4 (Hz)	m = 1 n = 2 38.4 (Hz)	m = 2 n = 2 61.4 (Hz)	m = 1 n = 3 78.8 (Hz)	m = 2 n = 3 99.8 (Hz)	m = 3 n = 3 138.2 (Hz)
Non-controlled	2.41 s	0.97 s	0.61 s	0.47 s	0.37 s	0.26 s
Controlled	0.85 s	0.34 s	0.21 s	0.39 s	0.29 s	0.25 s

[a]Plate size: 40 cm × 40 cm
[b]Gain = 300
[c]System damping = 1%

limited by materials, equipment, stability, and operation conditions. It seems obvious that a stronger control effect could be obtained if the amplitude of feedback voltage is kept at a constant maximum and only the sign of the voltage is changed with the direction of the velocity (Lyapunov control—an on–off control). In this section, Lyapunov control of a simply supported plate, with four segmented sensors/actuators, is studied and compared with the proportional voltage feedback.

In order to facilitate the solution procedures and to investigate the controlled responses, an initial displacement is imposed and the initial modal damping is assumed zero—an initial value problem. Thus, only the damping ratios contributed by the feedbacks are used when calculating the transient responses. The maximum feedback voltage is set at 240 V for Lyapunov control, a constant voltage. The maximum initial feedback voltage in the proportional feedback is also set at 240 V, which corresponds to a feedback gain of 100. (Note that the voltage amplitude varies in the proportional feedback.) A plate (40 cm × 40 cm) is coupled with four

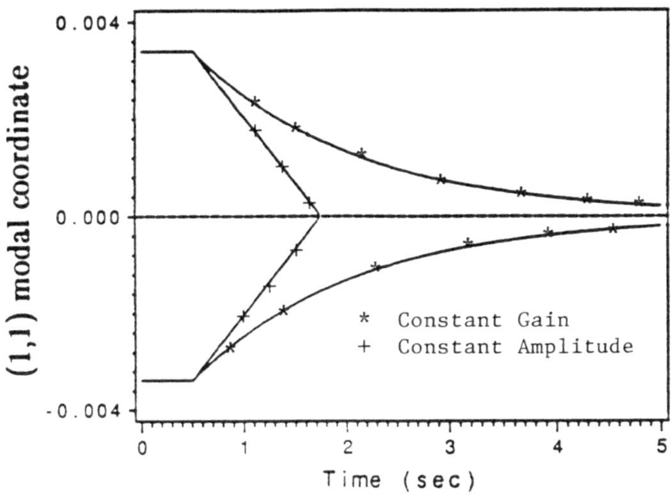

Fig. 7.15 Controlled time-history envelopes of the (1,1) modal coordinate

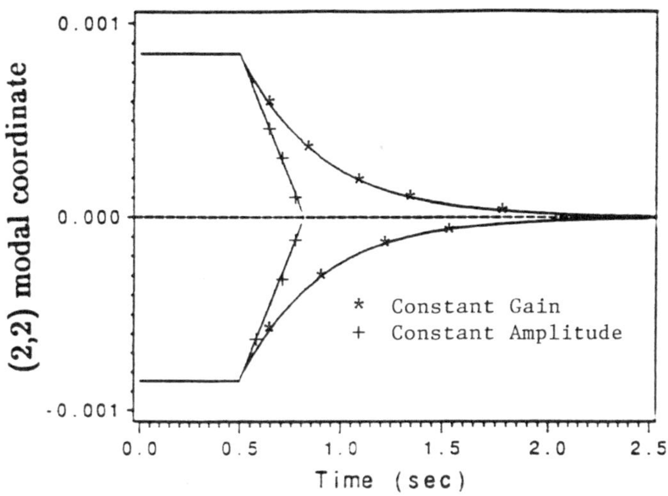

Fig. 7.16 Controlled time-history envelopes of the (2,2) modal coordinate

collocated piezoelectric sensors/actuators. All other physical properties and control conditions are the same as used in the preceding section. Transient responses of two modes, (1,1) and (2,2), are calculated and the oscillation envelopes are plotted in Figs. 7.15 and 7.16. Ten-percent setting times of the first six modes are also calculated and summarized in Table 7.5. In general, Lyapunov control, the constant amplitude, gives better control responses because the maximum voltages are always used in the feedback control loop.

Table 7.5 Modal ten-percent settling times (in seconds) of two control algorithms

w/ or w/o control	Modes					
	m = 1 n = 1 15.4 (Hz)	m = 1 n = 2 38.4 (Hz)	m = 2 n = 2 61.4 (Hz)	m = 1 n = 3 78.8 (Hz)	m = 2 n = 3 99.8 (Hz)	m = 3 n = 3 138.2 (Hz)
Proportional voltage	3.65 s	1.46	0.90	>5	>5	>5
Constant voltage	1.17 s	0.47	0.30	2.02	1.55	>5

[a]Plate: 40 cm × 40 cm
[b]Max. Voltage: 240
[c]No system damping
[d]>: longer

Comparing Tables 7.4 with 7.5, one can observe that the odd modes also serve as *breaking points* for amplitude controls. Within each ***mode group***, the 10% settling time decreases as the mode increases.

7.8 Summary

In the development of active piezoelectric/elastic structures, it was noted that a fully (symmetrically) distributed piezoelectric sensor/actuator could lead to minimum, or zero, sensing/control effects for anti-symmetrical modes of structures, especially with symmetrical boundary conditions. One method of improving the performance is to segment the symmetrically distributed sensor/actuator layers into a number of collocated sub-segments, i.e., the *segmentation* technique in Sect. 4.6, Chap. 4. In this chapter, mathematical models and analytical solutions of a simply supported plate with a single-piece distributed sensor/actuator and four-piece quarterly segmented sensors/actuators were derived. ***Modal sensitivities*** and ***modal feedback factors*** for the two sensor/actuator configurations are defined, and modal displacement and velocity feedbacks are formulated. Theoretical derivations and solutions suggested the following conclusions:

1) A single-piece symmetrical distributed sensor layer has sensing deficiencies (observation deficiencies) for all even modes because the locally generated positive and negative charges are canceled out on the whole effective sensor surface.
2) A single-piece symmetrical distributed actuator layer is also ineffective for controlling all even modes (controllability deficiency) due to reasons similar to those stated above.
3) Quarterly segmented sensors and actuators can sense and control most of the natural modes, except the quadruple modes, of the plate. The sensing and

control effects for all odd modes are identical to those of the single-piece sensor/actuator configuration.

4) Segmenting distributed sensor and actuator layers into a number of sub-segments does improve the observability and controllability of the plate system.

In the parametric studies, there were four analyses presented: 1) Free Vibration Analysis in which natural frequencies and mode shapes were presented; 2) Parametric Study in which controlled damping ratios were investigated based on a number of system parameters, e.g., feedback gains, mode orders, plate sizes, plate thickness, effective actuator areas, and other material properties; 3) Time-History Analyses which presented transient responses of the plate with and without feedback controls; 4) Control Algorithms in which two control algorithms, namely 1) the proportional feedback and 2) Lyapunov control, were implemented and their control effectiveness compared. The parametric studies suggested that:

1) The controlled modal damping ratio increases as the feedback gain/voltage increases. The odd modes serves as ***breaking points*** of ***mode groups*** in which the associated even modes have similar control effects as the leading odd mode in a ***mode group***. Similar characteristics were also observed in time-history analyses.
2) The control effects decrease as the effective actuator area is reduced or the plate becomes thicker. The effect due to increased bending stiffness of the plate is more significant than that introduced by the increased moment arm of the control moments.
3) The vibration control effects by the Lyapunov control are better than those via the proportional feedback control. Note that the Lyapunov control scheme could introduce unstable oscillations due to the sudden change of feedback voltage.

In general, the segmented sensor/actuator design improves the system observability/controllability for even modes without degrading the sensing/control merits for all odd modes of the simply supported plate. Since the lower modes are more important than the higher modes in structural monitoring and control, only several lower modes are considered in this study. Note that finer segmentation of sensors/actuators are possible and might provide better structural observability/controllability.

Note that the feedback gain is an absolute amplification ratio of sensor signals. Higher feedback gains generally lead to better control effects. However, in practice, the highest feedback voltage is restricted by a breakdown voltage, around 10–30 V/μm (d.c./a.c. voltage), for piezoelectric PVDF polymers. In addition, an abrupt change of feedback control resulting from high feedback gains could introduce unstable oscillations. Piezoelectric ceramics could be used to enhance the sensing and control effects. However, the material properties of piezoceramics can not be neglected in the analysis. Note that this study was carried out based on analytical solutions which need to be further verified via laboratory experiments.

Appendix

Mode shapes for the first six modes, Figs. 7.17, 7.18, 7.19, 7.20, 7.21 and 7.22, of a simply supported plate. Note that there are symmetrical modes for the modes: (2,1), (3,1), (3,2), etc.

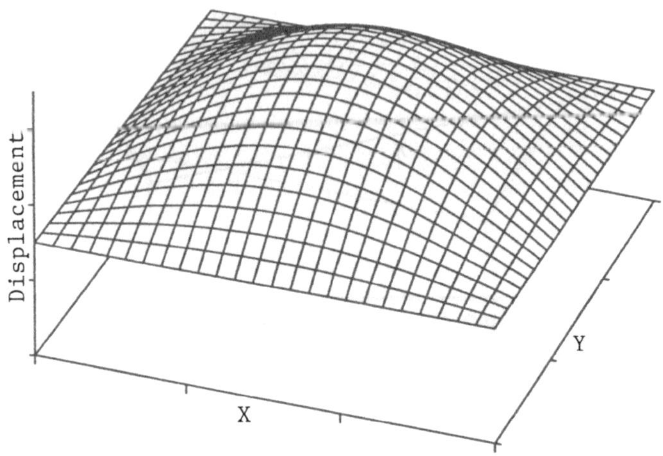

Fig. 7.17 The (1,1)-th plate mode shape

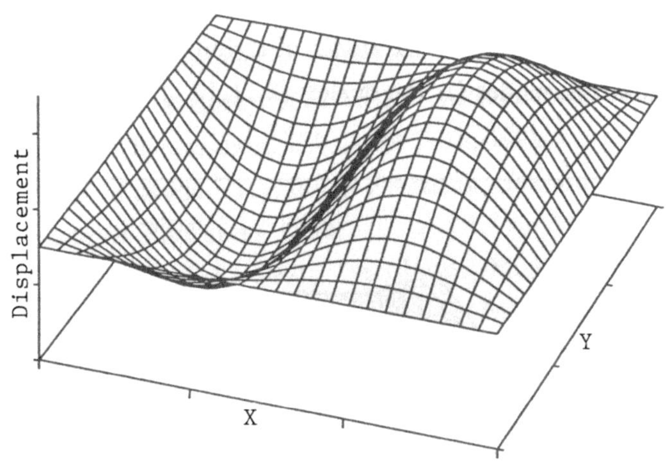

Fig. 7.18 The (2,1)-th plate mode shape

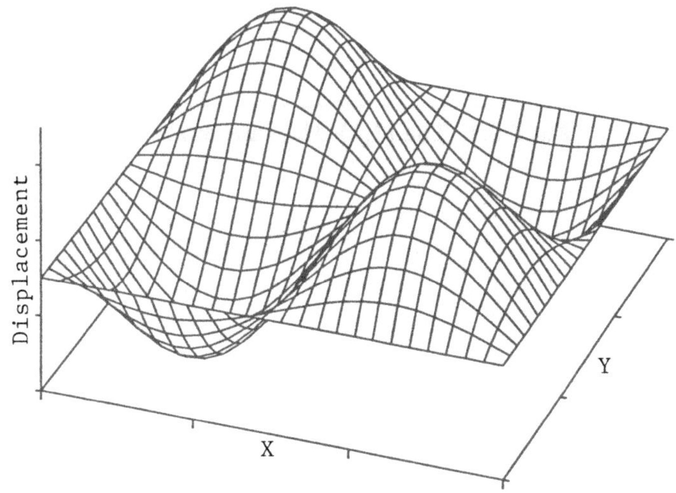

Fig. 7.19 The (2,2)-th plate mode shape

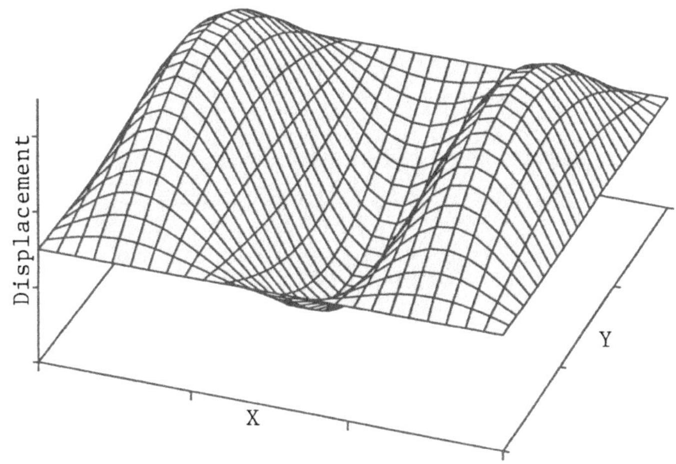

Fig. 7.20 The (3,1)-th plate mode shape

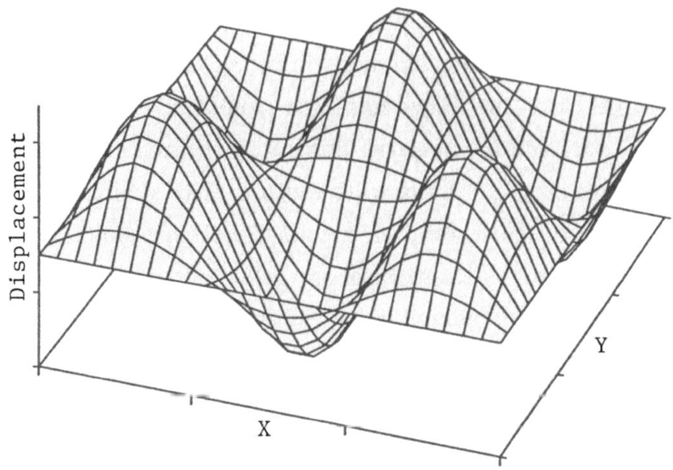

Fig. 7.21 The (3,2)-th plate mode shape

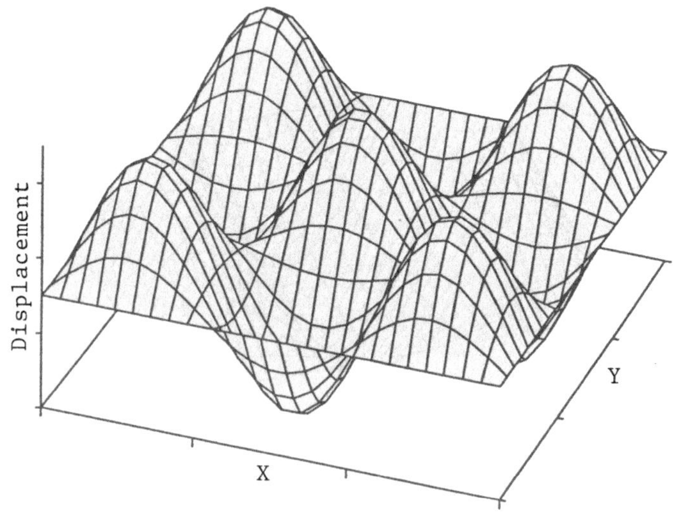

Fig. 7.22 The (3,3)-th plate mode shape

References

Burke, S. and Hubbard, J.E., 1990, Distributed Transducer Control Design for Thin Plates, *Electro–Optical Materials for Switches, Coatings, Sensor Optics, and Detectors* (1990), SPIE Vol.1307, pp.222–231.

Dimitriadis, E.K., Fuller, C.R., and Rogers, C.A., 1991, Piezoelectric Actuators for Distributed Vibration Excitation of Thin Plates, ASME *Journal of Vibration and Acoustics*, Vol.113, No.1, pp.100–107.

Fu, Haiqi, 1990, *Active Vibration Control of a Simply Supported Plate Using Segmented Piezoelectric Sensors and Actuators*, MSME Thesis, Department of Mechanical Engineering, University of Kentucky, Lexington, KY.

Lee, C. K., 1990, Theory of Laminated Piezoelectric Plates for the Design of Distributed Sensors/ Actuators. Part–1: Governing Equations and Reciprocal Relationships, *Journal of Acoustic Society of America*, Vol.87, No.3, pp.1144–1158.

Ricketts, D., 1981, Model for a Piezoelectric Polymer Flexural Plate Hydrophone, *Journal of Acoustic Society of America*, Vol.70, No.4, pp.929–935.

Ricketts, D., 1989, The Frequency of Flexible Vibration of Completely Free Composite Piezoelectric Polymer Plates, *Journal of Acoustic Society of America*, Vol.80, No.3, pp.2432–2439.

Soedel, W., 1976, Shells and Plates Loaded by Dynamic Moments with Special Attention to Rotating Point Moments, *Journal of Sound and Vibration*, Vol.48, No.2, pp.179–188.

Soedel, W., 1981, *Vibrations of Shells and Plates*, Marcel Dekker, New York.

Tzou, H.S., 1991a, *Active Elastic/Piezoelectric Shells: Theory and Applications*, (Monograph), Institute of Space and Astronautical Science, Kanagawa, Japan.

Tzou, H.S., 1991b, Distributed Modal Identification and Vibration Control of Continua: Theory and Applications, *ASME Journal of Dynamic Systems, Measurements, and Control*, Vol.113, No.3, pp.494–499, September 1991.

Tzou, H.S., 1992, A New Distributed Sensation and Control Theory for "Intelligent" Shells, *Journal of Sound and Vibration*, Vol.153, No.2, pp.335–350.

Tzou, H.S. and Fu, H.Q., "A Study of Segmentation of Distributed Sensors and Actuators, Part 1, Theoretical Analysis," *Journal of Sound & Vibration*, Vol.172, No.2, pp.247–260, April 1994a.

Tzou, H.S. and Fu, H.Q., "A Study of Segmentation of Distributed Sensors and Actuators, Part 2, Parametric Study and Vibration Controls," *Journal of Sound & Vibration*, Vol.172, No.2, pp.261–275, April 1994b.

Tzou, H.S. and Fukuda, T., 1991, *Piezoelectric Smart Systems Applied to Robotics, Micro–Systems, Identification, and Control*, Workshop Notes, IEEE Robotics and Automation Society, 1991 IEEE International Conference on Robotics and Automation, Sacramento, CA, April 7–12, 1991.

Tzou, H.S. and Fukuda, T., 1992, *Precision Sensors, Actuators, and Systems*, Kluwer Academic Publishers, 1992. (7/Plate–Segment.BkPzSm3).

Tzou, H.S. & Tseng, C.I., 1991, Distributed Modal Identification and Vibration Control of Continua: Piezoelectric Finite Element Formulation and Analysis, *ASME* Journal of Dynamic Systems, Measurements, and Control, Vol.113, No.3, pp.500–505, September 1991.

Chapter 8
Convolving Shell Sensors and Actuators Applied to Rings

Spillover problems can occur when modal cross couplings introduced by the residual modes coming into feedback in a closed-loop distributed control system. **Observation spillover** is due to the infinite summation of feedback control forces in the closed-loop control equations, and this spillover can introduce unstable dynamic responses in undamped structural systems (Meirovitch & Baruh, 1983). Thus, it is highly desirable that sensors only monitor those modes needed to be controlled such that observation spillover is prevented. In reality, however, sensors not only respond to desirable controlled modes, but also those uncontrolled residual modes. There are several techniques of reducing the observation spillover. Conventional practice is to place sensors, spatially **discrete** sensors, at modal nodes or nodal lines of the residual modes. The difficulty with this arrangement is that it is very difficult, if not impossible, to avoid all uncontrolled residual modes. Another common approach is to pre-filter the sensor data using a comb filter with phase-locked loops (Gustafson & Speyer, 1976). The phase-locked loop filter reduces the observation spillover in the frequency domain. This technique requires that 1) the controlled modal frequencies are precisely known; 2) there is a reasonable separation from the nearby residual modes; and 3) the signal-to-noise ratio is sufficiently high. The other method is to use spatially distributed ***modal sensors*** which respond only to a structural mode or a group of modes. In addition, feedback control forces for controlled modes could also appear in the governing equations of other uncontrolled modes resulting from modal interactions among all participating modes, i.e., the **control spillover**. (This is due to the surface integration of control forces on mode shapes in these equations.) Generic distributed sensing and control concepts based on distributed piezoelectric actuators using the shaping technique are presented in this chapter.

There are two sensor/actuator design principles: **1)** ***the segmentation technique*** **and 2)** ***the shaping technique***, as discussed in Sect. 4.6 of Chap. 4 (Tzou, 1991). Applications of the generic theories to flexible ring structures are demonstrated in this chapter and to cylindrical shell structures in the next chapter. This chapter focuses on sensing/control effectiveness of distributed convolving orthogonal

H. (HS) Tzou, *Piezoelectric Shells*, Solid Mechanics and Its Applications 247,
https://doi.org/10.1007/978-94-024-1258-1_8

sensors/actuators of circular ring shells based on the shaping technique. Lee and Moon (1990) proposed a distributed piezoelectric modal sensor designed for a flexible beam and a one-dimensional plate. Collins et al. (1991) proposed spatially shaped distributed piezoelectric sensors using a sinc function for monitoring beam oscillations. Tzou et al. also derived a generic distributed piezoelectric sensor/actuator theory (1991) and a thin piezoelectric finite element formulation (1990) for distributed monitoring and control of shells. Busch-Vishniac (1990) proposed spatially distributed transducers with sensor and actuator applications. In this chapter, detailed ***sensor*** and ***actuator (control) electromechanics*** of generic distributed shell sensors/actuators are studied. Spatially distributed convolving shell sensors and actuators are proposed and their performances evaluated. New spatially distributed convolving sensors/actuators for modal sensing/control of ring structures are proposed, and their detailed electromechanical behaviors are analyzed. Distributed convolving sensors are discussed first, followed by distributed convolving actuators (Zhong, 1991). Modal sensing and control effects are evaluated.

8.1 Sensor Mechanics

In this section, ***sensor mechanics*** of generic spatially distributed shell convolving sensors is discussed, which can be simplified to a broad class of distributed sensors based on four system parameters, i.e., two Lamé parameters and two radii of in-plane coordinate axes. A local voltage amplitude as a function of mechanical strains experienced in the piezoelectric shell continuum is derived first. An output amplitude of a distributed shell sensor involving surface charge average is formulated next. In a distributed modal sensor design, piezoelectric layers can be shaped and convolved, i.e., change of polarity, in order to match the sensor shape with the mode shape based on the shaping technique. It is assumed that the generic thin distributed piezoelectric shell continuum is bonded on an elastic shell continuum and the piezoelectric continuum serves as a distributed sensor. (Note that the distributed piezoelectric layers can also be embedded in a composite laminate.) A number of assumptions are employed in the later analyses: 1) piezoelectric layers are perfectly bonded with the elastic shell; 2) physical properties of the bonding material are negligible; 3) the laminated shell is still thin; and 4) the electrodes, with negligible physical properties, are perfectly conductive.

8.1.1 Thin Distributed Piezoelectric Shell Continua

It is assumed that a generic piezoelectric shell continuum is used as a distributed shell sensor. An output signal is measured across the top and bottom electroded surfaces, i.e., the transverse direction, Fig. 8.1. In later derivations,

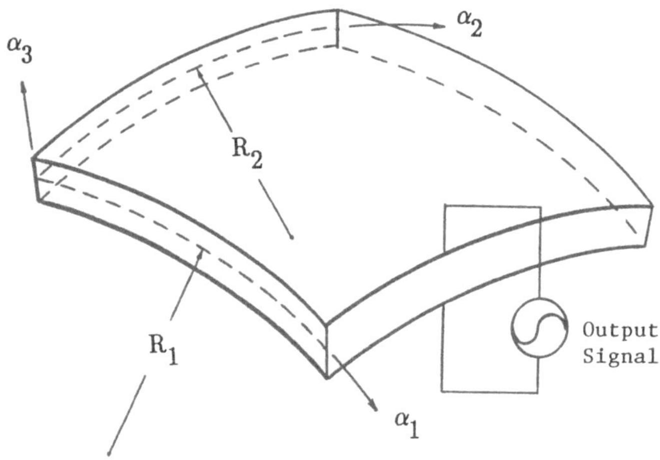

Fig. 8.1 A spatially distributed piezoelectric thin shell layer

electromechanics of a symmetrical hexagonal piezoelectric shell continuum is first discussed. A theory of distributed shell convolving sensors is derived next.

The charge equation of electrostatics, in the transverse direction, of a piezoelectric thin shell is, Sect. 2.4 of Chap. 2, (Tzou & Zhong, 1990):

$$\frac{\partial}{\partial \alpha_3}\left[(e_{31}S_{11} + e_{32}S_{22} + \varepsilon_{33}E_3)A_1A_2(1 + \frac{\alpha_3}{R_1})(1 + \frac{\alpha_3}{R_2})\right] = 0, \qquad (8.1.1a)$$

where e_{31} and e_{32} are the piezoelectric stress constants (for a symmetrical hexagonal piezoelectric material, $e_{31} = e_{32}$); S_{ii} is the normal strain in the i-th direction; ε_{33} is the dielectric constant; E_3 is the transverse electric field; A_i is the Lamé parameter; R_i is the radius of curvature of the i-th coordinate axis; α_3 is the transverse distance measured from the neutral surface. It is assumed that the piezoelectric layer is insensitive to in-plane shear deformations. The charge equation of electrostatics indicates the thickness variation is zero. (Note that a distributed sensor theory is derived from the charge equation of electrostatics of a **piezoelectric** thin shell in this chapter. The direct piezoelectric equation of linear piezoelectricity was used to derive the sensor theory in Sect. 4.2 of Chap. 4.) Note that there are two "sensor" configurations, 1) a piezoelectric thin shell continuum and 2) a piezoelectric layer coupled with an elastic shell, discussed later. (The first configuration merges to the second configuration when practical limitations are considered and assumptions imposed.)

Considering the charge in the transverse direction of the thin piezoelectric shell continuum, one can reduce the charge equation to

$$(e_{31}S_{11} + e_{32}S_{22} + \varepsilon_{33}E_3)A_1A_2(1 + \frac{\alpha_3}{R_1})(1 + \frac{\alpha_3}{R_2}) = C^*, \qquad (8.1.1b)$$

where C^* is the electric displacement which can be estimated using the electric boundary conditions. The equation of electrostatics, Eq. (8.1.1b), can be used for both piezoelectric shell continua and laminated piezoelectric layers. Note that $\alpha_3 \ll R_i$, thus,

$$\left(1 + \frac{\alpha_3}{R_i}\right) \cong 1. \tag{8.1.2}$$

In sensor applications, it is assumed that C^* is zero in an open-circuit boundary condition; more explicitly, there is no externally applied electric boundary condition or charge flows. Thus,

$$e_{31}S_{11} + e_{31}S_{22} + \varepsilon_{33}E_3 = 0, \tag{8.1.3}$$

It is assumed that the effective axes of the piezoelectric layer are aligned with the principal axes of the shell, i.e., the skew angle is zero. Integrating Eq. (8.1.3) over the piezoelectric layer thickness yields

$$\int_{\alpha_3} (e_{31}S_{11} + e_{31}S_{22})d\alpha_3 + \varepsilon_{33}\int_{\alpha_3} E_3 d\alpha_3 = 0. \tag{8.1.4}$$

Note that $\int_{\alpha_3} E_3 d\alpha_3 = \phi_3$ is the electric potential difference between the top and bottom surface electrodes. Thus,

$$\int_{\alpha_3} (e_{31}S_{11} + e_{31}S_{22})d\alpha_3 + \varepsilon_{33}\phi_3 = 0. \tag{8.1.5}$$

A local voltage amplitude $\phi_3(\alpha_1^*, \alpha_2^*)$ at a given point (α_1^*, α_2^*) can be derived as

$$\phi_3(\alpha_1^*, \alpha_2^*) = -\frac{1}{\varepsilon_{33}}\int_{\alpha_3} e_{31}\left[S_{11}(\alpha_1^*, \alpha_2^*) + S_{22}(\alpha_1^*, \alpha_2^*)\right]d\alpha_3. \tag{8.1.6}$$

This equation indicates that a local (**discrete**) output signal is a function of local strains; the signal is developed on a finite-area electrode centering around the point, Sect. 4.2 of Chap. 4. (Note that the output signal is primarily contributed by two normal strains.) For a **distributed** piezoelectric shell continuum with an effective surface electrode S^e, the signal output ϕ_3^S becomes

$$\begin{aligned}&\int_{S^e} \phi_3^S A_1A_2\left(1 + \frac{\alpha_3}{R_1}\right)\left(1 + \frac{\alpha_3}{R_2}\right)d\alpha_1 d\alpha_2\\ &= -\frac{\int_{S^e}\int_{\alpha_3} (e_{31}S_{11} + e_{31}S_{22})A_1A_2\left(1 + \frac{\alpha_3}{R_1}\right)\left(1 + \frac{\alpha_3}{R_2}\right)d\alpha_1 d\alpha_2 d\alpha_3}{\varepsilon_{33}}.\end{aligned} \tag{8.1.7}$$

Note that $\frac{\alpha_3}{R_1} \ll 1$ and $\frac{\alpha_3}{R_2} \ll 1$. Thus, taking a surface average over the entire electrode area $\mathbf{S^e}$ and rearranging the signal equation, one can derive

$$\phi_3^S = -\frac{\int_{S^e}\int_{\alpha_3}(e_{31}S_{11} + e_{31}S_{22})A_1A_2 d\alpha_1 d\alpha_2 d\alpha_3}{\varepsilon_{33}S^e}. \tag{8.1.8}$$

Note that $\int_{S^e} A_1A_2(1+\frac{\alpha_3}{R_1})(1+\frac{\alpha_3}{R_2})d\alpha_1 d\alpha_2 = S^e$, the effective electrode area.

For a thin shell continuum, (mechanical) normal strains can be further divided into membrane strains S_{ii}^o and bending strains k_{ii}, i.e.,

$$S_{11} = S_{11}^o + \alpha_3 k_{11}, \tag{8.1.9}$$

$$S_{22} = S_{22}^o + \alpha_3 k_{22}. \tag{8.1.10}$$

Thus, the output signal becomes

$$\phi_3^S = -\frac{1}{\varepsilon_{33}S^e}\int_{S^e}\int_{\alpha_3}\left[e_{31}\left(S_{11}^o + S_{22}^o\right) + \alpha_3(k_{11} + k_{22})\right]\cdot A_1A_2 d\alpha_1 d\alpha_2 d\alpha_3. \tag{8.1.11}$$

For a **piezoelectric shell continuum**, the thickness integration is from −h/2 to +h/2 where h is the shell thickness. However, for a **distributed piezoelectric shell layer** laminated on (or embedded in) an elastic shell, the thickness integration is respect to the thickness of the piezoelectric sensor layer; the thickness could be either uniform or non-uniform; this will be discussed in the next section.

Note that the overall signal output of the shell continuum is contributed by both membrane and bending strains experienced in the piezoelectric layer. The membrane strains S_{ii}^o and bending strains k_{ii} can be further expressed as a function of three displacements u_1, u_2, and u_3 in three axial directions (Tzou & Zhong, 1990). (Note that this case is similar to that in Sect. 4.2 of Chap. 4.)

1) Membrane Strains

$$S_{11}^o = \frac{1}{A_1}\frac{\partial u_1}{\partial \alpha_1} + \frac{u_2}{A_1A_2}\frac{\partial A_1}{\partial \alpha_2} + \frac{u_3}{R_1}, \tag{8.1.12}$$

$$S_{22}^o = \frac{1}{A_2}\frac{\partial u_2}{\partial \alpha_2} + \frac{u_1}{A_1A_2}\frac{\partial A_2}{\partial \alpha_1} + \frac{u_3}{R_2}, \tag{8.1.13}$$

2) Bending Strains

$$k_{11} = \frac{1}{A_1}\frac{\partial}{\partial \alpha_1}\left(\frac{u_1}{R_1} - \frac{1}{A_1}\frac{\partial u_3}{\partial \alpha_1}\right) + \frac{1}{A_1A_2}\left(\frac{u_2}{R_2} - \frac{1}{A_2}\frac{\partial u_3}{\partial \alpha_2}\right)\frac{\partial A_1}{\partial \alpha_2}, \tag{8.1.14}$$

$$k_{22} = \frac{1}{A_2}\frac{\partial}{\partial \alpha_2}\left(\frac{u_2}{R_2} - \frac{1}{A_2}\frac{\partial u_3}{\partial \alpha_2}\right) + \frac{1}{A_1 A_2}\left(\frac{u_1}{R_1} - \frac{1}{A_1}\frac{\partial u_3}{\partial \alpha_1}\right)\frac{\partial A_2}{\partial \alpha_1} \tag{8.1.15}$$

Note that the Lamé parameters A_i's and radii of curvature R_i's are geometry dependent, e.g., $A_1 = A_2 = 1$ and $R_1 = R_2 = \infty$ for a rectangular plate. (Detailed reduction procedures and examples were presented in Chaps. 3 and 4). Thus, the sensor equation can be further simplified based on these four parameters defined by the geometries.

As discussed above, the sensor signal is contributed by the bending strains and the membrane strains. Accordingly, two sensor sensitivities: 1) a ***transverse modal sensitivity*** and 2) a ***membrane modal sensitivity*** can be defined and these two sensitivities will be applied to distributed piezoelectric sensors. In general, the transverse modal sensitivity is defined for the transverse natural modes and the membrane modal sensitivity for the in-plane natural modes. Detailed contribution of each component will be evaluated in case studies. For a single layer piezoelectric shell continuum, the transverse modal sensitivity is zero because the neutral surface of the shell is located in the middle of the shell thickness. The thickness integration from −h/2 to +h/2 gives a zero output in the transverse sensitivity. Thus, a piezoelectric shell continuum or a distributed piezoelectric shell layer on the neutral surface of a structure only responds to in-plane contractions and expansions, and consequently it can be used only as a ***membrane sensor***. (The membrane modal sensitivity can exist if the membrane strains are not zero.) Note that the generic piezoelectric shell is fully covered with conducting electrodes on the top and bottom surfaces. In the later derivation, the surface electrodes are spatially shaped and the electric polarization is also altered in order to design generic distributed shell modal sensors based on the shaping technique.

8.1.2 Spatially Distributed Piezoelectric Convolving Sensor

Distributed piezoelectric shell layers can be either embedded or surface bonded with an elastic shell, and these layers are used as distributed sensors. For a spatially distributed piezoelectric shell convolving sensor, a weighting (or window) function $W(\alpha_1, \alpha_2)$ and a polarity function $\mathrm{sgn}[U_3(\alpha_1, \alpha_2)]$ can be added to the generic shell signal equation. Note that sgn(·) denotes a signum function used to change the piezoelectric polarity, which sgn(·) = 1 when (·) > 0, 0 when (·) = 0, and −1 when (·) < 0; $U_3(\alpha_1, \alpha_2)$ denotes a modal function. In the later derivation, two weighting functions are discussed. First, it is assumed that the thickness of the piezoelectric shell is a spatial function, i.e., $h^s = W_t(\alpha_1, \alpha_2)$ where the subscript t denotes the ***thickness shaping***, Fig. 8.2a. Secondly, it is assumed that the thickness is constant and the electrode shape $\mathbb{Z}(\cdot)$ is defined as $\mathbb{Z}(\alpha_1, \alpha_2) = W_s(\alpha_1, \alpha_2)$ where the subscript s denotes the ***surface shaping***, Fig. 8.2b.

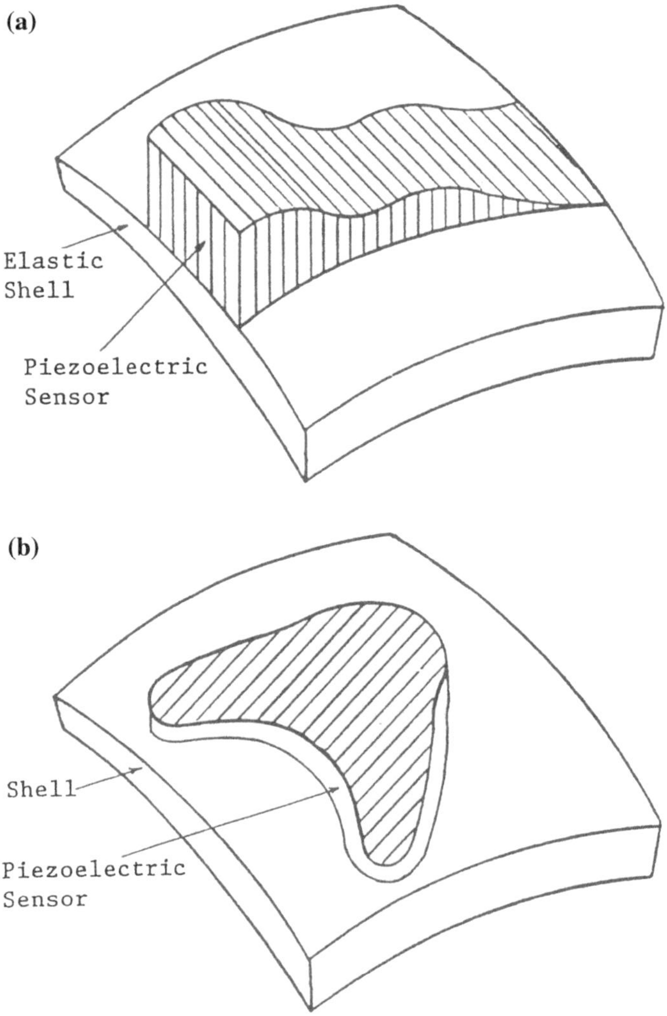

Fig. 8.2 Spatially distributed shell sensors. **a** Spatial thickness shaping, **b** Spatial surface shaping

1) Spatial Thickness Shaping

In the first case, where the piezoelectric shell thickness is a spatial function $W_t(\alpha_1, \alpha_2)$, the sensor equation becomes

$$\begin{aligned}
&\int_{S^e} \phi_3^S A_1 A_2 (1+\frac{\alpha_3}{R_1})(1+\frac{\alpha_3}{R_2}) d\alpha_1 d\alpha_2 \\
&= -\frac{e_{31}}{\varepsilon_{33}} \int_{\alpha_1}\int_{\alpha_2} \mathrm{sgn}[U_3(\alpha_1,\alpha_2)] \Big\{ W_t(\alpha_1,\alpha_2)\big(S_{11}^o + S_{22}^o\big) \\
&\quad + \int_{r_1}^{W_t(\alpha_1,\alpha_2)+r_1} \alpha_3 (k_{11}+k_{22}) d\alpha_3 \Big\} \cdot (1+\frac{\alpha_3}{R_1})(1+\frac{\alpha_3}{R_2}) A_1 A_2 d\alpha_1 d\alpha_2,
\end{aligned} \tag{8.1.16}$$

where r_1 is the distance measured from the shell neutral surface to the bottom of the piezoelectric sensor layer. Note that the first term inside the brace is contributed by the membrane strains and the second by the bending strains. The total output signal is contributed by the sum of membrane and bending strains. Recall that the output voltage is measured at an open-circuit condition. In addition, $R_i \gg \alpha_3$—an infinitesimal thickness; thus, $(1+\frac{\alpha_3}{R_i}) \cong 1$.

2) Spatial Surface Shaping

In the second case, the piezoelectric shell thickness is assumed constant. An electrode shape (or sensor shape) can be designed by using a spatial (shape) weighting function $W_s(\alpha_1, \alpha_2)$, i.e., $\mathbb{Z}(\alpha_1, \alpha_2) = W_s(\alpha_1, \alpha_2)$.

$$\begin{aligned}&\int_{\alpha_1}\int_{\alpha_2} \phi_3^S (1+\frac{\alpha_3}{R_1})(1+\frac{\alpha_3}{R_2}) A_1 A_2 d\alpha_1 d\alpha_2 \\ &= -\frac{1}{\varepsilon_{33}}\int_{\alpha_1}\int_{\alpha_2} W_s(\alpha_1,\alpha_2)\mathrm{sgn}[U_3(\alpha_1,\alpha_2)]\int_{\alpha_3}(e_{31}S_{11}+e_{31}S_{22}) \\ &\quad \cdot (1+\frac{\alpha_3}{R_1})(1+\frac{\alpha_3}{R_2}) A_1A_2 d\alpha_1 d\alpha_2 d\alpha_3.\end{aligned} \tag{8.1.17a}$$

Again, the curvature effects can be neglected, i.e., $(1+\frac{\alpha_3}{R_i}) \cong 1$. For a one-dimensional (1D) structure, e.g., beam, ring, arch, etc., the mechanical strains are functions of only one coordinate, say α_1. Thus, membrane strains S_{11}^o, S_{22}^o and bending strains k_{11}, k_{22} can be described as functions of the coordinate α_1, i.e., $W_s(\alpha_1, \alpha_2) \cong W_s(\alpha_1)$. Thus, the 1D sensor signal equation can be written as

$$\begin{aligned}\phi_3^S &= -\frac{e_{31}}{\varepsilon_{33} S^e}\int_{\alpha_1} \mathrm{sgn}[U_3(\alpha_1)]\Big[h^S(S_{11}^o + S_{22}^o) \\ &\quad + \frac{1}{2}(r_2^2 - r_1^2)(k_{11}+k_{22})\Big] A_1 A_2 \int_0^{W_S(\alpha_1)} d\alpha_2 d\alpha_1 \\ &= -\frac{e_{31}}{\varepsilon_{33} S^e}\int_{\alpha_1} W_S(\alpha_1)\mathrm{sgn}[U_3(\alpha_1)]\Big[h^S(S_{11}^o + S_{22}^o) \\ &\quad + \frac{1}{2}h^S\left(h+h^S\right)(k_{11}+k_{22})\Big] A_1 A_2 d\alpha_1\end{aligned}, \tag{8.1.17b}$$

where h^s is the thickness of the sensor layer; r_2 is the distance measured from the neutral surface of the shell to the top surface of the distributed piezoelectric sensor layer. Note that $r_2 - r_1 = h^s$, and $(r_2^2 - r_1^2) = h^s(h+h^s)$. Next, system equations and natural modes of circular ring structures are defined; then the generic sensor shaping theory is applied to distributed convolving piezoelectric ring sensors.

8.2 Circular Ring Shells

In this section, dynamic equations and natural modes of circular ring structures are studied, which establish the fundamentals for further analyses in distributed convolving sensors and actuators. Note that the **piezoelectric** ring equations will be derived from the generic equations of piezoelectric thin shells, Chap. 2. Elastic ring equations can be obtained by neglecting all electric membrane forces and moments in the piezoelastic ring equations.

8.2.1 Electromechanical Equations of a Piezoelectric Ring

It is assumed that a circular piezoelectric ring has a constant radius R, a thickness h, and a width b, Fig. 8.3. The fundamental form is

$$(ds)^2 = (R)^2(d\theta)^2 + (1)^2(dx)^2, \tag{8.2.1}$$

where θ defines the circumferential direction and x the width direction. α_3 is normal to the neutral surface defined by x and θ axes. Thus, the Lamé parameters are $A_1 = R, A_2 = 1$ where $\alpha_1 = \theta$ and $\alpha_2 = x$. The radii are $R_1 = R$ and $R_2 = \infty$.

Substituting the four parameters into the thin piezoelectric shell equations and simplifying them, one can derive the electromechanical equations for rings:

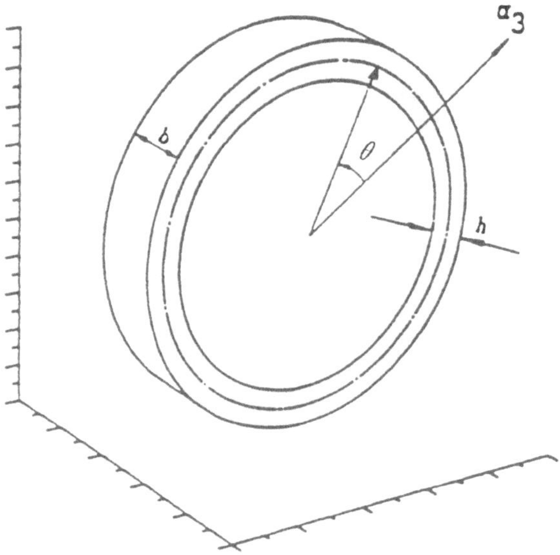

Fig. 8.3 A circular ring shell

$$\frac{1}{R}\frac{\partial\left(N_{\theta\theta}^{m}-N_{\theta\theta}^{e}\right)}{\partial\theta}+\frac{Q_{\theta3}^{m}}{R}=\rho h\frac{\partial^{2}u_{\theta}}{\partial t^{2}}, \tag{8.2.2}$$

$$\frac{1}{R}\frac{\partial Q_{\theta3}^{m}}{\partial\theta}-\frac{N_{\theta\theta}^{m}-N_{\theta\theta}^{e}}{R}=\rho h\frac{\partial^{2}u_{3}}{\partial t^{2}}, \tag{8.2.3}$$

where

$$Q_{\theta3}^{m}=\frac{1}{R}\frac{\partial\left(M_{\theta\theta}^{m}-M_{\theta\theta}^{e}\right)}{\partial\theta}. \tag{8.2.4}$$

Note that external mechanical excitations F_i's are not considered. The relations between strains and displacements can be simplified to

$$S_{\theta\theta}^{o}=\frac{1}{R}\left(\frac{\partial u_{\theta}}{\partial\theta}+u_{3}\right), \tag{8.2.5}$$

$$k_{\theta\theta}=\frac{1}{R^{2}}\left(\frac{\partial u_{\theta}}{\partial\theta}-\frac{\partial^{2}u_{3}}{\partial\theta^{2}}\right). \tag{8.2.6}$$

The mechanical membrane force and bending moment are

$$N_{\theta\theta}^{m}=KS_{\theta\theta}^{o}=\frac{K}{R}\left(\frac{\partial u_{\theta}}{\partial\theta}+u_{3}\right), \tag{8.2.7}$$

$$M_{\theta\theta}^{m}=Dk_{\theta\theta}=\frac{D}{R^{2}}\left(\frac{\partial u_{\theta}}{\partial\theta}-\frac{\partial^{2}u_{3}}{\partial\theta^{2}}\right). \tag{8.2.8}$$

Recall that the membrane stiffness K and bending stiffness D are defined as

$$K=\frac{Yh}{1-\mu^{2}}, \tag{8.2.9}$$

$$D=\frac{Yh^{3}}{12(1-\mu^{2})}. \tag{8.2.10}$$

The electric membrane force and bending moment are

$$N_{\theta\theta}^{e}=\int_{\alpha_3}e_{31}E_{3}d\alpha_{3}, \tag{8.2.11}$$

$$M_{\theta\theta}^{e}=\int_{\alpha_3}e_{31}E_{3}\alpha_{3}d\alpha_{3}. \tag{8.2.12}$$

The charge equation of electrostatics is

$$\frac{\partial(e_{31}S_{\theta\theta}+\varepsilon_{33}E_3)}{\partial\alpha_3}=0. \tag{8.2.13}$$

Based on the charge equation of electrostatics, one can define the electric boundary condition with an external surface charge Q_3 as

$$e_{31}S_{\theta\theta}+\varepsilon_{33}E_3=-Q_3. \tag{8.2.14}$$

Using the electric boundary condition, one can derive the electric field E_3 as a function of $S_{\theta\theta}$ and an external charge Q_3. Writing E_3 as a function of $S_{\theta\theta}$ and Q_3, substituting E_3 into the electric membrane force and bending moment, one can derive $N^e_{\theta\theta}$ and $M^e_{\theta\theta}$ as a function of the combined direct (induced by the internal elasticity) and converse (induced by the external boundary condition) effects.

$$\begin{aligned} N^e_{\theta\theta} &= -\int_{\alpha_3}\frac{e_{31}}{\varepsilon_{33}}\left[e_{31}\left(S^o_{\theta\theta}+\alpha_3 k_{\theta\theta}\right)+Q_3\right]d\alpha_3 \\ &= -\int_{-h/2}^{h/2}\frac{e_{31}}{\varepsilon_{33}}\left[e_{31}\left(S^o_{\theta\theta}+\alpha_3 k_{\theta\theta}\right)+Q_3\right]d\alpha_3 \\ &= -\frac{e_{31}^2 h}{\varepsilon_{33}}S^o_{\theta\theta}-\frac{e_{31}h}{\varepsilon_{33}}Q_3. \end{aligned} \tag{8.2.15}$$

$$\begin{aligned} M^e_{\theta\theta} &= -\int_{\alpha_3}\frac{e_{31}}{\varepsilon_{33}}\left[e_{31}\left(S^o_{\theta\theta}+\alpha_3 k_{\theta\theta}\right)+Q_3\right]\alpha_3 d\alpha_3 \\ &= -\int_{-h/2}^{h/2}\frac{e_{31}}{\varepsilon_{33}}\left[e_{31}\left(S^o_{\theta\theta}+\alpha_3 k_{\theta\theta}\right)+Q_3\right]\alpha_3 d\alpha_3 \\ &= -\frac{e_{31}^2}{\varepsilon_{33}}\frac{h^3}{12}k_{\theta\theta}. \end{aligned} \tag{8.2.16}$$

(Note that all external mechanical excitations and the boundary charge Q_3 can be neglected in a free-vibration analysis.) Substituting the mechanical and electric membrane forces and moments into Eqs. (8.2.2) and (8.2.3) and using the strain and displacement relations, one can derive

$$\left(D+\frac{e_{31}^2}{\varepsilon_{33}}\frac{h^3}{12}\right)\frac{1}{R^4}\left(\frac{\partial^2 u_\theta}{\partial\theta^2}-\frac{\partial^3 u_3}{\partial\theta^3}\right)+\left(K+\frac{e_{31}^2 h}{\varepsilon_{33}}\right)\frac{1}{R^2}\left(\frac{\partial^2 u_\theta}{\partial\theta^2}+\frac{\partial u_3}{\partial\theta}\right)=\rho h\frac{\partial^2 u_\theta}{\partial t^2}, \tag{8.2.17}$$

$$(D+\frac{e_{31}^2}{\varepsilon_{33}}\frac{h^3}{12})\frac{1}{R^4}(\frac{\partial^3 u_\theta}{\partial\theta^3}-\frac{\partial^4 u_3}{\partial\theta^4})-(K+\frac{e_{31}^2 h}{\varepsilon_{33}})\frac{1}{R^2}(\frac{\partial u_\theta}{\partial\theta}+u_3)-\frac{N_{\theta\theta}^e}{R}=\rho h\frac{\partial^2 u_3}{\partial t^2}, \tag{8.2.18}$$

where one can further define the modified membrane stiffness K' and bending stiffness D' for the **piezoelectric** circular rings:

$$K'=K+\frac{e_{31}^2 h}{\varepsilon_{33}}, \tag{8.2.19}$$

$$D'=D+\frac{e_{31}^2}{\varepsilon_{33}}\frac{h^3}{12}. \tag{8.2.20}$$

It is observed that both the bending and membrane stiffness constants of **piezoelectric** rings are higher than those of the **elastic** rings due to the piezoelectric constants. Governing equations for **elastic** rings and **laminated** rings are discussed next. Note that $N_{\theta\theta}^e/R$ is still preserved in the transverse equation; this can be used as a control force in open- or closed-loop actuation and controls, see Sect. 3.8 of Chap. 3.

1) Equations of Motion of an Elastic Ring

For an **elastic** ring, i.e., no piezoelectric effect, the ring equations can be further simplified to

$$\frac{D}{R^4}(\frac{\partial^2 u_\theta}{\partial\theta^2}-\frac{\partial^3 u_3}{\partial\theta^3})+\frac{K}{R^2}(\frac{\partial^2 u_\theta}{\partial\theta^2}+\frac{\partial u_3}{\partial\theta})=\rho h\frac{\partial^2 u_\theta}{\partial t^2}, \tag{8.2.21}$$

$$\frac{D}{R^4}(\frac{\partial^3 u_\theta}{\partial\theta^3}-\frac{\partial^4 u_3}{\partial\theta^4})-\frac{K}{R^2}(\frac{\partial u_\theta}{\partial\theta}+u_3)=\rho h\frac{\partial^2 u_3}{\partial t^2}. \tag{8.2.22}$$

2) Equations of Motion of a Laminated Ring

A **laminated** ring made of a steel ring and two piezoelectric polymeric polyvinylidene fluoride (PVDF) layers is shown in Fig. 8.4. It is assumed that the elastic ring is laminated with a distributed convolving sensor layer on one surface and with a distributed convolving actuator layer on the other surface. Note that material properties (e.g., mass and stiffness) of piezoelectric sensor/actuator layers are negligible. Thus, the governing equations of **elastic** rings will be used in the sensor/actuator analyses, Sects. 8.3 and 8.7.

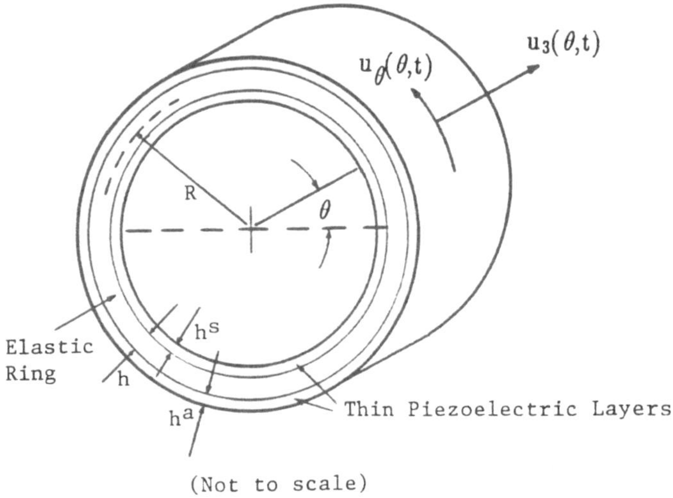

Fig. 8.4 A laminated circular ring shell

8.2.2 *Natural Frequencies and Mode Shapes*

In this section, natural frequencies and modal shapes of a **piezoelectric** ring structure are studied. Note that to obtain the analysis for an **elastic** ring, one can simply neglect all the electric terms in the piezoelectric equations. In the **laminated** ring configuration, governing equations and natural modes of the elastic ring are used, because the material properties of the sensor layers are not considered. (This will be discussed later.)

For a free vibration analysis, all external loads, both mechanical and electric, are zero. It is assumed that all points of the shell oscillate harmonically at one of the natural frequencies. Thus, the displacements $u_\theta(\theta, t)$ and $u_3(\theta, t)$ can be assumed in a harmonic form:

$$u_\theta(\theta, t) = U_{\theta n}(\theta)e^{j\omega_n t}, \tag{8.2.23}$$

$$u_3(\theta, t) = U_{3n}(\theta)e^{j\omega_n t}, \tag{8.2.24}$$

where ω_n is a natural frequency (rad/s). Substituting these into Eqs. (8.2.17) and (8.2.18) yields a set of ordinary differential equations (ODE's):

$$\frac{D'}{R^4}\left(\frac{d^2U_{\theta n}}{d\theta^2} - \frac{d^3U_{3n}}{d\theta^3}\right) + \frac{K'}{R^2}\left(\frac{d^2U_{\theta n}}{d\theta^2} + \frac{dU_{3n}}{d\theta}\right) + \rho h\omega_n^2 U_{\theta n} = 0, \tag{8.2.25}$$

$$\frac{D'}{R^4}\left(\frac{d^3U_{\theta n}}{d\theta^3} - \frac{d^4U_{3n}}{d\theta^4}\right) - \frac{K'}{R^2}\left(\frac{dU_{\theta n}}{d\theta} + U_{3n}\right) + \rho h\omega_n^2 U_{3n} = 0. \tag{8.2.26}$$

Note that the membrane stiffness K' and bending stiffness D' include both Young's modulus and the direct piezoelectric effects. In this section, it is assumed that the ring is free floating in space. The solutions (mode shape functions) of Eqs. (8.2.25) and (8.2.26) are assumed as

$$U_{\theta n}(\theta) = A_n \sin n(\theta - \phi^*), \tag{8.2.27a}$$

$$U_{3n}(\theta) = B_n \cos n(\theta - \phi^*), \tag{8.2.27b}$$

where A_n and B_n are arbitrary modal amplitudes; n is an integer, the mode number; ϕ^* is an arbitrary phase angle which has to be included since the ring does not show a preference for the orientation of its modes (Soedel, 1981). The orientation is determined by the distribution of the external loads. Substituting the mode shape functions into the electromechanical ring equations yields

$$\begin{bmatrix} \rho h\omega_n^2 - \frac{n^2 D'}{R^4} - \frac{n^2 K'}{R^2} & -\frac{n^3 D'}{R^4} - \frac{n K'}{R^2} \\ -\frac{n^3 D'}{R^4} - \frac{n K'}{R^2} & \rho h\omega_n^2 - \frac{n^4 D'}{R^4} - \frac{K'}{R^2} \end{bmatrix} \begin{Bmatrix} A_n \\ B_n \end{Bmatrix} = \{0\}. \tag{8.2.28}$$

For non-trial solutions, the modal amplitudes $\begin{Bmatrix} A_n \\ B_n \end{Bmatrix} \neq 0$, thus,

$$\det \begin{bmatrix} \rho h\omega_n^2 - \frac{n^2 D'}{R^4} - \frac{n^2 K'}{R^2} & -\frac{n^3 D'}{R^4} - \frac{n K'}{R^2} \\ -\frac{n^3 D'}{R^4} - \frac{n K'}{R^2} & \rho h\omega_n^2 - \frac{n^4 D'}{R^4} - \frac{K'}{R^2} \end{bmatrix} = 0. \tag{8.2.29}$$

Expanding the determinant and simplifying yields the ***characteristic equation***

$$\omega_n^4 - K_1\omega_n^2 + K_2 = 0, \tag{8.2.30}$$

where

$$K_1 = \frac{n^2+1}{R^2\rho h}\left(\frac{n^2 D'}{R^2} + K'\right), \tag{8.2.31a}$$

$$K_2 = \frac{n^2(n^2-1)^2}{R^6(\rho h)^2} D'K'. \tag{8.2.31b}$$

The natural frequencies are

$$\omega_n^2 = \frac{K_1}{2}\left(1 \pm \sqrt{1 - 4\frac{K_2}{K_1^2}}\right). \quad (8.2.32a)$$

Thus, there are two ***component natural frequencies*** for each mode number n, i.e.,

$$\omega_{n1}^2 = \frac{K_1}{2}\left(1 - \sqrt{1 - 4\frac{K_2}{K_1^2}}\right), \quad (8.2.32b)$$

$$\omega_{n2}^2 = \frac{K_1}{2}\left(1 + \sqrt{1 - 4\frac{K_2}{K_1^2}}\right). \quad (8.2.32c)$$

Note that the *component frequency* ω_{n1}, with a negative sign, denotes the ***transverse component natural frequency*** governing the transverse ring oscillation, and ω_{n2} denotes the ***circumferential component natural frequency*** governing the circumferential ring oscillation. For those *component natural frequencies* at lower n mode numbers ($n \leq 10$), modal oscillation amplitudes of the transverse component modes and circumferential component modes are coupled (Soedel, 1981). The amplitude ratio of transverse modes B_n over circumferential modes A_n is

$$\frac{B_{n_1}}{A_{n_1}} \cong -n, \quad (8.2.33a)$$

and

$$\frac{B_{n_2}}{A_{n_2}} \cong \frac{1}{n}, \quad (8.2.33b)$$

where the subscript n_1 denotes the transverse oscillation at ω_{n1} and n_2 the circumferential oscillation at ω_{n2}. Note that the transverse oscillation amplitude B_{n_1} is greater than the circumferential oscillation amplitude A_{n_1} at ω_{n1}, and less than it at ω_{n2}. The modified membrane stiffness K' and bending stiffness D' were defined in Sect. 8.2.1. Again, the (elastic) membrane stiffness K and bending stiffness D are used for **elastic** rings or **laminated** rings with negligible PVDF sensor/actuator layers.

Table 8.1 Natural frequencies of a steel free floating ring

n	f_{n1} (Hz) (transverse)	f_{n2} (Hz) (circumference)
0	0.0	0.173×10^5
1	0.0	0.245×10^5
2	268	0.387×10^5
3	758	0.548×10^5
4	1453	0.714×10^5
5	2350	0.883×10^5

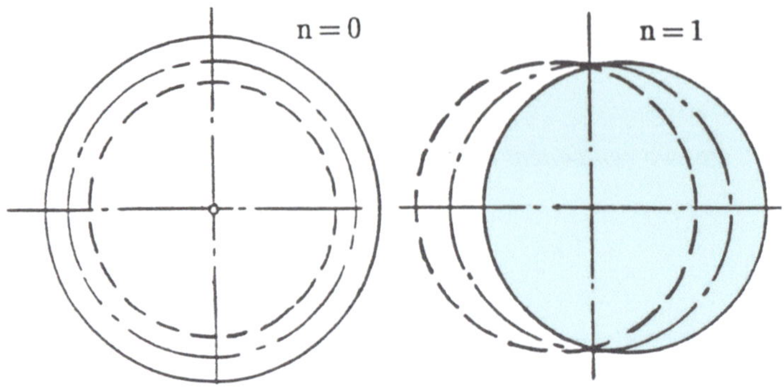

Fig. 8.5 Mode shapes of a circular ring, n = 0 and n = 1 modes

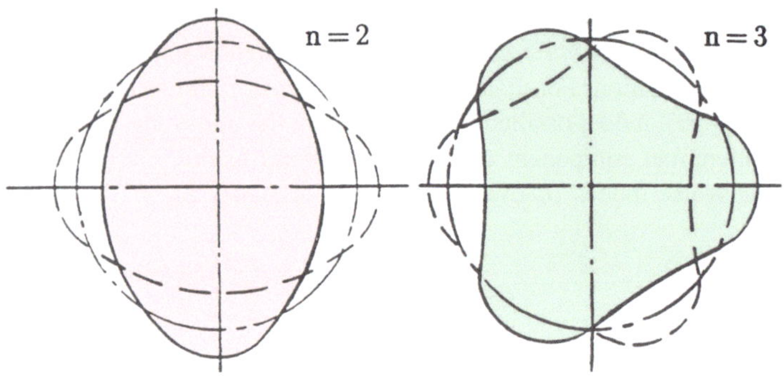

Fig. 8.6 Mode shapes of a circular ring, n = 2 and n = 3 modes

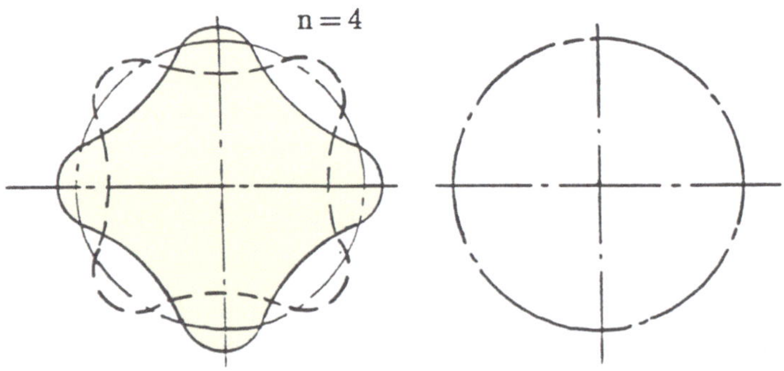

Fig. 8.7 Mode shapes of a circular ring, n = 4 and etc. modes

8.2.3 Natural Frequencies and Mode Shapes

In this section, natural frequencies and mode shapes of a free-floating ring are presented. A steel ring structure with a radius of 50 mm, width 10 mm, and thickness 1 mm is used here. Natural frequencies of the ring shell are summarized in Table 8.1; mode shapes are plotted in Figs. 8.5, 8.6 and 8.7.

In Table 8.1, the first column denotes the mode number, the second column gives the *transverse component natural frequency* for the n-th mode, and the third column gives the corresponding *circumferential component natural frequency*. It is clear that the circumferential component frequencies are much higher than those transverse component frequencies.

It is observed that n = 0 is a breathing mode; n = 1 is a rigid body mode; a set of transverse and circumferential *component* modes occur when n $\geq$ 2. Note that the component natural frequency for the transverse oscillation is ω_{n1} and the circumferential oscillation is ω_{n2}.

8.3 Distributed Convolving Ring Sensors

A closed elastic ring shell, Fig. 8.3, is a special case of an arch, which has a constant radius. (Note that a beam is a special case of an open ring with infinite curvature.) In this section, spatially distributed convolving sensors for ring structures are proposed. Analytical solutions for cosine shaped convolving piezoelectric ring sensors are then derived. It is assumed that the piezoelectric sensor layer on the ring structure, Fig. 8.4, has a constant thickness in the later analyses.

Based on the modal expansion technique, Sect. 3.5 of Chap. 3, the circumferential and transverse displacements for a free-floating circular ring can be expressed as

$$u_\theta(\theta, t) = \sum_{n=0}^{\infty} \eta_n(t) A_n \sin(n\theta - \phi^*), \tag{8.3.1}$$

$$u_3(\theta, t) = \sum_{n=0}^{\infty} \eta_n(t) B_n \cos(n\theta - \phi^*), \tag{8.3.2}$$

where $\eta_n(t)$ denotes the n-th modal participation factor, or the modal coordinate; ϕ^* is an arbitrary phase angle; A_n and B_n are the modal amplitudes. In the following derivation, it is assumed that a reference point is defined so that the phase angle $\phi^* = 0$.

As discussed previously, for a circular ring with free boundary conditions, the first mode n = 0 is a breathing mode and the second mode n = 1, is a translational rigid-body mode. For n $\geq$ 2, sets of a ***transverse component mode*** and a ***circumferential component mode*** appear. It should be noted that for a given integer n, there are two *component* natural frequencies ω_{n1} and ω_{n2} in which the former corresponds to a *transverse component mode*, i.e., the transverse oscillation dominates, and the latter to a *circumferential component mode*, i.e., the circumferential oscillation dominates (Soedel, 1981). As discussed previously, for *transverse* *component modes*, the modal oscillation amplitudes of the transverse and circumferential oscillation components are coupled by a factor when $n \leq 10$:

$$\frac{B_{n_1}}{A_{n_1}} \cong -n \quad \text{or} \quad B_{n_1} \cong -nA_{n_1}. \tag{8.3.3}$$

Similarly, for *circumferential component modes*, the ratio between the transverse oscillation component and the circumferential component is

$$\frac{B_{n_2}}{A_{n_2}} \cong \frac{1}{n} \quad \text{or} \quad A_{n_2} \cong nB_{n_2}. \tag{8.3.4}$$

Substituting the modal equations and strain expressions into the sensor equation, Eq. (8.1.17b), gives

$$\begin{aligned}\phi_3^S = -\frac{e_{31}}{\varepsilon_{33}S^e}\sum_{n=0}^{\infty}\eta_n(t)\left\{h^S(nA_{n_2}+B_{n_2})+\frac{1}{2R}\left(r_2^2-r_1^2\right)\left(nA_{n_1}+n^2B_{n_1}\right)\right\} \\ \cdot\int_0^{2\pi}\mathrm{sgn}(W_s(\theta))W_s(\theta)\cos(n\theta)d\theta\end{aligned}, \tag{8.3.5}$$

where A_{n_2} and B_{n_2} denote the modal amplitudes of ***circumferential component modes***; A_{n_1} and B_{n_1} denote the modal amplitudes of ***transverse component modes***. Thus, the first part, membrane strains, is primarily contributed by *circumferential component modes* with amplitudes A_{n_2} and B_{n_2}; the second part, bending strains, is contributed by *transverse component modes* with amplitudes A_{n_1} and B_{n_1}. The modal amplitudes, either A_{n_1}/B_{n_1} or A_{n_2}/B_{n_2}, are coupled by a constant as discussed previously. Equation (8.3.5) denotes the distributed sensor equation for spatially shaped piezoelectric sensors with spatially convolving electrodes.

8.3.1 *Shape Design*

Based on the modal orthogonality, appropriate shaped and convolved sensors only respond to the modal amplitudes of selected modes without the spillover of other residual modes. Thus, observation spillover problem can be prevented. Since the k-th mode shape is orthogonal to all other mode shapes, the shape function $W_{sk}(\alpha_1)$ can be designed as a *cosine function*, i.e.,

$$W_{sk}(\theta) = b'\cos(k\theta), \tag{8.3.6}$$

where the subscript s denotes the sensor; the subscript k is the k-th mode; b′ is a weighting factor usually assumed that b′ = b—the ring width. Substituting $W_{sk}(\theta)$ into the sensor equation and using $(r_2^2 - r_1^2) = h^s(h + h^s)$, one can derive

$$\phi_3^S = -\frac{e_{31}\pi b'}{\varepsilon_{33} S_k^e} \operatorname{sgn}(\cos k\theta) \left[h^S (kA_{k_2} + B_{k_2}) + \frac{1}{2R} h^S (h + h^S) (kA_{k_1} + k^2 B_{k_1}) \right] \eta_k(t), \tag{8.3.7}$$

where $\eta_k(t)$ is the modal participation factor. k_1 denotes the *transverse component mode* and k_2 the *circumferential component mode* for the n = k natural mode. A_{k_1} and B_{k_1} are respectively the circumferential and transverse modal oscillation amplitudes of the k-th transverse natural mode with a *transverse component* natural frequency ω_{k1}. A_{k_2} and B_{k_2} are the circumferential and transverse modal amplitudes of the k-th circumferential natural mode with a *circumferential component* natural frequency ω_{k2}.

8.3.2 Modal Sensitivities

Thus, two modal sensitivities: 1) a ***transverse modal sensitivity*** and 2) a ***circumferential modal sensitivity*** can be defined. (Note that the membrane modal sensitivity used for a piezoelectric shell continuum is defined as the circumferential modal sensitivity in ring sensors.) Each of the sensitivities can be further divided into two component sensitivities defined in terms of either the transverse or the circumferential oscillation amplitudes. Note that, again, these two amplitudes are coupled by a constant as discussed in Eqs. (8.3.3) and (8.3.4).

The ***transverse modal sensitivity*** S_t^t defined by the transverse oscillation amplitude B_{k_1} of the distributed cosine shaped convolving sensor is

$$S_t^t = \frac{\phi_3^S}{B_{k_1}\eta_k(t)} = -\frac{e_{31}\pi b'}{\varepsilon_{33} S_k^e} \left[\frac{1}{2R} h^S (h + h^S) (k^2 - 1) \right]. \tag{8.3.8a}$$

The transverse modal sensitivity S_c^t defined by the circumferential oscillation amplitude A_{k_1} of the distributed sensor is

$$S_c^t = \frac{\phi_3^S}{A_{k_1}\eta_k(t)} = -\frac{e_{31}\pi b'}{\varepsilon_{33} S_k^e} \left[\frac{1}{2R} h^S (h + h^S) (k^3 - k) \right]. \tag{8.3.8b}$$

The ***circumferential modal sensitivity*** S_c^t defined by the transverse oscillation amplitude B_{k_2} is

$$S_t^c = \frac{\phi_3^S}{B_{k_2}\eta_k(t)} = -\frac{e_{31}\pi b'}{\varepsilon_{33} S_k^e} \left[h^S (k^2 + 1) \right]; \tag{8.3.9a}$$

and the circumferential modal sensitivity S_t^c defined by the circumferential oscillation amplitude A_{k_2} is

$$S_c^c = \frac{\phi_3^S}{A_{k_2}\eta_k(t)} = -\frac{e_{31}\pi b'}{\varepsilon_{33}S_k^e}\left[h^S\left(k + \frac{1}{k}\right)\right]. \tag{8.3.9b}$$

Note that $h^s = r_2 - r_1$, and $(r_2^2 - r_1^2) = h^s(h + h^s)$. These two sensor sensitivities for the distributed convolving ring sensors are to be analyzed. Thus, for bending oscillations where transverse modes dominate, S^t should be used to estimate oscillation amplitudes. On the other hand, for circumferential oscillations, S^c is used.

8.4 Evaluation of Distributed Ring Sensors

A steel ring shell with a radius of 50 mm, width 10 mm, and thickness 1 mm is used in this study. Piezoelectric polymeric polyvinylidene fluoride material (25 μm) is spatially shaped as distributed modal sensors. Material and geometric properties are summarized in Table 8.2. Detailed sensor electromechanics and parametric studies for *transverse* and *circumferential component modes* are analyzed and results presented in this section. Contributions from two modal oscillation amplitudes, e.g., the in-plane circumferential component and the out-of-plane transverse component, are analyzed and compared. Variations of ring and sensor thicknesses are also investigated.

8.4.1 Cosine Shaped Sensors

As discussed previously, for a free-floating ring, the first mode k = 0 is a breathing mode and the second mode k = 1, is a translational rigid-body mode (Soedel, 1981). For k $\geq$ 2, sets of transverse and circumferential *component* modes, with distinct *component* natural frequencies, appear. Thus, cosine shaped distributed piezoelectric convolving sensors are primarily designed for k $\geq$ 2 and sensitivity analyses of the distributed sensors are also evaluated for k $\geq$ 2. Spatially

Table 8.2 Material and geometric properties

Properties	Steel	PVDF	Units
b	1.00×10^{-2}	1.00×10^{-2}	m
R	5.00×10^{-2}		m
h	1.00×10^{-3}	2.50×10^{-5}	m
ρ	7.80×10^{3}	1.80×10^{3}	kg/m^3
Y	2.10×10^{11}	1.60×10^{9}	N/m^2
μ	0.30	0.29	
d_{31}, d_{32}		6.00×10^{-12}	C/N
d_{33}		13.00×10^{-12}	C/N
$\varepsilon/\varepsilon_o$		10	
ε_o		8.85×10^{-10}	F/m

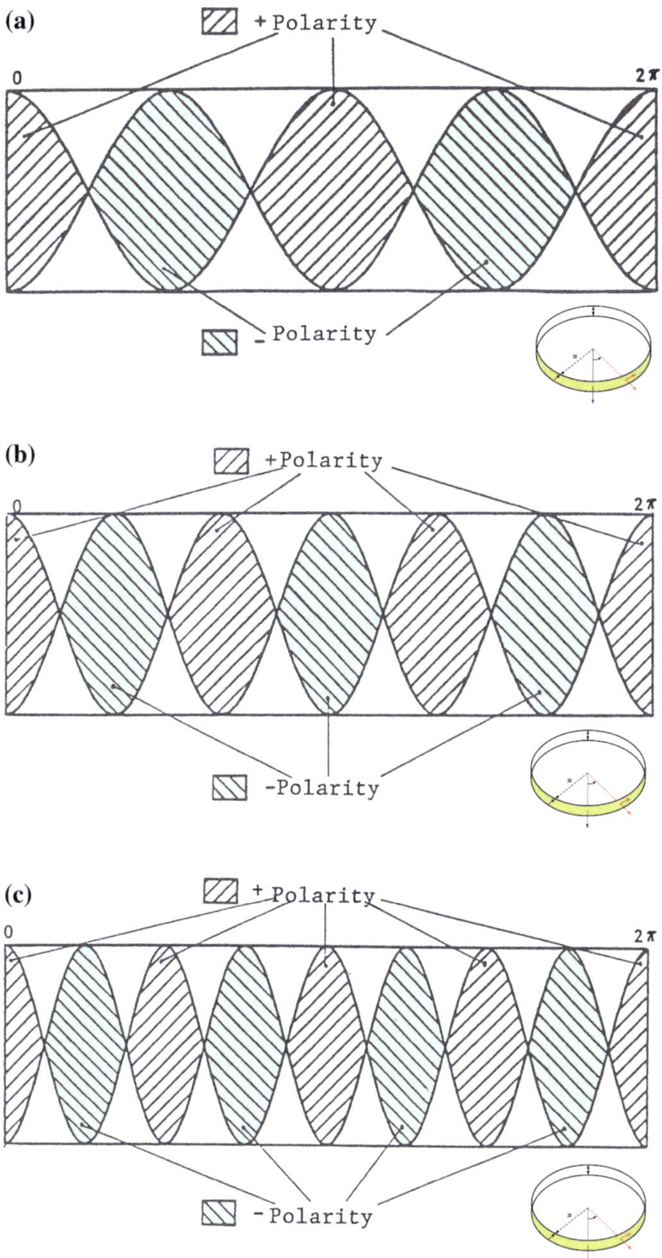

Fig. 8.8 Spatially distributed cosine shaped sensors. **a** Cosine shaped sensor for $k = 2$. **b** Cosine shaped sensor for $k = 3$. **c** Cosine shaped sensor for $k = 4$

distributed cosine shaped convolving sensors for k = 2, 3, and 4 modes are shown in Fig. 8.8a–c. Note that the sensors are defined from 0 to 2π and it is cut at $\theta = 0$. Also, the polarity changes are also illustrated.

Transverse modal sensitivity and circumferential modal sensitivity of the distributed cosine shaped convolving sensor are calculated and plotted for $k \geq 2$ modes. As discussed previously, there are two *component* natural modes for each k value, i.e., a *transverse component mode* and a *circumferential component mode*, with distinct natural frequencies. Each mode consists of two modal oscillation amplitudes: 1) an in-plane circumferential oscillation component and 2) an out-of-plane transverse oscillation component. These two oscillation components are coupled by a constant for $k \leq 10$ (Soedel, 1981). Thus, for an output signal measured, the voltage can be used to estimate both the transverse oscillation amplitude and the circumferential oscillation amplitude for a specific *component* natural mode at its natural frequency.

8.4.2 *Modal Sensitivity Analysis*

Transverse modal sensitivities, with two component sensitivities, for natural modes k = 2, 3, …, 10 are plotted in Fig. 8.9a. Circumferential modal sensitivities, with two component sensitivities, of the same modes are plotted in Fig. 8.9b. In each figure two oscillation components are plotted respectively. It is assumed that the output sensitivities are estimated at the same strain amplitudes for all evaluated natural modes.

Figure 8.9a suggests that for a measured signal, the inferred transverse oscillation amplitude, bending effect, is larger than the in-plane circumferential oscillation, membrane effect, for the transverse component natural modes. This is true because the transverse oscillation amplitude has to be much more significant in transverse natural modes. Figure 8.9b also leads to a similar conclusion for the circumferential component natural modes.

Transverse and circumferential modal sensitivities are also evaluated when the thickness of the ring and the sensor layer are changed. Figure 8.10a and b show the sensitivities of a ring (5 mm thick) with a 25 μm sensor layer. Figure 8.11a and b show the sensitivities of a ring (1 mm thick) with a 40 μm piezoelectric sensor layer. (Note that the original ring was 1 mm thick coupled with a 25 μm piezoelectric sensor layer.)

In general, membrane strains, the circumferential component, in the distributed sensor should be the same regardless the ring thickness. However, bending strains, the transverse component, increase when the ring becomes thicker. Higher strains in the sensor layer generate higher output signals. Comparing Figs. 8.10a and 8.9a, one can observe that the transverse sensitivity increases when the ring thickness

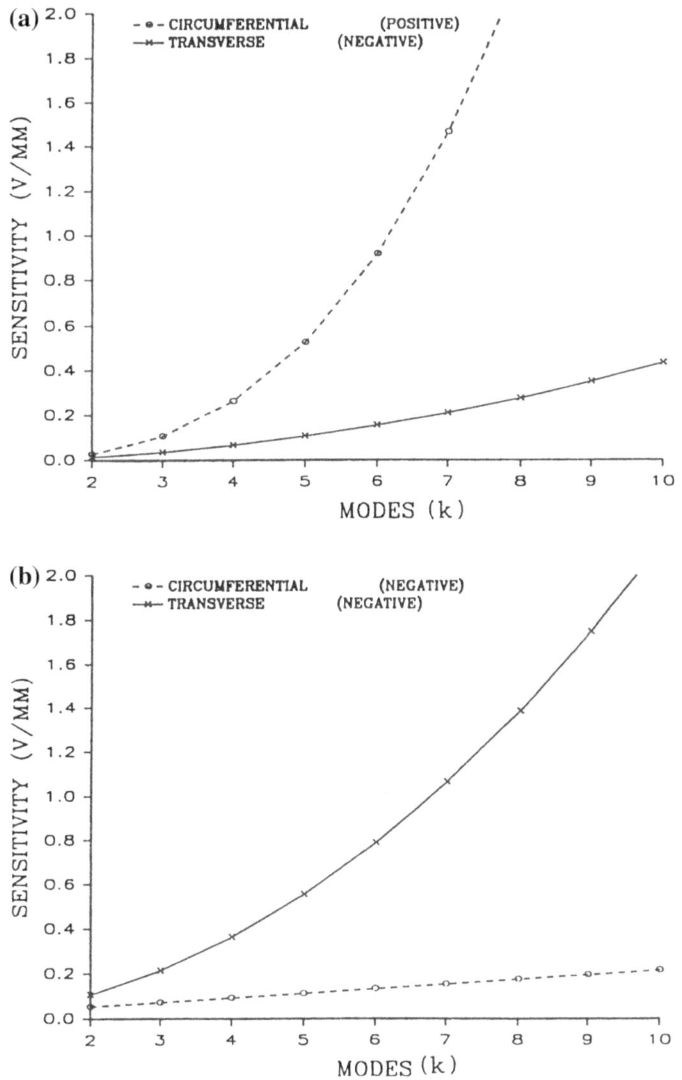

Fig. 8.9 Modal sensitivities (1 mm ring, 25 μm sensor). **a** Transverse modal sensitivity (1 mm ring, 25 μm sensor). **b** Circumferential modal sensitivities (1 mm ring, 25 μm sensor)

increases. However, the circumferential sensitivities in Figs. 8.10b and 8.9b are identical because the membrane strains remained unchanged. These results suggested that the membrane (circumferential) modal sensitivity is independent of ring thickness and the transverse modal sensitivity is a linear function of ring thickness. Figure 8.11a and b show the sensitivities increase when the sensor layer becomes thicker, at the same oscillation amplitudes.

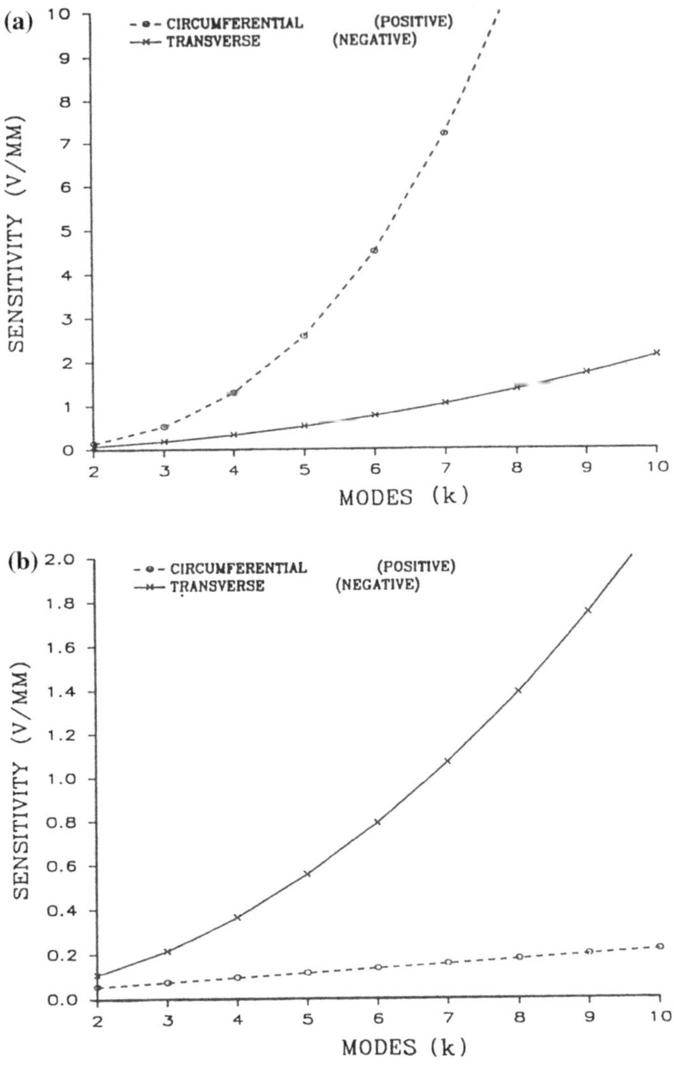

Fig. 8.10 Modal sensitivities (5 mm ring, 25 μm sensor). **a** Transverse modal sensitivity (5 mm ring, 25 μm sensor). **b** Circumferential modal sensitivities (5 mm ring, 25 μm sensor)

Distributed sensor mechanics and distributed convolving sensors have been presented. Design and performance of distributed convolving cosine sensors for circular ring structures have been studied. Next, distributed shell actuators and generic control concepts are discussed. Applications of the theory to distributed convolving actuators for a generic structure and a circular ring structure are demonstrated.

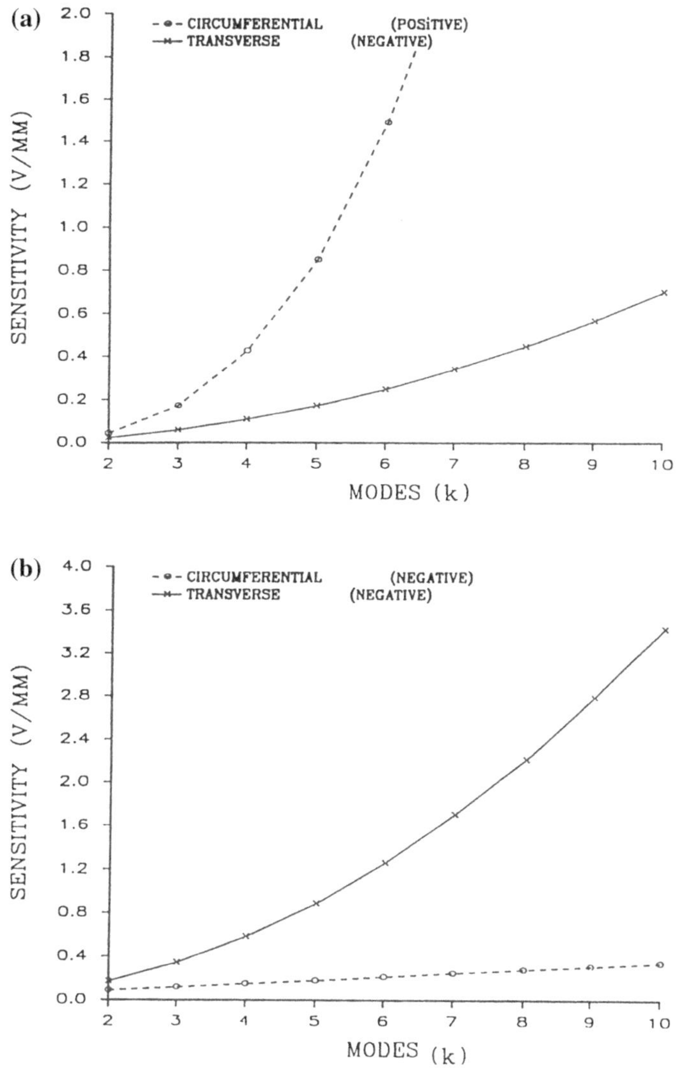

Fig. 8.11 Modal sensitivities (1 mm ring, 40 μm sensor). **a** Transverse modal sensitivity (1 mm ring, 40 μm sensor). **b** Circumferential modal sensitivities (1 mm ring, 40 μm sensor)

8.5 Distributed Vibration Controls

As discussed in Sect. 3.5 of Chap. 3, electric membrane forces and bending moments in the electromechanical equations of piezoelectric shells can be used to actively control the piezoelectric shells. The electric forces and moments can be manipulated by either open-loop or closed-loop schemes. In open-loop controls, a reference signal is required. In closed-loop controls, sensor signals representing the dynamic state of the continuum are used in a feedback control loop. Detailed derivations presented in Sect. 3.5 of Chap. 3 were generic, and they accommodated all three principal directions. In this section, the generic theories are reduced to the transverse vibration of continua or structures—the most practical case.

As discussed in Sect. 3.5, there are two kinds of excitation forces for the structures composed of piezoelectric materials. One is the conventional mechanical force and the other is the electric force introduced by the converse piezoelectric effect. If Love's operator of the converse piezoelectric effect $L_i^c(\phi_3)$ is designed as a function of mechanical motions, such as displacement, velocity, or acceleration, a feedback control system can be established. In these cases, voltages fed back to the piezoelectric actuators induce electric forces/moments which can be used to counteract and control the structural vibrations. Three generic feedback algorithms: 1) the displacement feedback, 2) the velocity feedback, and 3) the acceleration feedback were proposed and corresponding governing equations derived in Sect. 3.5. In this section, the transverse vibration (α_3 direction) is considered as the primary motion of interest.

8.5.1 Displacement Feedback

In the displacement feedback, it is assumed that Love's operator $L_3^c(\phi_3)$ contributed by the converse piezoelectric effect is a function of the transverse displacement, i.e.,

$$L_3^c\{\phi_3\} = \mathsf{F}_1^*\{u_3(\alpha_1, \alpha_2, t)\}, \tag{8.5.1a}$$

where F_i^* denotes a generic spatial function. The feedback can be further expressed in a modal coordinate expression:

$$L_3^c\{\phi_3\} = \sum_{m=1}^{\infty} \mathcal{G}_{3m}^{df}(\alpha_1, \alpha_2)\eta_m(t), \tag{8.5.1b}$$

where $\mathcal{G}_{3m}^{df}(\alpha_1, \alpha_2)$ is the *displacement modal feedback function*, a spatially distributed function. The modal control force defined for the k-th mode is

$$\hat{F}_k = \frac{1}{\rho h N_k} \int_{\alpha_1} \int_{\alpha_2} \left\{ \sum_{m=1}^{\infty} \mathcal{G}_{3m}^{df}(\alpha_1, \alpha_2)\eta_m(t) U_{3k}(\alpha_1, \alpha_2) \right\} A_1 A_2 d\alpha_1 d\alpha_2. \tag{8.5.2}$$

Substituting the modal control force into the modal equation gives

$$\begin{aligned} &\ddot{\eta}_k + \frac{c}{\rho h}\dot{\eta}_k + \omega_k^2 \eta_k \\ &= \frac{1}{\rho h N_k} \sum_{m=1}^{\infty} \int_{\alpha_1} \int_{\alpha_2} \left[\sum_{m=1}^{\infty} \mathcal{G}_{3m}^{df}(\alpha_1, \alpha_2)\eta_m(t) U_{3k}(\alpha_1, \alpha_2) \right] A_1 A_2 d\alpha_1 d\alpha_2, \end{aligned} \tag{8.5.3}$$

where the external mechanical forces are not considered, i.e., $F_3 = 0$, and

$$N_k = \int_{\alpha_1}\int_{\alpha_2} U_{3k}^2 A_1 A_2 d\alpha_1 d\alpha_2. \tag{8.5.4}$$

8.5.2 *Velocity Feedback*

Following the same procedures, one can further define the governing modal equations for the velocity feedback:

$$L_3^c\{\phi_3\} = \mathsf{F}_2^*\{\dot{u}_3(\alpha_1,\alpha_2,t)\}, \tag{8.5.5a}$$

$$L_3^c\{\phi_3\} = \sum_{m=1}^{\infty} \mathcal{G}_{3m}^{vf}(\alpha_1,\alpha_2)\dot{\eta}_m(t). \tag{8.5.5b}$$

and

$$\begin{aligned}
&\ddot{\eta}_k + \frac{c}{\rho h}\dot{\eta}_k + \omega_k^2\eta_k \\
&= \frac{1}{\rho h N_k}\sum_{m=1}^{\infty}\int_{\alpha_1}\int_{\alpha_2}\left[\mathcal{G}_{3m}^{vf}(\alpha_1,\alpha_2)\dot{\eta}_m(t)U_{3k}(\alpha_1,\alpha_2)\right]A_1A_2d\alpha_1d\alpha_2,
\end{aligned} \tag{8.5.6}$$

where the modal feedback control force for the velocity feedback is defined as

$$\hat{F}_k = \frac{1}{\rho h N_k}\int_{\alpha_1}\int_{\alpha_2}\left\{\sum_{m=1}^{\infty}\mathcal{G}_{3m}^{vf}(\alpha_1,\alpha_2)\dot{\eta}_m(t)U_{3k}(\alpha_1,\alpha_2)\right\}A_1A_2d\alpha_1d\alpha_2. \tag{8.5.7}$$

8.5.3 *Acceleration Feedback*

The acceleration feedback can be defined in the same manner.

$$L_3^c\{\phi_3\} = \mathsf{F}_3^*\{\ddot{u}_3(\alpha_1,\alpha_2,t)\}, \tag{8.5.8a}$$

$$L_3^c\{\phi_3\} = \sum_{m=1}^{\infty} \mathcal{G}_{3m}^{af}(\alpha_1,\alpha_2)\ddot{\eta}_m(t). \tag{8.5.8b}$$

and

$$\ddot{\eta}_k + \frac{c}{\rho h}\dot{\eta}_k + \omega_k^2\eta_k = \frac{1}{\rho h N_k}\sum_{m=1}^{\infty}\int_{\alpha_1}\int_{\alpha_2}\left[\mathcal{G}_{3m}^{af}(\alpha_1,\alpha_2)\ddot{\eta}_m(t)U_{3k}(\alpha_1,\alpha_2)\right]A_1A_2d\alpha_1 d\alpha_2. \tag{8.5.9}$$

where

$$\hat{F}_k = \frac{1}{\rho h N_k}\int_{\alpha_1}\int_{\alpha_2}\left\{\sum_{m=1}^{\infty}\mathcal{G}_{3m}^{af}(\alpha_1,\alpha_2)\ddot{\eta}_m(t)U_{3k}(\alpha_1,\alpha_2)\right\}A_1A_2d\alpha_1 d\alpha_2. \tag{8.5.10}$$

Note that a generic shell control system can be described by an infinite number of second order modal equations. The overall system response is contributed by all participating modes. However, only a finite number of modes are considered in practical applications. As discussed in Sect. 3.5 of Chap. 3 and in Sect. 7.1 of Chap. 7, control spillovers could occur when feedback control forces for controlled modes appear in the governing equations of other uncontrolled modes due to modal interactions among all participating modes. Generic distributed control concepts based on distributed convolving piezoelectric actuators are presented next.

In addition, three ***modal feedback functions*** $\mathcal{G}_{3m}^{df}(\alpha_1,\alpha_2), \mathcal{G}_{3m}^{vf}(\alpha_1,\alpha_2)$, and $\mathcal{G}_{3m}^{af}(\alpha_1,\alpha_2)$ are spatially distributed functions. If these modal feedback gain functions are defined so that they are orthogonal to a mode or a group of modes, these modes can be filtered out from being activated or controlled in the feedback control system. This concept will be discussed next.

8.6 Distributed Convolving Actuators and Controls

Note that control spillovers resulting from the cross coupling control forces can be observed in the modal equations of all three feedback algorithms. Based on the modal orthogonality, one can design the *spatial modal feedback functions*, i.e., $\mathcal{G}_{3m}^{df}(\alpha_1,\alpha_2), \mathcal{G}_{3m}^{vf}(\alpha_1,\alpha_2)$, and $\mathcal{G}_{3m}^{af}(\alpha_1,\alpha_2)$, with spatially distributed characteristics such that they are orthogonal to other natural modes, i.e., $k \neq p$. Thus, the modal feedback gain functions can be defined as a product of a constant modal gain $\mathcal{G}_{3n}^{f}$ and the n-th mode shape function $U_{3n}(\alpha_1,\alpha_2)$, i.e.,

$$\mathcal{G}_{3n}^{df}(\alpha_1,\alpha_2) = \mathcal{G}_{3n}^{df}U_{3n}(\alpha_1,\alpha_2), \tag{8.6.1a}$$

$$\mathcal{G}_{3n}^{vf}(\alpha_1,\alpha_2) = \mathcal{G}_{3n}^{vf}U_{3n}(\alpha_1,\alpha_2), \tag{8.6.1b}$$

$$\mathcal{G}_{3n}^{af}(\alpha_1,\alpha_2) = \mathcal{G}_{3n}^{af}U_{3n}(\alpha_1,\alpha_2), \tag{8.6.1c}$$

where $\mathcal{G}_{3n}^{df}$, $\mathcal{G}_{3n}^{vf}$, and $\mathcal{G}_{3n}^{af}$ are weighting factors (gain constants). By doing this, the control forces appear only in the equations of controlled modes, but not in the equations of uncontrolled modes. Based on this concept, one can further define the modal equations with the distributed modal actuators.

8.6.1 Displacement Feedback

Based on the modal orthogonality concept and Eq. (8.6.1a), the modal control force can be reduced to

$$\hat{F}_k = \frac{\mathcal{G}_3^{df}}{\rho h N_k} \eta_n(t) \int_{\alpha_1} \int_{\alpha_2} U_{3n}(\alpha_1, \alpha_2) U_{3k}(\alpha_1, \alpha_2) A_1 A_2 d\alpha_1 d\alpha_2, \tag{8.6.2}$$

such that

$$\text{i)} \quad \hat{F}_k = \frac{\mathcal{G}_3^{df}}{\rho h N_n} \eta_n(t) \int_{\alpha_1} \int_{\alpha_2} U_{3n}^2 A_1 A_2 d\alpha_1 d\alpha_2 \quad (\text{for } k = n), \tag{8.6.3a}$$

$$\text{ii)} \quad \hat{F}_k = 0 \quad (\text{for } k \neq n). \tag{8.6.3b}$$

Thus, the modal equation is decoupled from the residual modes, i.e.,

$$\Rightarrow \quad \ddot{\eta}_n + \frac{c}{\rho h} \dot{\eta}_n + \left(\omega_n^2 - \frac{\mathcal{G}_3^{df}}{\rho h N_n} \int_{\alpha_1} \int_{\alpha_2} U_{3n}^2 A_1 A_2 d\alpha_1 d\alpha_2 \right) \eta_n = 0. \tag{8.6.4}$$

Decoupled modal equations for the velocity and acceleration feedback controls can be derived in the same manner.

8.6.2 Velocity Feedback

$$\hat{F}_k = \frac{\mathcal{G}_3^{vf}}{\rho h N_k} \dot{\eta}_n(t) \int_{\alpha_1} \int_{\alpha_2} U_{3n}(\alpha_1, \alpha_2) U_{3k}(\alpha_1, \alpha_2) A_1 A_2 d\alpha_1 d\alpha_2; \tag{8.6.5}$$

and

$$\text{i)} \quad \hat{F}_k = \frac{\mathcal{G}_3^{vf}}{\rho h N_n} \dot{\eta}_n(t) \int_{\alpha_1} \int_{\alpha_2} U_{3n}^2 A_1 A_2 d\alpha_1 d\alpha_2 \quad (\text{for } k = n), \tag{8.6.6a}$$

$$\text{ii)}\quad \hat{F}_k = 0 \quad (\text{for } k \neq n). \tag{8.6.6b}$$

$$\Rightarrow \quad \ddot{\eta}_n + \frac{1}{\rho h}\left(c - \frac{\mathcal{G}_3^{vf}}{N_n}\int_{\alpha_1}\int_{\alpha_2} U_{3n}^2 A_1 A_2 d\alpha_1 d\alpha_2\right)\dot{\eta}_n + \omega_n^2 \eta_n = 0. \tag{8.6.7}$$

8.6.3 Acceleration Feedback

$$\hat{F}_k = \frac{\mathcal{G}_3^{af}}{\rho h N_k}\ddot{\eta}_n(t)\int_{\alpha_1}\int_{\alpha_2} U_{3n}(\alpha_1,\alpha_2)U_{3k}(\alpha_1,\alpha_2)A_1 A_2 d\alpha_1 d\alpha_2; \tag{8.6.8}$$

and

$$\text{i)}\quad \hat{F}_k = \frac{\mathcal{G}_3^{af}}{\rho h N_n}\ddot{\eta}_n(t)\int_{\alpha_1}\int_{\alpha_2} U_{3n}^2 A_1 A_2 d\alpha_1 d\alpha_2 \quad (\text{for } k = n), \tag{8.6.9a}$$

$$\text{ii)}\quad \hat{F}_k = 0 \quad (\text{for } k \neq n). \tag{8.6.9b}$$

$$\Rightarrow \quad \left(1 - \frac{\mathcal{G}_3^{af}}{\rho h N_n}\int_{\alpha_1}\int_{\alpha_2} U_{3n}^2 A_1 A_2 d\alpha_1 d\alpha_2\right)\ddot{\eta}_n + \frac{c}{\rho h}\dot{\eta}_n + \omega_n^2 \eta_n = 0. \tag{8.6.10}$$

Again, $\mathcal{G}_i^{df}$, $\mathcal{G}_i^{vf}$, and $\mathcal{G}_i^{af}$ are weighting factors (gain constants) which are independent of spatial coordinates and time. $U_{3n}(\alpha_1, \alpha_2)$ denotes the n-th mode shape function in the transverse direction. (Note that only the transverse oscillation is considered in the formulation, although the procedures can be extended to encompass all three principal directions.) As a demonstration of distributed modal actuators, spatially distributed convolving actuators are designed for a laminated circular ring and a laminated cylindrical shell (in Chap. 9).

8.7 Distributed Convolving Ring Actuators

As discussed previously, ring structures belong to the family of one-dimensional structures, such as arches, beams, rods, etc. Thus, one-dimensional piezoelectric modal convolving actuators can be established by designing either spatially convolving electrodes on a fully distributed piezoelectric layer or a spatially convolved and shaped actuator layer(s). Note that only the transverse electric field is considered and externally applied mechanical loads are neglected. Again, a laminated circular ring is used in the analysis. The ring has a thickness h and the actuator has a thickness h^a. Material properties of the actuator layer are neglected in the analysis. Modal excitation force, Eq. (3.5.7a), can be simplified as:

$$\hat{F}_k = \frac{1}{\rho h N_k} \int_{\alpha_1} \int_{\alpha_2} L_3^c\{\phi_3(\alpha_1)\} U_{3k} A_1 A_2 d\alpha_1 d\alpha_2. \tag{8.7.1}$$

In one-dimensional structures, the mode shape U_{3k} is a function of only one coordinate α_1, e.g., the circumferential direction in rings and arches, the longitudinal direction in beams and rods, etc. If the electrode area of the actuator is designed as an actuator shape function $W_{ak}(\alpha_1)$ where the subscript a is the actuator and k the k-th mode, the operator of the converse piezoelectric effect $L_3^c\{\phi_3(\alpha_1)\}$ is zero on the uncovered area(s) and is a constant on the area covered by the electrode, i.e., ϕ_3 = constant. Thus, the integration in Eq. (8.7.1) can be rewritten as

$$\begin{aligned} \hat{F}_k &= \frac{L_3^c\{\phi_3\}}{\rho h N_k} \int_{\alpha_1} \int_0^{W_{ak}(\alpha_1)} U_{3k} A_1 A_2 d\alpha_1 d\alpha_2 \\ &= \frac{L_3^c\{\phi_3\}}{\rho h N_k} \int_{\alpha_1} W_{ak}(\alpha_1) U_{3k} A_1 A_2 d\alpha_1. \end{aligned} \tag{8.7.2}$$

Equation (8.7.2) denotes the effective control force resulting from a spatially convolving piezoelectric actuator designed for one-dimensional structures. The spatially distributed electrode area acts as a (orthogonal) weighting function on the mode shape. Designing the electrode area resembling a specific mode shape makes the distributed actuator sensitive only to this particular mode and insensitive to all the other modes. Multiple modal actuations can be achieved by linear shape combination of all participating modes.

8.7.1 *Spatially Convolving Piezoelectric Ring Actuators*

In this section, distributed modal actuators designed for flexible ring structures are proposed and their performances evaluated. The equation of motion of a ring structure was discussed in Sect. 8.3. Because the radius of curvature R_θ, i.e., $R_\theta = R$, is not equal to infinity, the electric bending moment and membrane force induced by the converse piezoelectric effect are preserved in the system equation and they can be used for structural excitation and controls. In distributed structural controls, the electric membrane force and bending moments can be designated either open-loop or closed-loop. In closed-loop feedback controls, these forces/moments can be related to the velocity or displacement signals such that damping ratios or resonant frequencies can be manipulated. The modal participation equation of the ring structure is

$$\ddot{\eta}_k + \frac{c}{\rho h}\dot{\eta}_k + \omega_k^2\eta_k = \frac{1}{\rho h N_k}\int_\theta \left[W_{ak}(\theta)\frac{N_{\theta\theta}^e}{R} - \frac{1}{R^2}\frac{\partial^2\left[W_{ak}(\theta)M_{\theta\theta}^e\right]}{\partial\theta^2}\right]U_{3k}R d\theta, \tag{8.7.3}$$

where $W_{ak}(\theta)$ is the k-th actuator shape function based on the modal orthogonality; N_k is defined by the mode shape function:

$$N_k = Rb\int_0^{2\pi}\cos^2(n\theta)d\theta = \pi Rb. \tag{8.7.4}$$

In a feedback system, the electric loads, i.e., the electric membrane forces and bending moments, are treated as control forces. For spatially convolving actuators, these force/moment components can be expressed as a product of a spatially shaped weighing function $W_{ak}(\theta)$ and the electric forces or moments. Since the actuator is surface laminated, the electric membrane force and moment, Sect. 4.3.1 of Chap. 4, can be derived as

$$N_{\theta\theta}^e = \int_{\alpha_3} e_{31}E_3 d\alpha_3 = e_{31}\int_{\alpha_3} E_3 d\alpha_3 = e_{31}\phi_3, \tag{8.7.5a}$$

$$M_{\theta\theta}^e = e_{31}\phi_3 \cdot r_\theta^a = \frac{e_{31}(h+h^a)}{2}\phi_3, \tag{8.7.5b}$$

where h^a is the actuator thickness and h is the ring thickness. Substituting Eqs. (8.7.4), (8.7.5a) and (8.7.5b) as well as the mode shape function into Eq. (8.7.3) gives

$$\ddot{\eta}_k + \frac{c}{\rho h}\dot{\eta}_k + \omega_k^2\eta_k = \frac{e_{31}\phi_3}{\pi Rb\rho h}\int_0^{2\pi}\left[W_{ak}(\theta) - \frac{h+h^a}{2R} - \frac{\partial^2 W_{ak}(\theta)}{\partial\theta^2}\right]\cos(n\theta)d\theta. \tag{8.7.6}$$

Note that the first term in the bracket comes from the in-plane membrane force and the second is due to the bending effect. The weighting function $W_{ak}(\theta)$ (spatially convolving electrode area) is designed based on a mode shape function:

$$W_{ak}(\theta) = b''\cos(k\theta). \tag{8.7.7}$$

b'' is a weighting constant. For convenience, let $b'' = b$, the ring width. In this case, only the k-th mode can be controlled. The k-th modal equation of the ring with cosine shaped actuator can be written as

$$\begin{aligned}&\ddot{\eta}_k + \frac{c}{\rho h}\dot{\eta}_k + \omega_k^2\eta_k \\ &= \frac{e_{31}\phi_3}{\pi R\rho h}\left[1 + \frac{k^2(h+h^a)}{2R}\right]\int_0^{2\pi}\cos(k\theta)\cos(n\theta)d\theta \\ &= \frac{e_{31}\phi_3}{R\rho h}\left[1 + \frac{k^2(h+h^a)}{2R}\right].\end{aligned} \tag{8.7.8}$$

Again, the first term is related to the in-plane membrane control effect, and the second is due to the bending control effect. The velocity and displacement feedback can be setup via the input voltage ϕ_3.

8.7.1.1 Displacement Feedback

In the case of displacement feedback, the voltage is

$$\phi_3 = -\mathcal{G}_r^{df}\eta_k(t), \tag{8.7.9a}$$

where $\mathcal{G}_r^{df}$ is the total gain in the feedback loop, which can be written as:

$$\mathcal{G}_r^{df} = \mathcal{S}_r^{tr}\cdot\mathcal{G}_{am}, \tag{8.7.9b}$$

where $\mathcal{G}_{am}$ is the amplifier gain and $\mathcal{S}_r^{tr}$ is a sensor sensitivity. Substituting Eq. (8.7.9a) into Eq. (8.7.8) yields

$$\ddot{\eta}_k + \frac{c}{\rho h}\dot{\eta}_k + \left\{\omega_k^2 + \frac{e_{31}}{R}\left[1 + \frac{k^2(h+h^a)}{2R}\right]\mathcal{G}_r^{df}\right\}\eta_k = 0, \tag{8.7.10}$$

or

$$\ddot{\eta}_k + 2\zeta\omega_k\dot{\eta}_k + \left\{\omega_k^2 + \frac{e_{31}}{R}\left[1 + \frac{k^2(h+h^a)}{2R}\right]\mathcal{G}_r^{df}\right\}\eta_k = 0. \tag{8.7.11}$$

Recall that the damping ratio $\zeta = c/(2\rho h\omega_k)$. Note that the modal resonant frequency is controlled in the displacement feedback.

8.7.1.2 Velocity Feedback

For the velocity feedback, the signal taken from a sensor is differentiated, amplified, and then applied to the piezoelectric actuator.

$$\phi_3 = -\mathcal{G}_r^{vf}\dot{\eta}_k(t), \tag{8.7.12a}$$

where $\mathcal{G}_r^{vf}$ is the total gain of the feedback loop and it can be further expressed as

$$\mathcal{G}_r^{vf} = S_r^{tr} \cdot \mathcal{G}_{am} \cdot \mathcal{G}_{de}, \tag{8.7.12b}$$

where S_r^{tr} is the sensitivity of the sensor layer discussed earlier. $\mathcal{G}_{de}$ and $\mathcal{G}_{am}$ are differentiator and amplifier gains, respectively. Substituting Eq. (8.7.12a) into Eq. (8.7.8) yields

$$\ddot{\eta}_k + \frac{1}{\rho h}\left\{c + \frac{e_{31}}{R}\left[1 + \frac{k^2(h + h^a)}{2R}\right]\mathcal{G}_r^{vf}\right\}\dot{\eta}_k + \omega_k^2\eta_k = 0. \tag{8.7.13}$$

The equivalent controlled modal damping ratio for the ring structure is

$$\zeta_k' = \frac{1}{2\rho h\omega_k}\left\{c + \frac{e_{31}}{R}\left[1 + \frac{k^2(h + h^a)}{2R}\right]\mathcal{G}_r^{vf}\right\}. \tag{8.7.14}$$

It is observed that the velocity feedback can be used to increase the system damping. In later analyses, only the velocity feedback is considered and evaluated.

8.8 Control Effectiveness of Convolving Ring Actuators

In this section, control effectiveness of distributed cosine shaped actuators for ring structures is evaluated. The physical model was presented in Sect. 8.3 and the material properties were listed in Table 8.2. The shape of the convolving ring actuators are described by cosine functions, i.e., $W(\theta) = b\cos(k\theta)$, which have the same shapes as those plotted in Fig. 8.8a–c for k = 2, 3, 4.

8.8.1 Comparison of Bending and Membrane Effects

In order to compare the effective control effects introduced by electric membrane force and bending moment, the laminated ring structure (shown in Fig. 8.4) is considered, in which h = 1 mm, h^a = 0.025 mm, R = 50 mm. Bending contributions are calculated and summarized in Table 8.3. Note that the membrane effect is assumed unity.

$$\Delta_r = \frac{k^2(h + h^a)}{2R} = \frac{(1 + 0.025)k^2}{2 \times 50} = 0.01025k^2. \tag{8.8.1}$$

Table 8.3 Comparison of membrane and bending control effects

k	0	1	2	3	4	5
Δ_r	0	1.025%	4.100%	9.225%	16.400%	25.625%

It is observed that the ratio contributed by the bending control effect varies from 1.025% for the first mode up to 25.625% for the fifth mode. The contribution of bending effects increases as the mode number increases. These calculations suggest that the membrane control effect is the dominating control action for the lower modes when $n \leq 10$, and the bending control becomes significant when $n > 10$, $n = k$. Note that both membrane and bending control effects are considered in the later analyses.

8.8.2 Modal Damping Controls

It is assumed that the circular ring is made of steel and the distributed convolving actuators are made of polymeric piezoelectric polyvinylidene fluoride (PVDF), Fig. 8.4. Note that the actuator layer is bonded on the top surface and the sensor layer on the bottom surface. Material properties of the PVDF and bonding material are neglected. It is also assumed that the velocity signals are acquired from the distributed convolving sensors such that observation spillover is prevented. In this section, only the velocity feedback is considered and controlled damping ratios are evaluated.

The controlled damping ratios of the first three transverse modes ($n = 2, 3, 4$) are plotted versus the (normalized) system feedback gains with a unit of V/mm (voltage per unit displacement), Fig. 8.12a–c. The initial damping ratio is set at 1%. Recall that the $n = 0$ mode is a breathing mode and the $n = 1$ mode is a rigid body mode. The distributed actuators are designed for $n \geq 2$ modes.

It is observed that the control damping ratios linearly increase with the control gains, and decrease as the mode number increases. Note that the convergence of a modal time history is determined by a combined effect of modal natural frequency and damping ratio, i.e., $e^{-\zeta_n \omega_n}$, Sect. 7.7 of Chap. 7. The higher mode oscillations usually converge much faster than the lower modes. It is also observed that the electric membrane force contributes the majority of total control effects and this effect decreases as the mode number increases. The resultant control effect is a sum of the membrane control effect and the bending control effect. The control effect from the electric bending moments is relatively less sensitive than that from the electric membrane forces for different modes. Damping controls of a 0.5 mm circular ring with the 25 μm actuator and a 1 mm ring with a 40 μm piezoelectric actuator are respectively plotted in Figs. 8.13 and 8.14.

When the elastic ring becomes thinner, the control effect increases rather significantly, about five times. It is observed that the reduction of control moment arm, from (1.025/2) mm to (0.525/2) mm, is insignificant because the control action primarily comes from the in-plane membrane control forces. Consequently, the control effect to the flexible ring becomes much more prominent. The control contribution from the 40 μm-actuator is relatively insignificant. (Note that it may not be the case when the primary control action comes from the electric bending moments, e.g., for n > 10 modes.) In general, for the control effectiveness, the change of structural flexibility outweighs the moment arm change if the distributed actuator is made of PVDF materials, as discussed in Sect. 7.7 of Chap. 7. It should be noted that the gain used in these figures is the **system gain** of the total control system. In a real control system design, it includes the sensitivity of the distributed modal sensor, the gain of the amplifier, and the gain of the derivative circuit (to acquire the velocity information).

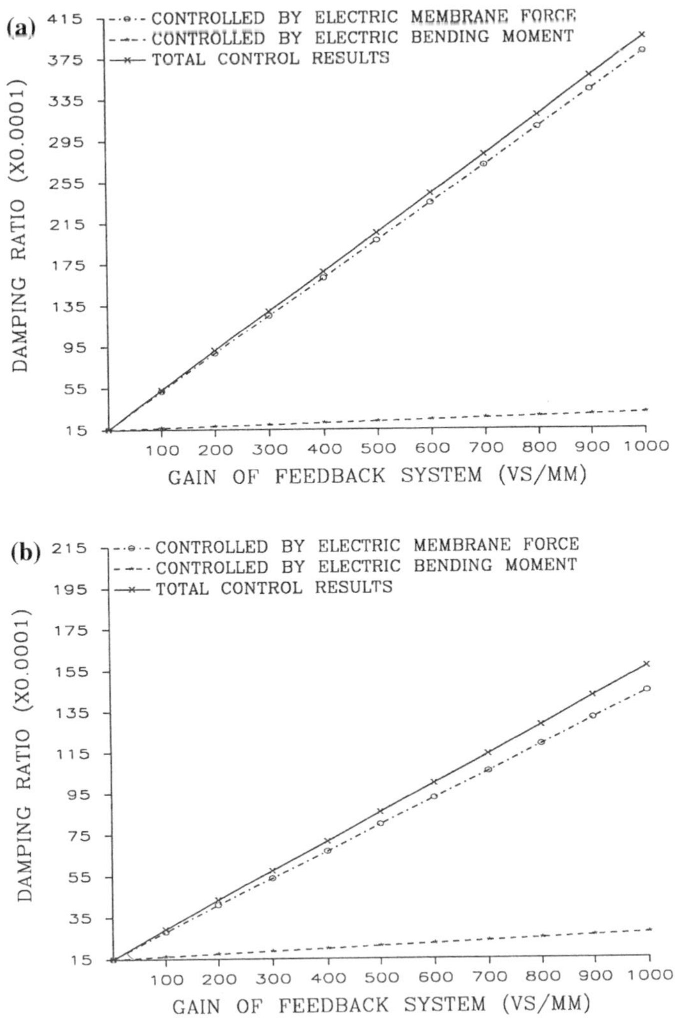

Fig. 8.12 Damping control of a circular ring (1 mm, 25 μm), n = 2, 3, 4

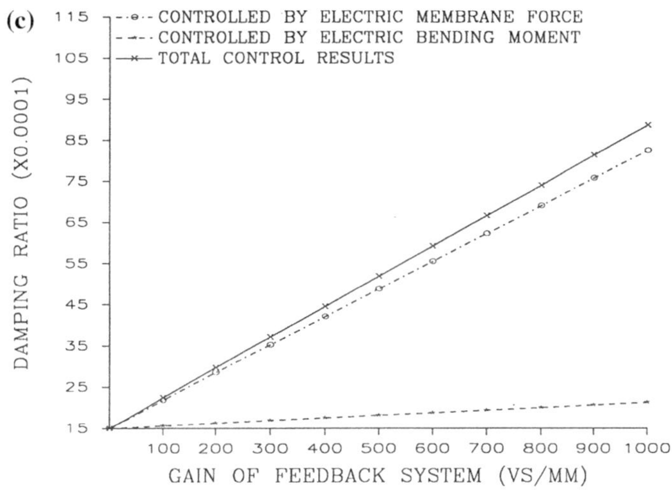

Fig. 8.12 (continued)

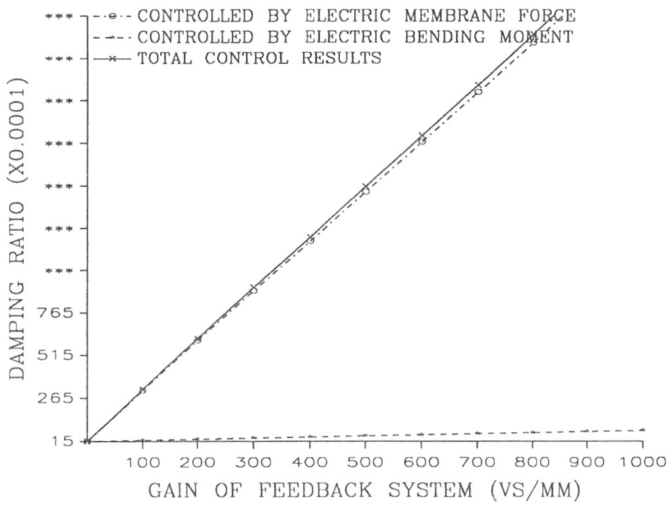

Fig. 8.13 Damping control of a circular ring (0.5 mm, 25 μm), n = 2

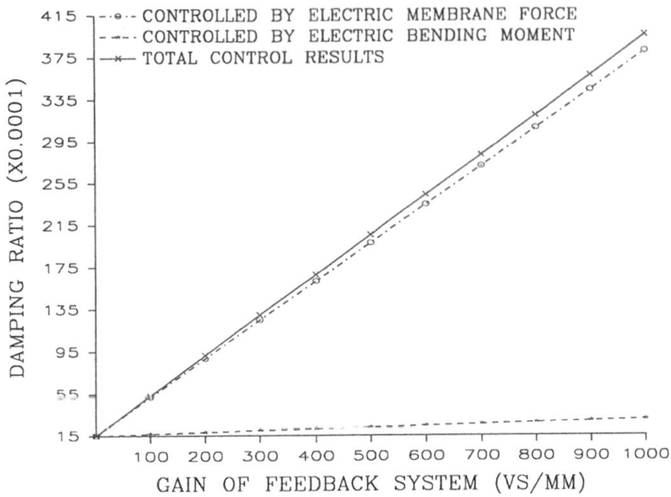

Fig. 8.14 Damping control of a circular ring (1 mm, 40 μm), n = 2

8.9 Summary

In this chapter, generic distributed piezoelectric shell convolving sensors and actuators were proposed and detailed electromechanical behaviors (***sensor*** and ***actuator mechanics***) were analyzed. It was observed that the sensor output is contributed by membrane strains and bending strains experienced in the sensor layer. Two sensor sensitivities: 1) a ***transverse modal sensitivity*** and 2) a ***membrane modal sensitivity*** can be defined accordingly. In general, the transverse modal sensitivity is defined for out-of-plane transverse natural modes and the membrane modal sensitivity for in-plane natural modes. Proper design of distributed sensor shape and convolution can provide modal filtering to prevent observation spillover in distributed structural control systems.

Cosine shaped piezoelectric convolving modal sensors and actuators, based on the shaping technique, were designed and analyzed for ring shell structures. A ***transverse modal sensitivity*** for transverse component natural modes and a ***circumferential modal sensitivity*** (equivalent to the *membrane modal sensitivity* in shell sensors) for circumferential component natural modes were defined and results plotted. Parametric studies suggested that the transverse modal sensitivity increases when the ring structure becomes thicker because the bending strains in the sensor layer increase. However, circumferential modal sensitivity remained unchanged because the membrane strains were independent of the ring thickness. It was also observed that both modal sensitivities increase when the piezoelectric sensor layer becomes thicker.

Control effectiveness of distributed modal actuators was evaluated. Analyses suggested that the primary control action comes from the in-plane membrane control forces; the contribution from the electric bending moment was relatively insignificant for natural modes n < 10. In addition, structural flexibility was observed to be of importance in feedback controls. It was observed that control effect increases as the structural stiffness decreases. Increasing control moment arm (as the ring becomes thicker) of the actuator layer seems insignificant in overall control effects for lower natural modes.

Theoretical and parametric studies of sensor and actuator mechanics, carried out in this study, provide a better understanding of the electromechanics and functionalities of distributed piezoelectric sensors and actuators. Proper selections of piezoelectric sensor thickness, shape, and convolution can provide spatial modal filtering and prevent observation and control spillovers in distributed structural control systems. Other studies of ring shells and motor applications have also been documented (Hu, Li and Tzou, 2013, 2014; Smithmaitrie and Tzou, 2005; Tzou, Zhong and Natori, 1993; Tzou and Ye, 1994).

References

Busch-Vishniac, I.J., 1990, "Spatially Distributed Transducers: Part-2," *ASME Journal of Dynamic Systems, Measurements, and Control*, Vol. (112), pp. 381–390.

Colins, S.A., Miller, D.W., von Flotow, A.H., 1991, "Piezoelectric Spatial Filters for Active Vibration Control," *Recent Advances in Active Control of Sound and Vibration*, Technomic, pp. 219–234.

Gustafson, D. and Speyer, J., 1976, "Linear Minimum Variable Filters Applied to Carrier Tracking," *IEEE Transactions on Automatic Control*, Vol. (AC-21), pp. 65–73.

Hu, S.D., LI, H. and Tzou, H.S., 2013, "Flexoelectric Responses of Circular Rings," *ASME Journal of Vibration and Acoustics*, 135(2), 00.021003. 2013.

Hu, S.D., LI, H. and Tzou, H.S., 2014, "Comparison of Flexoelectric and Piezoelectric Dynamic Signal Responses on Flexible Rings," *Journal of Intelligent Structures and Material Systems*, 25(7): pp. 832–844, 2014.

Lee, C.K. and Moon, F.C., 1990, "Modal Sensors/Actuators," ASME *Journal of Applied Mechanics*, Vol. (57), pp. 434–441.

Meirovitch, L. and Baruh, H., 1983, "On the Problem of Observation Spillover in Self-Adjoint Distributed-Parameter Systems," *J. of Optimization Theory and Applications*, Vol. (39), No. (2), pp. 269–291.

Smithmaitrie, P. and Tzou, H.S., 2005, "Electro-dynamics, Micro-actuation and Design of Ultrasonic Curvilinear Arc Stators," *Journal of Sound & Vibration*, Vol. 284, pp. 635–650.

Soedel, W., 1981, Vibrations of Plates and Shells, Dekker, N.Y.

Tzou, H.S., 1991, "Distributed Modal Identification and Vibration Control of Continua," ASME *Journal of Dynamic Systems, Measurements, and Control*, Vol. (113), No. (3), pp. 494–499.

Tzou, H.S. and Tseng, C.I., 1990, "Distributed Piezoelectric Sensor/Actuator Design for Dynamic Measurement/Control of Distributed Parameter Systems: A Finite Element Approach," *Journal of Sound and Vibration*, Vol. (138), No. (1), pp. 17–34.

Tzou H.S. and Ye, Y., 1994, "Piezothermoelasticity and Precision Control of Piezoelectric Systems: Theory and Finite Element Analysis," *ASME Transactions, Journal of Vibration & Acoustics*, Vol. 116, No. 4, pp. 489–495, October 1994.

Tzou, H.S. and Zhong, J.P., 1990, "Electromechanical Dynamics of Piezoelectric Shell Distributed Systems: Theory (I) and Applications (II)," *Robotics Research-1990*, ASME-DSC-Vol. 26, pp. 199–211. 1990 ASME WAM, Dallas, TX, November 27–31, 1990.

Tzou, H.S., Zhong, J.P., and Natori, M.C., 1993, "Sensor Mechanics of Distributed Shell Convolving Sensors Applied to Flexible Rings," ASME Transactions, *Journal of Vibration & Acoustics*, Vol. 115, No. 1, pp. 40–46.

Zhong, J.P., 1991, *A Study on Piezoelectric Shell Dynamics Applied to Structural Identification and Control*, Ph.D. Thesis, Department of Mechanical Engineering, University of Kentucky, 1991.

Chapter 9
Sensing and Control of Cylindrical Shells

Generic theories for distributed convolving modal sensors and actuators have been proposed and their sensing/control effectiveness on ring shells evaluated in Chap. 8. In this chapter, distributed piezoelectric sensors and actuators designed for structural sensing and control of cylindrical shells are presented. Free vibration analysis of cylindrical shells is presented first and then distributed sensing and control via distributed piezoelectric sensors and actuators. Detailed electromechanics and performances of the sensors and actuators laminated on the cylindrical shells are discussed (Zhong, 1991). Note that there are three cylindrical shells discussed in this chapter: 1) a piezoelectric cylindrical shell, 2) an elastic cylindrical shell, and 3) a piezoelectric laminated cylindrical shell. The differences between the first two cases are: 1) the electric and electromechanical coupling terms and 2) a charge equation of electrostatics. The **laminated** cylindrical shell is made of an elastic shell whose inner and outer surfaces are laminated with distributed piezoelectric layers. These piezoelectric layers are used as a distributed sensor and an actuator respectively.

Free vibration analysis is conducted for a piezoelectric cylindrical shell. Natural frequencies for an elastic cylindrical shell can be easily derived by neglecting all electric terms. Since the material properties of the piezoelectric layers are neglected in the **laminated** shell case, governing equations of the **elastic** cylindrical shell are used. (Note that damping controls of a piezoelectric cylindrical shell via in-plane electric membrane forces were studied in Chap. 3, Sect. 3.8.)

9.1 Cylindrical Shell

A cylindrical shell is a special case of the generic shell continuum defined by the three tri-orthogonal axes α_1, α_2, and α_3, in Chaps. 2 and 4. The cylindrical shell can be defined by a cylindrical coordinate system with x, θ, and α_3 axes, in which x defines the longitudinal direction (length), θ the circumferential direction, and α_3 the transverse direction, Fig. 9.1a, and its distributed sensing and control, Fig. 9.1b.

H. (HS) Tzou, *Piezoelectric Shells*, Solid Mechanics and Its Applications 247,
https://doi.org/10.1007/978-94-024-1258-1_9

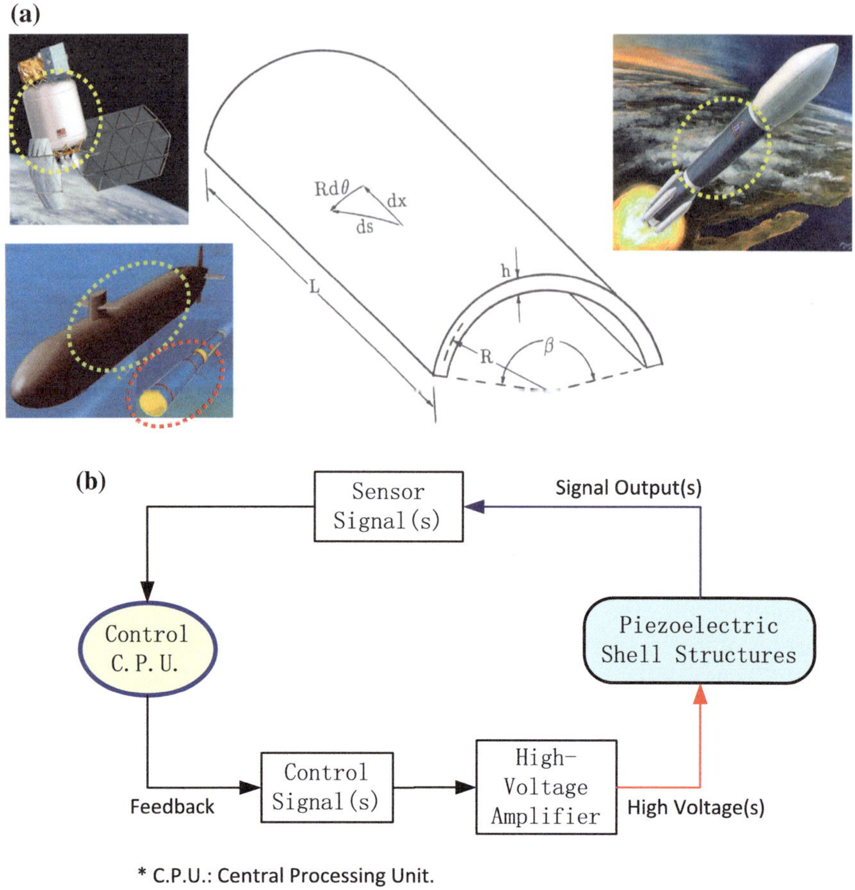

Fig. 9.1 Sensing and control of cylindrical shells. **a** A cylindrical shell. **b** Distributed sensing and feedback control

In this section, piezoelectric electromechanical equations of **piezoelectric** cylindrical shells will be derived based on the system equations of the generic piezoelectric thin shell. Dynamic equations for an **elastic** cylindrical shell can be obtained by eliminating all electric related terms in the governing equations for piezoelectric cylindrical shells.

As discussed previously, the fundamental form for the cylindrical shell is

$$(ds)^2 = (1)^2(dx)^2 + (R)^2(d\theta)^2. \tag{9.1.1}$$

Thus, the Lamé parameters are $A_1 = 1$ and $A_2 = R$. Observing the cylindrical coordinate system, one can obtain radii of curvatures: $R_1 = \infty$ and $R_2 = R$.

9.1.1 Governing Piezoelectric Equations

Substituting the Lamé parameters and the radii of curvatures into the piezoelectric thin shell equations and simplifying them accordingly, one can derive the electromechanical equations of the **piezoelectric** cylindrical shell in u_x, u_θ, and u_3 displacements:

$$\frac{\partial\left(N_{xx}^m - N_{xx}^e\right)}{\partial x} + \frac{1}{R}\frac{\partial N_{\theta x}^m}{\partial \theta} - \rho h \ddot{u}_x = 0, \tag{9.1.2}$$

$$\begin{aligned} &\frac{\partial N_{x\theta}^m}{\partial x} + \frac{1}{R}\frac{\partial\left(N_{\theta\theta}^m - N_{\theta\theta}^e\right)}{\partial \theta} + \frac{1}{R}\frac{\partial M_{x\theta}^m}{\partial x} \\ &+ \frac{1}{R^2}\frac{\partial\left(M_{\theta\theta}^m - M_{\theta\theta}^e\right)}{\partial \theta} - \rho h \ddot{u}_\theta = 0 \end{aligned}, \tag{9.1.3}$$

$$\begin{aligned} &\frac{\partial^2\left(M_{xx}^m - M_{xx}^e\right)}{\partial x^2} + \frac{2}{R}\frac{\partial^2 M_{\theta x}^m}{\partial x \partial \theta} + \frac{1}{R^2}\frac{\partial^2\left(M_{\theta\theta}^m - M_{\theta\theta}^e\right)}{\partial \theta^2} \\ &- \frac{N_{\theta\theta}^m - N_{\theta\theta}^e}{R} - \rho h \ddot{u}_3 = 0 \end{aligned}. \tag{9.1.4}$$

The charge equation of electrostatics is

$$\frac{\partial}{\partial \alpha_3}\left(e_{31} S_{xx} + e_{31} S_{\theta\theta} + \varepsilon_{33} E_3\right) R\left(1 + \frac{\alpha_3}{R}\right) = 0. \tag{9.1.5}$$

Note that the governing equations for an **elastic** cylindrical shell can be derived by eliminating all electric related terms in Eqs. (9.1.3)–(9.1.5). This set of governing equations will be further reduced to a conventional form, in terms of displacements u_x, u_θ and u_3, when the mechanical and electric forces/moments are defined next.

9.1.2 Simplified Governing Equations and Free Vibration Analysis

In this section, conventional governing equations of a cylindrical shell structure are formulated first; then, natural frequencies and mode shapes are derived. It is assumed that the shell is simply supported on all edges. The analysis is performed for a piezoelectric laminated cylindrical shell, i.e., an elastic shell coupled with distributed sensor/actuator layers, Fig. 9.2.

As discussed previously, the equations of the corresponding **elastic** shell can be obtained by eliminating the electric terms. For a laminated shell, material

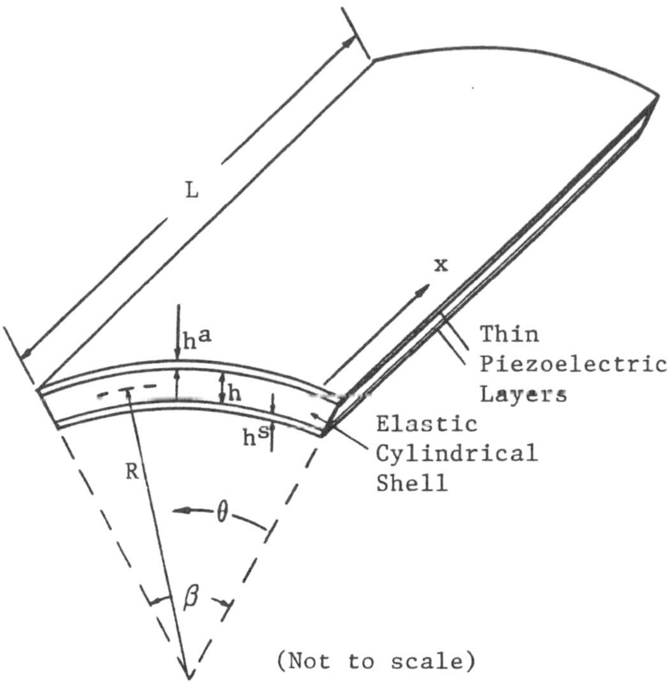

Fig. 9.2 A laminated simply supported cylindrical shell

properties, e.g., mass and stiffness, of the piezoelectric layers are neglected so that the **elastic** shell equations are used. Procedures for calculating natural frequencies and natural modes are similar for all three cases. (Note that the primary differences are in assumptions and related simplifications.) Numerical solutions are calculated for an elastic steel cylindrical shell which is used as the (master) elastic structure in the laminated shell case in which distributed sensors and actuators are evaluated.

From the charge equation of electrostatics and an external applied charge Q_3 (constant), the electric field induced by the direct and converse piezoelectric effects can be written as a function of mechanical normal strains and the external charge (Chaps. 4 and 8):

$$E_3 = -\frac{1}{\varepsilon_{33}}\left[e_{31}(S_{xx} + S_{\theta\theta}) + Q_3\right]. \tag{9.1.6}$$

(Recall that the piezoelectric shell has a hexagonal symmetrical structure and the in-plane shear effect is neglected.) Note that the external charge boundary condition is not considered in the free vibration analysis. The strains are composed of membrane strains and bending strains, i.e. $S_{ij} = (\overset{\circ}{S}_{ij} + \alpha_3 k_{ij})$, Substituting this into the electric membrane forces and bending moments, defined in Chap. 2, and integrating from (−h/2) to (h/2), one can derive

$$N^e_{xx} = \int_{\alpha_3} e_{31} E_3 d\,\alpha_3 = -\frac{e_{31}^2 h}{\varepsilon_{33}}\left(S^o_{xx} + S^o_{\theta\theta}\right) - \frac{e_{31} h}{\varepsilon_{33}} Q_3, \tag{9.1.7a}$$

$$N^e_{\theta\theta} = \int_{\alpha_3} e_{31} E_3 d\,\alpha_3 = -\frac{e_{31}^2 h}{\varepsilon_{33}}\left(S^o_{xx} + S^o_{\theta\theta}\right) - \frac{e_{31} h}{\varepsilon_{33}} Q_3, \tag{9.1.7b}$$

$$M^e_{xx} = \int_{\alpha_3} e_{31} E_3 \alpha_3 d\,\alpha_3 = -\frac{e_{31}^2 I}{\varepsilon_{33}}(k_{xx} + k_{\theta\theta}), \tag{9.1.8a}$$

$$M^e_{\theta\theta} = \int_{\alpha_3} e_{31} E_3 \alpha_3 d\,\alpha_3 = -\frac{e_{31}^2 I}{\varepsilon_{33}}(k_{xx} + k_{\theta\theta}). \tag{9.1.8b}$$

Thus,

$$N^m_{xx} - N^e_{xx} = \left(K + \frac{e_{31}^2 h}{\varepsilon_{33}}\right) S^o_{xx} + \left(K\mu + \frac{e_{31}^2 h}{\varepsilon_{33}}\right) S^o_{\theta\theta} + \frac{e_{31} h}{\varepsilon_{33}} Q_3, \tag{9.1.9a}$$

$$N^m_{\theta\theta} - N^e_{\theta\theta} = \left(K\mu + \frac{e_{31}^2 h}{\varepsilon_{33}}\right) S^o_{xx} + \left(K + \frac{e_{31}^2 h}{\varepsilon_{33}}\right) S^o_{\theta\theta} + \frac{e_{31} h}{\varepsilon_{33}} Q_3 \tag{9.1.9b}$$

$$M^m_{xx} - M^e_{xx} = \left(D + \frac{e_{31}^2 I}{\varepsilon_{33}}\right) k_{xx} + \left(D\mu + \frac{e_{31}^2 I}{\varepsilon_{33}}\right) k_{\theta\theta}, \tag{9.1.10a}$$

$$M^m_{\theta\theta} - M^e_{\theta\theta} = \left(D\mu + \frac{e_{31}^2 I}{\varepsilon_{33}}\right) k_{xx} + \left(D + \frac{e_{31}^2 I}{\varepsilon_{33}}\right) k_{\theta\theta}. \tag{9.1.10b}$$

Recall that *D* and *K* are the bending and membrane stiffness respectively. I is the area moment of inertia per unit width, i.e., $I = h^3/12$. Substituting the four system parameters, two Lamé parameters and two radii, into the strain-displacement relations of a thin shell continuum, defined in Sect. 2.4 of Chap. 2, gives the strain-displacement relations for the cylindrical shells. Replacing the strains in Eqs. (9.1.9a & b)–(9.1.10a & b) and substituting the force and moments into the governing equations, Eqs. (9.1.2)–(9.1.4), one can obtain

$$\begin{aligned} &\left(K + \frac{e_{31}^2 h}{\varepsilon_{33}}\right) \frac{\partial^2 u_x}{\partial x^2} + \left(K\mu + \frac{e_{31}^2 h}{\varepsilon_{33}}\right) \frac{1}{R}\left(\frac{\partial^2 u_\theta}{\partial x \partial \theta} + \frac{\partial u_3}{\partial x}\right) \\ &+ \frac{1}{R}\frac{K(1-\mu)}{2}\left(\frac{\partial^2 u_\theta}{\partial x \partial \theta} + \frac{1}{R}\frac{\partial^2 u_x}{\partial \theta^2}\right) - \rho h \ddot{u}_x = 0 \end{aligned}, \tag{9.1.11}$$

$$\frac{K(1-\mu)}{2}\left(\frac{\partial^2 u_\theta}{\partial x^2}+\frac{1}{R}\frac{\partial^2 u_x}{\partial x\partial\theta}\right)+\frac{1}{R}\left(K\mu+\frac{e_{31}^2 h}{\varepsilon_{33}}\right)\frac{\partial^2 u_x}{\partial x\partial\theta}$$
$$+\frac{1}{R^2}\left(K+\frac{e_{31}^2 h}{\varepsilon_{33}}\right)\left(\frac{\partial^2 u_\theta}{\partial\theta^2}+\frac{\partial u_3}{\partial\theta}\right)+\frac{1}{R^2}\frac{D(1-\mu)}{2}\left(\frac{\partial^2 u_\theta}{\partial x^2}\right.$$
$$\left.-2\frac{\partial^3 u_3}{\partial x^2\partial\theta}\right)-\frac{1}{R^2}\left(D\mu+\frac{e_{31}^2 I}{\varepsilon_{33}}\right)\frac{\partial^3 u_3}{\partial x^2\partial\theta} \qquad (9.1.12)$$
$$+\frac{1}{R^4}\left(D+\frac{e_{31}^2 I}{\varepsilon_{33}}\right)\left(\frac{\partial^2 u_\theta}{\partial\theta^2}-\frac{\partial^3 u_3}{\partial\theta^3}\right)-\rho h\ddot{u}_\theta=0$$

$$-\left(D+\frac{e_{31}^2 I}{\varepsilon_{33}}\right)\frac{\partial^4 u_3}{\partial x^4}+\left(D\mu+\frac{e_{31}^2 I}{\varepsilon_{33}}\right)\frac{1}{R^2}\left(\frac{\partial^3 u_\theta}{\partial x^2\partial\theta}-\frac{\partial^4 u_3}{\partial x^2\partial\theta^2}\right)$$
$$+\frac{2}{R^2}\frac{D(1-\mu)}{2}\left(\frac{\partial^3 u_\theta}{\partial x^2\partial\theta}-2\frac{\partial^4 u_3}{\partial x^2\partial\theta^2}\right)-\frac{1}{R^2}\left(D\mu+\frac{e_{31}^2 I}{\varepsilon_{33}}\right)\frac{\partial^4 u_3}{\partial x^2\partial\theta^2}$$
$$+\frac{1}{R^4}\left(D+\frac{e_{31}^2 I}{\varepsilon_{33}}\right)\left(\frac{\partial^3 u_\theta}{\partial\theta^3}-\frac{\partial^4 u_3}{\partial\theta^4}\right)-\frac{1}{R}\left[\left(K\mu+\frac{e_{31}^2 h}{\varepsilon_{33}}\right)\frac{\partial u_x}{\partial x}\right. \qquad (9.1.13)$$
$$\left.-\frac{1}{R}\left(K+\frac{e_{31}^2 h}{\varepsilon_{33}}\right)\left(\frac{1}{R}\frac{\partial u_\theta}{\partial\theta}+\frac{u_3}{R}\right)\right]+\frac{e_{31}h}{\varepsilon_{33}R}Q_3-\rho h\ddot{u}_3=0$$

Note that due to a non-zero curvature in the circumferential θ-direction, $\frac{e_{31}hQ_3}{\varepsilon_{33}R}$ is still preserved in the transverse equation; this can be used as a control force in open or closed-loop controls. However, $\frac{e_{31}hQ_3}{\varepsilon_{33}R}$ is neglected in the free vibration analysis presented later. External mechanical forces are not considered in these expressions. Simplifying the above equations yields

$$K_1\frac{\partial^2 u_x}{\partial x^2}+K_2\frac{\partial^2 u_\theta}{\partial x\partial\theta}+K_3\frac{\partial^2 u_x}{\partial\theta^2}+K_4\frac{\partial u_3}{\partial x}-\rho h\ddot{u}_x=0, \qquad (9.1.14)$$

$$K_5\frac{\partial^2 u_\theta}{\partial x^2}+K_6\frac{\partial^2 u_\theta}{\partial\theta^2}+K_7\frac{\partial^2 u_x}{\partial x\partial\theta}-K_8\frac{\partial^3 u_3}{\partial x^2\partial\theta}-K_9\frac{\partial^3 u_3}{\partial\theta^3}$$
$$+K_{10}\frac{\partial u_3}{\partial\theta}-\rho h\ddot{u}_\theta=0, \qquad (9.1.15)$$

$$-K_{11}\frac{\partial^4 u_3}{\partial x^4}-K_{12}\frac{\partial^4 u_3}{\partial x^2\partial\theta^2}-K_{13}\frac{\partial^4 u_3}{\partial\theta^4}+K_{14}\frac{\partial^3 u_\theta}{\partial x^2\partial\theta}+K_{15}\frac{\partial^3 u_\theta}{\partial\theta^3}$$
$$-K_{16}\frac{\partial u_x}{\partial x}-K_{17}\left(\frac{\partial u_\theta}{\partial\theta}+u_3\right)-\rho h\ddot{u}_3=\frac{e_{31}h}{\varepsilon_{33}R}Q_3, \qquad (9.1.16)$$

where

$$K_1 = K + \frac{e_{31}^2 h}{\varepsilon_{33}}, \tag{9.1.17a}$$

$$K_2 = \frac{1}{R}\left[\frac{K(1+\mu)}{2} + \frac{e_{31}^2 h}{\varepsilon_{33}}\right], \tag{9.1.17b}$$

$$K_3 = \frac{K(1-\mu)}{2R^2}, \tag{9.1.17c}$$

$$K_4 = \frac{1}{R}\left(K\mu + \frac{e_{31}^2 h}{\varepsilon_{33}}\right), \tag{9.1.17d}$$

$$K_5 = \frac{1}{2}\left[K(1-\mu) + \frac{D(1-\mu)}{R^2}\right], \tag{9.1.17e}$$

$$K_6 = \frac{1}{R^2}\left[K + \frac{e_{31}^2 h}{\varepsilon_{33}} + \frac{1}{R^2}\left(D + \frac{e_{31}^2 I}{\varepsilon_{33}}\right)\right], \tag{9.1.17f}$$

$$K_7 = \frac{1}{R}\left[\frac{K(1+\mu)}{2} + \frac{e_{31}^2 h}{\varepsilon_{33}}\right], \tag{9.1.17g}$$

$$K_8 = \frac{1}{R^2}\left(D + \frac{e_{31}^2 I}{\varepsilon_{33}}\right), \tag{9.1.17h}$$

$$K_9 = \frac{1}{R^4}\left(D + \frac{e_{31}^2 I}{\varepsilon_{33}}\right), \tag{9.1.17i}$$

$$K_{10} = \frac{1}{R^2}\left(K + \frac{e_{31}^2 I}{\varepsilon_{33}}\right), \tag{9.1.17j}$$

$$K_{11} = D + \frac{e_{31}^2 I}{\varepsilon_{33}}, \tag{9.1.17k}$$

$$K_{12} = \frac{2}{R^2}\left(D + \frac{e_{31}^2 I}{\varepsilon_{33}}\right), \tag{9.1.17l}$$

$$K_{13} = \frac{1}{R^4}\left(D + \frac{e_{31}^2 I}{\varepsilon_{33}}\right), \tag{9.1.17m}$$

$$K_{14} = \frac{1}{R^2}\left(D + \frac{e_{31}^2 I}{\varepsilon_{33}}\right), \tag{9.1.17n}$$

$$K_{15} = \frac{1}{R^4}\left(D + \frac{e_{31}^2 h}{\varepsilon_{33}}\right), \tag{9.1.17o}$$

$$K_{16} = \frac{1}{R}\left(K\mu + \frac{e_{31}^2 h}{\varepsilon_{33}}\right), \tag{9.1.17p}$$

$$K_{17} = \frac{1}{R^2}\left(K + \frac{e_{31}^2 h}{\varepsilon_{33}}\right). \tag{9.1.17q}$$

It can be observed that $K_2 = K_7$, $K_4 = K_{16}$, $K_8 = K_{14} = 1/2(K_{12})$, $K_9 = K_{13}$. Note that piezoelectric coefficients also contribute to the stiffness constants and so increase the natural frequencies. The additional stiffness resulting from the direct piezoelectric effect is about one percent of the elastic part of a piezoelectric cylindrical shell made of polymeric polyvinylidene fluoride (PVDF). For laminated cylindrical shells, if polymeric PVDF is used for sensor/actuator material and steel for elastic material, the contributions from the direct piezoelectric effects in a free vibration analysis are also negligible. However, for piezoelectric ceramics, this contribution would need to be reevaluated.

In a free vibration analysis, all external mechanical and electric excitations are assumed to be zero. It is assumed that the shell oscillates harmonically at a natural frequency. The harmonic oscillations can be assumed as (Soedel, 1981)

$$u_x(x,\theta,t) = U_x(x,\theta)e^{j\omega t}, \tag{9.1.18a}$$

$$u_\theta(x,\theta,t) = U_\theta(x,\theta)e^{j\omega t}, \tag{9.1.18b}$$

$$u_3(x,\theta,t) = U_3(x,\theta)e^{j\omega t}, \tag{9.1.18c}$$

where $U_i(x,\theta)$ is a modal function or mode shape function, ω is the natural frequency, and $j = \sqrt{-1}$. Substituting Eqs. (9.1.18a)–(9.1.18c) into Eqs. (9.1.14)–(9.1.16) and neglecting $\frac{e_{31} h Q_3}{\varepsilon_{33} R}$, one can derive

$$K_1 \frac{\partial^2 U_x}{\partial x^2} + K_2 \frac{\partial^2 U_\theta}{\partial x \partial \theta} + K_3 \frac{\partial^2 U_x}{\partial \theta^2} + K_4 \frac{\partial U_3}{\partial x} + \rho h \omega^2 U_x = 0, \tag{9.1.19}$$

$$\begin{aligned} &K_5 \frac{\partial^2 U_\theta}{\partial x^2} + K_6 \frac{\partial^2 U_\theta}{\partial \theta^2} + K_7 \frac{\partial^2 U_x}{\partial x \partial \theta} - K_8 \frac{\partial^3 U_3}{\partial x^2 \partial \theta} - K_9 \frac{\partial^3 U_3}{\partial \theta^3}, \\ &+ K_{10} \frac{\partial U_3}{\partial \theta} + \rho h \omega^2 U_\theta = 0 \end{aligned} \tag{9.1.20}$$

$$\begin{aligned}&-K_{11}\frac{\partial^4 U_3}{\partial x^4}-K_{12}\frac{\partial^4 U_3}{\partial x^2\partial\theta^2}-K_{13}\frac{\partial^4 U_3}{\partial\theta^4}+K_{14}\frac{\partial^3 U_\theta}{\partial x^2\partial\theta}\\&+K_{15}\frac{\partial^3 U_\theta}{\partial\theta^3}-K_{16}\frac{\partial U_x}{\partial x}-K_{17}\left(\frac{\partial U_\theta}{\partial\theta}+U_3\right)+\rho h\omega^2 U_3=0\end{aligned}. \quad (9.1.21)$$

For a simply supported cylindrical shell panel, the mode shape functions can be assumed as a product of sine and cosine functions:

$$U_x(x,\theta)=A\cos\frac{m\pi x}{L}\sin\frac{n\pi\theta}{\beta}, \quad (9.1.22a)$$

$$U_\theta(x,\theta)=B\sin\frac{m\pi x}{L}\cos\frac{n\pi\theta}{\beta}, \quad (9.1.22b)$$

$$U_3(x,\theta)=C\sin\frac{m\pi x}{L}\sin\frac{n\pi\theta}{\beta}, \quad (9.1.22c)$$

where A, B, C are modal oscillation amplitudes. These mode shape functions have to satisfy the given boundary conditions, i.e., displacements and moments are zero on all four boundaries. Substituting assumed mode shape functions into Eqs. (9.1.19)–(9.1.21) yields

$$\left[\rho h\omega^2-K_1\left(\frac{m\pi}{L}\right)^2-K_3\left(\frac{n\pi}{\beta}\right)^2\right]A-K_2\frac{m\pi}{L}\frac{n\pi}{\beta}B+K_4\frac{m\pi}{L}C=0, \quad (9.1.23)$$

$$\begin{aligned}&-K_7\left(\frac{m\pi}{L}\frac{n\pi}{\beta}\right)A+\left[\rho h\omega^2-K_5\left(\frac{m\pi}{L}\right)^2-K_6\left(\frac{n\pi}{\beta}\right)^2\right]B\\&+\left[K_8\left(\frac{m\pi}{L}\right)^2\left(\frac{n\pi}{\beta}\right)+K_9\left(\frac{n\pi}{\beta}\right)^3+K_{10}\frac{n\pi}{\beta}\right]C=0\end{aligned}, \quad (9.1.24)$$

$$\begin{aligned}&K_{16}\frac{m\pi}{L}A+\left[K_{14}\frac{n\pi}{\beta}\left(\frac{m\pi}{L}\right)^2+K_{15}\left(\frac{n\pi}{\beta}\right)^3+K_{17}\left(\frac{n\pi}{\beta}\right)\right]B\\&+\left[\rho h\omega^2-K_{11}\left(\frac{m\pi}{L}\right)^4-K_{12}\left(\frac{m\pi}{L}\right)^2\left(\frac{n\pi}{\beta}\right)^2\right.\\&\left.-K_{13}\left(\frac{n\pi}{\beta}\right)^4-K_{17}\right]C=0\end{aligned}, \quad (9.1.25)$$

or

$$\begin{bmatrix}\rho h\omega^2-k_{11} & k_{12} & k_{13}\\ k_{21} & \rho h\omega^2-k_{22} & k_{23}\\ k_{31} & k_{32} & \rho h\omega^2-k_{33}\end{bmatrix}\begin{bmatrix}A\\B\\C\end{bmatrix}=0, \quad (9.1.26)$$

where

$$k_{11} = K_1\left(\frac{m\pi}{L}\right)^2 + K_3\left(\frac{n\pi}{\beta}\right)^2, \tag{9.1.27a}$$

$$k_{12} = k_{21} = -K_2\frac{m\pi}{L}\frac{n\pi}{\beta}, \tag{9.1.27b}$$

$$k_{13} = k_{31} = K_4\frac{m\pi}{L}, \tag{9.1.27c}$$

$$k_{22} = K_5\left(\frac{m\pi}{L}\right)^2 + K_6\left(\frac{n\pi}{\beta}\right)^2, \tag{9.1.27d}$$

$$k_{23} = k_{32} = K_8\left(\frac{m\pi}{L}\right)^2\frac{n\pi}{\beta} + K_9\left(\frac{n\pi}{\beta}\right)^3 + K_{10}\frac{n\pi}{\beta}, \tag{9.1.27e}$$

$$k_{33} = K_{11}\left(\frac{m\pi}{L}\right)^4 + K_{12}\left(\frac{m\pi}{L}\right)^2\left(\frac{n\pi}{\beta}\right)^2 + K_{13}\left(\frac{n\pi}{\beta}\right)^4 + K_{17}. \tag{9.1.27f}$$

For nontrivial solutions, i.e., $\begin{Bmatrix} A \\ B \\ C \end{Bmatrix} \neq 0$, the determinant of the coefficient matrix in Eq. (9.1.26) must be zero. Thus, a characteristic equation can be derived as

$$\omega^6 + \tilde{\gamma}_1\omega^4 + \tilde{\gamma}_2\omega^2 + \tilde{\gamma}_3 = 0, \tag{9.1.28}$$

where

$$\tilde{\gamma}_1 = -\frac{1}{\rho h}(k_{11} + k_{22} + k_{33}), \tag{9.1.29a}$$

$$\tilde{\gamma}_2 = \frac{1}{(\rho h)^2}\left(k_{11}k_{33} + k_{22}k_{33} + k_{11}k_{22} - k_{23}^2 - k_{12}^2 - k_{13}^2\right), \tag{9.1.29b}$$

$$\tilde{\gamma}_3 = \frac{1}{(\rho h)^3}\left(k_{11}k_{23}^2 + k_{22}k_{13}^2 + k_{33}k_{12}^2 + 2k_{12}k_{23}k_{13} - k_{11}k_{22}k_{33}\right). \tag{9.1.29c}$$

Solving the characteristic equation, one can obtain three ***component natural frequencies*** of the mn-th mode:

$$\omega_{1\,mn}'^{\;2} = -\frac{2}{3}\sqrt{\tilde{\gamma}_1^2 - 3\tilde{\gamma}_2}\cos\frac{\tilde{\alpha}}{3} - \frac{\tilde{\gamma}_1}{3}, \tag{9.1.30a}$$

$$\omega_{2\,mn}'^{\;2} = -\frac{2}{3}\sqrt{\tilde{\gamma}_1^2 - 3\tilde{\gamma}_2}\cos\frac{\tilde{\alpha} + 2\pi}{3} - \frac{\tilde{\gamma}_1}{3} \tag{9.1.30b}$$

$$\omega'^{\,2}_{3\,mn} = -\frac{2}{3}\sqrt{\tilde{\gamma}_1^2 - 3\tilde{\gamma}_2}\cos\frac{\tilde{\alpha}+4\pi}{3} - \frac{\tilde{\gamma}_1}{3}, \tag{9.1.30c}$$

where

$$\tilde{\alpha} = \cos^{-1}\left[\frac{27\tilde{\gamma}_3 + 2\tilde{\gamma}_1^3 - 9\tilde{\gamma}_1\tilde{\gamma}_2}{2\sqrt{\left(\tilde{\gamma}_1^2 - 3\tilde{\gamma}_2\right)^3}}\right]. \tag{9.1.31}$$

Usually, the lowest *component natural frequency* is associated with the transverse component or transverse oscillation; the second is the in-plane longitudinal (x) component; the third is the in-plane circumferential (θ) component. However, depending on the relative length and the curvature angle (i.e., the aspect ratio) of the cylindrical shell panel, the second and the third could be reversed. The relative modal amplitude of the i-th *component* natural frequency normalized with respect to C_{imn} can be obtained from Eq. (9.1.26).

$$\left[\begin{array}{c|c} \rho h \omega'^{\,2}_{imn} - k_{11} & k_{12} \\ \hline k_{21} & \rho h \omega'^{\,2}_{imn} - k_{22} \end{array}\right]\begin{bmatrix} A_{imn} \\ B_{imn} \end{bmatrix} = -C_{imn}\begin{bmatrix} k_{13} \\ k_{23} \end{bmatrix}, \tag{9.1.32}$$

where i = 1, 2, 3. Using Cramer's rule, one can define the amplitude ratios

$$\frac{A_{imn}}{C_{imn}} = -\frac{\begin{vmatrix} k_{13} & k_{12} \\ k_{23} & \rho h \omega'^{2}_{imn} - k_{22} \end{vmatrix}}{\tilde{D}_{imn}}, \tag{9.1.33a}$$

$$\frac{B_{imn}}{C_{imn}} = -\frac{\begin{vmatrix} \rho h \omega'^{2}_{imn} - k_{11} & k_{13} \\ k_{21} & k_{23} \end{vmatrix}}{\tilde{D}_{imn}}, \tag{9.1.33b}$$

where

$$\tilde{D}_{imn} = \left|\begin{array}{c|c} \rho h \omega'^{\,2}_{imn} - k_{11} & k_{12} \\ \hline k_{21} & \rho h \omega'^{\,2}_{imn} - k_{22} \end{array}\right|. \tag{9.1.34}$$

Thus, the relative amplitude ratios of the i-th *component* natural frequency are defined as

$$\frac{A_{imn}}{C_{imn}} = -\frac{k_{13}\left(\rho h \omega'^{\,2}_{imn} - k_{22}\right) - k_{12}k_{23}}{\left(\rho h \omega'^{2\,2}_{imn} - k_{11}\right)\left(\rho h \omega'^{\,2}_{imn} - k_{22}\right) - k_{12}^2}, \tag{9.1.35a}$$

$$\frac{B_{imn}}{C_{imn}} = -\frac{k_{23}\left(\rho h \omega'^{\,2}_{imn} - k_{11}\right) - k_{21}k_{13}}{\left(\rho h \omega'^{\,2}_{imn} - k_{11}\right)\left(\rho h \omega'^{\,2}_{imn} - k_{22}\right) - k_{12}^2}. \tag{9.1.35b}$$

Since three are three *component* natural frequencies, three ratios can be defined. Note that the modal oscillation amplitudes can also be normalized with respect to either A_{imn} or B_{imn}. The three natural modes are

$$\Longrightarrow \begin{bmatrix} U_{xi} \\ U_{\theta i} \\ U_{3i} \end{bmatrix} = \begin{bmatrix} A_{imn} \cos\frac{m\pi x}{L} \sin\frac{n\pi\theta}{\beta} \\ B_{imn} \sin\frac{m\pi x}{L} \cos\frac{n\pi\theta}{\beta} \\ C_{imn} \sin\frac{m\pi x}{L} \sin\frac{n\pi\theta}{\beta} \end{bmatrix}, \quad (9.1.36)$$

where the C_{imn} is an arbitrary constant. For simplicity, the modal amplitudes A_{imn}, B_{imn}, and C_{imn} are simply represented as A_i, B_i, and C_i in later derivations.

9.1.3 *Natural Frequencies and Modes of a Laminated Cylindrical Shell*

It is assumed that a simply supported steel cylindrical shell (with a thickness h) is laminated with piezoelectric PVDF layers on the top and bottom surfaces. It is also assumed that the top layer is an actuator layer (thickness h^a) and the bottom a sensor layer (thickness h^S). (Two dimensional shaping of distributed sensors and actuators will be discussed in the next four sections.) As discussed previously, material properties of the PVDF layers and the bonding materials are negligible. Thus, natural frequencies are calculated for the simply supported (elastic) steel cylindrical

Table 9.1 Dimensional and material properties of the laminated cylindrical shell

Properties	Steel	PVDF	Units
L	1.00×10^{-1}	1.000×10^{-2}	m
R	5.00×10^{-2}		m
h	1.00×10^{-3}	2.500×10^{-5}	m
β	60°, 90°, 120°		
ρ	7.8×10^3	1.800×10^3	Kg/m^3
Y	2.10×10^{11}	1.600×10^9	N/m^2
μ	0.300	0.290	
d_{31}		6.000×10^{-12}	C/N
d_{32}		6.000×10^{-12}	C/N
d_{33}		13.000×10^{-12}	C/N
$\varepsilon/\varepsilon_o$		10	
ε_o		8.85×10^{-10}	F/m

Table 9.2 Natural frequencies (Hz) of a 60° cylindrical shell

n	m					
	0	1	2	3	4	5
0	$.56941 \times 10^1$	$.16074 \times 10^5$	$.16463 \times 10^5$	$.16630 \times 10^5$	$.16961 \times 10^5$	$.17617 \times 10^5$
1	$.75842 \times 10^3$	$.34961 \times 10^4$	$.85061 \times 10^4$	$.11934 \times 10^5$	$.14164 \times 10^5$	$.15970 \times 10^5$
2	$.34505 \times 10^4$	$.38334 \times 10^4$	$.56362 \times 10^4$	$.84064 \times 10^4$	$.11311 \times 10^5$	$.14149 \times 10^5$
3	$.79476 \times 10^4$	$.82045 \times 10^4$	$.90960 \times 10^4$	$.10743 \times 10^5$	$.13035 \times 10^5$	$.15788 \times 10^5$
4	$.14244 \times 10^5$	$.14490 \times 10^5$	$.15258 \times 10^5$	$.16594 \times 10^5$	$.18511 \times 10^5$	$.20977 \times 10^5$
5	$.22340 \times 10^5$	$.22586 \times 10^5$	$.23330 \times 10^5$	$.24592 \times 10^5$	$.26383 \times 10^5$	$.28704 \times 10^5$

Table 9.3 Natural frequencies (Hz) of a 90° cylindrical shell

n	m					
	0	1	2	3	4	5
0	$.56941 \times 10^1$	$.16074 \times 10^5$	$.16463 \times 10^5$	$.16630 \times 10^5$	$.16961 \times 10^5$	$.17617 \times 10^5$
1	$.26814 \times 10^3$	$.56856 \times 10^4$	$.11362 \times 10^5$	$.14039 \times 10^5$	$.15521 \times 10^5$	$.16792 \times 10^5$
2	$.14548 \times 10^4$	$.27201 \times 10^4$	$.65880 \times 10^4$	$.10126 \times 10^5$	$.12845 \times 10^5$	$.15135 \times 10^5$
3	$.34505 \times 10^4$	$.38334 \times 10^4$	$.56362 \times 10^4$	$.84064 \times 10^4$	$.11311 \times 10^5$	$.14149 \times 10^5$
4	$.62485 \times 10^4$	$.65181 \times 10^4$	$.75433 \times 10^4$	$.94424 \times 10^4$	$.11946 \times 10^5$	$.14795 \times 10^5$
5	$.98458 \times 10^4$	$.10097 \times 10^5$	$.10921 \times 10^5$	$.12411 \times 10^5$	$.14534 \times 10^5$	$.17180 \times 10^5$

Table 9.4 Natural frequencies (Hz) of a 120° cylindrical shell

n	m					
	0	1	2	3	4	5
0	$.56941 \times 10^1$	$.16074 \times 10^5$	$.16463 \times 10^5$	$.16630 \times 10^5$	$.16961 \times 10^5$	$.17617 \times 10^5$
1	$.10536 \times 10^3$	$.75056 \times 10^4$	$.13030 \times 10^5$	$.15037 \times 10^5$	$.16106 \times 10^5$	$.17133 \times 10^5$
2	$.75842 \times 10^3$	$.34961 \times 10^4$	$.85061 \times 10^4$	$.11934 \times 10^5$	$.14164 \times 10^5$	$.15970 \times 10^5$
3	$.18782 \times 10^4$	$.27404 \times 10^4$	$.60004 \times 10^4$	$.94327 \times 10^4$	$.12288 \times 10^5$	$.14773 \times 10^5$
4	$.34505 \times 10^4$	$.38334 \times 10^4$	$.56362 \times 10^4$	$.84064 \times 10^4$	$.11311 \times 10^5$	$.14149 \times 10^5$
5	$.54736 \times 10^4$	$.57561 \times 10^4$	$.68907 \times 10^4$	$.98623 \times 10^4$	$.11582 \times 10^5$	$.14463 \times 105$

shell panel. The dimensions and material properties are summarized in Table 9.1. Note that there are three shell deepness considered here, i.e., 60°, 90° and 120° curvature orientations, to evaluate the effect of shell deepness. Natural frequencies for the three shell curvatures are summarized in Tables 9.2, 9.3, and 9.4, respectively. Natural mode shapes of the first nine modes are plotted in Figs. 9.3, 9.4, 9.5, 9.6, 9.7, 9.8, 9.9, 9.10, 9.11 and 9.12.

Natural frequencies and mode shapes of cylindrical shells were analyzed. Distributed sensing and control of the cylindrical shell via the coupled polymeric piezoelectric PVDF sensor/actuator are discussed next.

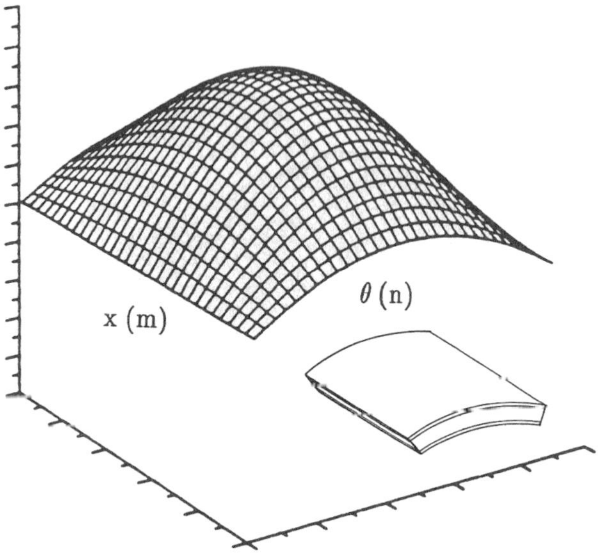

Fig. 9.3 The (1,1) mode shape of a simply supported cylindrical shell

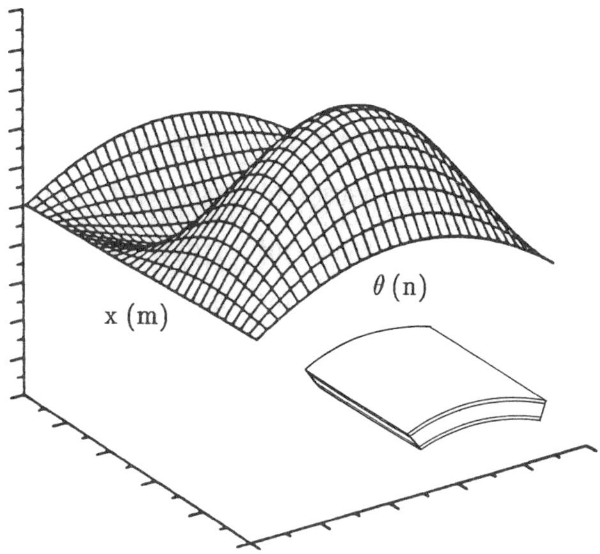

Fig. 9.4 The (2,1) mode shape of a simply supported cylindrical shell

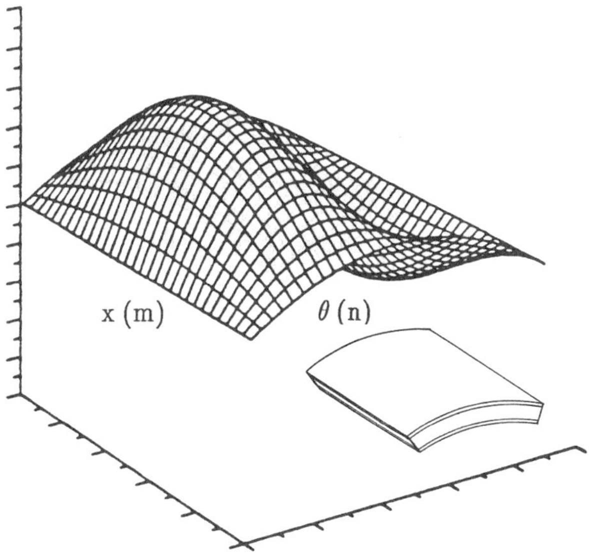

Fig. 9.5 The (1,2) mode shape of a simply supported cylindrical shell

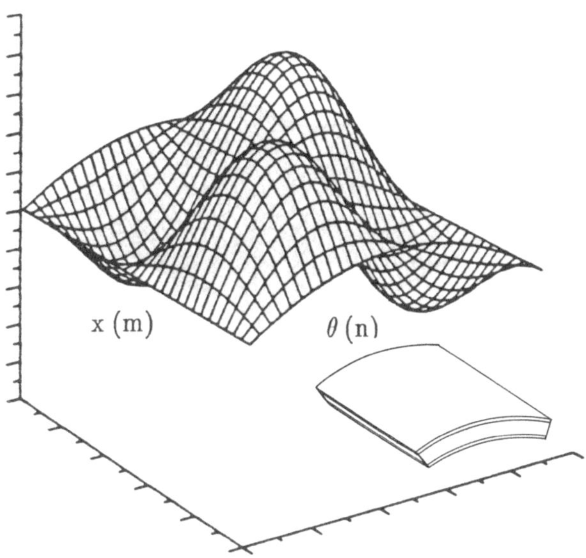

Fig. 9.6 The (2,2) mode shape of a simply supported cylindrical shell

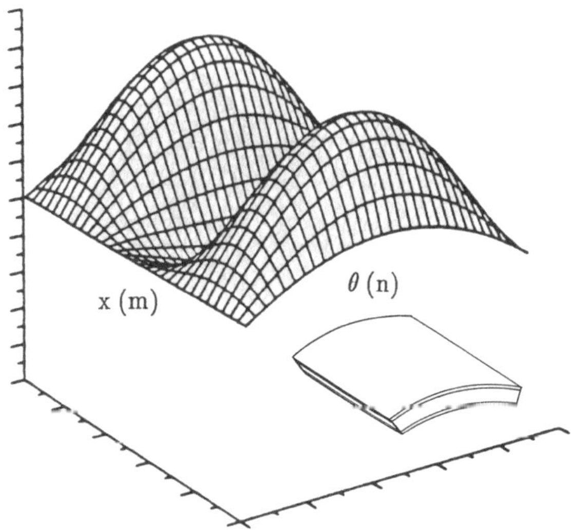

Fig. 9.7 The (3,1) mode shape of a simply supported cylindrical shell

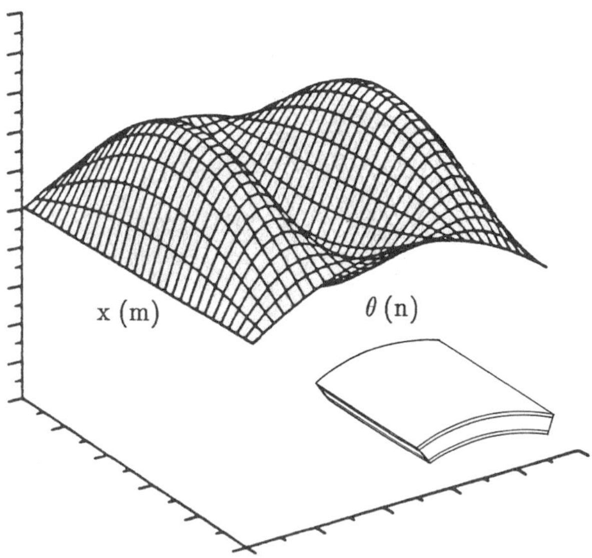

Fig. 9.8 The (1,3) mode shape of a simply supported cylindrical shell

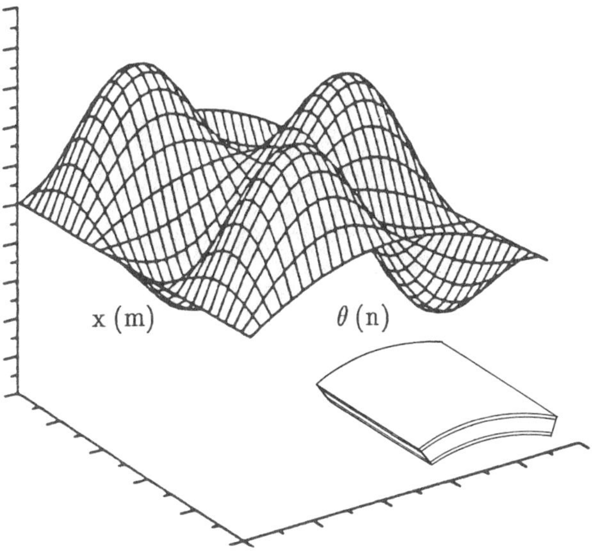

Fig. 9.9 The (3,2) mode shape of a simply supported cylindrical shell

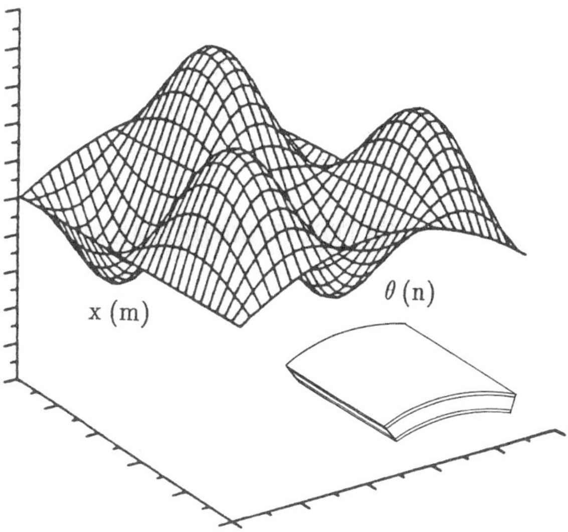

Fig. 9.10 The (2,3) mode shape of a simply supported cylindrical shell

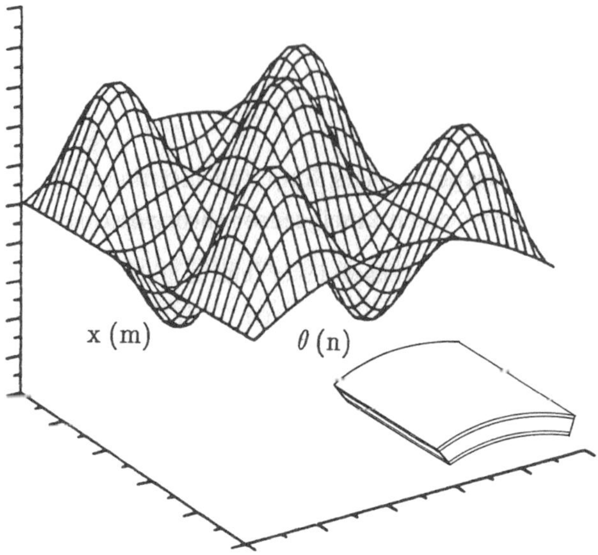

Fig. 9.11 The (3,3) mode shape of a simply supported cylindrical shell

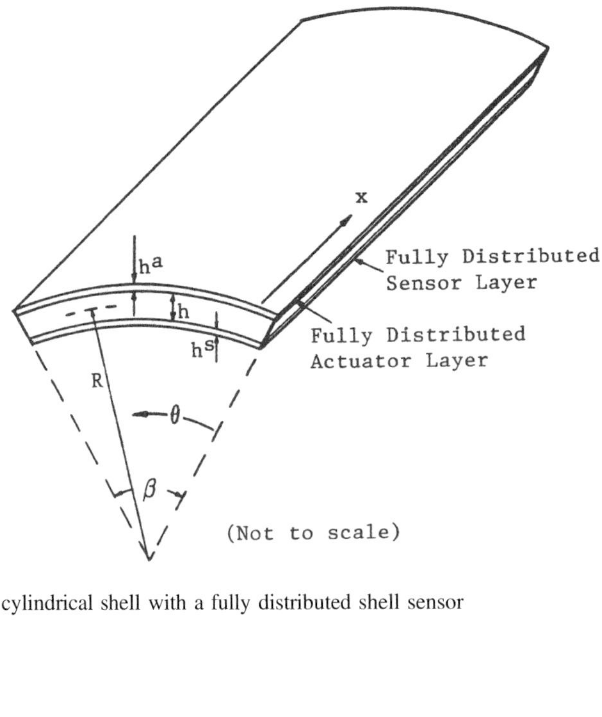

Fig. 9.12 A cylindrical shell with a fully distributed shell sensor

9.2 Distributed Sensors for Cylindrical Shells

Generic sensor electromechanics of convolving piezoelectric shell sensors was discussed; application of the theory to distributed convolving cosine shaped sensors for circular rings was also demonstrated in Chap. 8. In this section, two dimensional distributed sensors for cylindrical shells are proposed and their performances evaluated. There are two distributed sensors proposed for the cylindrical shells: 1) a fully distributed sensor and 2) a diagonally distributed stripe sensors. Each of the sensors is to be discussed. Note that the laminated cylindrical shell (with thickness h, length L and angle of rotation β) is used in the modeling and analysis.

9.2.1 Estimation of Sensor Signals

In this section, output signals from the distributed sensors of cylindrical shells are estimated for the generic shell sensor theory discussed in Chaps. 4 and 8. It is assumed that the distributed sensor layer has an uniform thickness h^S. Sensor sensitivities corresponding to the three principal motions are defined. Displacements in three principal directions can be expressed in the modal expansion forms:

$$u_x(x,\theta,t) = \sum_{m=1}^{\infty}\sum_{n=1}^{\infty} \eta_{mn}(t) A_{mn} \cos\frac{m\pi x}{L} \sin\frac{n\pi\theta}{\beta}, \tag{9.2.1a}$$

$$u_\theta(x,\theta,t) = \sum_{m=1}^{\infty}\sum_{n=1}^{\infty} \eta_{mn}(t) B_{mn} \sin\frac{m\pi x}{L} \cos\frac{n\pi\theta}{\beta}, \tag{9.2.1b}$$

$$u_3(x,\theta,t) = \sum_{m=1}^{\infty}\sum_{n=1}^{\infty} \eta_{mn}(t) C_{mn} \sin\frac{m\pi x}{L} \sin\frac{n\pi\theta}{\beta}. \tag{9.2.1c}$$

A_{mn}, B_{mn}, and C_{mn} are constants. For the mn-th natural mode, the relative oscillation amplitudes A_{mn}, B_{mn}, and C_{mn} are defined in Sect. 9.1.2. The strain-displacement relations are

$$S_{xx} = \frac{\partial u_x}{\partial x} + \alpha_3 \frac{\partial \beta_x}{\partial x}, \tag{9.2.2a}$$

$$S_{\theta\theta} = \frac{1}{R}\frac{\partial u_\theta}{\partial \theta} + \frac{u_3}{R} + \frac{\alpha_3}{R}\frac{\partial \beta_\theta}{\partial \theta}, \tag{9.2.2b}$$

$$S_{x\theta} = \frac{\partial u_\theta}{\partial x} + \frac{1}{R}\frac{\partial u_x}{\partial \theta} + \alpha_3\left(\frac{\partial \beta_\theta}{\partial x} + \frac{1}{R}\frac{\partial \beta_x}{\partial \theta}\right), \tag{9.2.2c}$$

and the rotation angles β_x and β_θ are

$$\beta_x = -\frac{\partial u_3}{\partial x}, \tag{9.2.3a}$$

$$\beta_\theta = \frac{u_\theta}{R} - \frac{1}{R}\frac{u_3}{\partial \theta}. \tag{9.2.3b}$$

Substituting the membrane and bending strains experienced in the distributed sensor layer into the generic sensor equation and expressing displacements in the modal expansion forms, one can derive the (averaged) sensor output signal (Chap. 4, Sect. 4.2):

$$\begin{aligned}\phi_3^S = &-\frac{e_{31}R}{\varepsilon_{33}S^e}\sum_{m=1}^{\infty}\sum_{n=1}^{\infty}\left\{h^S\left(\frac{1}{R}C - \frac{m\pi}{L}A - \frac{n\pi}{\beta R}B\right)\right.\\ &\left. + \frac{h^S(h+h^S)}{2}\left[\left(\frac{m\pi}{L}\right)^2 C - \frac{1}{R^2}\frac{n\pi}{\beta}B + \frac{1}{R^2}\left(\frac{n\pi}{\beta}\right)^2 C\right]\right\}, \\ &\cdot \eta_{mn}(t)\int_{S^e}\sin\frac{m\pi x}{L}\sin\frac{n\pi\theta}{\beta}dx\,d\theta\end{aligned} \tag{9.2.4}$$

where A, B, and C are modal amplitudes defined in Sect. 9.1.2. (Note that $A = A_{mn}$, $B = B_{mn}$, and $C = C_{mn}$, for simplicity.) S^e denotes the effective sensor electrode surface. It can be observed that the signal is a function of all participating modes represented by the modal participation factor, or the mn-th modal coordinate, η_{mn}. Each modal signal is determined by two strain components: 1) a membrane component (leading by h^S) and 2) a bending component (leading by $[h^S(h+h^S)/2]$). Note that depending on the location of the sensor layer and the vibration modes of the shell, one of these two components may be absent. For example, the sensor layer experiences only membrane strains when it is embedded on the neutral surface or the elastic shell only experiences membrane oscillations. On the other hand, the sensor layer experiences only bending strains if the layer is surface coupled or embedded away from the neutral surface while the shell is experiencing only bending modes. (In general, the neutral surface of an elastic shell only exhibits membrane actions.) Otherwise, the output signal is contributed by both strain effects, although the contribution of each component could be different.

Note that the sensor signal is surface averaged. As discussed in Chaps. 4 and 8, sensor shape and/or electrode can be spatially shaped and convolved to achieve modal filtering and monitoring capabilities, via *the shaping technique*. Two (surface) distributed sensors, i.e., 1) a fully distributed and 2) a diagonally distributed, for cylindrical shells are presented next. Performances of these sensors are evaluated. Distributed actuation and control characteristics are presented later.

9.2.2 *Fully Distributed Shell Sensor*

In this case, it is assumed that a piezoelectric layer fully covers the bottom surface of the cylindrical shell and is used as a distributed sensor, Fig. 9.12.

The (averaged) sensor signal of the fully distributed shell sensor is

$$\begin{aligned}\phi_3^S &= -\frac{e_{31}R}{\varepsilon_{33}S^e}\sum_{m=1}^{\infty}\sum_{n=1}^{\infty}\left\{h^S\left(\frac{1}{R}C-\frac{m\pi}{L}A-\frac{n\pi}{\beta R}B\right)\right.\\&\left.+\frac{h^S(h+h^S)}{2}\left[\left(\frac{m\pi}{L}\right)^2C-\frac{1}{R^2}\frac{n\pi}{\beta}B+\frac{1}{R^2}\left(\frac{n\pi}{\beta}\right)^2C\right]\right\}\eta_{mn}(t)\\&\cdot\int_0^L\int_0^\beta \sin\frac{m\pi x}{L}\sin\frac{n\pi\theta}{\beta}dx\,d\theta,\\&=\frac{4e_{31}}{\pi^2\varepsilon_{33}}\sum_{m=1}^{\infty}\sum_{n=1}^{\infty}\frac{1}{mn}\left\{h^S\left(\frac{1}{R}C-\frac{m\pi}{L}A-\frac{n\pi}{\beta R}B\right)+\frac{h^S(h+h^S)}{2}\right.\\&\left.\cdot\left[\left(\frac{m\pi}{L}\right)^2C-\frac{1}{R^2}\frac{n\pi}{\beta}B+\frac{1}{R^2}\left(\frac{n\pi}{\beta}\right)^2C\right]\right\}\eta_{mn}(t)\hat{\delta}_{mn}\end{aligned}\tag{9.2.5}$$

where

$$\begin{aligned}\hat{\delta}_{mn}&=\frac{1}{4}[1-\cos(m\pi)][1-\cos(n\pi)]\\&=\begin{cases}1, & \text{if both m and n are odd integers;}\\0, & \text{if m or n is even integers.}\end{cases}\end{aligned}\tag{9.2.6}$$

It is observed that the output voltage ϕ_3^S is equal to zero for all even modes (either m or n is an even integer) since this voltage is an average voltage. The even modes are anti-symmetrical modes: there are positive and negative strains in the sensor layer are of equal amplitudes, and thus the resulting signals cancel out each other in the surface integration.

Recall that there are three ***component natural frequencies*** (i.e., ω'_{1mn}, ω'_{2mn}, and ω'_{3mn}) associated with the mn-th mode, Sect. 9.1.2. It is assumed that the lowest component frequency is the mn-th transverse mode at the natural frequency ω_{3mn}; the second is the mn-th in-plane longitudinal (x) mode at ω_{xmn}; the third is the mn-th in-plane circumferential (θ) mode at $\omega_{\theta mn}$. (Note that depending on the relative length and curvature angle of the cylindrical shell, the last two *component* modes could be reversed.) Dividing the sensor output by the modal participation factor $\eta_{mn}(t)$ and the transverse modal amplitude C_i gives the sensitivity defined by the transverse modal oscillation amplitude for a selected frequency. Thus, three sensor sensitivities: 1) a transverse sensitivity, 2) an in-plane longitudinal (x direction) sensitivity, and 3) an in-plane circumferential (θ direction) sensitivity can be defined for the fully distributed piezoelectric sensor. Each of the sensitivities can be further defined in terms of the transverse oscillation amplitudes.

The ***transverse*** sensitivity S^{tr}_{fpmn} defined by the transverse oscillation amplitude C_1 is

$$S^{tr}_{f\,pmn} = \frac{\phi_3^S}{C_1 \eta_{mn}} = \frac{4e_{31}}{\pi^2 \varepsilon_{33}} \frac{1}{mn} \frac{h^S(h+h^S)}{2} \cdot \left(\left(\frac{m\pi}{L}\right)^2 - \frac{1}{R^2} \frac{n\pi}{\beta} \frac{B_1}{C_1} + \frac{1}{R^2} \left(\frac{n\pi}{\beta}\right)^2 \right) \hat{\delta}_{mn}. \tag{9.2.7a}$$

Note that, for simplicity, the amplitudes B_1 and C_1 are defined for the transverse component natural frequency ω_{3mn} of the mn-th mode.

The in-plane ***longitudinal*** (x direction) sensitivity $S^{in\,x}_{f\,pmn}$ is

$$S^{in\,x}_{f\,pmn} = \frac{\phi_3^S}{C_2 \eta_{mn}} = \frac{4e_{31}}{\pi^2 \varepsilon_{33}} \frac{h^S}{mn} \left(\frac{1}{R} - \frac{m\pi}{L} \frac{A_2}{C_2} - \frac{n\pi}{\beta R} \frac{B_2}{C_2} \right) \hat{\delta}_{mn}. \tag{9.2.7b}$$

The amplitudes A_2, B_2 and C_2 are defined for the mnth component natural frequency ω_{xmn}.

The in-plane ***circumferential*** (θ direction) sensitivity $S^{in\,\theta}_{f\,pmn}$ is

$$S^{in\,\theta}_{f\,pmn} = \frac{\phi_3^S}{C_3 \eta_{mn}} = \frac{4e_{31}}{\pi^2 \varepsilon_{33}} \frac{h^S}{mn} \left(\frac{1}{R} - \frac{m\pi}{L} \frac{A_3}{C_3} - \frac{n\pi}{\beta R} \frac{B_3}{C_3} \right) \hat{\delta}_{mn}. \tag{9.2.7c}$$

The amplitudes A_3, B_3 and C_3 are defined for the mn-th component natural frequency $\omega_{\theta mn}$. $\frac{A_i}{C_i}$ and $\frac{B_i}{C_i}$ (i = 1, 2, 3) are defined by Eqs. (9.1.35a & b) in Sect. 9.1. Since the sensor is only sensitive to the odd modes, there is no observation spillover if controlled modes are all odd modes. Note that three sensitivities are defined with respect to the transverse modal oscillating amplitude; these sensitivities could also be defined (or normalized) with respect to the other two modal oscillating amplitudes, i.e., A_i and B_i, as discussed in Sect. 8.3 of Chap. 8. Detailed analyses and evaluation are presented later.

9.2.3 *Diagonal Stripe Shell Sensors*

A diagonal stripe sensor is a narrow piezoelectric stripe diagonally distributed across two far corners of the cylindrical shell. The other alternative is to deposit a narrow conducting electrode diagonally on a fully distributed piezoelectric layer, Fig. 9.13.

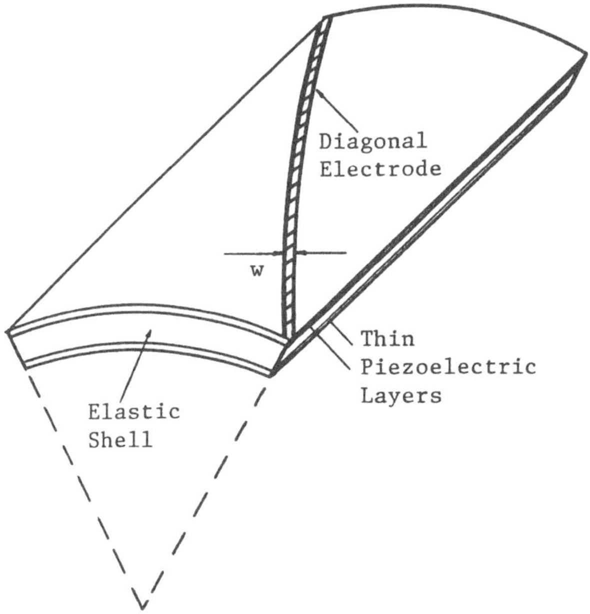

Fig. 9.13 A cylindrical shell with a diagonal stripe sensor

The output signal of the stripe sensor can be defined as

$$\begin{aligned}\phi_3^S &= -\frac{e_{31}R}{\varepsilon_{33}S^e}\sum_{m=1}^{\infty}\sum_{n=1}^{\infty}\left\{h^S\left(\frac{1}{R}C-\frac{m\pi}{L}A-\frac{n\pi}{\beta R}B\right)\right.\\&\left.+\frac{h^S(h+h^S)}{2}\left(\left(\frac{m\pi}{L}\right)^2C-\frac{1}{R^2}\frac{n\pi}{\beta}B+\frac{1}{R^2}\left(\frac{n\pi}{\beta}\right)^2C\right)\right\}\eta_{mn}(t)\\&\cdot\int_0^L\int_{\frac{\beta}{L}x}^{\frac{\beta}{L}x+w}\left(\sin\frac{m\pi x}{L}\sin\frac{n\pi\theta}{\beta}\right)dx\,d\theta,\\&=\frac{e_{31}R}{\varepsilon_{33}S^e}\sum_{m=1}^{\infty}\sum_{n=1}^{\infty}\left\{h^S\left(\frac{1}{R}C-\frac{m\pi}{L}A-\frac{n\pi}{\beta R}B\right)\right.\\&\left.+\frac{h^S(h+h^S)}{2}\left(\left(\frac{m\pi}{L}\right)^2C-\frac{1}{R^2}\frac{n\pi}{\beta}B+\frac{1}{R^2}\left(\frac{n\pi}{\beta}\right)^2C\right)\right\}\eta_{mn}(t)\frac{\beta}{n\pi}\\&\cdot\int_0^L\left(\sin\frac{m\pi x}{L}\cos\left[\frac{n\pi}{\beta}\left(\frac{\beta}{L}x+w\right)\right]-\sin\frac{m\pi x}{L}\cos\frac{n\pi x}{L}\right)dx\end{aligned}\tag{9.2.8}$$

where w is the electrode width of the diagonal stripe sensor. S^e denotes the effective electrode area of the piezoelectric sensor layer, and $S^e \approx w\sqrt{L^2+(R\beta)}$. Applying the modal orthogonality and simplifying the above equation, one can obtain

$$
\begin{aligned}
\phi_3^S = &-\frac{e_{31}R}{\varepsilon_{33}S^e}\sum_{m=1}^{\infty}\sum_{n=1}^{\infty}\left\{h^S\left(\frac{1}{R}C-\frac{m\pi}{L}A-\frac{n\pi}{\beta R}B\right)\right.\\
&\left.+\frac{h^S\left(h+h^S\right)}{2}\left(\left(\frac{m\pi}{L}\right)^2C-\frac{1}{R^2}\frac{n\pi}{\beta}B+\frac{1}{R^2}\left(\frac{n\pi}{\beta}\right)^2C\right)\right\}.\\
&\cdot\eta_{mn}(t)\frac{\beta}{n\pi}\sin\frac{n\pi w}{\beta}\int_0^L\left(\sin\frac{m\pi x}{L}\sin\frac{n\pi x}{L}\right)dx
\end{aligned}
\tag{9.2.9}
$$

Applying the modal orthogonality again and carrying out the integration, one can derive the sensor signal for the m = n modes:

$$
\begin{aligned}
\phi_3^S = &-\frac{e_{31}}{\varepsilon_{33}}\frac{RL\beta}{2\pi w\sqrt{L^2+(\beta R)^2}}\sum_{m=1}^{\infty}\sum_{n=1}^{\infty}\frac{1}{n}\sin\frac{n\pi w}{\beta}\left\{h^S\left(\frac{1}{R}C-\frac{m\pi}{L}A\right.\right.\\
&\left.\left.-\frac{n\pi}{\beta R}B\right)+\frac{h^S\left(h+h^S\right)}{2}\left(\left(\frac{m\pi}{L}\right)^2C-\frac{1}{R^2}\frac{n\pi}{\beta}B+\frac{1}{R^2}\left(\frac{n\pi}{\beta}\right)^2C\right)\right\},\\
&\cdot\eta_{mn}(t)\hat{\delta}_{mn}
\end{aligned}
\tag{9.2.10}
$$

where

$$
\hat{\delta}_{mn}=\begin{cases}0, & \text{if } m\neq n;\\ 1, & \text{if } m=n.\end{cases}
\tag{9.2.11}
$$

Thus, the diagonal stripe sensor is sensitive only to all m = n modes, and insensitive to all $m \neq n$ modes. If controlled modes are m = n modes, there is no observation spillover from $m \neq n$ modes. For a diagonal stripe sensor surface bonded on a cylindrical shell, there are also three sensitivities: 1) a transverse sensitivity, 2) an in-plane longitudinal (x direction) sensitivity, and 3) an in-plane circumferential (θ direction) sensitivity. Each of the sensitivities can be further defined in terms of the transverse oscillation amplitudes.

Thus, the ***transverse*** sensitivity $S^{tr}_{d\,pmn}$ of the stripe sensor defined by the transverse oscillation amplitude is

$$
\begin{aligned}
S^{tr}_{d\,pmn} &= \frac{\phi_3^S}{C_1\eta_{mn}(t)}\\
&=\frac{e_{31}}{\varepsilon_{33}}\frac{RL\beta}{2\pi b\sqrt{L^2+(\beta R)^2}}\frac{1}{n}\sin\frac{n\pi b}{\beta}\frac{h^S\left(h+h^S\right)}{2}.\\
&\cdot\left(\left(\frac{m\pi}{L}\right)^2-\frac{1}{R^2}\frac{n\pi}{\beta}\frac{B_1}{C_1}+\frac{1}{R^2}\left(\frac{n\pi}{\beta}\right)^2\right)\hat{\delta}_{mn}
\end{aligned}
\tag{9.2.12a}
$$

The in-plane ***longitudinal*** (x direction) sensitivity $S^{tr\,x}_{d\,pmn}$ is

$$
\begin{aligned}
S^{tr\,x}_{d\,pmn} &= \frac{\phi_3^S}{C_2 \eta_{mn}(t)} \\
&= \frac{e_{31}}{\varepsilon_{33}} \frac{RL\beta}{2\pi b\sqrt{L^2+(\beta R)^2}} \frac{1}{n} \sin\frac{n\pi b}{\beta} h^S \cdot \\
&\cdot \left(\frac{1}{R} - \frac{m\pi}{L}\frac{A_2}{C_2} - \frac{n\pi}{\beta R}\frac{B_2}{C_2} \right) \hat{\delta}_{mn}
\end{aligned}
\tag{9.2.12b}
$$

The in-plane ***circumferential*** (θ direction) sensitivity $S^{tr\,\theta}_{d\,pmn}$ is

$$
\begin{aligned}
S^{tr\,\theta}_{d\,pmn} &= \frac{\phi_3^S}{C_3 \eta_{mn}(t)} \\
&= \frac{e_{31}}{\varepsilon_{33}} \frac{RL\beta}{2\pi b\sqrt{L^2+(\beta R)^2}} \frac{1}{n} \sin\frac{n\pi b}{\beta} h^S \cdot \\
&\cdot \left(\frac{1}{R} - \frac{m\pi}{L}\frac{A_3}{C_3} - \frac{n\pi}{\beta R}\frac{B_3}{C_3} \right) \hat{\delta}_{mn}
\end{aligned}
\tag{9.2.12c}
$$

Thus, two distributed piezoelectric sensors: 1) a fully distributed sensor and 2) a diagonal stripe sensor for cylindrical shells were proposed and their sensitivities defied. Sensor sensitivities, defined by the transverse oscillation amplitude, for the two distributed sensors were defined. Note that these three sensitivities can also be defined by the oscillation amplitudes of the other two oscillating amplitudes of the mn-th natural mode, as discussed in Sect. 9.2.2 and in Sect. 8.3 of Chap. 8. Numerical solutions of the distributed sensor sensitivities are presented next.

9.3 Evaluation of Distributed Cylindrical Shell Sensors

As discussed in Sect. 9.2, two distributed sensors were proposed and their sensitivities analyzed. In this section, sensitivities and parametric studies of these two sensors are evaluated. It is assumed that the sensors are made of polymeric polyvinylidene fluoride (PVDF) with a thickness of 25 μm. The cylindrical shell is made of steel (1 mm thick) with three curvature angles, i.e., 60°, 90°, and 120° to account for the shell deepness or curvature influences. Material properties were summarized in Table 9.1. Free vibration analysis of the cylindrical shell was presented in Sect. 9.1.

9.3.1 Fully Distributed Shell Sensor

It was observed that the fully distributed shell sensor is insensitive to all even modes due to signal cancellations from positive and negative strain regions. The sensitivities of transverse modes, in-plane x-direction modes, and in-plane θ-direction modes of odd modes are analyzed. The odd modes are considered for both n and m.

Figure 9.14a–c show the sensitivities of transverse modes, in-plane x-direction modes, and in-plane θ-direction modes. There are three cylindrical shells, i.e., 60°, 90°, and 120°, plotted in each figure. It is observed that the shallow shell (60° curvature) gives higher sensitivities than those of deep shells (90° and 120° curvatures). This is because the strain level is higher for the shallow shell when a unit transverse oscillation is imposed to all three shell curvatures.

Thickness effects of the sensor and the cylindrical shell are also evaluated. Figure 9.15a–c show three sensitivities calculated for a 40 μm thick PVDF distributed sensor. As shown in analytical solutions, all three sensitivities increase.

Figure 9.16a–c give the sensitivities for a 5 mm-thick cylindrical shell with a 25 μm piezoelectric sensor. It is observed that the in-plane membrane sensitivities remain identical to those of the 1 mm-thick shell and the transverse sensitivity increases. As explained for distributed convolving ring sensors, the membrane strains experienced in the distributed sensors should be the same regardless of the elastic shell thickness. However, the bending strain increases as the sensor is placed farther away from the neutral surface. These phenomena can also be directly observed from the analytical solutions.

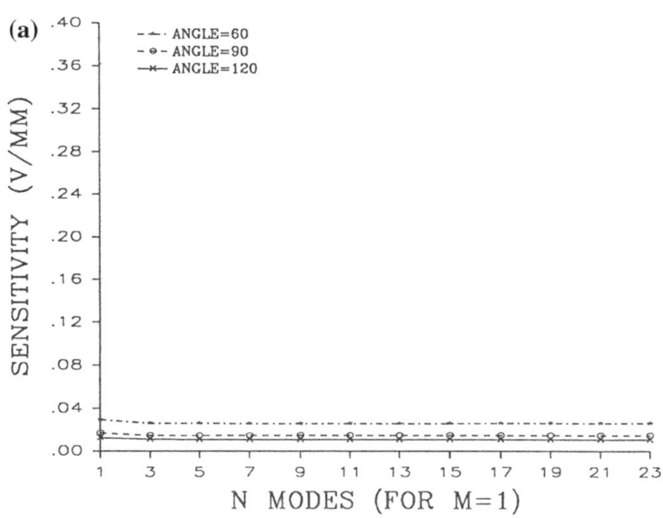

Fig. 9.14 Three sensitivities of the fully distributed sensor (1 mm, 25 μm). **a** Transverse sensitivity (1 mm, 25 μm). **b** In-plane x sensitivity (1 mm, 25 μm). **c** In-plane θ sensitivity sensor (1 mm, 25 μm)

(b)

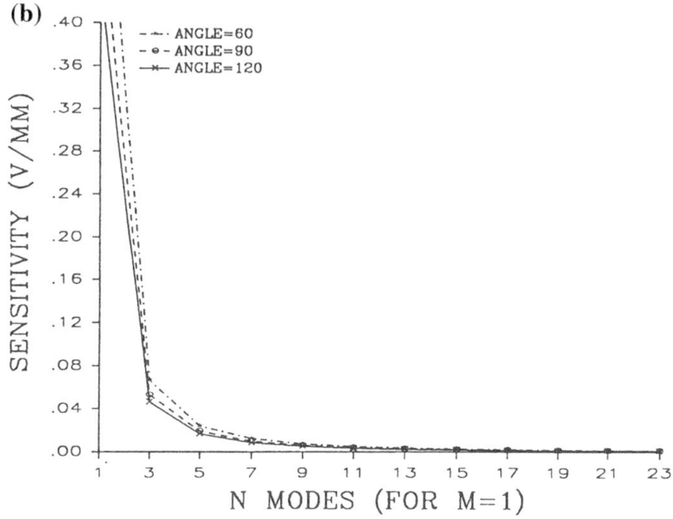

(c)

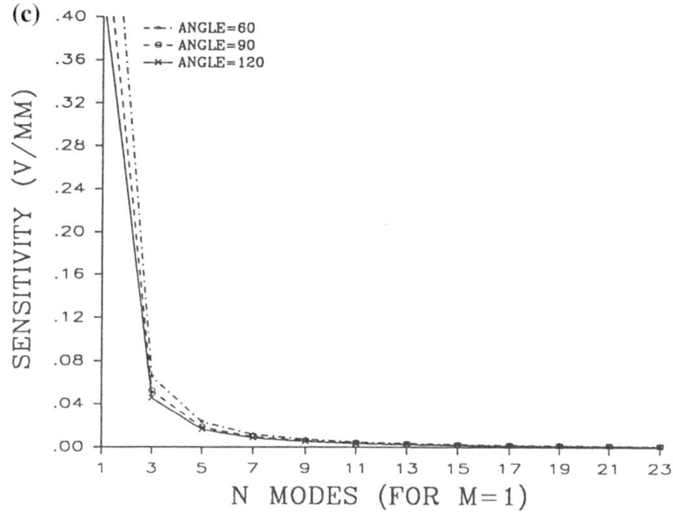

Fig. 9.14 (continued)

(a)
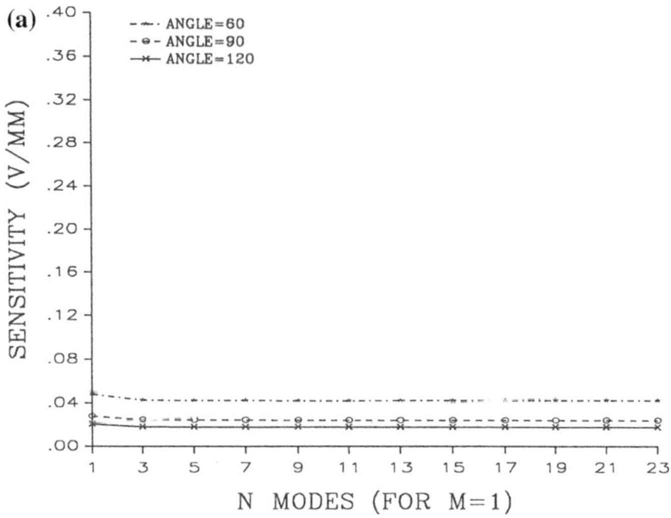

(b)
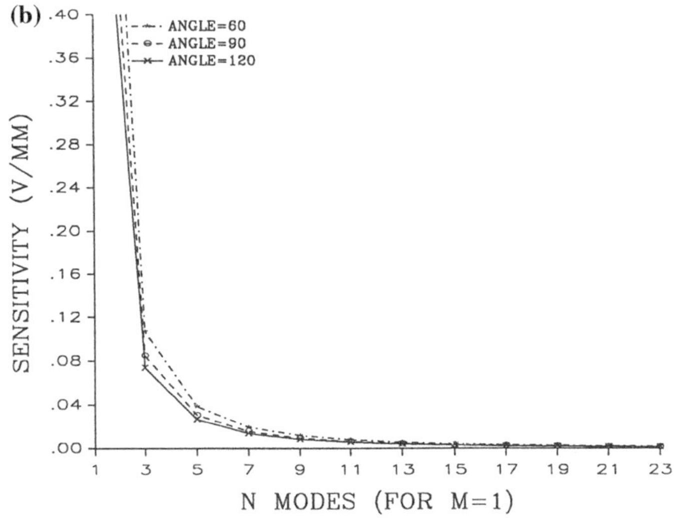

Fig. 9.15 Three sensitivities of the fully distributed sensor (1 mm, 40 μm). **a** Transverse sensitivity (1 mm, 40 μm). **b** In-plane x sensitivity (1 mm, 40 μm). **c** In-plane θ sensitivity (1 mm, 40 μm)

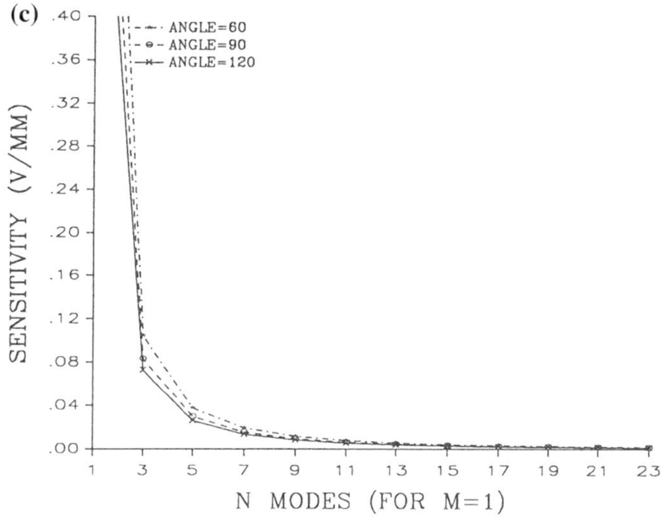

Fig. 9.15 (continued)

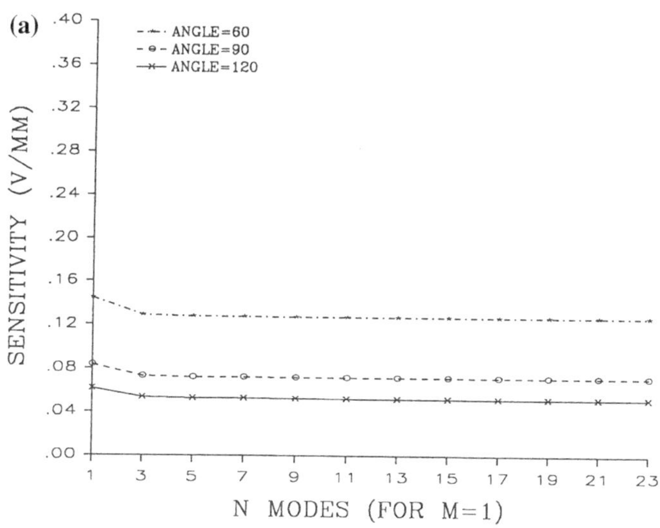

Fig. 9.16 Three sensitivities of the fully distributed sensor (5 mm, 25 μm). **a** Transverse sensitivity (5 mm, 25 μm). **b** x sensitivity (5 mm, 25 μm). **c** θ sensitivity (5 mm, 25 μm)

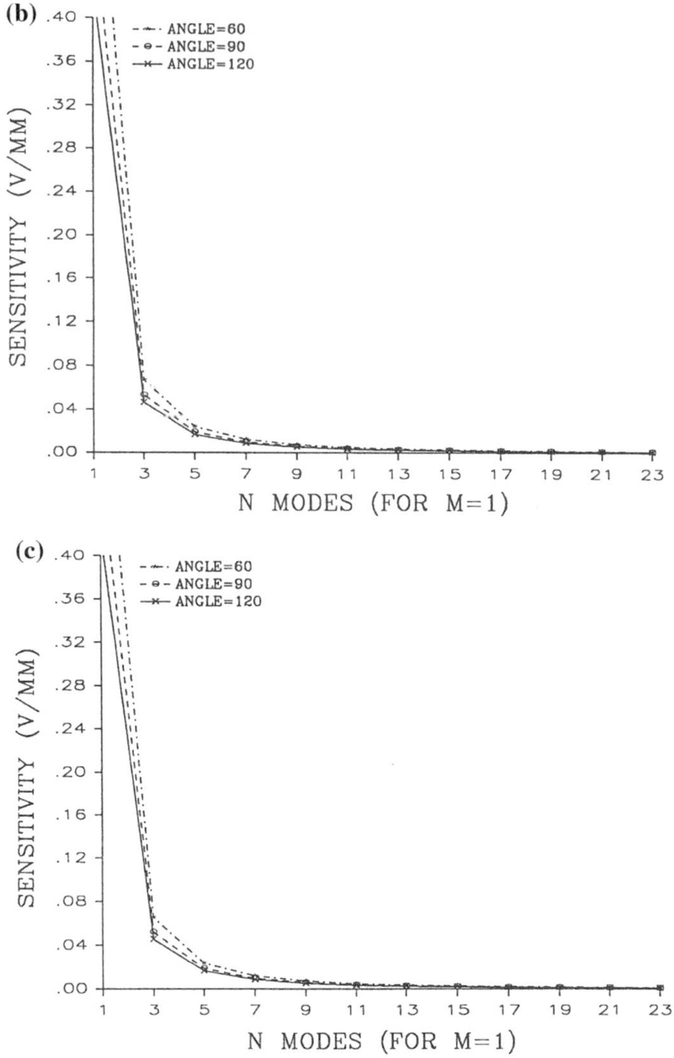

Fig. 9.16 (continued)

9.3.2 Diagonal Stripe Shell Sensors

Three sensitivities of diagonally distributed stripe sensors for three cylindrical shells are plotted in Fig. 9.17a–c. Recall that the diagonal stripe sensors are sensitive to m = n modes and insensitive to other $m \neq n$ modes. Sensitivities of transverse modes, in-plane x direction modes, and in-plane θ direction modes are calculated and analyzed. It is observed that the sensitivities for shallow shells are higher than those of deep shells because the strains are higher if the same magnitude of oscillation is considered.

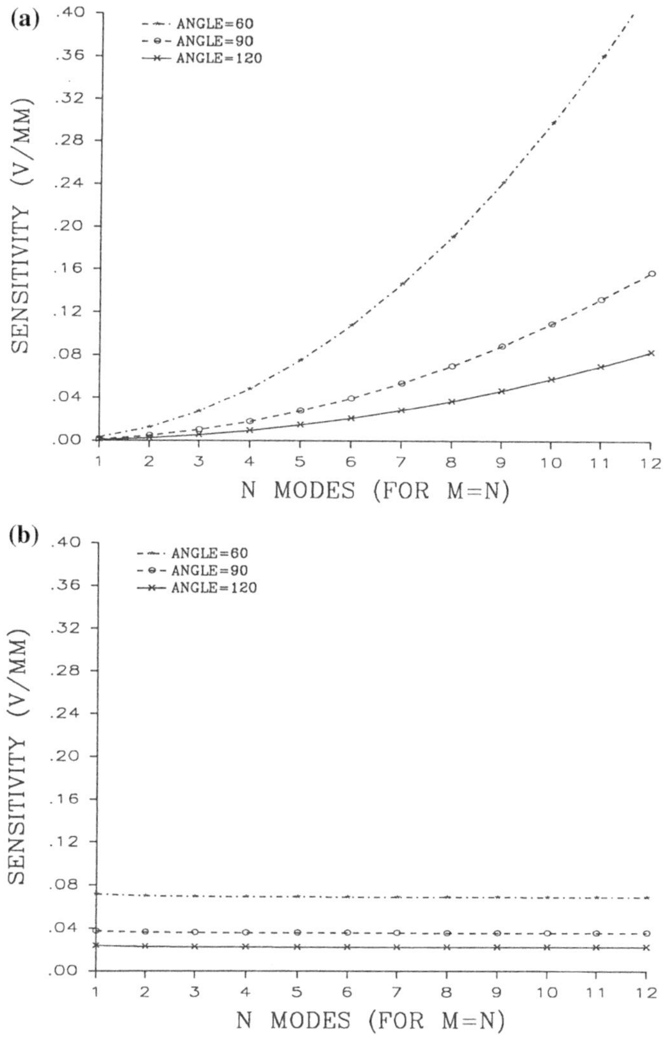

Fig. 9.17 Three sensitivities of the diagonal stripe sensor (1 mm, 25 μm). **a** Transverse sensitivity (1 mm, 25 μm). **b** In-plane x sensitivity (1 mm, 25 μm). **c** In-plane θ sensitivity (1 mm, 25 μm)

Thickness changes for both the piezoelectric sensor and the cylindrical shell are considered. Figure 9.18a–c give the sensitivities of 40 μm diagonal stripe sensors. Sensor sensitivities increase as the sensor thickness increases. Three sensitivities of a 5 mm cylindrical shell with a 25 μm piezoelectric sensor are plotted in Fig. 9.19a–c. Again, the transverse sensitivity increases due to the increase of bending strains. However, the other in-plane sensitivities remain identical as those of the 1 mm shell because the membrane strains remain unchanged regardless the thickness change.

(c)

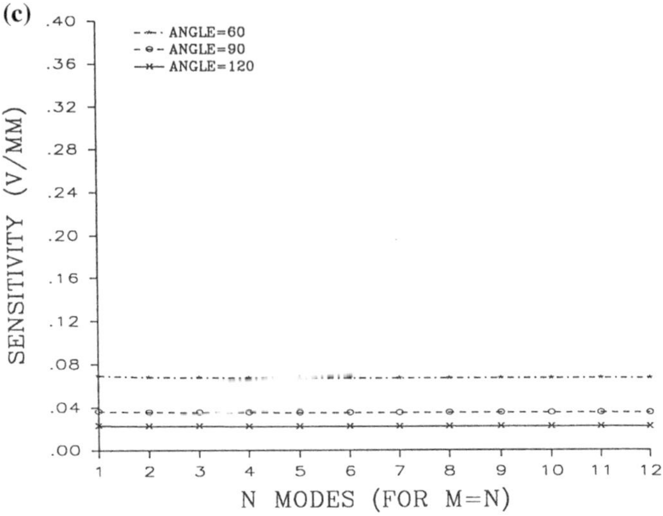

Fig. 9.17 (continued)

(a)

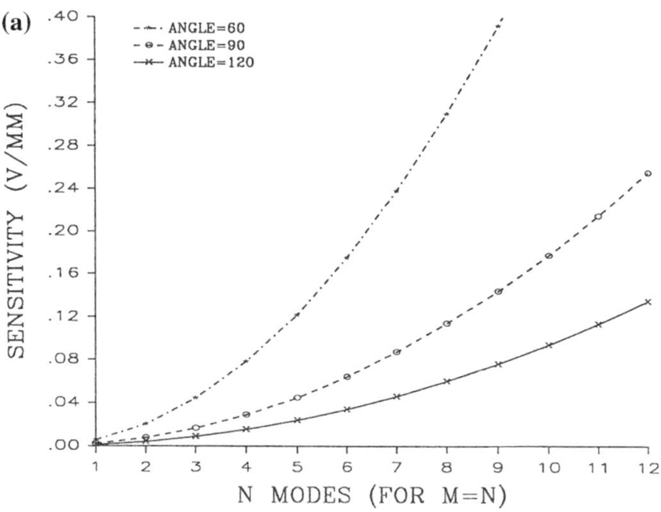

Fig. 9.18 Three sensitivities of the diagonal stripe sensor (1 mm, 40 μm). **a** Transverse sensitivity (1 mm, 40 μm). **b** In-plane x sensitivity (1 mm, 40 μm). **c** In-plane θ sensitivity (1 mm, 40 μm)

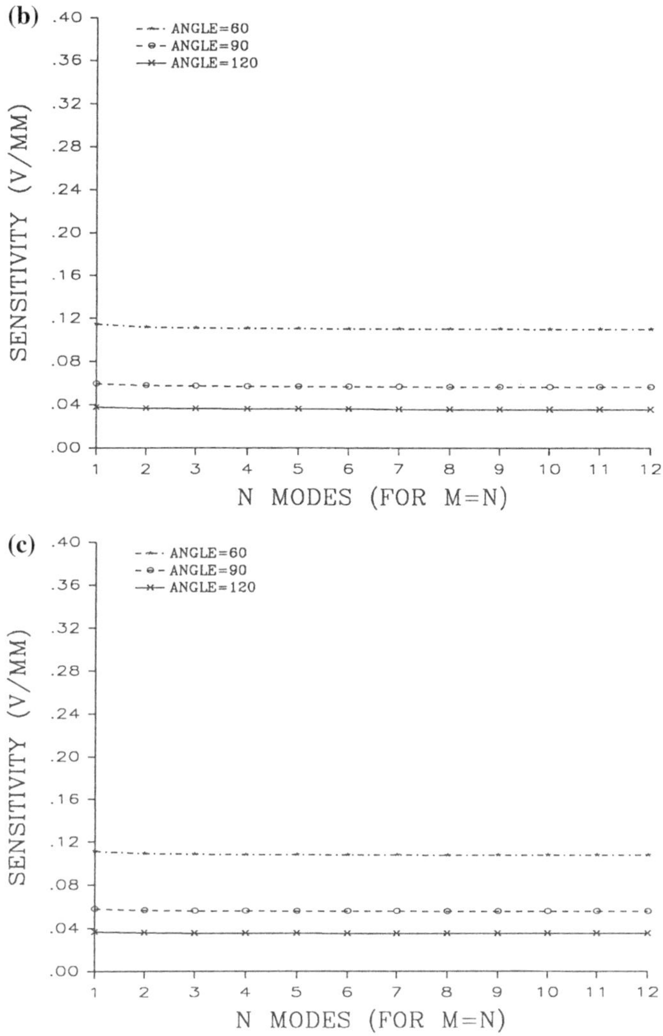

Fig. 9.18 (continued)

In summary, the fully distributed sensor is designed for all odd modes and the diagonally distributed stripe sensor is for all m = n modes. Output signal is contributed by the membrane strains and bending strains. Depending on the location of the sensor and the shell vibration modes, the contribution of these strains can be different. In general, membrane strains contribute to the in-plane sensitivities and the bending strains to the transverse sensitivity of distributed shell sensors. Distributed actuation and control behaviors of these two kinds, i.e., a fully distributed and a stripe, of actuators are discussed next.

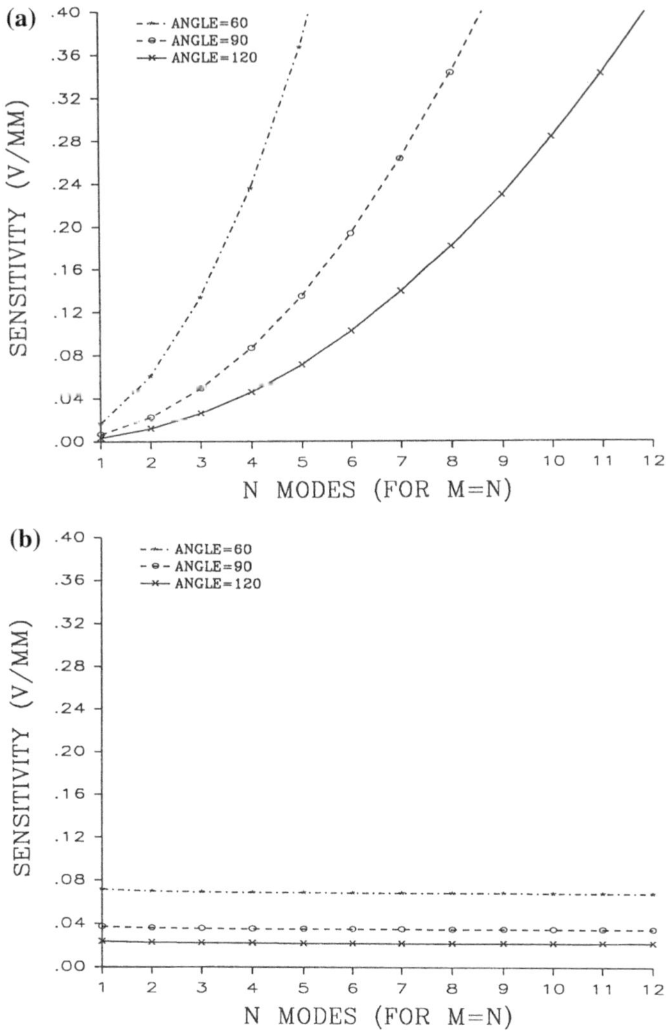

Fig. 9.19 Three sensitivities of the diagonal stripe sensor (5 mm, 25 μm). **a** Transverse sensitivity (5 mm, 25 μm). **b** In-plane x sensitivity (5 mm, 25 μm). **c** In-plane θ sensitivity (5 mm, 25 μm)

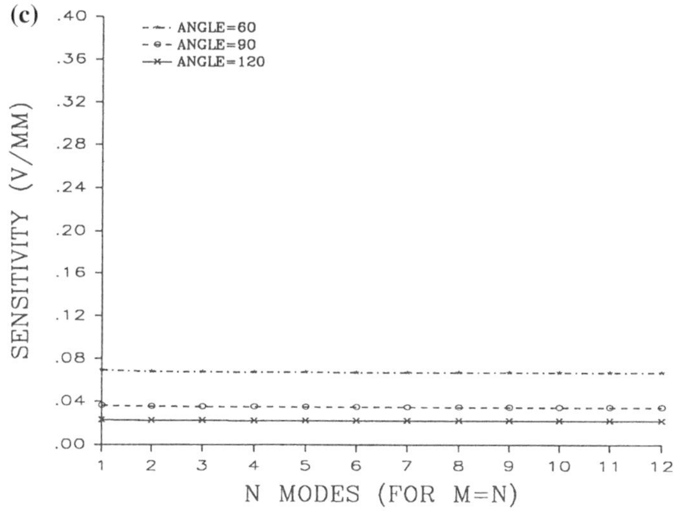

Fig. 9.19 (continued)

9.4 Two-Dimensional Distributed Actuators for Cylindrical Shells

There are two fundamental designs for two-dimensional distributed actuators: 1) a thickness variation and 2) a spatially distribution, Sect. 4.6 of Chap. 4, from which spatially distributed actuations can be generated, Chap. 8. In this section, spatially distributed actuators for cylindrical shells are proposed and their performances evaluated.

It is assumed that a distributed piezoelectric actuator layer is laminated on the top surface of a simply supported cylindrical shell (thickness h). The piezoelectric actuator layer (thickness h^a) is made of a polymeric PVDF material and its material properties are neglected in the analysis. Again, the shell is made of steel with a thickness of 1 mm. Note that there are three angular curvatures, i.e., 60°, 90°, and 120° from which the effect of shallow and deep shells can be evaluated. From Chaps. 3, 4 and 8, one can derive the transverse modal equation for a simply supported cylindrical shell with distributed actuators as

$$\ddot{\eta}_{mn} + \frac{c}{\rho h}\dot{\eta}_{mn} + \omega_{mn}^2 \eta_{mn} = \frac{1}{\rho h N_{mn}} \int_x \int_\theta \left(\frac{N_{\theta\theta}^e}{R} - \frac{\partial^2 M_{xx}^e}{\partial x^2} - \frac{1}{R^2}\frac{\partial^2 M_{\theta\theta}^e}{\partial \theta^2} \right) \sin\frac{m\pi x}{L} \sin\frac{n\pi\theta}{\beta} R\,dx\,d\theta, \tag{9.4.1}$$

and

$$N_{mn} = \int_0^L \int_0^\beta \left(\sin\frac{m\pi x}{L} \sin\frac{n\pi\theta}{\beta} \right)^2 R dx d\theta = \frac{LR\beta}{4}. \tag{9.4.2}$$

Since the actuator is laminated on the surface of the cylindrical shell, the electric membrane force $N^e_{\theta\theta}$ and bending moments (M^e_{xx} and $M^e_{\theta\theta}$) can be defined from the derivations in Sect. 4.3 of Chap. 4. (Note that only the transverse vibration $u_3(x, \theta, t) = \sum_{m=1}^{\infty} \sum_{n=1}^{\infty} \eta_{mn}(t) C_{mn} \sin\frac{m\pi x}{L} \sin\frac{n\pi\theta}{\beta}$ is considered.) It is intended to use the electric membrane force and the control moments to actively control the transverse vibration of shells. Again, two actuator configurations: 1) a fully distributed and 2) a diagonal stripe are considered.

9.4.1 Fully Distributed Actuator

A fully distributed actuator is a PVDF layer covering the whole top surface of the cylindrical shell, Fig. 9.12. The resultant effect of the distributed actuator is equivalent to four counteracting moments acting on the four boundary edges, and these moments can be defined by step functions, Sect. 7.3 of Chap. 7. Assuming a feedback voltage ϕ_3^a is injected into the actuator, using the equivalent control-moment approaches, Sect. 7.3, and Eq. (9.4.1), one can derive the control forces and moments in the modal domain:

Membrane control force:

$$\int_0^L \int_0^\beta \frac{1}{R} N^e_{\theta\theta} \sin\frac{m\pi x}{L} \sin\frac{n\pi\theta}{\beta} R dx d\theta = \frac{4 e_{31} L\beta}{\pi^2} \frac{1}{mn} \phi_3^a \hat{\delta}_{mn}. \tag{9.4.3}$$

Control moments:

$$\int_x \int_\theta \frac{\partial^2 M^e_{xx}}{\partial x^2} \sin\frac{m\pi x}{L} \sin\frac{n\pi\theta}{\beta} R dx d\theta = -\frac{e_{31}(h + h^a)}{2} \frac{4m\beta R}{nL} \phi_3^a \hat{\delta}_{mn}, \tag{9.4.4a}$$

$$\int_x \int_\theta \frac{1}{R^2} \frac{\partial^2 M^e_{\theta\theta}}{\partial \theta^2} \sin\frac{m\pi x}{L} \sin\frac{n\pi\theta}{\beta} R dx d\theta = -\frac{e_{31}(h + h^a)}{2} \frac{4nL}{m\beta R} \phi_3^a \hat{\delta}_{mn}, \tag{9.4.4b}$$

where

$$\begin{aligned} \hat{\delta}_{mn} &= \frac{1}{4}[1 - \cos(m\pi)][1 - \cos(n\pi)] \\ &= \begin{cases} 1, & \text{If both of m and n are odd integers ;} \\ 0, & \text{for other m and n.} \end{cases} \end{aligned} \tag{9.4.5}$$

Note that the resultant control force and moments are zero for all even modes due to voltage cancellations of anti-symmetrical modes in a symmetrical structure as explained in Chap. 7. Substituting the resultant force and moment expressions into the modal equation gives the independent modal control equation:

$$\Longrightarrow \ddot{\eta}_{mn} + \frac{c}{\rho h}\dot{\eta}_{mn} + \omega_{mn}^2\eta_{mn} = \frac{16e_{31}}{LR\beta\rho h}\left\{\frac{L\beta}{\pi^2}\frac{1}{mn} + \frac{h+h^a}{2}\left(\frac{m\beta R}{nL} + \frac{nL}{m\beta R}\right)\right\}\phi_3^a\hat{\delta}_{mn}. \tag{9.4.6}$$

Modal equations for the displacement and velocity feedback algorithms will be derived next.

1) Displacement Feedback

For a displacement feedback, the voltage signal is directly amplified and applied to the actuator.

$$\phi_3^a = -\mathcal{G}_{fp}^{df}\eta_{mn}(t), \tag{9.4.7}$$

where $\mathcal{G}_{fp}^{df}$ is the system feedback gain and it is defined as the product of the sensor sensitivity S_{fp} and an amplifier gain $\mathcal{G}_{am}$:

$$\mathcal{G}_{fp}^{df} = S_{fp}\mathcal{G}_{am}. \tag{9.4.8}$$

For odd modes, substituting Eq. (9.4.7) into Eq. (9.4.6) yields

$$\ddot{\eta}_{mn} + \frac{c}{\rho h}\dot{\eta}_{mn} + \left\{\omega_{mn}^2 + \frac{16e_{31}}{LR\beta}\left(\frac{L\beta}{\pi^2}\frac{1}{mn} + \frac{h+h^a}{2}\left(\frac{m\beta R}{nL} + \frac{nL}{m\beta R}\right)\mathcal{G}_{fp}^{df}\hat{\delta}_{mn}\right\}\eta_{mn} = 0; \tag{9.4.9a}$$

or defined in the modal damping ratio:

$$\Longrightarrow \ddot{\eta}_{mn} + 2\zeta\omega_{mn}\dot{\eta}_{mn} + \left\{\omega_{mn}^2 + \frac{16e_{31}}{LR\beta}\left(\frac{L\beta}{\pi^2}\frac{1}{mn} + \frac{h+h^a}{2}\left(\frac{m\beta R}{nL} + \frac{nL}{m\beta R}\right)\mathcal{G}_{fp}^{df}\hat{\delta}_{mn}\right\}\eta_{mn} = 0. \tag{9.4.9b}$$

Note that feedback terms primarily contribute to the stiffness part of the modal equation

2) Velocity Feedback

In a velocity feedback, the control voltage applied to the distributed piezoelectric actuator can be described as

$$\phi_3^a = -\mathcal{G}_{fp}^{vf}\dot{\eta}_{mn}(t), \tag{9.4.10}$$

where $\mathcal{G}_{fp}^{vf}$ is the system feedback gain which is defined as the product of the sensor sensitivity S_{fp}, the amplifier gain $\mathcal{G}_{am}$, and the gain of a differentiator circuit $\mathcal{G}_{de}$:

$$\mathcal{G}_{fp}^{vf} = S_{fp} \cdot \mathcal{G}_{am} \cdot \mathcal{G}_{de}. \tag{9.4.11}$$

For odd modes, substituting Eq. (9.4.10) into Eq. (9.4.6) yields

$$\Longrightarrow \ddot{\eta}_{mn} + \frac{1}{\rho h}\left\{c + \frac{16e_{31}}{LR\beta}\left(\frac{L\beta}{\pi^2}\frac{1}{mn} + \frac{h+h^a}{2}\left(\frac{m\beta R}{nL} + \frac{nL}{m\beta R}\right)\right)\mathcal{G}_{fp}^{vf}\hat{\delta}_{mn}\right\}\dot{\eta}_{mn} + \omega_{mn}^2\eta_{mn} = 0 \tag{9.4.12}$$

Dividing the equivalent damping constant by $2\omega_{mn}$ gives an equivalent controlled modal damping ratio.

$$\begin{aligned}
\zeta'_{mn} &= \frac{1}{2\rho h\omega_{mn}}\left\{c + \frac{16e_{31}}{LR\beta}\left(\frac{L\beta}{\pi^2}\frac{1}{mn} + \frac{h+h^a}{2}\left(\frac{m\beta R}{nL} + \frac{nL}{m\beta R}\right)\right)\mathcal{G}_{fp}^{vf}\hat{\delta}_{mn}\right\} \\
&= \zeta_{mn} + \frac{8e_{31}}{LR\beta\rho h\omega_{mn}}\left\{\frac{L\beta}{\pi^2}\frac{1}{mn} + \frac{h+h^a}{2}\left(\frac{m\beta R}{nL} + \frac{nL}{m\beta R}\right)\right\}\mathcal{G}_{fp}^{vf}\hat{\delta}_{mn}. \\
&= \zeta_{mn} + \frac{8e_{31}}{LR\beta\rho h\omega_{mn}}\frac{1}{mn}\left\{\frac{L\beta}{\pi^2} + \frac{h+h^a}{2}\left(\frac{(m\beta R)^2 + (nL)^2}{L\beta R}\right)\right\}\mathcal{G}_{fp}^{vf}\hat{\delta}_{mn}
\end{aligned} \tag{9.4.13}$$

As discussed previously, the velocity feedback can be used to adjust the modal damping ratio and the displacement feedback can be used to change the modal frequency. It should be pointed out, again, that the sensor signal is directly contributed by the strains and indirectly contributed by the displacements. Even so, the conventional "displacement" feedback term is used. Regarding the velocity feedback, the signal is actually a time derivative of mechanical strains, and it is indirectly defined by velocities. Still, the conventional "velocity" feedback term is used. Note that only the velocity feedback is evaluated in the numerical analyses, Sect. 9.5.

9.4.2 *Distributed Diagonal Stripe Actuator*

It is assumed that a narrow piezoelectric stripe is diagonally attached to two far corners of the cylindrical shell. (The other alternative is to use a fully distributed piezoelectric layer with a narrow diagonally distributed electrode.) Figure 9.13 illustrates the diagonal stripe actuator.

The diagonal stripe actuator can be treated as a diagonal line actuator and be mathematically described by a Kronecker delta function (Soedel, 1981). In this case, because of the narrowness of the stripe actuator, the effect of the electric moment is neglected. The electric membrane control force can be expressed as:

$$N_{\theta\theta}^{e} = \frac{\sqrt{(R\beta)^2 + L^2}}{R\beta} e_{31}\phi_3^a \delta\left(x - \frac{L}{\beta}\theta\right), \tag{9.4.14}$$

where $\delta(\cdots)$ is a delta function, i.e., $\delta(\cdots) = 0$ when $x \neq L\theta/\beta$; $\delta(\cdots) = 1$ when $x = L\theta/\beta$.

The modal equation becomes

$$\Longrightarrow \quad \ddot{\eta}_{mn} + \frac{c}{\rho h}\dot{\eta}_{mn} + \omega_{mn}^2 \eta_{mn} = \frac{2e_{31}}{\rho h R^2 \beta}\sqrt{(R\beta)^2 + L^2}\,\phi_3^a \tilde{\delta}_{mn}, \tag{9.4.15}$$

where

$$\tilde{\delta}_{mn} = \begin{cases} 1, & \text{if } m = n; \\ 0, & \text{if } m \neq n. \end{cases} \tag{9.4.16}$$

This indicates that a diagonal strip actuator can excite or actuate only m = n modes and is insensitive to m ≠ n modes. Again, modal equations for the displacement and velocity feedbacks are presented next.

1) Displacement Feedback

For a displacement feedback, the signal is directly amplified and applied to the actuator.

$$\phi_3^a = -\mathcal{G}_{dp}^{df}\eta_{mn}(t), \tag{9.4.17}$$

where the system gain is defined by a sensor sensitivity and an amplifier gain:

$$\mathcal{G}_{dp}^{df} = S_{dp}\mathcal{G}_{am}. \tag{9.4.18}$$

The governing modal equation for the m = n modes becomes

$$\ddot{\eta}_{mn} + \frac{c}{\rho h}\dot{\eta}_{mn} + \left(\omega_{mn}^2 + \frac{2e_{31}}{\rho h R^2 \beta}\sqrt{(R\beta)^2 + L^2}\,\mathcal{G}_{dp}^{df}\,\tilde{\delta}_{mn}\right)\eta_{mn} = 0; \qquad (9.4.19a)$$

or

$$\Longrightarrow \ddot{\eta}_{mn} + 2\zeta\omega_{mn}\dot{\eta}_{mn} + \left(\omega_{mn}^2 + \frac{2e_{31}}{\rho h R^2 \beta}\sqrt{(R\beta)^2 + L^2}\,\mathcal{G}_{dp}^{df}\,\tilde{\delta}_{mn}\right)\eta_{mn} = 0. \qquad (9.4.19b)$$

2) Velocity Feedback

For a velocity feedback, the displacement signal from the sensor is differentiated, amplified, and then fed back to the actuator.

$$\phi_3^a = -\mathcal{G}_{dp}^{vf}\,\dot{\eta}_{mn}(t), \qquad (9.4.20)$$

where the system gain $\mathcal{G}_{dp}^{vf}$ is defined as:

$$\mathcal{G}_{dp}^{vf} = S_{dp}\,\mathcal{G}_{am}\,\mathcal{G}_{de}, \qquad (9.4.21)$$

where S_{dp} is the normalized modal sensitivity of the diagonal stripe sensor, Eq. (9.2.10). For m = n modes, substituting Eq. (9.4.20) into Eq. (9.4.15) yields

$$\Longrightarrow \ddot{\eta}_{mn} + \frac{1}{\rho h}\left(c + \frac{2e_{31}}{R^2\beta}\sqrt{(R\beta)^2 + L^2}\,\mathcal{G}_{dp}^{vf}\,\hat{\delta}_{mn}\right)\dot{\eta}_{mn} + \omega_{mn}^2\eta_{mn} = 0. \qquad (9.4.22)$$

The controlled modal damping ratio is defined as

$$\begin{aligned}\zeta'_{mn} &= \frac{1}{2\rho h\omega_{mn}}\left(c + \frac{2e_{31}}{R^2\beta}\sqrt{(R\beta)^2 + L^2}\,\mathcal{G}_{dp}^{vf}\,\hat{\delta}_{mn}\right) \\ &= \zeta_{mn} + \frac{e_{31}}{R^2\rho h\beta\omega_{mn}}\sqrt{(R\beta)^2 + L^2}\,\mathcal{G}_{dp}^{vf}\,\hat{\delta}_{mn}.\end{aligned} \qquad (9.4.23)$$

Again, the modal frequency can be changed in the displacement feedback, and the damping ratio be controlled in the velocity feedback. Note that only the velocity feedback is considered in the later analyses.

9.5 Evaluation of Distributed Actuators

The laminated cylindrical shell defined earlier is used in the numerical analysis. The cylindrical shell is made of a 1 mm steel and the piezoelectric actuators are made of thin polymeric PVDF layers. Note, again, actuator material properties are neglected in the analysis. Control effects of the two actuators, i.e., the fully distributed actuator and the diagonal stripe actuator, are studied.

As expressed in the governing equation, there are totally three electric actuation terms. i.e., $N^e_{\theta\theta}$, M^e_{xx}, and $M^e_{\theta\theta}$, which can be used to control the simply supported cylindrical shell. It is observed that $N^e_{\theta\theta}$ and $M^e_{\theta\theta}$ are effective for the circumferential oscillations and M^e_{xx} for the longitudinal oscillations. (Note that $N^e_{\theta\theta}$ is preserved due to the non-zero radius of curvature in the θ-direction.) Contributions of these electric force and moment terms are evaluated next.

9.5.1 *Evaluation of Bending Effects*

Recall that there are two major control components: 1) the membrane control force and 2) the control moments. In this section, the relative effectiveness of these components is compared. Using a cylindrical shell with a dimension of $\beta = 60°$, $L = 100$ mm, $R = 50$ mm, $h = 1$ mm, $h^a = 0.025$ mm as an example, one can calculate the electric membrane force and bending moments and evaluate their control contributions to the overall control effect. The membrane component in Eq. (9.4.13) is

$$\frac{L\beta}{\pi^2} = 10.61. \tag{9.5.1}$$

Define Δ_c to represent the bending component in Eq. (9.4.13):

$$\Delta_c = \frac{h + h^a}{2}\left(\frac{(m\beta R)^2 + (nL)^2}{L\beta R}\right) = 9.8 \times 10^{-5}\left(2742m^2 + 10^4 n^2\right). \tag{9.5.2}$$

For $m = 1$, $n = 1$, $\Delta_c = 1.25$ and $m = 1$, $n = 3$, $\Delta_c = 9.09$. Thus, except for the (1,1) mode, control effects introduced by the electric bending moments must be included in the analysis. Detailed control contributions of the membrane and bending effects are studied next.

9.5.2 *Evaluation of the Fully Distributed Actuator*

Damping controls, via the fully distributed actuator, of the cylindrical shell are presented in this section. Control effects resulting from the electric control membrane force and bending moments of the first four odd modes, (1,1), (1,3), (3,1), and

(3,3), are separately plotted and compared. Note that the fully distributed actuator is effective only for the odd modes and ineffective for the even modes, due to voltage cancellations. Control effects due to thickness variations of the PVDF layer and the cylindrical shell are also evaluated. Note that the cylindrical shell is 1 mm thick and the piezoelectric actuator is 25 μm. The controlled damping ratios of the (1,1) mode contributed by the membrane force is given in Fig. 9.20a, by control moments in Fig. 9.20b, and the resultant controlled damping ratios in Fig. 9.20c. In each figure, the damping ratio is plotted versus the normalized system feedback gain in Volts/mm or V/mm. There are three shell curvatures, i.e., 60°, 90°, and 120°, in each figure. In addition, Fig. 9.21a–c show the damping variations for the (1,3) mode, Fig. 9.22a–c the (3,1) mode, and Fig. 9.23a–c the (3,3) mode.

In general, the controlled damping ratio increases (linearly) as the system feedback gain increases, although the increment is relatively small. Note that the initial damping ratio was assumed 1% for all modes. The control effects from electric membrane forces, electric bending moments, and total control effect are illustrated in separate figures. The control effect from electric membrane forces is larger than that from electric bending moments, but they are on the same order. The total control effect is equal to the sum of control effects from electric membrane forces and bending moments. For the (1,1) mode, the control effect contributed by the in-plane membrane force (90%) is much more significant than that by the bending moment (10%). As the mode increases, the membrane force contribution decreases more significantly than the moment contribution. However, as discussed in Chap. 7, the convergence of a modal time history is determined by the product of modal damping ratio and natural frequency, i.e., $e^{-\zeta_{mn}\omega_{mn}t}$. Usually, high frequency components converge much faster than the lower frequency components. Thus, lower frequency components are usually observable in the overall system oscillation. It is also observed that the control effects for deep shells (120° curvature) are better than those for the shallow shells (60° curvature), except for the (1,3) mode. This is because the natural frequency of 120° cylindrical shell is higher that of the 60° shell, Tables 9.2, 9.3 and 9.4 in Sect. 9.1.

Calculations were also made for a thinner cylindrical shell (0.5 mm) and a thicker piezoelectric actuator (40 μm). Control effects for the (1,1) mode of these two cases are plotted in Fig. 9.24a–c (0.5 mm shell) and Fig. 9.25a–c (40 μm actuator). It is observed that the control effects are better for the 0.5 mm cylindrical shell because the shell is more flexible and thus is easier to control. In addition, a thicker piezoelectric actuator also contributes better control effects.

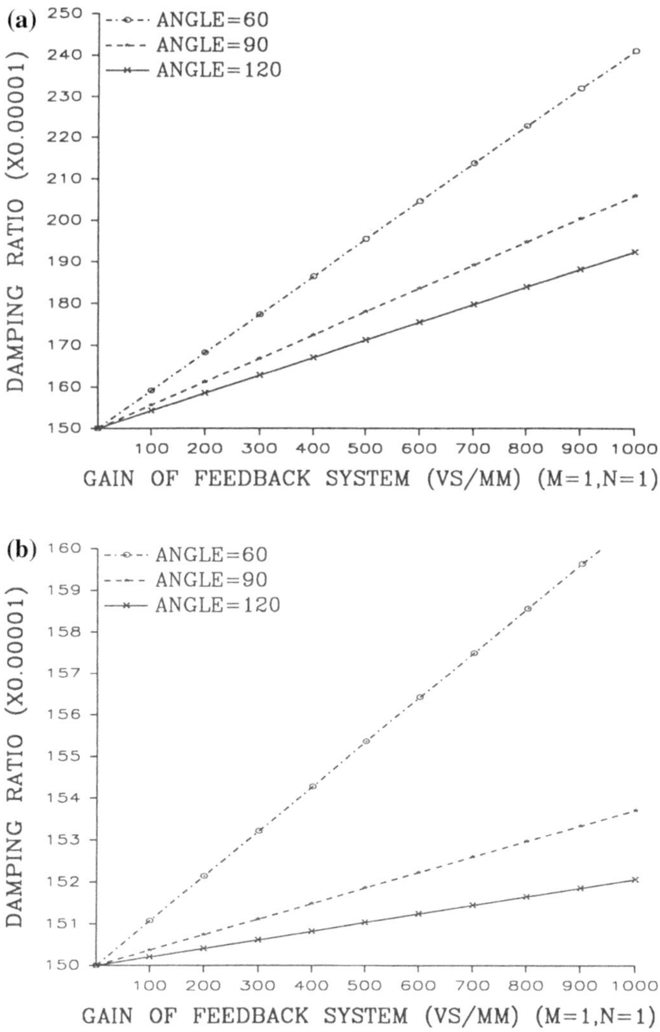

Fig. 9.20 Damping control of the fully distributed actuator, (1,1) mode & (1 mm, 25 μm). **a** Damping control (membrane), (1,1) mode & (1 mm, 25 μm). **b** Damping control (bending), (1,1) mode & (1 mm, 25 μm). **c** Total damping control (resultant)

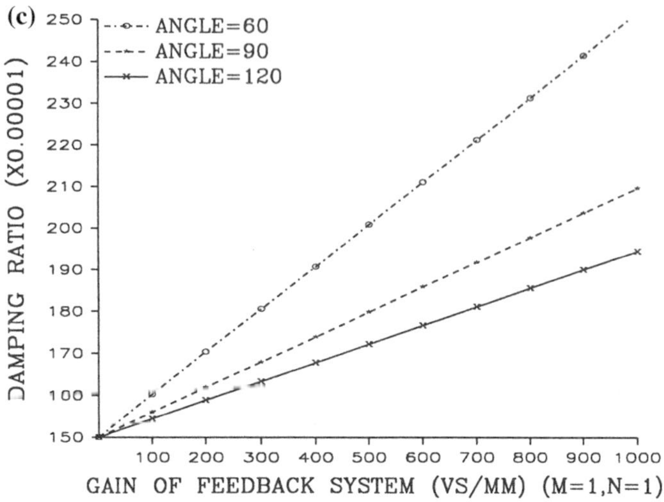

Fig. 9.20 (continued)

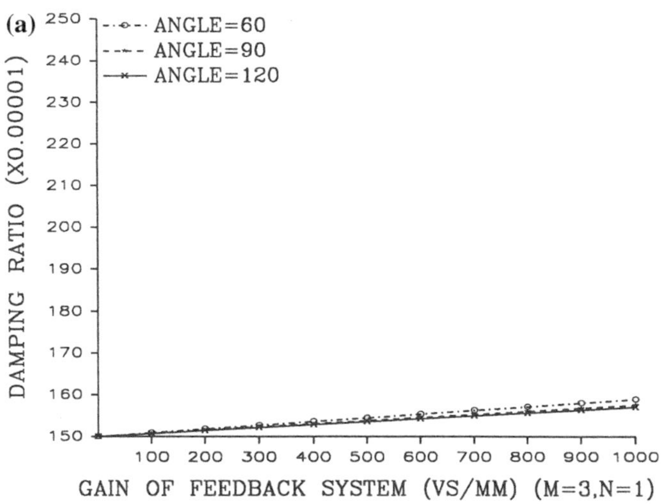

Fig. 9.21 Damping control of the fully distributed actuator, (3,1) mode & (1 mm, 25 μm). **a** Damping control (membrane), (3,1) mode & (1 mm, 25 μm). **b** Damping control (bending), (3,1) mode & (1 mm, 25 μm). **c** Total damping control (resultant)

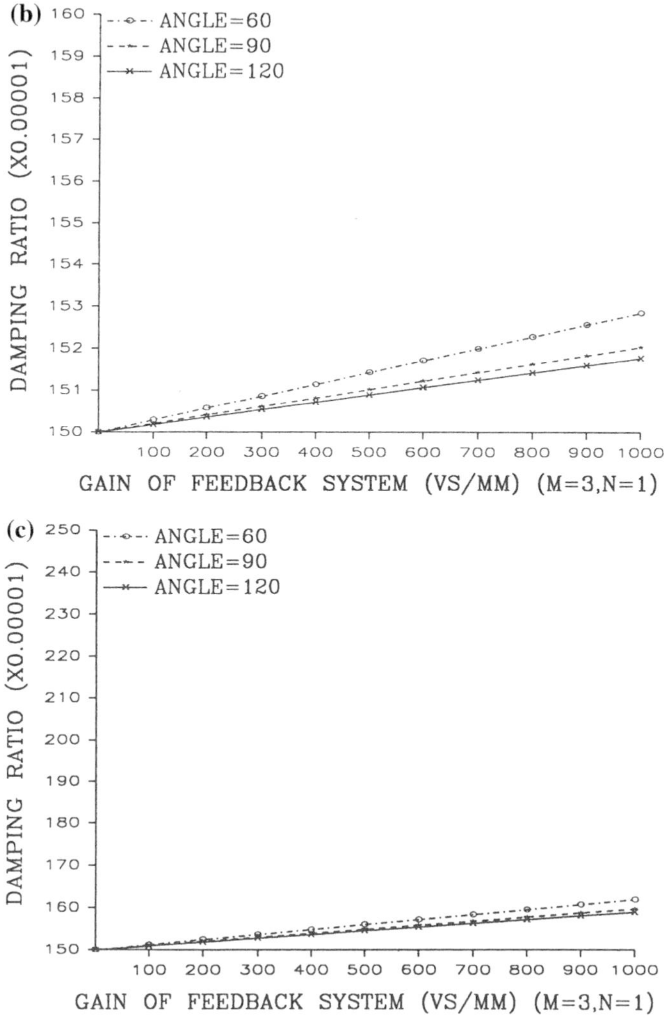

Fig. 9.21 (continued)

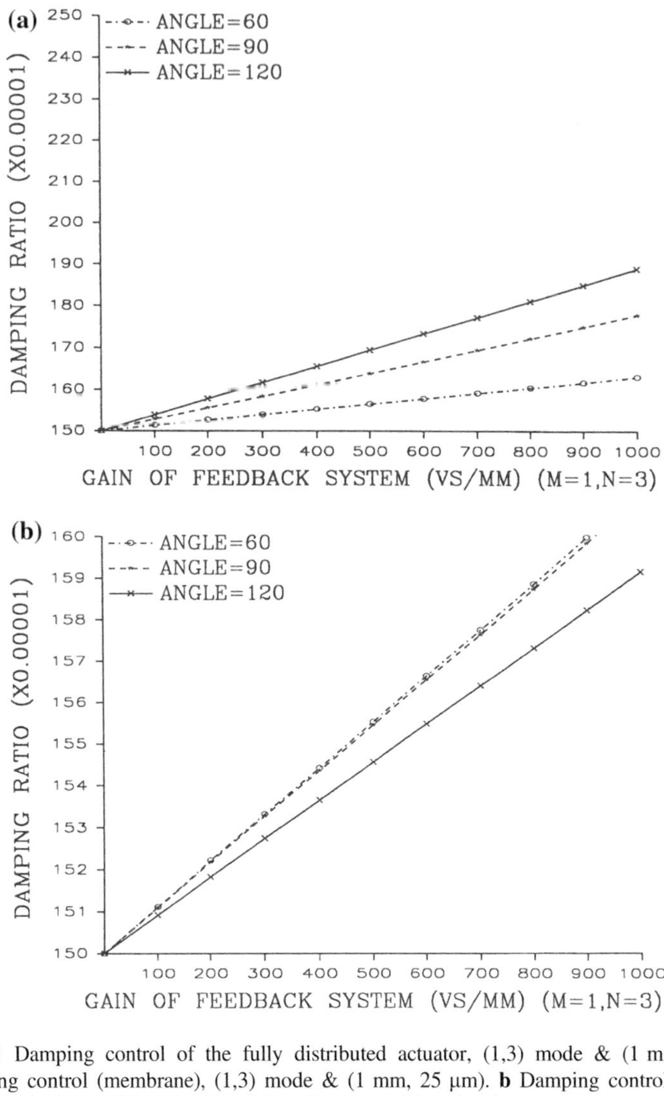

Fig. 9.22 Damping control of the fully distributed actuator, (1,3) mode & (1 mm, 25 μm). **a** Damping control (membrane), (1,3) mode & (1 mm, 25 μm). **b** Damping control (bending), (1,3) mode & (1 mm, 25 μm). **c** Total damping control (resultant)

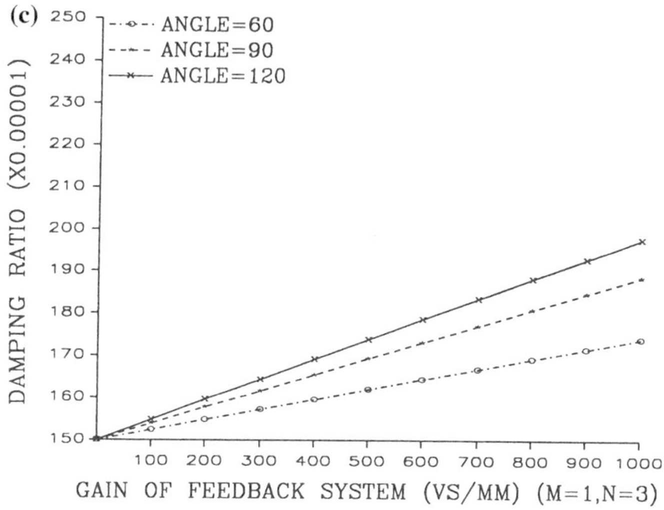

Fig. 9.22 (continued)

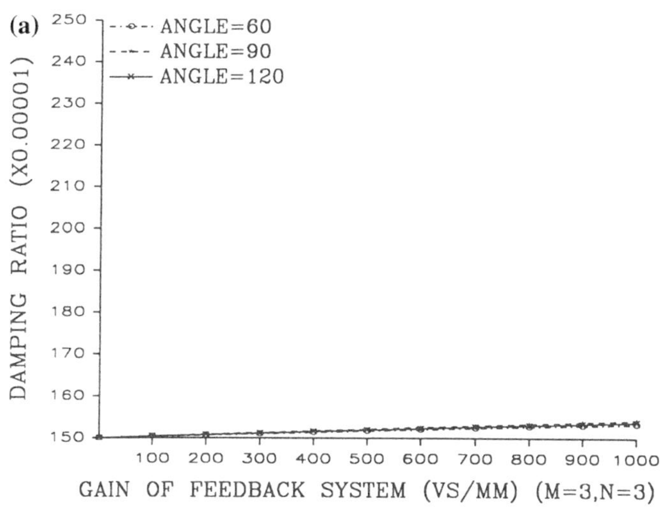

Fig. 9.23 Damping control of the fully distributed actuator, (3,3) mode & (1 mm, 25 μm). **a** Damping control (membrane), (3,3) mode & (1 mm, 25 μm). **b** Damping control (bending), (3,3) mode & (1 mm, 25 μm). **c** Total damping control (resultant)

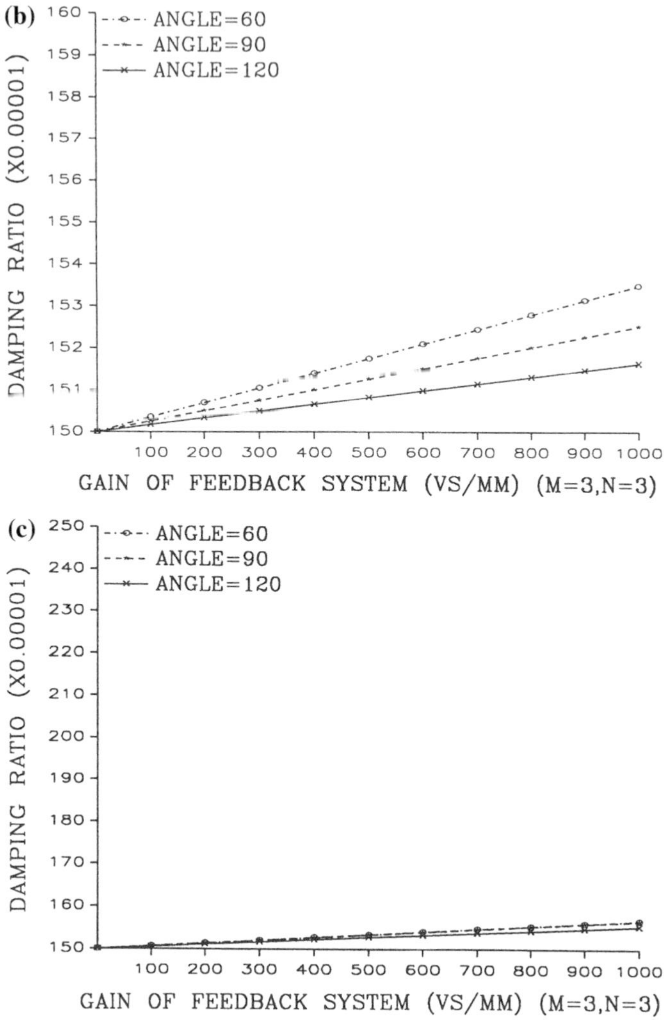

Fig. 9.23 (continued)

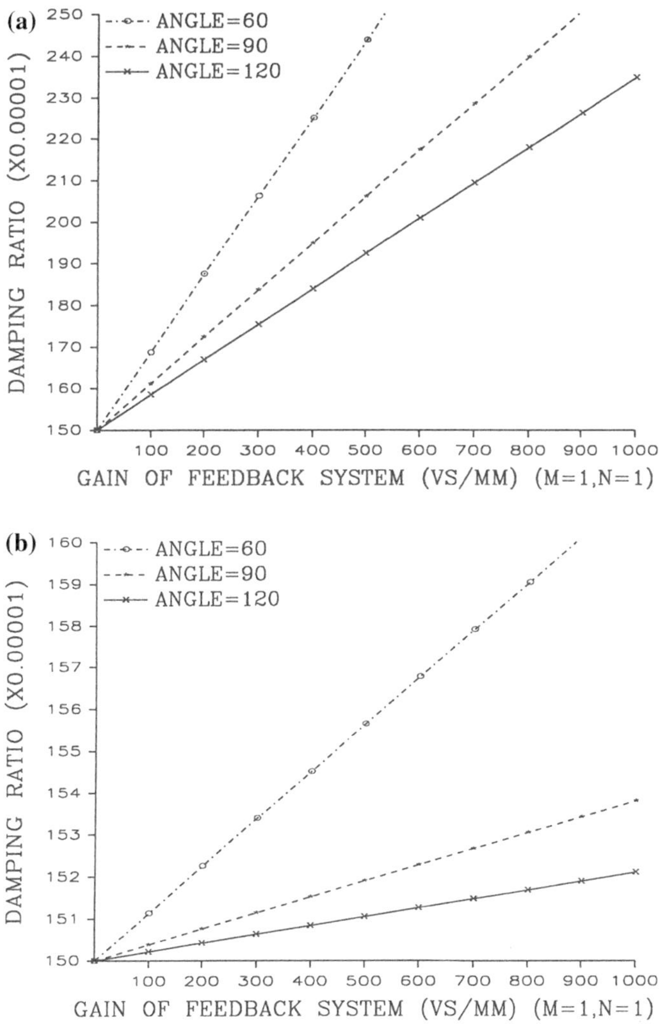

Fig. 9.24 Damping control of the fully distributed actuator, (1,1) mode & (0.5 mm, 25 μm). **a** Damping control (membrane), (1,1) mode & (0.5 mm, 25 μm). **b** Damping control (bending), (1,1) mode & (0.5 mm, 25 μm). **c** Total damping control (resultant)

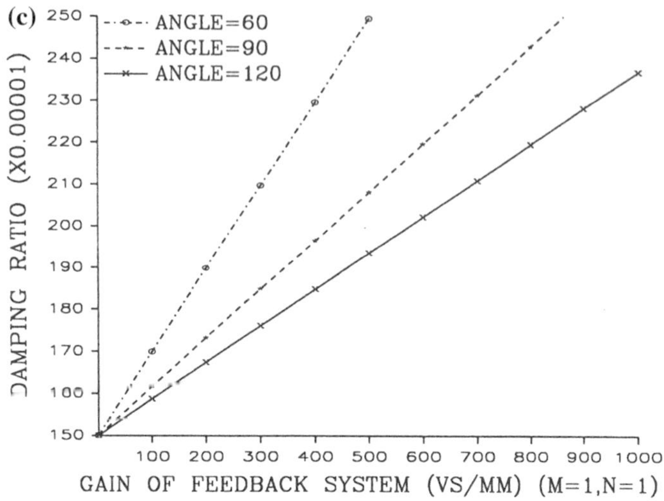

Fig. 9.24 (continued)

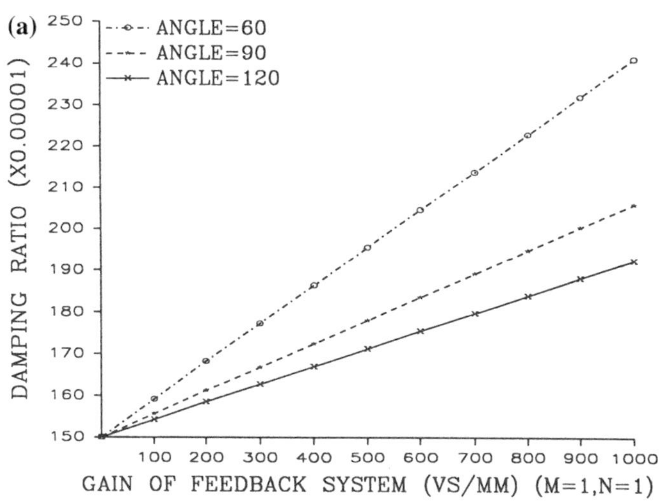

Fig. 9.25 Damping control of the fully distributed actuator, (1,1) mode & (1 mm, 40 μm). **a** Damping control (membrane), (1,1) mode & (1 mm, 40 μm). **b** Damping control (bending), (1,1) mode & (1 mm, 40 μm). **c** Total damping control (resultant)

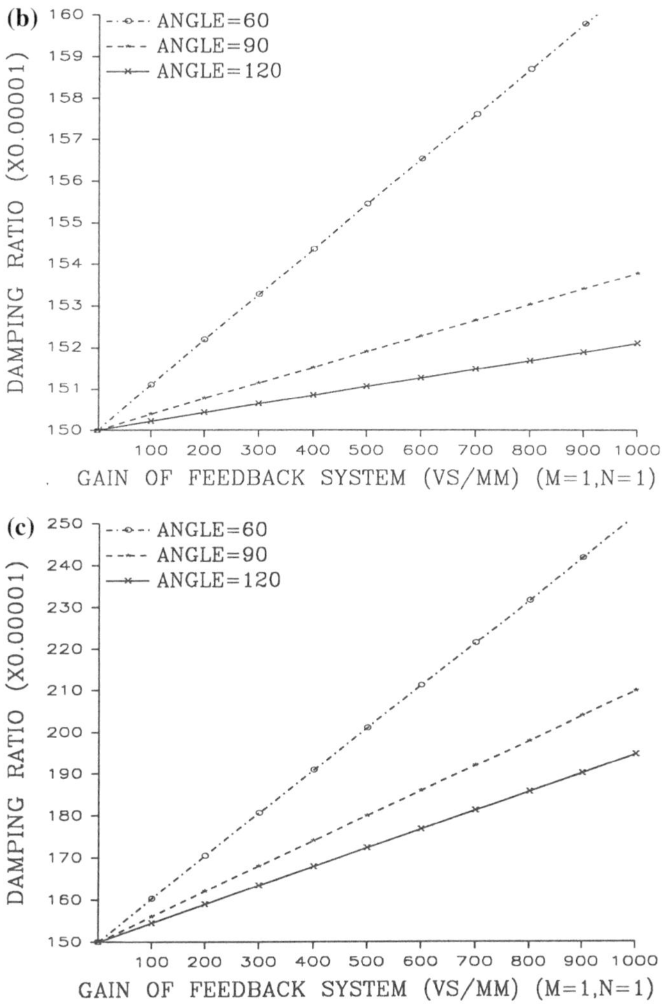

Fig. 9.25 (continued)

9.5.3 Evaluation of the Diagonal Stripe Actuator

Similar parametric studies are carried out for the distributed diagonal stripe actuators. Controlled damping ratios for the first three modes, (1,1), (2,2), and (3,3), of a 1 mm cylindrical shell with a 25 μm distributed stripe actuator are calculated versus the system feedback gains, Figs. 9.26, 9.27, and 9.28. (Note that the distributed diagonal stripe actuator is effective only to the m = n modes.) In addition, thickness

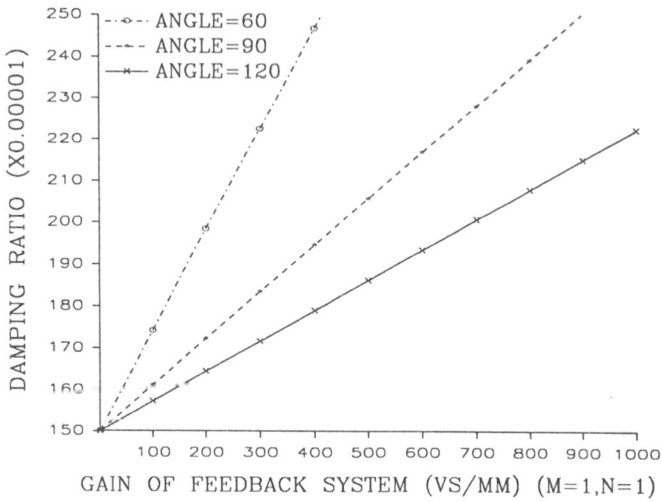

Fig. 9.26 Damping control of the distributed diagonal stripe actuator, (1,1) mode & (1 mm, 25 μm)

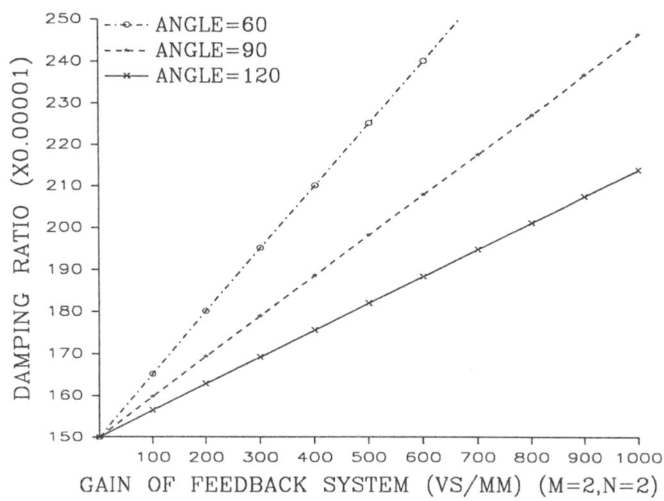

Fig. 9.27 Damping control of the distributed diagonal stripe actuator, (2,2) mode & (1 mm, 25 μm)

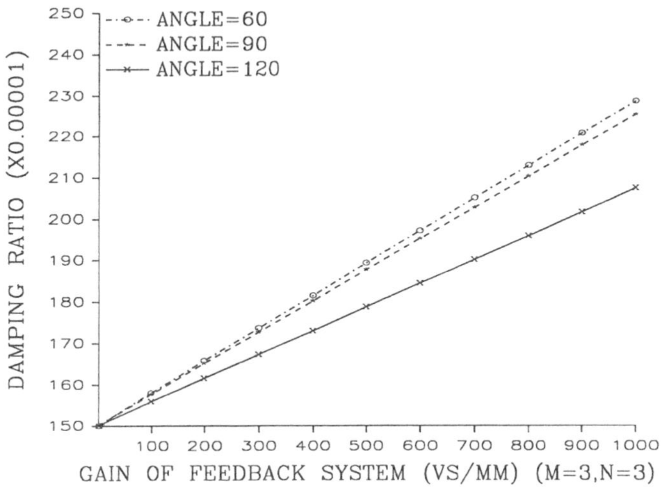

Fig. 9.28 Damping control of the distributed diagonal stripe actuator, (3,3) mode & (1 mm, 25 μm)

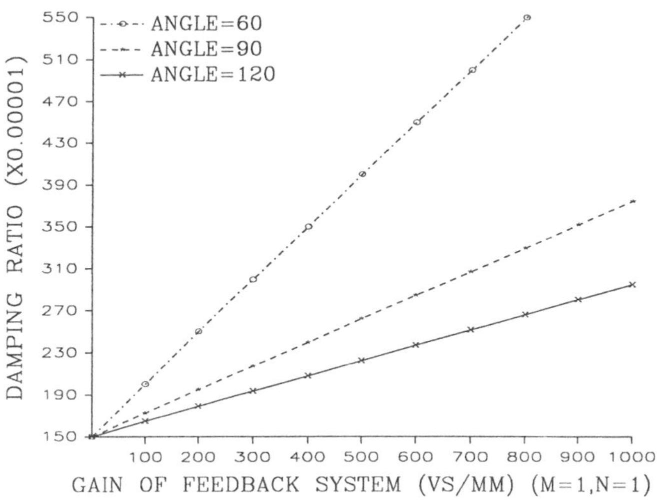

Fig. 9.29 Damping control of the distributed diagonal stripe actuator, (1,1) mode & (0.5 mm, 25 μm)

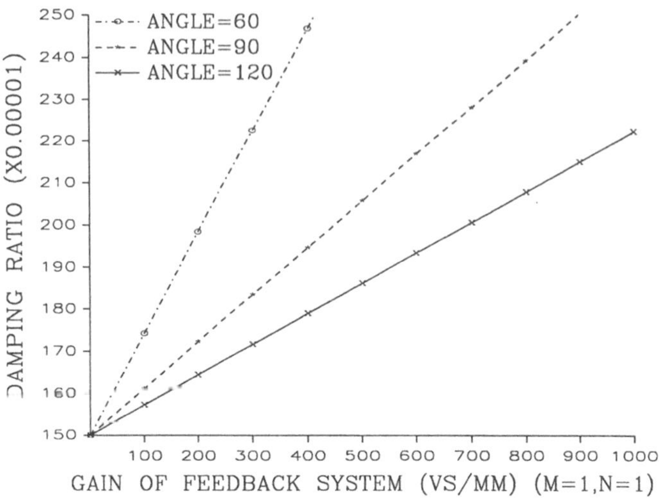

Fig. 9.30 Damping control of the distributed diagonal stripe actuator, (1,1) mode & (1 mm, 40 μm)

variations are also studied. Figure 9.29 shows the damping control of a 0.5 mm cylindrical with a 25 μm actuator and Fig. 9.30 shows the results of a 1.0 mm shell with a 40 μm actuator. Note that the electric bending control effect is neglected because of the narrowness of the stripe actuator.

Figures 9.26, 9.27 and 9.28 suggest that the controlled damping ratios decrease as the mode number increases. However, as discussed in Chap. 7, the overall convergence of a modal time history is determined by the product of the modal natural frequency and damping ratio. Still, higher mode oscillations usually converge much faster than the lower modes. Control effects are also enhanced when a thicker actuator is used or the cylindrical shell becomes thinner, i.e., more flexible. Note that the distributed stripe actuators are effective only to the m = n modes. The control effects are primarily contributed by the in-plane electric membrane control forces. The vibration control effects are more significant for larger feedback gains. The system feedback gain used in these figures is a total gain of the control system. In a real feedback control system, it includes the sensitivity of the distributed modal sensor, the amplifier gain, and the gain of derivative circuits.

9.6 Summary and Conclusions

In this chapter, distributed sensors and actuators for cylindrical shells were designed and their spatially distributed sensing/control effects were analyzed. Mathematical model and analytical solutions suggest that the fully distributed shell sensor is sensitive only to all odd modes and insensitive to all even modes. This is due to signal cancellations of positive and negative signals in opposite strain regions. The diagonal stripe sensor is sensitive only to the $m = n$ modes and insensitive to the $m \neq n$ modes. Three sensor sensitivities, i.e., transverse, in-plane longitudinal x and in-plane circumferential θ, were defined for each sensor and their normalized sensitivities evaluated. It was observed that the in-plane sensitivities are insensitive to thickness variations of elastic shells because the in-plane strains remain identical regardless of the thickness change. However, the transverse sensitivity increases as the shell becomes thicker due to an increase of bending strains.

Two distributed piezoelectric actuators, a fully distributed actuator and a diagonal stripe actuator, were designed and their actuation performances evaluated. For the fully distributed actuator, the control effects are lumped at the boundaries, as explained in Chap. 7, and can be defined by step functions. It shows that there are two electric control components: an in-plane membrane control force and a bending control moment. The study suggested that the in-plane control force contributes a majority of the control effects for lower natural modes, e.g., ≅90% for the (1,1) mode, of the cylindrical shell. However, this membrane control effect decreases as the mode number increases. The bending control effects basically remain about the same influence for all shell modes evaluated.

Studies of the distributed diagonal stripe actuator suggested that the control effect basically comes from the in-plane electric membrane control force. The bending effects were neglected because the stripe actuator was very narrow. The controlled damping ratios increase linearly with the feedback gains. These damping ratios decrease as the mode number increases. It should be noted that the overall convergence of a modal time history is determined by the combined effect of the modal natural frequency and damping ratio. Higher mode oscillations usually converge much faster than the lower modes. It was also observed that the control effects increase as the actuator becomes thicker or the shell becomes thinner due to increased control force/moment or reduced shell rigidity. Note that, again, the fully distributed actuator is effective only to all odd modes and the stripe actuator to all $m = n$ modes (Tzou et al. 1993, 1994; Tzou and Zhong 1993a, b, 1994). Additional studies of distributed sensing and control of various cylindrical shells (Hu, et al., 2013, Li, et al., 2012, 2015; DeHaven, et al., 2007; Tzou, Bao and Venkayya, 1996; Tzou and Zhong, 1996), spherical shells (Zhou and Tzou, 2000; Smithmaitrie and Tzou, 2004; Tzou, Smithmaitrie and Ding., 2002), conical shells (Tzou, Chai and Wang, 2003, Wang, et al, 2006; Chai, Tzou and Higuchi, 2006; Chai, DeHaven and Tzou, 2005), etc. have also been reported.

References

DeHaven, J.G., Han, Y. and Tzou, 2007, "Transition of Membrane/bending Neural Signals on Transforming Adaptive Shells," *Journal of Vibration and Control*, Vol.13, No.7, pp. 1007–1029.

Chai, W.K., DeHaven, J.G. and Tzou, H.S., 2005, "Distributed Micro-control Characteristics of Shallow Conical Shell Sections," *Journal of Vibration and Control*, Vol.11, No.11, pp. 1397–1411 .

Chai, W.K., Tzou, H.S. and Higuchi, K., 2006, "Micro-actuation Characteristics of Conical Shell Sections," *Journal of Sound & Vibration*. Vol.293, No1–2, pp. 286–298.

Hu, S.D., Li, H. and Tzou, H.S., 2013, "Distributed Neural Signals on Parabolic Cylindrical Shells," *Journal of Sound and Vibration* 332, 2984–3001.

Li, H.Y., Li, H. and Tzou, H.S., 2015, "Frequency Control of Beams and Cylinrical Shells with Light-Activated Shape Memory Polymers," ASME *Journal of Vibration and Acoustics*, 37, p. 011006.

Li, H., Hu, S D., Tzou, H.S., and Chen, Z.B., 2012, "Optimal Vibration Control of Conical Shells with Collocated Helical Sensor/Actuator Pairs," *Journal of Theoretical and Applied Mechanics*, 50(3), pp. 769–784.

Smithmaitrie, P. and Tzou, H.S., 2004, Micro-control actions of actuator patches laminated on hemispherical shells, *Journal of Sound and Vibration*, 277, 691–710.

Soedel, W., 1981, *Vibrations of Plates and Shells*, Dekker, N.Y.

Tzou, H.S., Bao, Y., Venkayya, V.B., 1996, "Study of Segmented Transducers Laminated on Cylindrical Shells, Part 2: Actuator Patches," *Journal of Sound & Vibration*, Vol.197, No.2, pp. 225–249, October 1996.

Tzou, H.S., Chai, W.K. and Wang, D.W., 2003, "Modal Voltages and Micro-Signal Analysis of Conical Shells of Revolution," *Journal of Sound and Vibration* 260, 589–609.

Tzou, H. S., Smithmaitrie, P. and Ding, J. H., 2002, "Sensor Electromechanics and Distributed Signal Analysis of Piezo(Electric)-Elastic Spherical Shells," *Mechanical Systems and Signal Processing (Journal of)*, Vol.16(2–3), pp. 185–199.

Tzou, H.S., Zhong, J.P. and Natori, M.C. "Sensor Mechanics of Distributed Shell Convolving Sensors Applied to Flexible Rings," ASME Transactions, *Journal of Vibration & Acoustics*, Vol.115, No.1, pp. 40–46, January 1993.

Tzou, H.S. and Zhong, J.P., "Adaptive Piezoelectric Shell Structures: Theory and Experiments," *Mechanical Systems and Signal Processing (Journal of),* Vol.7, No.4, pp.307–319, July 1993a.

Tzou, H.S. and Zhong, J.P., "Electromechanics and Vibrations of Piezoelectric Shell Distributed Systems: Theory and Applications," ASME Transactions, *Journal of Dynamic Systems, Measurement, and Control*, Vol.115, No.3, pp. 506–517, 09, 1993b.

Tzou, H.S. and Zhong, J.P., "A Linear Theory of Piezoelastic Shell Vibrations," *Journal of Sound & Vibration*, Vol.175, No.1, pp. 77–88, August 1994.

Tzou, H.S., Zhong, J.P. and Hollkamp, J.J. "Spatially Distributed Orthogonal Piezoelectric Shell Actuators: Theory and Applications," *Journal of Sound & Vibration*, Vol.177, No.3, pp. 363–378, October 1994.

Tzou and H.S. J. Zhong, J., 1996, "Spatially Filtered Vibration Control of Cylindrical Shells," *Shock and Vibration Journal*, (special issue of on Modern Vibration Engineering, ICVE'94), Vol.3, No.4, pp. 269–278, 1996.

Wang, D.W., Tzou, H.S., Arnold, S.M. and Lee, H.J., 2006, "Control of Static Shape, Dynamic Oscillation and Thermally Induced Vibration of Nozzles," ASME Transactions, *Journal of Pressure Vessel Technology*, Vol.128, pp. 357–363.

Zhong, J.P., 1991, *A Study on Piezoelectric Shell Dynamics Applied to Structural Identification and Control*, Ph.D. Thesis, Department of Mechanical Engineering, University of Kentucky, 1991.

Zhou, Y.H., and Tzou, H.S., 2000, "Active Control of Nonlinear Piezoelectric Circular Shallow Spherical Shells," *International Journal of Solids and Structures*, 37(12), pp. 1663–1677.

Chapter 10
Microscopic Actuations and Optimal Control of Parabolic Shells

With the focusing and transmitting capability of radial signals, open paraboloidal shells are commonly used as key structures and/or components for communication antennas, space mirrors and reflectors, solar collectors, etc. (Kildal, 1984; Sadowy, et al., 2001; Dragovan, 2000; Proce, et al., 2002; Fernandez-Garcia et al., 2010). Paraboloidal shells belong to the shell of revolution family, Sect. 3.3 in Chap. 3, and they usually have two unequal radii (Tzou, Ding and Hagiwara, 2002; Tzou and Ding, 2004; Yue, Deng and Tzou, 2008). Due to the weight limitation for transportation, these space structures usually are of high flexibility, low stiffness and lightly damped. Therefore, environmental excitations or variations, such as orbital maneuver, changing path, non-uniform heating, etc., can easily induce unexpected vibrations and consequently affect the precision and accuracy. Thus, active actuation and vibration control is desirable to suppress the structural vibration so as to assure its high precision and accuracy.

Based on the smart structures and structronics technology, piezoelectric materials have been used in sensor and actuator applications where high accuracy and precision are needed (Tzou and Anderson, 1992; Tzou, 1993). Over the past three decades, the piezoelectric actuation behaviors and vibration controls had expanded into shell structures, such as circular cylindrical (Qiu and Tani, 1995; Tzou, Bao, Venkayya, 1996; Washington, 1996), spherical (Zhou and Tzou, 2000; Smithmaitrie and Tzou, 2004), conical (Tzou, Wang and Chai, 2002; Li, Chen and Tzou, 2010), paraboloidal (Tzou, Ding and Hagiwara, 2002; Tzou and Ding, 2004; Yue, Deng and Tzou, 2008) and toroidal shells (Tzou, Chai and Wang, 2004), etc. Using piezoelectric materials as active actuators, static shape control of parabolic cylindrical antennas and corresponding effect on radiation pattern was investigated (Gupta, et al., 2005). However, researches of vibration control of parabolic cylindrical shells are sparse in open literatures, expect a recent study of the piezoelectric sensing behavior on parabolic cylindrical shell panels (Hu, Li and Tzou, 2013).

Thus, this chapter focuses on the spatially distributed actuation behavior of piezoelectric actuators on an open parabolic cylindrical shell panel with simply-supported boundary conditions (BCs). Control equations of the parabolic

H. (HS) Tzou, *Piezoelectric Shells*, Solid Mechanics and Its Applications 247,
https://doi.org/10.1007/978-94-024-1258-1_10

cylindrical panel laminated with a piezoelectric actuator are presented. A new transverse mode shape function used for the study of distributed sensing of parabolic cylindrical shells (Hu, Li and Tzou, 2013) is also applied to the modal actuation analysis here. The actuation effectiveness induced by an arbitrary piezoelectric patch is derived, and the spatially distribution of microscopic actuation effectiveness is also investigated when the actuator area shrunk to infinitesimal. The actuation effectiveness and its three components are evaluated in parametric studies with different actuator sizes and natural modes. Optimal actuator location(s) is also identified in case studies. Optimal control effectiveness was also evaluated (Hu, 2014).

10.1 Modal Control of Parabolic Cylindrical Panels

In this section, the governing control equation of a parabolic cylindrical panel laminated with an arbitrary piezoelectric actuator is derived. The modal control force induced by the actuator patch is formulated.

10.1.1 Governing Control Equations

A parabolic cylindrical shell panel is a single-curvature cylindrical panel whose circumferential direction is parabola-shaped with a non-constant radius. Figure 10.1a shows a parabolic cylindrical panel with an arbitrary actuator patch laminated on its outer surface. In the tri-orthogonal curvature coordinate system, x denotes the longitudinal direction, ψ denotes the meridional (or circumferential) direction with $\psi = 0$ at the ridge and α_3 denotes the transverse direction. Note that the two Lamé parameters are $A_x = 1$, $A_\psi = b/\cos^3\psi$ and two curvature radii are $R_x = \infty$, $R_\psi = b/\cos^3\psi$. The parabola parameter $b = a^2/2c = 2f$ where a is the projected span distance of the parabola; c is the meridian height; f is the focal distance, as illustrated in Fig. 10.1b. All four edges of the parabolic cylindrical panel are assumed simply-supported.

It is assumed that the piezoelectric patch is bonded perfectly on the outer surface of the elastic shell panel so as to deform consistently. When applied with an external electric voltage across the thickness of the piezoelectric patch, the actuator generates a control force onto the base structure due to the converse piezoelectric effect. Since the piezoelectric layer is so thin as compared with the shell panel, the elastic and mass properties of the piezoelectric patch do not affect the elastic shell (Tzou and Anderson, 1992; Tzou, 1993; Qiu and Tani, 1995; Tzou, Bao, Venkayya, 1996; Smithmaitrie and Tzou, 2004; Tzou, Wang and Chai, 2002; Li, Chen and Tzou, 2010).

The parabolic cylindrical shell panel usually vibrates in three directions, i.e., the transverse, longitudinal and meridional vibrations. However, the transverse

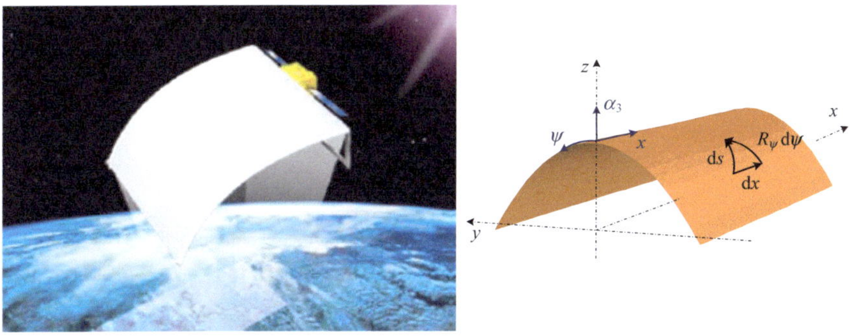

(Rahmat-Samii, et al, 2005)

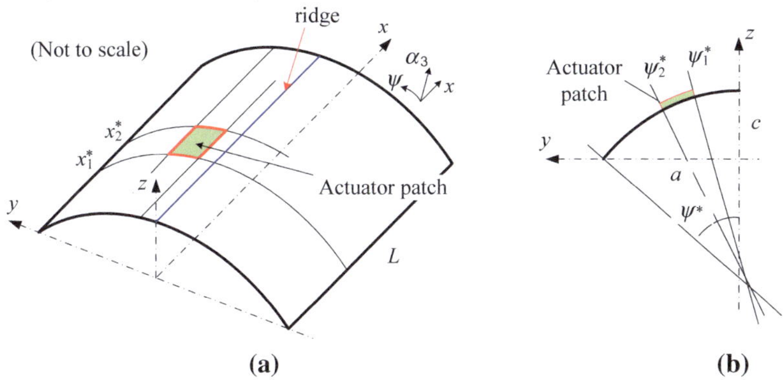

Fig. 10.1 A parabolic cylindrical shell panel with an actuator patch: **a** perspective view; **b** left-half of the front view

vibration is the dominant dynamic response of the shell in practical applications while the other two vibration components are often small and negligible (Tzou, 1993; Tzou, Ding and Hagiwara, 2002; Hu, Li and Tzou, 2013; Soedel, 2004). Thus, only the transverse vibration is considered in this study and consequently the transverse dynamic equation can be defined as (Hu, Li and Tzou, 2013)

$$
\begin{aligned}
\frac{b}{\cos^3\psi}\rho h\ddot{u}_3 = & \left(\frac{b}{\cos^3\psi}\frac{\partial^2 M_{xx}}{\partial x^2} + \frac{\partial^2 M_{\psi x}}{\partial x \partial \psi}\right) \\
& + \frac{\partial M_{x\psi}}{\partial x \partial \psi}\frac{\partial}{\partial \psi}\left(\frac{\cos^3\psi}{b}\frac{\partial M_{\psi\psi}}{\partial \psi}\right) - N_{\psi\psi} + \frac{b}{\cos^3\psi} q_3,
\end{aligned}
\tag{10.1.1}
$$

where N_{ij} is the membrane forces per unit length; M_{ij} is the bending moment per unit length; ρ is the mass density; h is the shell thickness; u_3 is the displacement in the transverse direction; q_3 is the external force applied to the transverse direction; the dot denotes the partial derivation with respect to time. Here, N_{ij} and M_{ij} are respectively the resultant forces and moments including two components related to the original elastic effect and the piezoelectric control effect. According to the piezoelectric actuation behavior on shell structures, the resultant forces/moments can be defined as (Tzou, 1993)

$$N_{xx} = N_{xx}^m - N_{xx}^a, \quad N_{\psi\psi} = N_{\psi\psi}^m - N_{\psi\psi}^a, \tag{10.1.2a, b}$$

$$M_{xx} = M_{xx}^m - M_{xx}^a, \quad M_{\psi\psi} = M_{\psi\psi}^m - M_{\psi\psi}^a, \tag{10.1.2c, d}$$

where the superscript "a" denotes the electric control effect resulting from the piezoelectric actuator; the superscript "m" denotes the original elastic (or mechanical) effect. When subject to an input control signal $\phi^a(\mathrm{x}, \psi, \mathrm{t})$, the induced control forces/moments are

$$N_{xx}^a = d_{31} Y_p \phi^a(x, \psi, t), \quad N_{\psi\psi}^a = d_{32} Y_p \phi^a(x, \psi, t), \tag{10.1.3a, b}$$

$$M_{xx}^a = r_x^a d_{31} Y_p \phi^a(x, \psi, t), \quad M_{\psi\psi}^a = r_\psi^a d_{32} Y_p \phi^a(x, \psi, t), \tag{10.1.3c, d}$$

where r_i^a ($i = x,\psi$) is the moment arm from panel's neutral surface to actuator's mid-surface; d_{3i} ($i = 1, 2$) is the piezoelectric strain constant; Y_p is Young's modulus of the piezoelectric actuator; $\phi^a(\mathrm{x}, \psi, \mathrm{t})$ is the imposed control signal. Note that the in-plane twisting control effects resulting from the piezoelectric effect are assumed zero for shells, i.e., $N_{x\psi}^a = M_{x\psi}^a = 0$, so that $M_{x\psi} = M_{x\psi}^m$ and $M_{\psi x} = M_{\psi x}^m$. Substituting Eqs. (10.1.2a)–(2d) into the transverse dynamic equation of parabolic cylindrical panels, Eq. (10.1.1), yields the governing control equation

$$\begin{aligned} &\frac{\partial}{\partial x}\left[\frac{\partial M_{xx}^m}{\partial x} + \frac{\cos^3\psi}{b}\frac{\partial M_{\psi x}^m}{\partial \psi}\right] - \frac{\cos^3\psi}{b} N_{\psi\psi}^m + \frac{\cos^3\psi}{b}\frac{\partial}{\partial \psi}\left[\frac{\partial M_{x\psi}^m}{\partial x} + \frac{\cos^3\psi}{b}\frac{\partial M_{\psi\psi}^m}{\partial \psi}\right] \\ &= \rho h \ddot{u}_3 - \left[-\frac{\partial^2 M_{xx}^a}{\partial x^2} - \frac{\cos^3\psi}{b}\frac{\partial}{\partial \psi}\left(\frac{\cos^3\psi}{b}\frac{\partial M_{\psi\psi}^a}{\partial \psi}\right) + \frac{\cos^3\psi}{b} N_{\psi\psi}^a\right] - q_3. \end{aligned} \tag{10.1.4}$$

The elastic bending moments and the membrane force can be expressed as functions of the transverse displacement. The modal control force is presented next.

10.1.2 Modal Control Force

According to the modal expansion method, the dynamic response of the parabolic cylindrical panel is composed of all participating modes (Tzou, 1993; Soedel, 2004):

$$u_3(x, \psi, t) = \sum_{m=1}^{\infty}\sum_{n=1}^{\infty} \eta_{mn}(t) U_{3\,mn}(x, \psi), \tag{10.1.5}$$

where m and n denote the vibration mode components in the x- and ψ-directions, respectively; η_{mn} is the modal participating factor; and U_{3mn} is the (m,n)-th mode shape function. Substituting mode shape functions into the simplified shell dynamic equation, using the orthogonal condition and imposing the damping effect, one can derive the independent modal control equation of the parabolic cylindrical shell panel as

$$\ddot{\eta}_{mn} + 2\zeta_{mn}\omega_{mn}\dot{\eta}_{mn} + \omega_{mn}^2\eta_{mn} = \hat{F}_{mn}^m + \hat{F}_{mn}^c, \tag{10.1.6}$$

where ζ_{mn} is the (m,n)-th damping ratio; ω_{mn} is the (m,n)-th natural frequency; $\hat{F}_{mn}^m$ and $\hat{F}_{mn}^c$ denote the modal mechanical excitation contributed by q_3 and the modal control force contributed by the electric input, respectively. Note that this study focuses on the modal control force $\hat{F}_{mn}^c$ which is defined as (Tzou, 1993)

$$\hat{F}_{mn}^c = \frac{1}{\rho h N_{mn}} \int_\psi \int_x \left[L_3^c(\phi_3) U_{3mn}\right] \frac{b}{\cos^3\psi} \mathrm{d}x\mathrm{d}\psi, \tag{10.1.7}$$

where $N_{mn} = \int_\psi \int_x U_{3mn}^2 b/\cos^3\psi \, \mathrm{d}x\mathrm{d}\psi$ and Love's control operator is a function of control signals ϕ^a:

$$L_3^c(\phi) = -\frac{\partial^2 M_{xx}^a}{\partial x^2} - \frac{\cos^3\psi}{b}\frac{\partial}{\partial\psi}\left(\frac{\cos^3\psi}{b}\frac{\partial M_{\psi\psi}^a}{\partial\psi}\right) + \frac{\cos^3\psi}{b} N_{\psi\psi}^a. \tag{10.1.8}$$

With the expression of mode shape function U_{3mn}, the modal control force induced by the actuator can be investigated.

10.2 Distributed Actuation Effectiveness

In this section, the expression of modal control force generated by an actuator patch is obtained and the microscopic actuation effectiveness induced by an infinitesimal actuator element is also derived in order to precisely reveal the spatially distributed microscopic actuation and control behavior on parabolic cylindrical shells.

10.2.1 Transverse Mode Shape Function

Due to the complicated analytical procedure and the dynamic equation of the parabolic cylindrical shell panel, analytical mode shape functions are usually difficult to obtain. With experienced "trial-and-errors", a *new* mode shape function satisfying the specified BCs is formulated (Hu, Li and Tzou, 2013):

$$U_{3mm} = C_{mm}\sqrt{\frac{\cos^3\psi^*}{\cos^3\psi}}\sin\left(\frac{m\pi x}{L}\right)\sin\left[\frac{n\pi(\psi+\psi^*)}{2\psi^*}\right] \tag{10.2.1}$$

where L is shell's longitudinal length; ψ is the meridional angle that varies from $-\psi^*$ to ψ^*; $2\psi^*$ is the total span angle or curvature angle (as shown in Fig. 10.1); and C_{mn} is an arbitrary modal amplitude. Equation (10.2.1) satisfies the simply-supported boundary condition that the transverse displacements and the bending moments are zero at all four boundaries. As panel's curvature varies along the meridional direction of the parabolic cylindrical panel, the amplitude distribution of the mode shape is also non-uniform. Furthermore, based on Eq. (10.2.1), when considering the mode shape distribution in the ψ-direction, the amplitude is larger near the edges rather than the ridge, because of the smaller denominator $\cos^3\psi$ towards two edges. This behavior will lead to distinct actuation effectiveness and behaviors discussed next.

10.2.2 Actuation Effectiveness of an Arbitrary Actuator Patch

As shown in Fig. 10.1, an arbitrary segmented actuator patch covers the shell's surface from x_1^* to x_2^* in the longitudinal direction and from ψ_1^* to ψ_2^* in the meridional direction. It is assumed that the actuator has a uniform thickness h^a. If the electrode resistance is neglected, the control voltage on the segmented patch is constant. Then, the effective control voltage actually applied to the structure can be defined as

$$\phi^a(x,\psi,t) = \phi^a(t)\left[u_s(x-x_1^*) - u_s(x-x_2^*)\right] \times \left[u_s(\psi-\psi_1^*) - u_s(\psi-\psi_2^*)\right], \tag{10.2.2}$$

where $u_s(\cdot)$ is the unit step function: $u_s(x{-}x_i^*) = 1$ when $x > x_i^*$, and $u_s(x{-}x_i^*) = 0$ when $x < x_i^*$. $\phi^a(\mathrm{t})$ is the imposed control voltage either in open-loop or closed-loop controls. Substituting the control forces and moments, Eqs. (10.1.3a)–(3d), into Eq. (10.1.7), one obtains the modal control force induced by the actuator patch:

$$\widehat{F}^c_{mn} = \frac{Y_p C_{mn} \phi^a(t)}{\rho h N_{mn}} \int_0^L \int_{-\psi^*}^{\psi^*} \left\{ d_{32} \left[u_s(x - x_1^*) - u_s(x - x_2^*) \right] \left[u_s(\psi - \psi_1^*) - u_s(\psi - \psi_2^*) \right] \right.$$
$$- r_x^a d_{31} \frac{b}{\cos^3 \psi} \frac{\partial}{\partial x} \left[\delta(x - x_1^*) - \delta(x - x_2^*) \right] \left[u_s(\psi - \psi_1^*) - u_s(\psi - \psi_2^*) \right]$$
$$- r_\psi^a d_{32} \frac{\cos^3 \psi}{b} \left[u_s(x - x_1^*) - u_s(x - x_2^*) \right] \frac{\partial}{\partial \psi} \left[\delta(\psi - \psi_1^*) - \delta(\psi - \psi_2^*) \right]$$
$$\left. + r_\psi^a d_{32} \frac{3 \cos^2 \psi \sin \psi}{b} \left[u_s(x - x_1^*) - u_s(x - x_2^*) \right] \left[\delta(\psi - \psi_1^*) - \delta(\psi - \psi_2^*) \right] \right\}$$
$$\times \sqrt{\frac{\cos^3 \psi^*}{\cos^3 \psi}} \sin\left(\frac{m\pi x}{L}\right) \sin\left[\frac{n\pi(\psi + \psi^*)}{2\psi^*}\right] \mathrm{d}\psi \mathrm{d}x, \tag{10.2.3}$$

where $\delta(\cdot)$ is the Dirac delta function and

$$\delta(x - x_i^*) = \infty \ \text{ when } x = x_i^*, \quad \text{and } \delta(x - x_i^*) = 0 \ \text{ when } x \neq x_i^* \tag{10.2.4}$$

Accordingly, the total modal control force contains three contributing components resulting from its control forces/moments:

$$\widehat{F}^c_{mn} = \frac{Y_p C_{mn} \phi^a(t) d_{3i}}{\rho} (\widehat{T}_{mn})_{\text{total}}$$
$$= \frac{Y_p C_{mn} \phi^a(t) d_{3i}}{\rho} \left[(\widehat{T}_{mn})_{\psi,\text{mem}} + (\widehat{T}_{mn})_{x,\text{bend}} + (\widehat{T}_{mn})_{\psi,\text{bend}} \right], \tag{10.2.5}$$

where $(\widehat{T}_{mn})_{\text{total}}$ denotes the total modal control force; $(\widehat{T}_{mn})_{\psi,\text{mem}}$, $(\widehat{T}_{mn})_{x,\text{bend}}$ and $(\widehat{T}_{mn})_{\psi,\text{bend}}$ denote the three force components induced by the membrane force in the meridional direction, the bending moment in the longitudinal direction and the bending moment in the meridional direction, respectively. Note that the contributing components $(\widehat{T}_{mn})_{\text{total}}$, $(\widehat{T}_{mn})_{\psi,\text{mem}}$, $(\widehat{T}_{mn})_{x,\text{bend}}$ and $(\widehat{T}_{mn})_{\psi,\text{bend}}$ only reflect actuator's spatial-dependent behaviors depending on actuator's placements. The true magnitude of the modal control force can be calculated once materials properties and control parameters are specified in Eq. (10.2.5). Accordingly, these three components can be respectively expressed as

$$(\widehat{T}_{mn})_{\psi,\text{mem}} = \frac{1}{h N_{mn}} \int_0^L \int_{-\psi^*}^{\psi^*} \left\{ \sqrt{\frac{\cos^3 \psi^*}{\cos^3 \psi}} \sin\left[\frac{n\pi(\psi + \psi^*)}{2\psi^*}\right] \right.$$
$$\left. \times \sin\left(\frac{m\pi x}{L}\right) \left[u_s(x - x_1^*) - u_s(x - x_2^*) \right] \times \left[u_s(\psi - \psi_1^*) - u_s(\psi - \psi_2^*) \right] \right\} \mathrm{d}\psi \mathrm{d}x, \tag{10.2.6}$$

$$(\widehat{T}_{mn})_{x,\mathrm{bend}} = \frac{r^a}{hN_{mn}} \int_0^L \int_{-\psi^*}^{\psi^*} \frac{b}{\cos^* \psi} \left\{ \sqrt{\frac{\cos^3 \psi^*}{\cos^3 \psi}} \sin\left[\frac{n\pi(\psi + \psi^*)}{2\psi^*}\right] \right.$$
$$\left. \times \sin\left(\frac{m\pi x}{L}\right) \frac{\partial}{\partial x} \left[\delta(x - x_1^*) - \delta(x - x_2^*)\right] \times \left[u_s(\psi - \psi_1^*) - u_s(\psi - \psi_2^*)\right] \right\} \mathrm{d}\psi \mathrm{d}x, \tag{10.2.7}$$

$$(\widehat{T}_{mn})_{\psi,\mathrm{bend}} = \frac{r^a}{hN_{mn}} \int_0^L \int_{-\psi^*}^{\psi^*} \left\{ \sqrt{\frac{\cos^3 \psi^*}{\cos^3 \psi}} \sin\left[\frac{n\pi(\psi + \psi^*)}{2\psi^*}\right] \right.$$
$$\times \sin\left(\frac{m\pi x}{L}\right) \left\{ \frac{3\cos^2 \psi \sin \psi}{b} \left[u_s(x - x_1^*) - u_s(x - x_2^*)\right] \left[\delta(\psi - \psi_1^*) - \delta(\psi - \psi_2^*)\right] \right\}$$
$$\left. - \frac{\cos^3 \psi}{b} \left[u_s(x - x_1^*) - u_s(x - x_2^*)\right] \times \frac{\partial}{\partial \psi} \left[\delta(\psi - \psi_1^*) - \delta(\psi - \psi_2^*)\right] \mathrm{d}\psi \mathrm{d}x. \tag{10.2.8}$$

Note that if the actuator location changes along the ψ-direction with the same spatial interval (i.e., $\Delta\psi = \psi_2^* - \psi_1^*$ is a constant), the actuator area also correspondingly varies with different positions due to the curvature effect. Apparently, the actual actuator size is the smallest at the ridge and gradually increases along the meridional direction toward the boundary when $\Delta\psi$ is constant. Therefore, the modal control force induced by the piezoelectric actuator needs to be normalized so that the actual effective control capability per actuator area can be demonstrated. The normalized modal control force is obtained by dividing the modal control force by the corresponding actuator area, i.e.,

$$(\widehat{F}^c_{mn})_{\mathrm{eff}} = \frac{\widehat{F}^c_{mn}}{S^a}, \tag{10.2.9}$$

where the actuator area is defined as

$$S^a = \int_{x_1^*}^{x_2^*} \int_{-\psi^*}^{\psi^*} \left[u_s(x - x_1^*) - u_s(x - x_2^*)\right] \times \left[u_s(\psi - \psi_1^*) - u_s(\psi - \psi_2^*)\right] \frac{b}{\cos^3 \psi} \mathrm{d}\psi \mathrm{d}x \tag{10.2.10}$$

Obviously, the effective actuation effect of modal control force per actuator area is more objective in practical applications. However, it is still an average result. By shrinking the actuator area to infinitesimal, one can obtain the limit of Eq. (10.2.9) as S^a approaches zero. Thus, the microscopic point modal control force $\widetilde{F}^c_{mn}(x, \psi)$ at an arbitrary position becomes

$$\widetilde{F}^c_{mn} = \frac{Y_p C_{mn} \phi^a(t)}{\rho h N_{mn}} \times \left\{ r^a_x d_{31} \left(\frac{m\pi}{L}\right)^2 \sin\left(\frac{m\pi x}{L}\right) \sqrt{\frac{\cos^3\psi^*}{\cos^3\psi}} \sin\left(\frac{n\pi(\psi+\psi^*)}{2\psi^*}\right) \right.$$
$$- r^a_\psi d_{32} \frac{\cos^3\psi}{b} \sin\left(\frac{m\pi x}{L}\right) \frac{d^2}{d\psi^2} \left[\frac{\cos^3\psi}{b} \sqrt{\frac{\cos^3\psi^*}{\cos^3\psi}} \sin\left(\frac{n\pi(\psi+\psi^*)}{2\psi^*}\right)\right]$$
$$- r^a_\psi d_{32} \sin\left(\frac{m\pi x}{L}\right) \frac{\cos^3\psi}{b} \frac{d}{d\psi} \left[3\cos^2\psi \sin\psi \sqrt{\frac{\cos^3\psi^*}{\cos^3\psi}} \sin\left(\frac{n\pi(\psi+\psi^*)}{2\psi^*}\right)\right]$$
$$\left. + d_{32} \frac{\cos^3\psi}{b} \sqrt{\frac{\cos^3\psi^*}{\cos^3\psi}} \sin\left(\frac{m\pi x}{L}\right) \sin\left(\frac{n\pi(\psi+\psi^*)}{2\psi^*}\right) \right\}. \tag{10.2.11}$$

Although impractical in engineering applications, the microscopic control force in Eq. (10.2.11) precisely predicts the spatial microscopic distribution of actuation effectiveness when the average effect is eliminated. Based on the spatial distribution of microscopic actuation effectiveness, the optimal locations with the maximal control effects can be clearly revealed. Similarly, the microscopic point actuation force can also be further divided into three contributing components:

$$\begin{aligned} \widetilde{F}^c_{mn} &= \frac{Y_p C_{mn} \phi^a(t) d_{3i}}{\rho} (\widetilde{T}_{mn})_{\text{total}} \\ &= \frac{Y_p C_{mn} \phi^a(t) d_{3i}}{\rho} \left[(\widetilde{T}_{mn})_{x,\text{bend}} + (\widetilde{T}_{mn})_{\psi,\text{bend}} + (\widetilde{T}_{mn})_{\psi,\text{mem}}\right], \end{aligned} \tag{10.2.12}$$

where $(\tilde{T}_{mn})_{\text{total}}$ denotes the total microscopic actuation effectiveness; $(\tilde{T}_{mn})_{x,\text{bend}}$, $(\tilde{T}_{mn})_{\psi,\text{bend}}$ and $(\tilde{T}_{mn})_{\psi,\text{mem}}$ denote the microscopic actuation effectiveness components induced by the bending moment in the x-direction, the bending moment in the ψ-direction and the membrane force in the ψ-direction, respectively. Apparently, these actuation components can be respectively expressed as

$$(\tilde{T}_{mn})_{x,\text{bend}} = \frac{r^a_x}{h N_m} \sqrt{\frac{\cos^3\psi^*}{\cos^3\psi}} \sin\left(\frac{n\pi(\psi+\psi^*)}{2\psi^*}\right) \left(\frac{m\pi}{L}\right)^2 \sin\left(\frac{m\pi x}{L}\right), \tag{10.2.13}$$

$$(\tilde{T}_{mn})_{\psi,\text{bend}} = \frac{r^a_\psi}{h N_m} \frac{\cos^3\psi}{b} \left\{ \frac{d^2}{d\psi^2} \left[\frac{\cos^3\psi}{b} \sqrt{\frac{\cos^3\psi^*}{\cos^3\psi}} \sin\left(\frac{n\pi(\psi+\psi^*)}{2\psi^*}\right)\right] \right.$$
$$\left. + \frac{d}{d\psi} \left[\frac{3\cos^2\psi \sin\psi}{b} \sqrt{\frac{\cos^3\psi^*}{\cos^3\psi}} \sin\left(\frac{n\pi(\psi+\psi^*)}{2\psi^*}\right)\right] \right\} \sin\left(\frac{m\pi x}{L}\right), \tag{10.2.14}$$

$$(\tilde{T}_{mn})_{\psi,\mathrm{mem}} = \frac{1}{hN_{mn}}\frac{\cos^3\psi}{b}\sqrt{\frac{\cos^3\psi^*}{\cos^3\psi}}\sin\left[\frac{n\pi(\psi+\psi^*)}{\psi^*}\right]\sin\left(\frac{m\pi x}{L}\right). \quad (10.2.15)$$

These individual contributing components of microscopic actuation effectiveness are evaluated in case studies.

10.3 System State Space Equations and Optimal Control

In order to evaluate the optimal vibration control of the parabolic cylindrical shell structronic system (Ray, 2003; Tzou and Ding, 2004; Chen and Shen, 1997; Li et al., 2012), dynamic control equations should be transformed into the state space formulation, so that all state variables can be effectively controlled via the modern control theory. The dynamic equation Eq. (10.1.6) can be rewritten into the state space form as

$$\dot{\mathbf{x}} = \mathbf{A}\mathbf{x} + \mathbf{B}\mathbf{u}, \quad (10.3.1a)$$

$$\mathbf{y} = \mathbf{C}\mathbf{x}, \quad (10.3.1b)$$

where $\mathbf{x}$ is the state-variable vector; $\mathbf{u}$ is the control input vector; $\mathbf{y}$ is the output vector; and $\mathbf{A}$, $\mathbf{B}$ and $\mathbf{C}$ are known coefficient matrices related to the system state, control input and system output, respectively. For the optimal control of parabolic cylindrical shell panels, the modal participating factor η_{mn} (t) and the modal velocity $\dot{\eta}_{mn}(t)$ are the system state variables, i.e., $\mathbf{x} = \{\eta_{mn}(t)\ \dot{\eta}_{mn}(t)\}^t$, where the superscript "t" denotes the transpose of a matrix or vector.

Since the modal control force $\hat{F}^c_{mn}$ is the only control input in this study, the control input vector is simplified to be a single voltage, i.e., $\mathbf{u} = \phi^a$. The related coefficient matrix is degraded to be a column vector $\mathbf{B} = \{0\ B_2\}^t$. The sensing signal generated by the piezoelectric patch is chosen as the system output. Simultaneously, the sensing signal ϕ^s is only dependent on the modal participating factor, and the system output matrix is also degraded into a row rector as $\mathbf{C} = \{C_1\ 0\}$. Actually, it is obvious that the element B_2 equals to the *modal actuation factor* of the piezoelectric actuator and C_1 the *modal sensitivity* of the piezoelectric sensor (Hu, 2014). Hence, the state space formulation of parabolic cylindrical shell system is written as

$$\begin{Bmatrix} \dot{\eta}_{mn}(t) \\ \ddot{\eta}_{mn}(t) \end{Bmatrix} = \begin{bmatrix} 0 & 1 \\ -\omega^2_{mn} & -2\zeta_{mn}\omega_{mn} \end{bmatrix} \begin{Bmatrix} \eta_{mn}(t) \\ \dot{\eta}_{mn}(t) \end{Bmatrix} + \begin{Bmatrix} 0 \\ A^a_{mn} \end{Bmatrix} \phi^a_{mn} + \begin{Bmatrix} 0 \\ 1 \end{Bmatrix} \hat{F}^m_{mn}, \quad (10.3.2a)$$

$$\phi_{mn}^{s} = \{ S_{mn} \quad 0 \} \left\{ \begin{array}{c} \eta_{mn}(t) \\ \dot{\eta}_{mn}(t) \end{array} \right\}. \tag{10.3.2b}$$

The relationship between the feedback control input **u** and system state **x** is determined using the LQ optimal control algorithm. Note that the state equation is usually a set of first-order equations. For better related to the original dynamics, the acceleration term is preserved in Eq. (10.3.2a).

10.3.1 LQ Optimal Control Procedure

In this section, the LQ optimal control procedure is presented to determine the optimal relationship between the control input and the system state. The criterion function of the LQ optimal control is defined as the integration of the quadratic form of the state **x** and a second quadratic form of the control **u** (Kwakernaak and Sivan,1972):

$$J = \int_{t_0}^{\infty} [\mathbf{x}^t(t)\mathbf{Q}(t)\mathbf{x}(t) + \mathbf{u}^t(t)\mathbf{R}\mathbf{u}(t)]\mathrm{d}t, \tag{10.3.3}$$

where **Q** is the positive semi-definite state weighting matrix; **R** is the positive-definite control weighting matrix; and t_0 is the initial time. The optimal relationship is searched to minimize this criterion function. In doing so, a better control performance can be achieved with less input control energy. In this study, the control weighting matrix **R** is a single parameter R and the state weight matrix **Q** is defined as a diagonal matrix

$$\mathbf{Q} = \begin{bmatrix} 1 & 0 \\ 0 & Q_{22} \end{bmatrix}, \tag{10.3.4}$$

where Q_{22} is a state weight coefficient denoting the weight of modal velocity relative to the modal participation factor. For the LQ optimal control, it is assumed that the feedback control input is linearly proportional to the system state

$$\mathbf{u} = -\mathbf{G}\mathbf{x}. \tag{10.3.5}$$

The optimal control gain **G** can be further defined with the weighting matrix **R**, system input vector **B** and a specific matrix **M**

$$\mathbf{G} = \mathbf{R}^{-1}\mathbf{B}^t\mathbf{M}, \tag{10.3.6}$$

where the superscript "−1" denotes the inverse of a matrix. The 2 × 2 symmetric matrix **M** is assumed that

$$\mathbf{M} = \begin{bmatrix} m_{11} & m_{12} \\ m_{21} & m_{22} \end{bmatrix}. \tag{10.3.7}$$

10.3.2 LQ Optimal Feedback Control Gain

The optimal feedback control gain can be obtained as

$$\mathbf{G} = \left[m_{21}B_2/R \quad m_{22}B_2/R \right]. \tag{10.3.8}$$

Based on the Riccati Equation in the modern control theory, the matrix elements m_{21} and m_{22} are finally solved:

$$m_{21} = \frac{\left(-\omega_{mn}^2 + \sqrt{\omega_{mn}^4 + B_2^2/R}\right)R}{B_2^2}, \tag{10.3.9a}$$

$$m_{22} = \frac{\left[-2\zeta_{mn}\omega_{mn} + \sqrt{\omega_{mn}^4 + (2m_{21} + c^2)B_2^2/R}\right]R}{B_2^2}. \tag{10.3.9b}$$

Until now, the feedback control gain based on the LQ optimal control algorithm is obtained.

Physically, the optimal control gain can be expressed in another form. As mentioned before, the sensing signal from the sensor is used as the feedback control signal after the treatment of the controller. Assume that $\mathbf{G}^*$ is the control gain matrix between the control signal $\phi_{mn}^a(t)$ and the sensing signal vector $\left\{\phi_{mn}^s(t), \dot{\phi}_{mn}^s(t)\right\}^t$. Thus, one can define the control signal (control input) as

$$\mathbf{u} = \phi_{mn}^a(t) = -\mathbf{G}^* \begin{Bmatrix} \phi_{mn}^s(t) \\ \dot{\phi}_{mn}^s(t) \end{Bmatrix} = -S_{mn}\mathbf{G}^*\mathbf{x}. \tag{10.3.10}$$

Based on the relationship between Eqs. (10.3.5) and (10.3.10), the control gain matrix $\mathbf{G}^*$ can be determined as

$$\mathbf{G}^* = \frac{\mathbf{G}}{S_{mn}}. \tag{10.3.11}$$

The solved $\mathbf{G}^*$ can be set in the controller so as to achieve the required optimal control gain $\mathbf{G}$. The LQ optimal vibration control results of a parabolic cylindrical panel using piezoelectric sensor and actuator are demonstrated in case studies (Hu, 2014).

10.4 Actuations Effectiveness

The modal control forces induced by an arbitrary actuator patch and an infinitesimal actuator element laminated on a parabolic cylindrical shell have been formulated previously. In this section, the total microscopic actuation effectiveness and the three contributing components, i.e., $(\tilde{T}_{mn})_{\text{total}}$, $(\tilde{T}_{mn})_{x,\text{bend}}$, $(\tilde{T}_{mn})_{\psi,\text{bend}}$ and $(\tilde{T}_{mn})_{\psi,\text{mem}}$ are evaluated and their spatial distributions are plotted in detail. The dominating behavior of actuation component(s) and the distinct curvature effect are discussed. For practical applications, the modal control forces and the contribution components of actuator patches, i.e., $(\widehat{T}_{mn})_{\text{total}}$, $(\widehat{T}_{mn})_{\psi,\text{mem}}$,$(\widehat{T}_{mn})_{x,\text{bend}}$ and $(\widehat{T}_{mn})_{\psi,\text{bend}}$ are evaluated, and different sizes of actuator patches are considered.

Being thin-film, lightweight, easily segmented/shaped and dynamically sensitive, flexible polyvinylidene fluoride (PVDF) polymeric materials can be conveniently applied to various in-situ distributed actuation applications of flexible structures. Thus, PVDF material is selected as the piezoelectric actuator in case studies. Note that, however, the previous actuation theory and control forces are applicable to all other piezoelectric actuators. Without loss of generality, it is further assumed that 1) both piezoelectric strain constants are identical, i.e., $d_{31} = d_{32}$, 2) the thickness of PVDF patch is a constant and the moment arm $r_x^a = r_\psi^a = r^a = (h + h^a)/2$.

To demonstrate the curvature effect of shell panels, two distinct curvatures of panels are selected. The ratio of the meridian height to the span radial distance of a parabola is 1:4 (shallow panel) and 1:1 (deep panel). Correspondingly, the half span angle ψ^* for each shape is 0.464 rad and 1.107 rad, respectively. However, these cases have the same meridional arc length L_ψ. Also, the meridional length and the longitudinal length have the same dimension (i.e., the expanded total areas of both cases are identical) in order to reveal the domination of force components. Material properties of the elastic parabolic cylindrical panel and PVDF actuator are listed in Table 10.1, where Y_p is Young's modulus of PVDF; ρ is the mass density of elastic panel; h is panel's thickness; h^a is the thickness of PVDF actuator; L and L_ψ denote the longitudinal length and the meridional arc length, respectively; d_{31} and d_{32} are the piezoelectric strain constants. Table 10.2 summarizes detailed geometry parameters of two panels with different curvatures.

Table 10.1 Properties of parabolic cylindrical panel and PVDF

Parameters	Value	Units
Y_p	1.6×10^9	N/m^2
ρ	7.8×10^3	kg/m^3
h	1×10^{-3}	m
h^a	50×10^{-6}	m
L, L_ψ	0.1	m
d_{31}, d_{32}	1.6×10^{-12}	C/N

10.4.1 Spatial Microscopic Actuations

Spatial modal actuation characteristics of two, i.e., a shallow and a deep, parabolic cylindrical shells are investigated in this section. In each case, the control force component $(\tilde{T}_{mn})_{\psi,\mathrm{mem}}$ resulting from the meridional membrane force, the force component $(\tilde{T}_{mn})_{x,\mathrm{bend}}$ resulting from the longitudinal bending moment, the force component $(\tilde{T}_{mn})_{\psi,\mathrm{bend}}$ resulting from the meridional bending moment and the total modal control force distribution $(\tilde{T}_{mn})_{\mathrm{total}}$ are evaluated.

10.4.1.1 Case-1: $a = 4c = 0.048$ m—Shallow

As noted in Table 10.2, structure parameters of the shallow parabolic cylindrical panel investigated in this case are: the meridian height $c = 0.012$ m, the span distance $a = 0.048$ m and the half span angle $\psi^* = 0.464$ rad. The modal control forces and their components of lower modes (1,1)–(2,2) are plotted in Figs. 10.2, 10.3, 10.4 and 10.5. In each figure, the top-left plot denotes the control force component $(\tilde{T}_{mn})_{\psi,\mathrm{mem}}$ resulting from the meridional membrane force; the top-right denotes the force component $(\tilde{T}_{mn})_{x,\mathrm{bend}}$ resulting from the longitudinal bending moment; the bottom-left denotes the force component $(\tilde{T}_{mn})_{\psi,\mathrm{bend}}$ resulting from the meridional bending moment; and the bottom-right denotes the total modal control force distribution $(\tilde{T}_{mn})_{\mathrm{total}}$, i.e., the summation of all contributing components. Additionally, these components are plotted with respected to the longitudinal (x) coordinate and the meridional (ψ) coordinate. The spatially distributed microscopic actuation behaviors of modal control forces induced by distributed microscopic actuator elements can be clearly observed in these figures.

Generally, Figs. 10.2, 10.3, 10.4 and 10.5 show that the modal control force component $(\tilde{T}_{mn})_{\psi,\mathrm{mem}}$ induced by the meridional membrane force dominates the total control effects and the components induced by the bending (both longitudinal and meridional) control moments are relatively insignificant for lower modes. However, with the increasing of mode numbers, the bending-contributed components (both the meridional and longitudinal) increase, while the membrane-contributed component decreases. Furthermore, when the mode number $m > n$, the longitudinal bending component dominantly contribute to the bending component; while when the mode number $m < n$, the meridional bending component

Table 10.2 Geometries of parabolic cylindrical panels

Case	1, shallow	2, deep
Span distance a (m)	0.0480	0.0338
Meridian height c (m)	0.0120	0.0338
Half span angle ψ^* (rad)	0.464	1.107

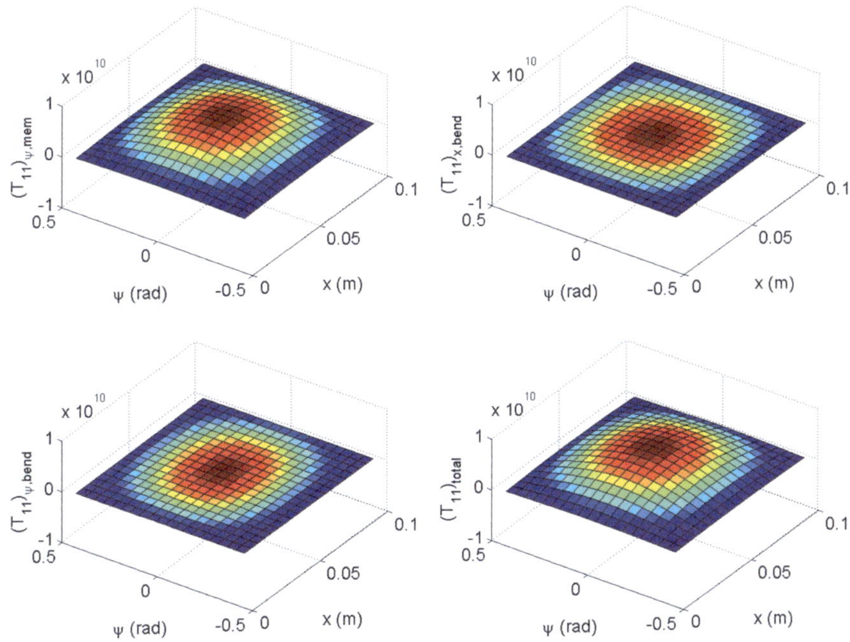

Fig. 10.2 Spatial distributions of microscopic actuation effectiveness of shallow panels, mode (1,1)

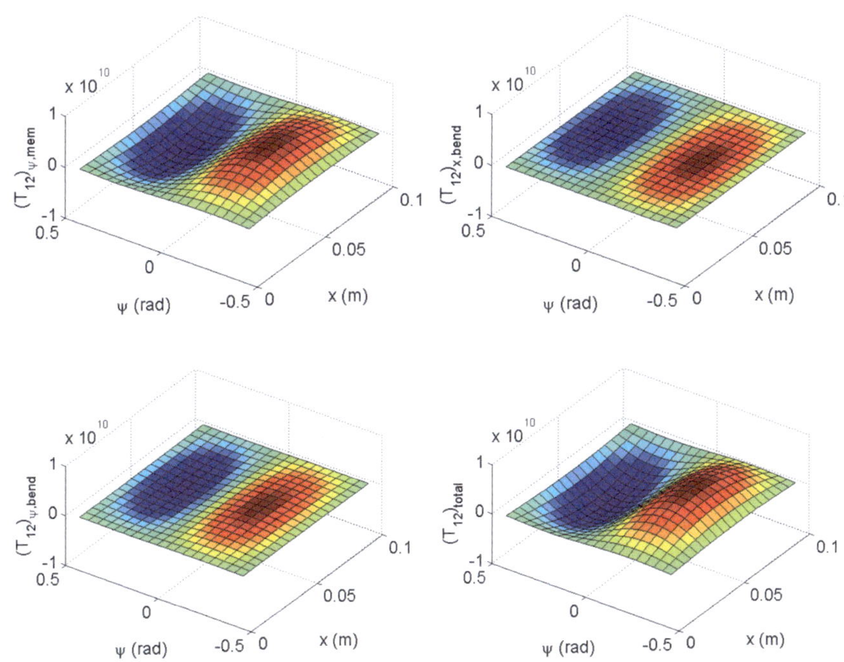

Fig. 10.3 Spatial distributions of microscopic actuation effectiveness of shallow panels, mode (1,2)

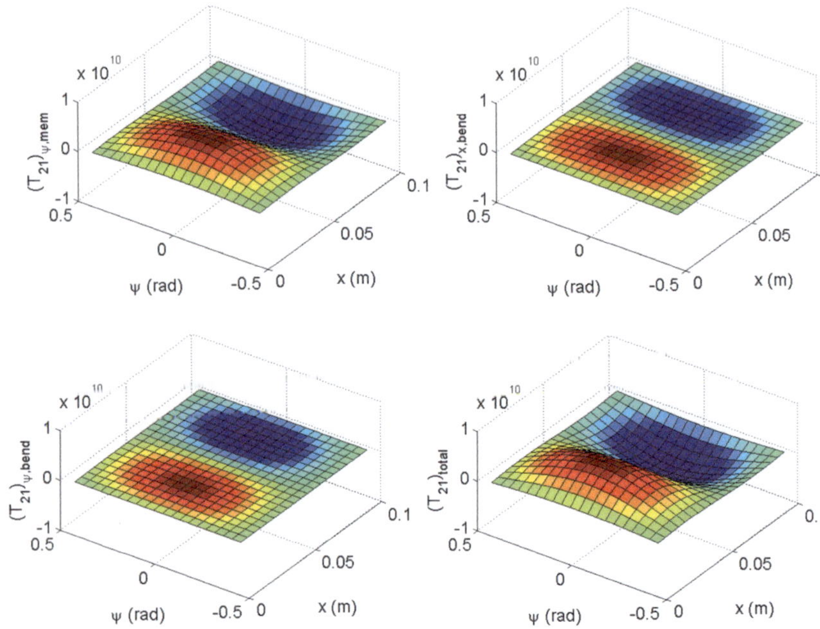

Fig. 10.4 Spatial distributions of microscopic actuation effectiveness of shallow panels, (2,1)

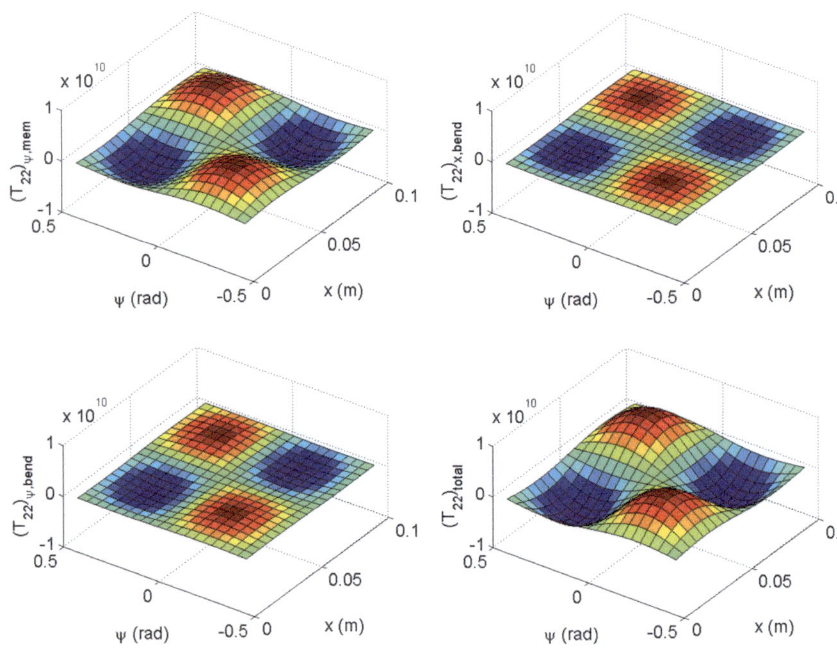

Fig. 10.5 Spatial distributions of microscopic actuation effectiveness of shallow panels, (2,2)

dominates. Note that there are positive and negative control forces induced by actuators. The positive forces induce positive control effects while the negative forces amplify the vibrations. However, since the sign of control signal can be manipulated through the control circuit, one only needs to concern the absolute magnitudes to infer the absolute control effectiveness. The maximal *peak* means that the largest control force can be generated with the same input control signal. Thus, it is the optimal location where the actuator should be placed to maximize the actuation and control effect.

In this study, only the optimal actuation location in the meridional direction is discussed, since the distribution behavior in the longitudinal direction can be inferred from the results of circular cylindrical shell panels (Tzou, 1993; Tzou, Bao and Venkayya, 1996). For the (m,1) mode group, as shown in Figs. 10.2 and 10.4, the ridge of the panel is the optimal region. For the (m,2) mode group, there are two identical optimal regions, Figs. 10.3 and 10.5. When the parabolic cylindrical panel is relatively shallow, the curvature effect (or the non-uniform curvature distribution) along the meridional direction is insignificant, and thus the distributed control effectiveness is similar to that of circular cylindrical shells (Tzou, 1993; Tzou, Bao and Venkayya, 1996). In order to reveal the curvature effect, a deeper shell panel (a:c = 1:1) is considered and its spatial actuation distributions are presented next.

10.4.1.2 Case-2: $a = c = 0.0338$ m—Deep

In this case, the meridian height c and the span distance a are both 0.0338 m, and the corresponding half span angle is $\psi^* = 1.107$ rad. The modal control forces and their components of modes (1,1)–(2,2) are plotted in Figs. 10.6, 10.7, 10.8 and 10.9 and component arranges are the same as those in **Case-1**. Again, the total control effect is dominantly contributed by the membrane-contributed control force component. Also, the meridional/longitudinal bending components increase with the corresponding mode number m/n.

The distinct modal actuation behavior resulting from the structural curvature effect, namely the non-uniform curvature in the meridional direction, can be clearly observed here. Taking Fig. 10.7 as an example, the maximal meridional membrane/bending components (i.e., $(\tilde{T}_{mn})_{\psi,\mathrm{mem}}$ and $(\tilde{T}_{mn})_{\psi,\mathrm{bend}}$) are located at the region very near the panel's ridge rather than two boundaries. Since the membrane control effect dominates for deep shells, actuator placed at the large curvature region would provide better control effect. Due to the drastic curvature variation, Figs. 10.6 and 10.8 show that a smaller region around the ridge of the panel is related to the optimal actuation for the (m,1) mode group. Figures 10.7 and 10.9 show that two identical optimal regions are also observed for the (m,2) mode group. Due to the anti-symmetric distribution in the ψ direction when the meridional vibration mode n is an even number, the control force in the ridge is zero. Although it is not the optimal region any more as in the odd modes, the location of the maximal control force exhibits a little shift towards the ridge where the curvature is larger. Actually,

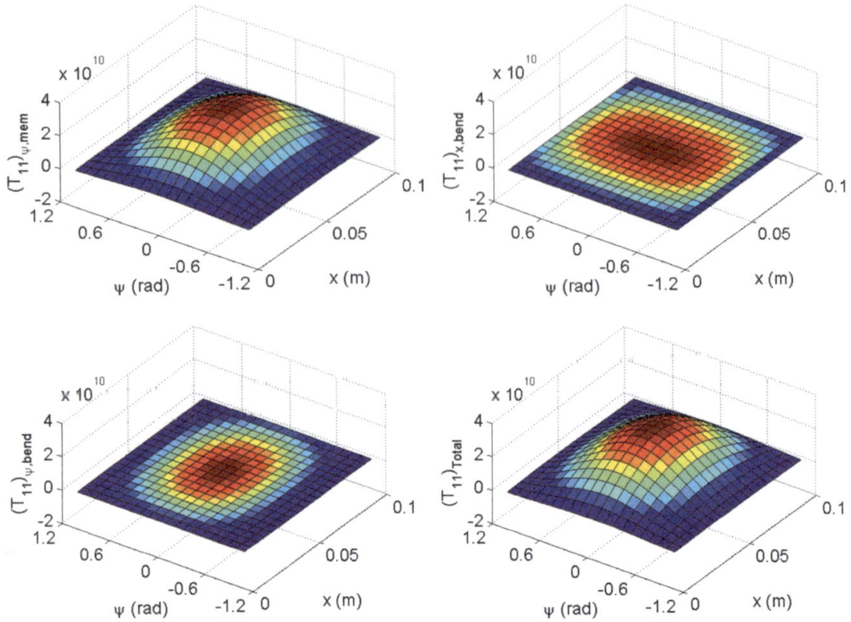

Fig. 10.6 Spatial distributions of microscopic actuation effectiveness of deep panels, mode (1,1)

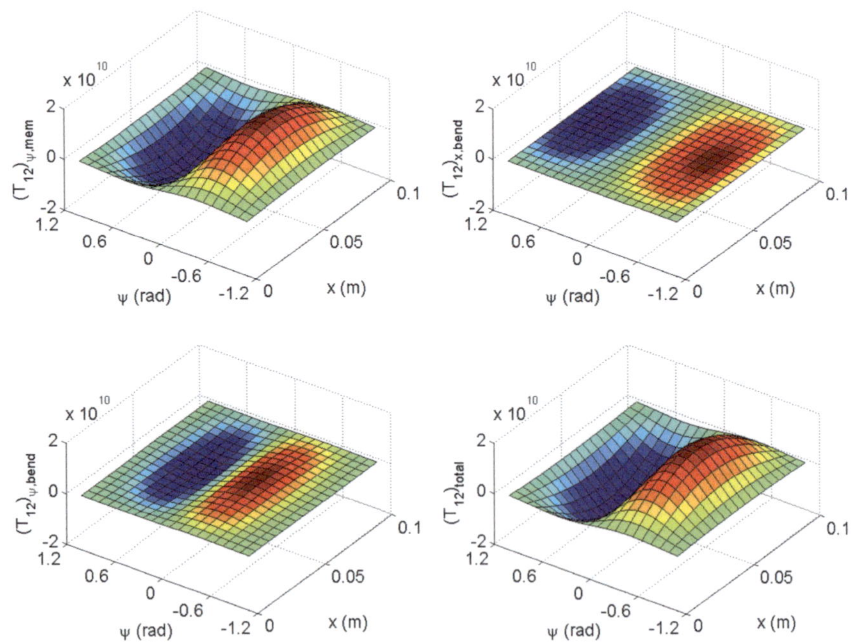

Fig. 10.7 Spatial distributions of microscopic actuation effectiveness of deep panels, mode (1,2)

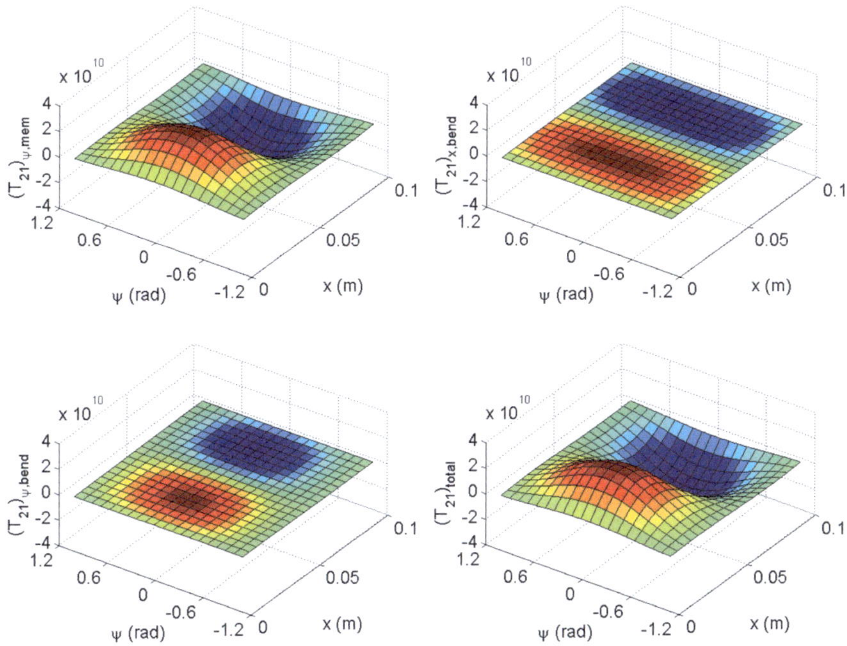

Fig. 10.8 Spatial distributions of microscopic actuation effectiveness of deep panels, mode (2,1)

the maximal peaks significantly incline to shell's ridge as shown in bottom-left plots of Figs. 10.7 and 10.9.

Thus, from the spatial distributions of microscopic actuation effectiveness of actuator elements, the optimal actuation location alters with the shell's curvature and the natural mode. However as a whole, the actuation can achieve the maximum where the meridional angle ψ is relatively smaller (i.e., larger curvature) near the ridge. The actuation location should be optimally selected to maximize the control effectiveness of specific natural mode(s) with the maximal control force; or on the contrary, zero control force could be selected to avoid the corresponding mode(s). In engineering applications, infinite actuator elements are not practical. Thus, actuator sizes are investigated next.

10.4.2 Actuator Size

In practical applications, it is crucial to control structural vibrations with limited number of actuators. Thus, the actuator size and its position are studied here. It is assumed that the piezoelectric layer is fully distributed along the meridional direction, and is segmented into a series of actuator patches as shown in Fig. 10.10. The distribution of modal control forces in the longitudinal direction behaves like a sine

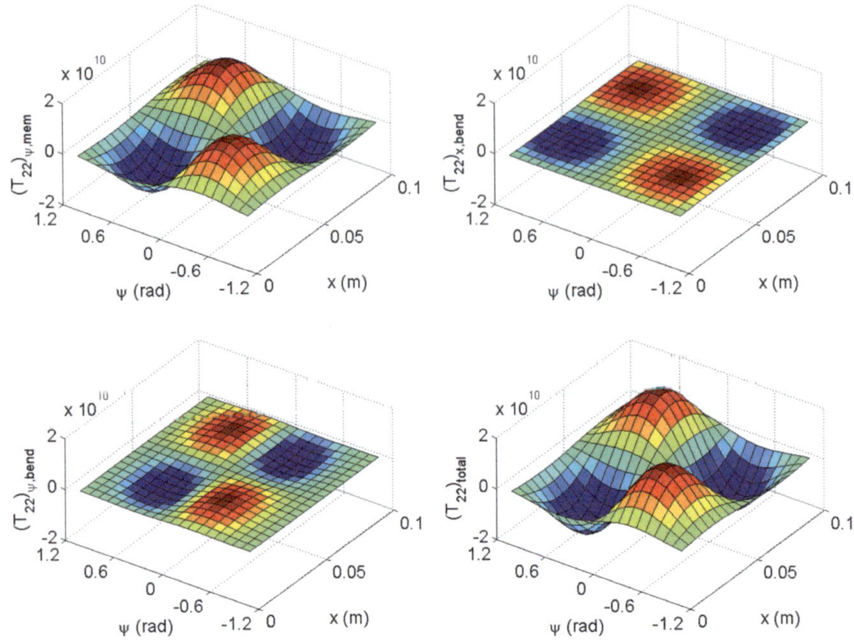

Fig. 10.9 Spatial distributions of microscopic actuation effectiveness of deep panels, mode (2,2)

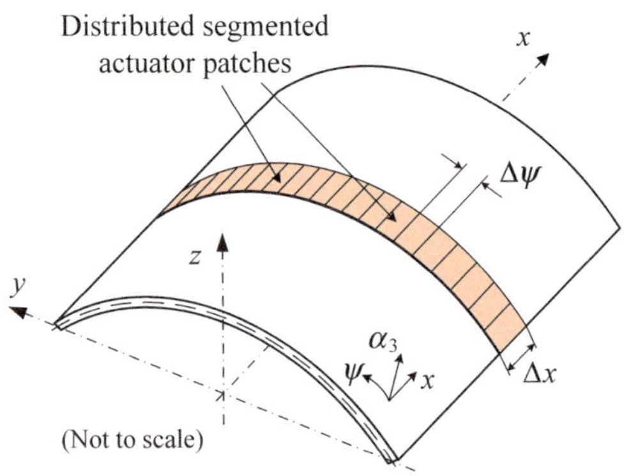

Fig. 10.10 Distributed segmented actuator patches

function and it exhibits similar characteristic as that of the circular cylindrical shell panel (Tzou, Bao and Venkayya, 1996). Therefore, this case mainly focuses on the distributed modal control behavior in the meridional direction in order to explore the parabolic curvature effect, as compared with that of open circular cylindrical shells.

The parabolic cylindrical panel with a geometry ratio $a{:}c$ = 1:1 is considered here. Note that the longitudinal actuator width is set to be $\Delta x = 0.005$ m. Three segmentation schemes that the actuator layer is respectively segmented into 40 patches, 20 patches, and 10 patches are considered to evaluate the actuator size effects. In each segmentation case, the meridional interval angle $\Delta\psi$ of all segmented actuator patches are the same. Firstly, the segmentation scheme of 40 actuator patches (i.e., $\Delta\psi = 0.0554$ rad) is studied. The total modal control force $(\tilde{T}_{mn})_{\text{total}}$ and its components $(\tilde{T}_{mn})_{\psi,\text{mem}}$, $(\tilde{T}_{mn})_{x,\text{bend}}$ and $(\tilde{T}_{mn})_{\psi,\text{bend}}$ are evaluated at various patch locations with modes $n = 1,2,3,4$ and these results are plotted in Fig. 10.11 where the horizontal coordinate denotes y axis as that in Fig. 10.10.

The 40-segment case provides a fine distribution of actuation characteristics and control component distributions along the meridional direction. The results are very similar to the microscopic behaviors presented previously. However, to segment such a small size is still impractical in engineering applications. Hence, the size of the actuator patch is doubly and quadruply enlarged and, thus the number of segmented patches is reduced to 20 and 10, respectively. Since the actuation effectiveness distributed in the meridional direction is varying, the most essential issue is the signal average and/or cancellation effect. Figures 10.12 and 10.13 illustrate the actuation characteristics at various segment positions. The patterns are similar to the 40-segment case, but the maximal actuation magnitude decreases when the size enlarged, due to signal average and/or cancellation.

In order to evaluate the variation of actuation effectiveness with respect to segment numbers, the maximal total actuation effectiveness of four segmentation cases with modes $n = 1,2,3,4$ are plotted and compared in Fig. 10.14. Generally

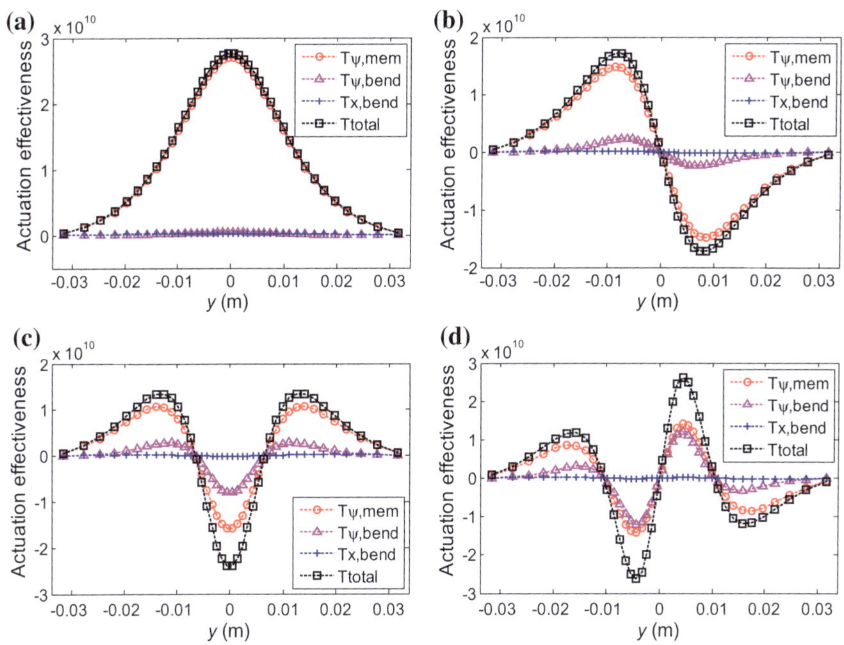

Fig. 10.11 Distributed actuation effectiveness of 40 segments: **a** the $(m,1)$ mode; **b** the $(m,2)$ mode; **c** the $(m,3)$ mode; **d** the $(m,4)$ mode

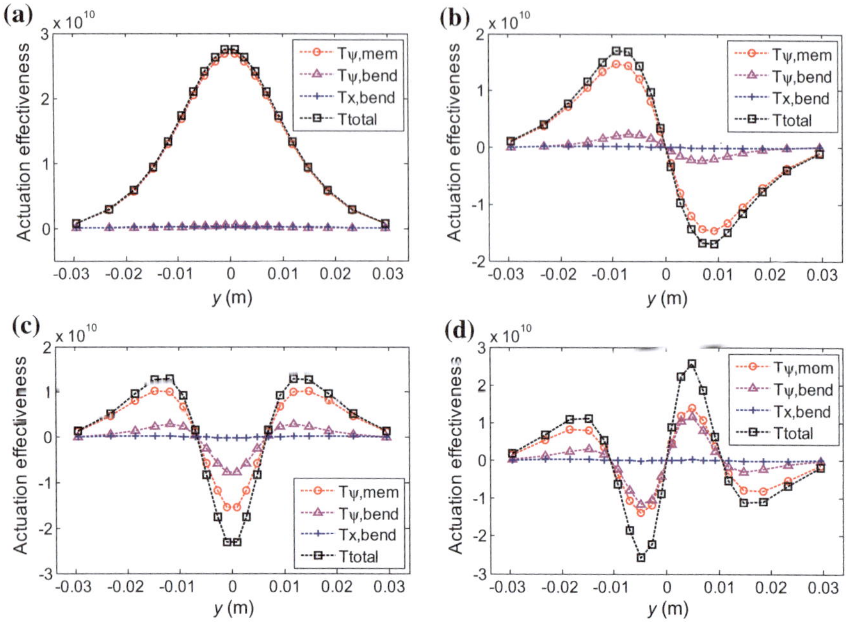

Fig. 10.12 Distributed actuation effectiveness of 20 segments: **a** the $(m,1)$ mode; **b** the $(m,2)$ mode; **c** the $(m,3)$ mode; **d** the $(m,4)$ mode

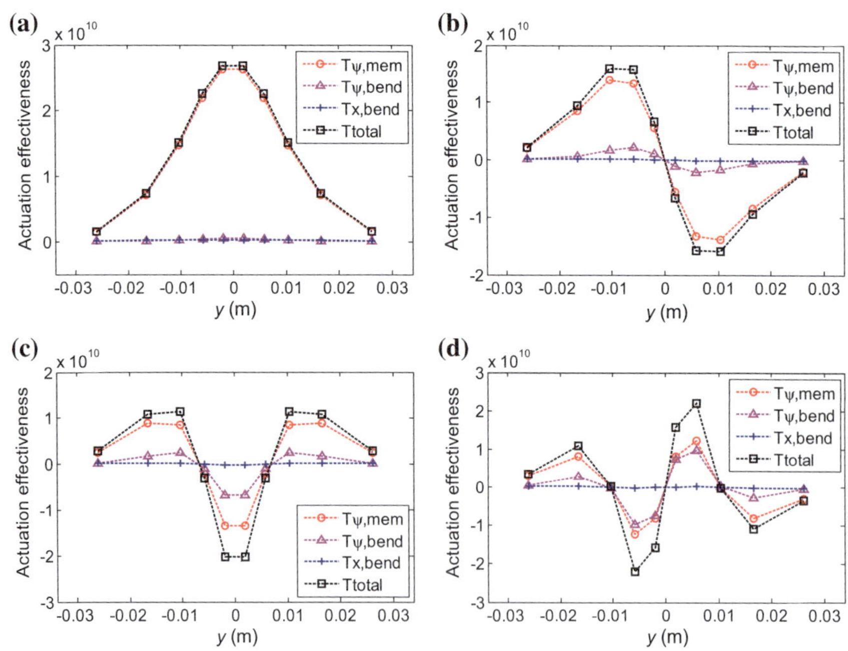

Fig. 10.13 Distributed actuation effectiveness of 10 segments: **a** the $(m,1)$ mode; **b** the $(m,2)$ mode; **c** the $(m,3)$ mode; **d** the $(m,4)$ mode

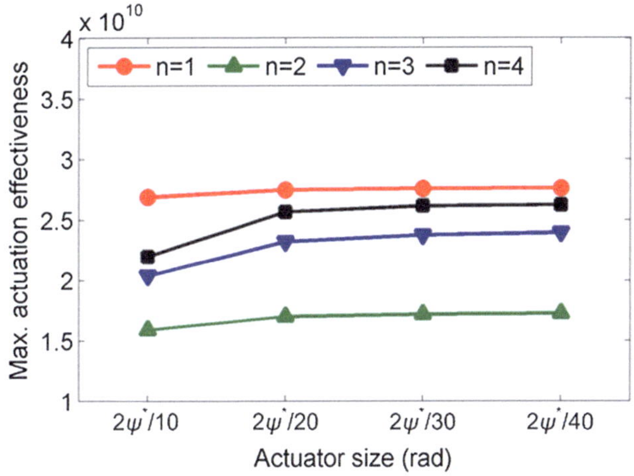

Fig. 10.14 Maximal actuation effectiveness versus actuator segmentation

speaking, the maximal total control force increases with the segment number for each natural mode. In detail, the maximal total control force increases significantly when the segment number increases from 10 to 20, but increases relatively slowly when the actuator segment is more than 20. Thus, a segmentation scheme of 20 segments (i.e., the meridional interval angle $\Delta\psi = 2\psi^*/20$ rad) is preferred since it leads to a faster enhancement of distributed actuation with limited and manageable signal and control channels.

10.5 Optimal Control

Optimal control effects of the parabolic cylindrical shell panel using collocated piezoelectric sensor and actuator are evaluated in case studies. Note that the piezoelectric sensor and actuator are laminated on the inner and outer surfaces of the shell panel respectively and they are connected with an optimal controller, Fig. 10.15. The generic optimal control derivation should be applicable to any kind of piezoelectric materials used. Due to properties of thin film, light weight and high dynamic sensitivity, polyvinylidene fluoride (PVDF) materials can be easily segmented into desirable shapes and they are convenient and applicable in various in-situ distributed sensing and control of structures, especially for shell structures. Thus, piezoelectric PVDF material is used as the sensor and actuator in this study. Material properties of the elastic parabolic cylindrical shell panel and the PVDF sensor/actuator are summarized in Table 10.3. Without loss of generality, it is assumed that the 1) the piezoelectric constants $d_{31} = d_{32}$ and $e_{31} = e_{32}$, and 2) PVDF sensor/actuator has a uniform thickness, i.e., $r_x^s = r_\psi^s = (h^s + h)/2$ and $r_x^a = r_\psi^a(h^a + h)/2$. In the following case studies, the dimension of PVDF sensor/actuator patches is selected to be $x = x_2^* - x_1^* = 0.02$ m in the longitudinal direction and $\Delta\psi = \psi_2^* - \psi_1^* = 0.1$ rad in the meridional direction.

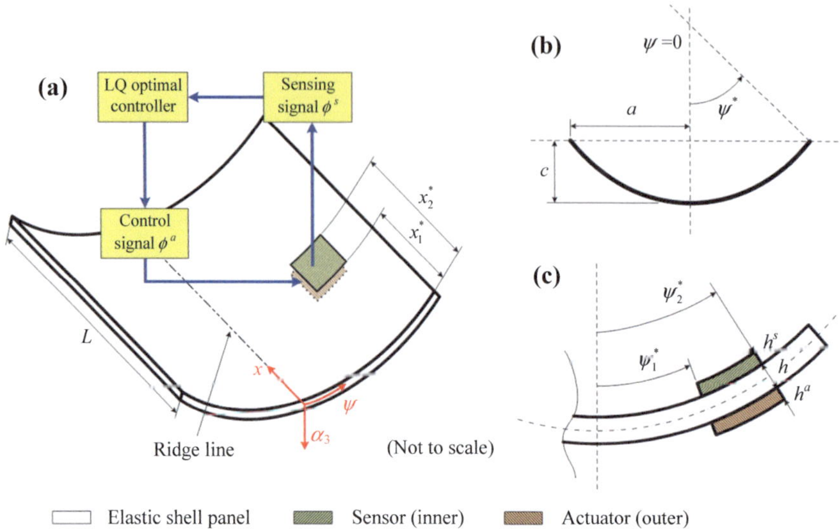

Fig. 10.15 A parabolic cylindrical shell system: **a** global configuration of the system; **b** geometry of a parabolic curve; **c** partial cross-section with sensor and actuator

Table 10.3 Detailed properties of elastic shell panel and PVDF sensor/actuator

Properties	Shell panel	Piezo-patches	Units
Half span distance	$a = 0.096$	/	m
Meridian height	$c = 0.024$	/	m
Longitudinal length	$L = 0.2$	/	m
Thickness	$h = 0.001$	$h^s = h^a = 5\times10^{-5}$	m
Young's modulus	$Y = 3.1 \times 10^9$	$Y_p = 1.6 \times 10^9$	N/m^2
Mass density	$\rho = 1190$	/	kg/m^3
Piezoelectric constant	/	$d_{31}, d_{32} = 6\times10^{-12}$	m/V
	/	$e_{31}, e_{32} = 9.6 \times 10^{-3}$	C/m^2
Dielectric permittivity	/	$\varepsilon_{33} = 8.85 \times 10^{-11}$	F/m

The optimal control responses for four independent natural modes (1,2), (1,3), (2,3) and (2,4) are considered in four cases. The natural frequencies of these four modes are 255, 223, 439 and 447 Hz, respectively. For each independent mode, assume that 1) the original modal damping ratio is 1%; 2) the initial transverse deflection of the reference position where the deflection reaches the maximum of the shell is 1 μm; and 3) the initial velocity is zero. The reference position with the maximal displacement is evaluated in these four cases. Note that the reference location of each mode is different, Sect. 10.4.1.

In order to achieve the greatest control effectiveness, the collocated piezoelectric sensor and actuator are selectively laminated on the region where the actuation factors reach the maximal values. Note that the laminating region of piezoelectric

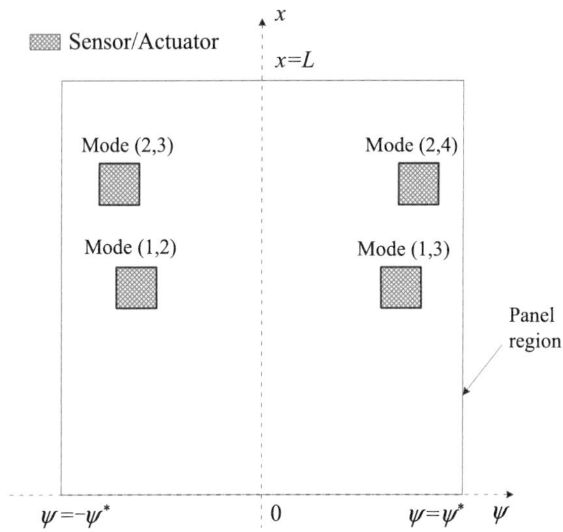

Fig. 10.16 Illustration of locations of PVDF patches for four modes

Table 10.4 Locations of PVDF patches related to four shell modes

Mode	x direction		ψ direction	
	x_1^* (m)	x_2^* (m)	ψ_1^* (rad)	ψ_2^* (rad)
(1,2)	0.09	0.11	−0.34	−0.24
(1,3)	0.09	0.11	0.27	0.37
(2,3)	0.14	0.16	−0.38	−0.28
(2,4)	0.14	0.16	0.31	0.41

patches depends on mode shapes that vary with different natural modes. The optimal actuation locations for each mode are beyond the scope of this study, so that the evaluation results of optimal laminating locations and relative modal sensitivities and actuation factors are used directly. (Detailed information can be found in (Hu, Li and Tzou, 2012; Hu 2014).) Locations of pairs of PVDF sensors/actuators for four independent modes are illustrated in Fig. 10.16 and their detailed coordinates are summarized in Table 10.4.

For the LQ optimal control study, the control input related parameter R and the performance related parameter Q_{22} should be designed in connection with the requirement of control performance and control cost. In case studies, these two parameters are designed according to the input limitation of PVDF materials that the control voltage should be limited within the range of less than 400 V so as to avoid the overload and/or damages. The detailed control responses of LQ optimal control algorithm and the role of design parameters R and Q_{22} of four shell modes are demonstrated as follows.

10.5.1 Case 1: Mode (1,2)

The optimal control response and relative control voltage of shell mode (1,2) are evaluated and illustrated in this sub-section. It is assumed that the initial displacement is 1.0×10^{-6} m and an initial damping of 1%. Snap-back time-domain modal responses with and without optimal control in the beginning 0.05 s are plotted in Fig. 10.17. The optimal design parameters are chosen to be $R = 1$ and $Q_{22} = 328{,}000$, so that the optimal control efficiency can be maximized with the limitation of maximal PVDF control voltage satisfied. Consequently, the optimal control gain is solved as $\mathbf{G}^{*} = [-2.5 \times 10^{-13}, 734.2]$. Figure 10.17 shows that the maximal displacement of the parabolic cylindrical shell panel is rapidly reduced as compared with the uncontrolled original vibratory response. The vibration is almost suppressed in the beginning 0.05 s. In fact, the damping ratio under the optimal control is enhanced from the original 1 to 4.86%.

The absolute control voltage of mode (1,2) is restricted to be less than 400 V as shown in Fig. 10.18 and it decreases with the vibration suppression. Note that the sign of voltage response is opposite to the modal displacement response, since it is based on the velocity feedback.

10.5.2 Case 2: Mode (1,3)

The snap-back time-domain displacement response of mode (1,3) under the optimal control is plotted in Fig. 10.19. In this case, the optimal design parameters are $R = 1$ and $Q_{22} = 364{,}000$ and the relative optimal control gain is $\mathbf{G}^{*} = [-2.9 \times 10^{-14}, -61.2]$. In doing so, the control voltage is exactly restricted in the targeted range as shown in Fig. 10.20. The controlled damping ratio reaches 7.79% and the vibration amplitude is controlled to be nearly zero within about eight vibratory cycles. Note, from Figs. 10.17 and 10.19, that the control efficiency is greatly enhanced in mode (1,3) than in mode (1,2) due to the enhanced modal

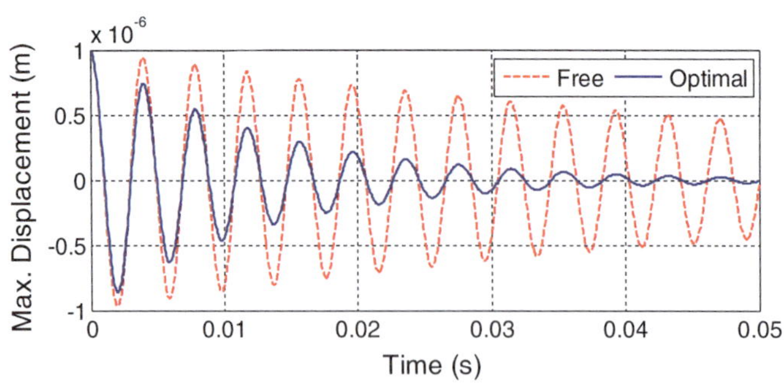

Fig. 10.17 Optimal control response of mode (1,2)

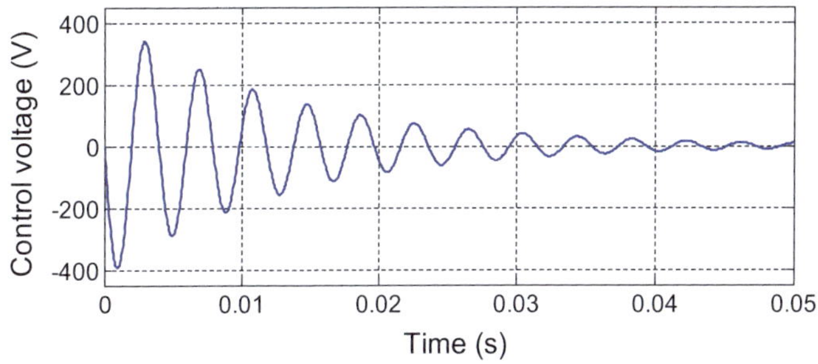

Fig. 10.18 Control voltage of mode (1,2)

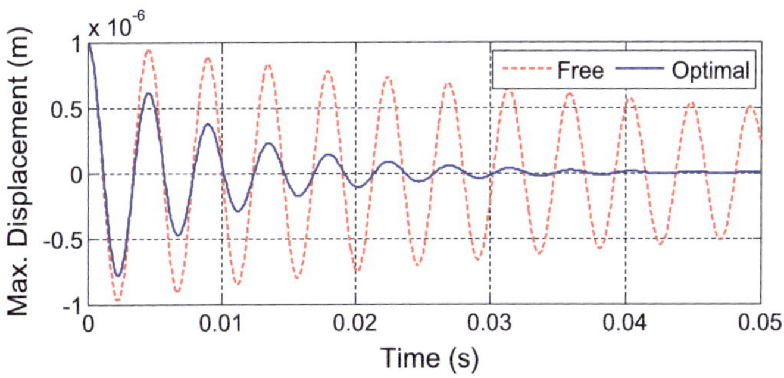

Fig. 10.19 Optimal control response of mode (1,3)

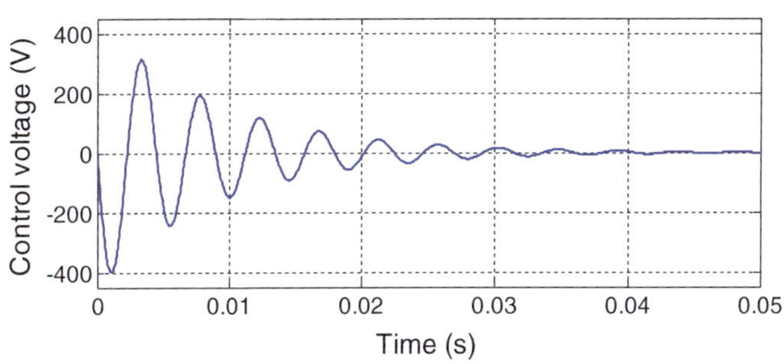

Fig. 10.20 Control voltage of mode (1,3)

actuation factor A^a_{mn}. In addition, the fact that the frequency of mode (1,3) is the lowest one among all shell natural frequencies of the parabolic cylindrical shell panel and it also contributes to the enhancement of control effect.

10.5.3 Case 3: Mode (2,3)

The optimal control response of the parabolic cylindrical shell panel, relative control voltage, design parameters and controlled damping ratio are evaluated for mode (2,3) in this case. Optimal design parameters are chosen as $R = 1$ and $Q_{22} = 199{,}000$. Thus, the modal damping ratio is 3.99% with the designed optimal control gain $\mathbf{G}^{*} = [-1.7 \times 10^{-14}, -36.7]$. The increase of wave number m in the longitudinal direction leads to the rise in both shell's natural frequency and actuator's actuation factor; however, they have opposite effects to the control efficiency. As a result, it is found that the damping ratio is reduced as compared with those in modes (1,2) and (1,3). Thus, it can be inferred that the vibration mode with a higher natural frequency is relatively more difficult to suppress. Figures 10.21 and 10.22 illustrate the time-domain responses of the maximal displacement of the parabolic cylindrical shell panel and the control voltage of mode (2,3), respectively.

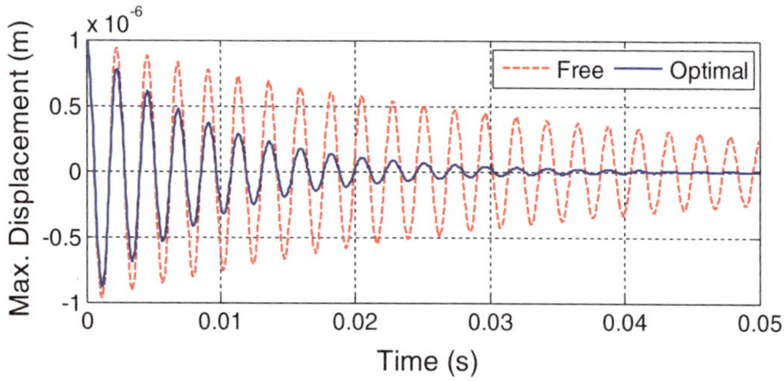

Fig. 10.21 Optimal control response of mode (2,3)

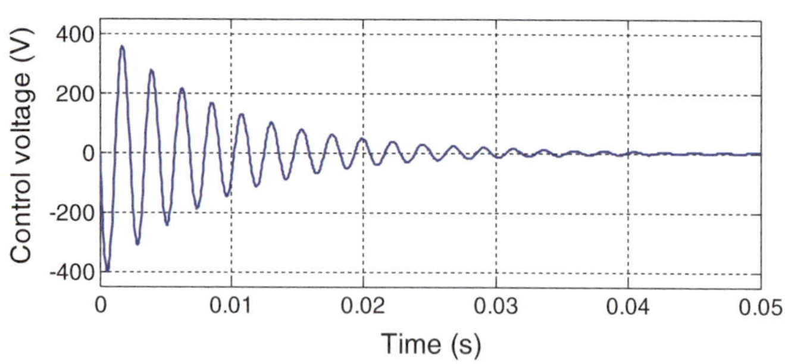

Fig. 10.22 Control voltage of mode (2,3)

10.5.4 Case 4: Mode (2,4)

The optimal control results of mode (2,4) are evaluated and illustrated in Figs. 10.23 and 10.24, where $R = 1$ and $Q_{22} = 190{,}000$. The optimal control gain is solved to be $\mathbf{G}^{*} = [-9.1 \times 10^{-15}, -16.7]$. As a result, the damping ratio under the optimal control reaches 4.52%. Modal displacement responses and control voltage illustrated in Figs. 10.23 and 10.24 are similar to those in Case 3, due to the similar natural frequencies. The slight enhancement of damping ratio owes to the increment of actuation factor of piezoelectric actuator because of the wave number n increasing from 3 to 4.

Looking into the optimal control gains $\mathbf{G}^{*}$ in four cases, one can find that the gain element related to the modal participation factor G_1^* is so insignificant that it is nearly ineligible as compared with the modal velocity G_2^*. Thus, it is concluded that the LQ optimal controller is effective for the vibration control of parabolic cylindrical shell panels and the feedback of modal velocity dominantly contributes to the remarkable enhancement of damping ratio.

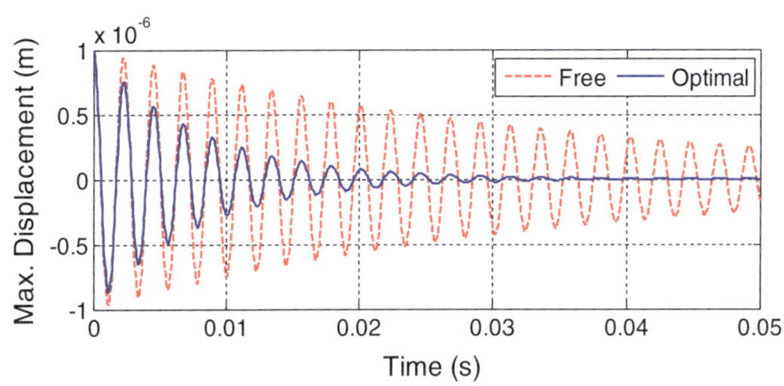

Fig. 10.23 Optimal control response of mode (2,4)

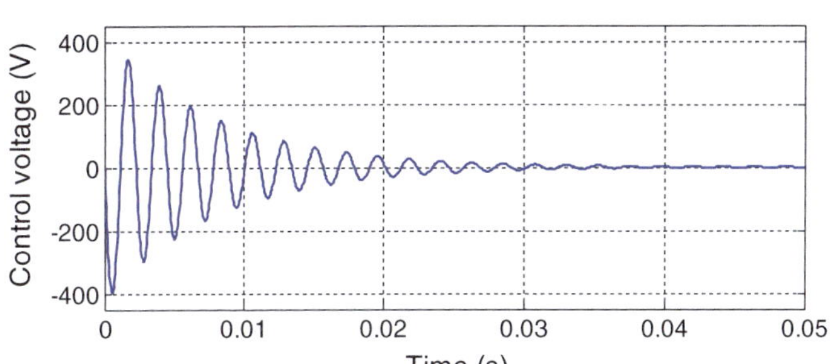

Fig. 10.24 Control voltage of mode (2,4)

10.5.5 Design Parameters R and $\mathbf{Q}_{22}$

Recall that **Q** is the positive semi-definite state weighting matrix and **R** is the positive-definite control weighting matrix in Eq. (10.3.3). By minimizing the criterion function, the LQ optimal control algorithm provides a method to achieve the maximal possible reduction of mechanical vibration with less control cost. In order to satisfy the requirement of specific engineering application, the weight between the control performance and the control cost can be adjusted through the proper selection of design parameters R and Q_{22}. In this section, different design parameters are chosen to reveal the role of R and Q_{22} in designing the LQ optimal controller.

The time responses of optimal control on parabolic cylindrical shell panels (mode (1,3)) with respect to the control cost weight R are illustrated in Fig. 10.25, where dotted line denotes the uncontrolled response and dashed-dotted line, dashed line and solid line denote the optimal control responses with R = 1, 2 and 4, respectively. Relative control voltages are plotted in Fig. 10.26.

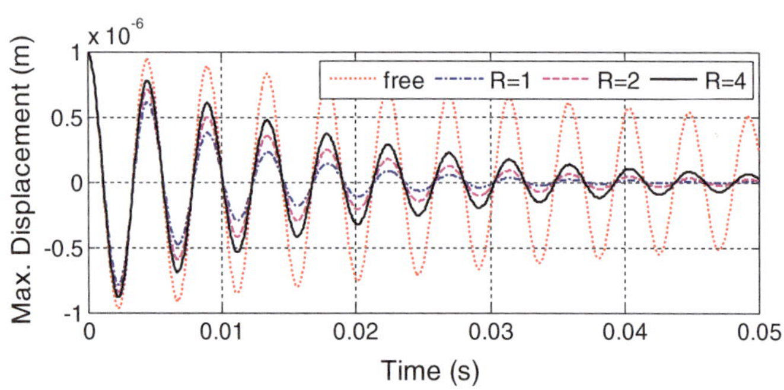

Fig. 10.25 Optimal control responses of mode (1,3) with different values of R

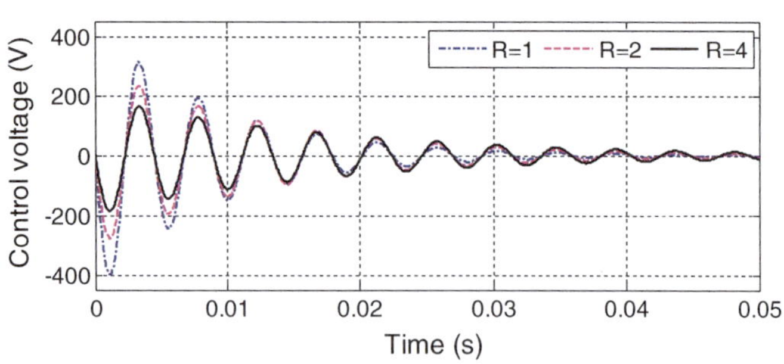

Fig. 10.26 Optimal control voltages of mode (1,3) with different values of R

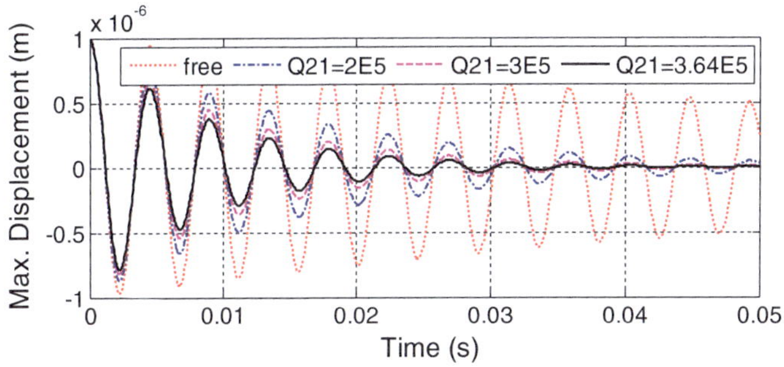

Fig. 10.27 Optimal control responses of mode (1,3) with different values of Q_{22}

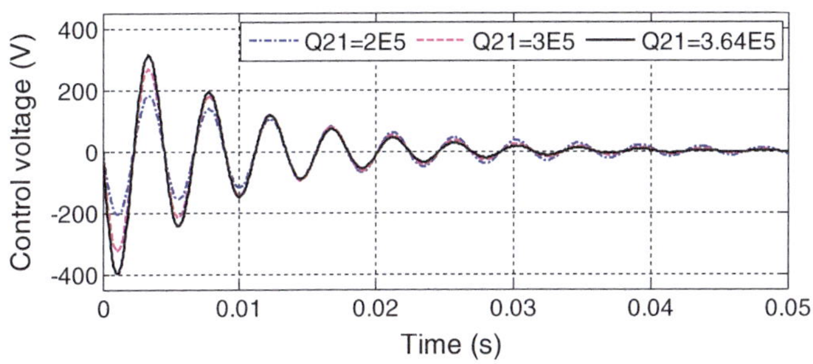

Fig. 10.28 Optimal control voltages of mode (1,3) with different values of Q_{22}

It is observed that for a constant Q_{22}, the increase of R reduces the initial control voltage and thus extends the vibration settling time. That's because the control energy cost is manually reduced due to the increase of control cost weight in the criterion function. On the contrary, when the state weight element Q_{22} increases (with a constant R), the initial control voltage is enhanced so that the shell vibration is suppressed faster as shown in Figs. 10.27 and 10.28. Therefore, relatively larger Q_{22} and/or smaller R are usually designed for faster vibration suppression while more control energies are relatively required, and vice versa.

10.6 Summary

Open parabolic cylindrical shells are important to radial signal collection, reflection and/or transmission applied to radar antennas, space reflectors, solar collectors, etc. The spatially distributed microscopic modal control effectiveness induced by piezoelectric actuators laminated on a simply-supported parabolic cylindrical shell panel was investigated in this study. Distinct distributed modal actuation behaviors of transverse vibrations of the shell were analyzed based on a *newly*-formulated mode shape function. The expression of modal control force induced by an actuator patch was derived. The spatially distributed microscopic actuation effectiveness induced by an infinitesimal actuator element was also derived to precisely illustrate the spatial distribution behavior.

The total control force of segmented actuators $(\widehat{T}_{mn})_{\text{total}}$ or that of an infinitesimal actuator $(\tilde{T}_{mn})_{\text{total}}$ consists of three actuation components: 1) a component $(\widehat{T}_{mn})_{\psi,\text{mem}}$ (or $(\tilde{T}_{mn})_{\psi,\text{mem}}$) contributed by the membrane force in the meridional direction, 2) a component $(\widehat{T}_{mn})_{\psi,\text{bend}}$ (or $(\tilde{T}_{mn})_{\psi,\text{bend}}$) contributed by the bending moment in the meridional direction and 3) a component $(\widehat{T}_{mn})_{x,\text{bend}}$ (or $(\widehat{T}_{mn})_{x,\text{bend}}$) contributed by the bending moment in the longitudinal direction. Their respective total modal control force and their three components were evaluated with different shell curvatures, i.e., deep and shallow, in two case studies. Their dominating component and the distinct curvature effect were evaluated. Analysis results suggest that the meridional membrane-contributed component dominates the total actuation effect for lower natural modes. However, the meridional/longitudinal bending-contributed components increase with the corresponding mode number m or n, while the membrane-contributed component decreases.

For the shallow parabolic shell whose curvature change is insignificant, the spatial distribution of actuation effectiveness is similar to that of the circular cylindrical shell panel. However, for the deep parabolic shell whose meridional curvature variation is significant, the curvature effect to distributed actuation behavior becomes evident. Thus, for odd modes, the maximal actuation effectiveness locates at the shell ridge, while for even modes, the peak/valley closest to the ridge is the optimal position to achieve the maximal actuation effect. The study of actuator sizes suggests that a segmentation scheme of the meridional interval angle $\Delta\psi = 2\psi^*/20$ rad (i.e., 20 actuator segments) is practical and feasible for engineering applications.

Optimal vibration control of parabolic cylindrical shell panels laminated with piezoelectric sensors and actuators with the LQ optimal control algorithm was discussed and case studies presented. A pair of piezoelectric patches laminated on the inner and outer surfaces of the shell panel served as the distributed sensor and actuator, respectively. The independent modal control equation was obtained and also defined in the state space form. Based on the sensing/actuation behaviors of the piezoelectric sensor/actuator, the LQ controller for each independent mode was designed, and the relative optimal control gain and the control voltage were solved.

The optimal control effects for four shell natural modes were evaluated in case studies. Numerical results show that the LQ optimal controller with the feedback control gain is effective to the vibration control of parabolic cylindrical shell panels. The damping ratio can be greatly enhanced for all shell modes evaluated, and the maximum reaches 7.79% for mode (1,3). The optimal control efficiency is dependent on natural frequency, sensing sensitivity, actuation factor, and optimal design parameters. In this study, the influence of design parameters R and Q_{22} was also evaluated. Relatively larger Q_{22} and/or smaller R results in faster reduction of modal oscillation with more control energy cost, and vice versa. These results would provide a design reference in practical applications. The size, shape and laminating location of piezoelectric sensor/actuator are also key design parameters. They will affect the control efficiency through directly adjusting the sensing sensitivity and actuation factor. Investigations of roles of these design parameters may be a future work on sensing and control of parabolic cylindrical shell panels.

Acknowledgements This research was supported, in part, by the Natural Science Foundation of China (No. 11172262), the 2nd phase of the "985 Project" at Zhejiang University and the State Key Laboratory of Fluid Power Transmission and Control (No. SKLoFP_CX_001).

References

Chen, C. Q., and Shen, Y. P., 1997, "Optimal Control of Active Structures with Piezoelectric Modal Sensors and Actuators." *Smart Materials and Structures*, 6(4), pp. 403–409.

Dragovan, M., "*The DART System for Far-IR/Submillimeter Space Missions*," JPL White Paper, 2000.

Fernandez-Garcia, A., Zarza, E., Valenzuela, L., and Perez, M., 2010, "Parabolic-Trough Solar Collectors and Their Applications," *Renewable and Sustainable Energy Reviews*, 14(7), pp. 1695–1721.

Gupta, V. K., Seshu, P., Issac, K. K., and Shevgaonkar, R. K., 2005, "Beam Steering and Shaping of Smart Cylindrical Antenna," *AIAA Journal*, 43(1), pp. 165–173.

Hu, S.D., 2014, *Electromechanical Coupling, Distributed Sensing and Control of Flexoelectric/Piezoelectric Smart Structures*, Ph.D. Thesis, Zhejiang University, China.

Hu, S. D., Li, H., and Tzou, H. S., 2012, "Distributed Actuation Effectiveness of Flexible Parabolic Cylindrical Panels," *Proceedings of ASME 2012 IDETC/CIE*, Chicago, USA, 2012.

Hu, S. D., Li, H., and Tzou, H. S., 2013, "Distributed Neural Signals on Parabolic Cylindrical Shells", *Journal of Sound and Vibration*, 332(12), pp. 2984–3001.

Kwakernaak, H., and Sivan, R., 1972, Linear Optimal Control Systems, Wiley-Interscience, New York.

Kildal, P. S., 1984, "Radiation Characteristics of the EISCAT VHF Parabolic Cylindrical Reflector Antenna," *IEEE Transactions on Antennas and Propagation*, 32(6), pp. 541–552.

Li, H., Chen, Z. B., and Tzou, H. S., 2010, "Distributed Actuation Characteristics of Clamped-Free Conical Shells Using Diagonal Piezoelectric Actuators," *Smart Materials and Structures*, 19(11), pp. 115015.

Li, H., Hu, S. D., Tzou, H. S., and Chen, Z. B., 2012, "Optimal Vibration Control of Conical Shells With Collocated Helical Sensor/Actuator Pairs," *Journal of Theoretical and Applied Mechanics*, 50(3), pp. 769–784.

Price, H., Lupfert, E., Kearney, D., Zarza, E., Cohen, G., Gee, R., and Mahoney, R., 2002, "Advances in Parabolic Trough Solar Power Technology," *Journal of Solar Energy Engineering-Transactions of the ASME*, 124(2), pp. 109–125.

Qiu, J., and Tani, J., 1995, "Vibration Control of a Cylindrical Shell Using Distributed Piezoelectric Sensors and Actuators," *Journal of Intelligent Material Systems and Structures*, 6(4), pp. 474–481.

Ray, M. C., 2003, "Optimal Control of Laminated Shells Using Piezoelectric Sensor and Actuator Layers," *AIAA Journal*, 41(6), pp. 1151–1157.

Sadowy, G., Berkun, A., Durden, S., Huang, J., Im, E., Lopez, B., Lou, M., Rahmat-Samii, Y., Liu, F. Y., and Rengarajan, S., 2001, "Technologies for the Next Generation of Spaceborne Precipitation Radars," IEEE Proceedings in Aerospace Conference, 4, pp. 1811–1823.

Soedel, W., 2004, *Vibrations of Shells and Plates*, Marcel Dekker Inc., New York, USA.

Smithmaitrie, P., and Tzou, H. S., 2004, "Micro-Control Actions of Actuator Patches Laminated on Hemispherical Shells," *Journal of Sound and Vibration*, 277(4–5), pp. 691-710.

Tzou, H. S., 1993, *Piezoelectric Shells: Distributed Sensing and Control of Continua*, Kluwer Academic Publishers, Dordrecht, Boston, London.

Tzou H.S., Anderson G.L., *Intelligent Structural System*. Dordrecht Boston London: Kluwer Academic Publishers. 1992.

Tzou, H. S., Bao, Y., and Venkayya, V. B., 1996, "Parametric Study of Segmented Transducers Laminated on Cylindrical Shells: 2. Actuator Patches," *Journal of Sound and Vibration*, 197(2), pp. 225–249.

Tzou, H. S., Chai, W. K., and Wang, D. W., 2004, "Micro-Control Actions and Location Sensitivity of Actuator Patches Laminated on Toroidal Shells Toroidal Shell," *Journal of Vibration and Acoustics-Transactions of the ASME*, 126(2), pp. 284–297. (ParbCydlActF. ShlParbC2).

Tzou, H. S., Ding, J. H., and Hagiwara, I., 2002, "Micro-Control Actions of Segmented Actuator Patches Laminated on Deep Paraboloidal Shells," *JSME International Journal Series C-Mechanical Systems Machine Elements and Manufacturing*, 45(1), pp. 8–15.

Tzou, H. S., and Ding, J. H., 2004, "Optimal Control of Precision Paraboloidal Shell Structronic Systems," Journal of Sound and Vibration, 276(1–2), pp. 273–291.

Tzou, H. S., Wang, D. W., and Chai, W. K., 2002, "Dynamics and Distributed Control of Conical Shells Laminated with Full and Diagonal Actuators," *Journal of Sound and Vibration*, 256(1), pp. 65–79.

Washington, G., 1996, "Smart Aperture Antennas," *Smart Materials and Structures*, 5(6), pp. 801–805.

Yue, H. H., Deng, Z. Q., and Tzou, H. S., 2008, "Optimal Actuator Locations and Precision Micro-Control Actions on Free Paraboloidal Membrane Shells," *Communications in Nonlinear Science and Numerical Simulation*, 13(10), pp. 2298–2307.

Zhou, Y. H., and Tzou, H., 2000, "Active Control of Nonlinear Piezoelectric Circular Shallow Spherical Shells," *International Journal of Solids and Structures*, 37(12), pp. 1663–1677.

Chapter 11
Linear/Nonlinear Piezoelectric Shell Energy Harvesters

Development of energy harvesting technologies, e.g., converting vibration energy to electric energy, has attracted much attention in recent years. Among the proposed technologies, piezoelectric energy harvesters can easily convert the vibration induced energy to useful electric energy with the electromechanical coupling of piezoelectricity (Kim, et al., 2011; Beeby et al., 2006; Liu and Chen, 2012; Anton and Sodano, 2007; Sodano, et al., 2004). Most piezoelectric energy harvesters have been designed based on various non-shell structures, such as cantilever beams and plates. Erturk and Inman studied the piezoelectric cantilever–beam energy harvester, derived voltage/power on the resistive load in the closed-circuit condition and also carried out laboratory experiments to validate the output powers (Erturk and Inman, 2008, 2009). Karami et al. (2011) presented a single-crystal energy harvesters based on a unimorph beam and compared the analytical results of the voltage on the resistive shunt with the experimental data. Kim (2005a, b) proposed an energy harvester based on a clamped unimorph piezoelectric plate structure and the results were validated in experiments (Kim, 2011).

Electromechanical coupling mechanism and distributed sensing of shell structures, which are commonly used in aerospace, civil and mechanical systems, also have captured researchers' attention for decades. Howard et al. studied the modal voltages in an open-circuit condition based on a generic double-curvature shell and applied the generic sensing theory to beams and conical shells (Howard, Chai, Tzou, 2001). Piezoelectric electromechanical coupling mechanisms based on other shells have also been studied, such as circular cylindrical panels, circular cylindrical shells, circular rings, conical shells and paraboloidal shells, etc. (Tzou, Bao, Venkaya, 1996a, b; Smithmaitrie and Tzou, 2004; Tzou et al., 1994; Zhang et al., 2014; Hu, Li, Tzou, 2013, 2014; Li, et al., 2012; Huang, et al., 2012). However, most of them only considered the signal generations in the **open-circuit** condition and the sensing theory becomes impractical when it is used to power electric devices whose impendence can no longer be ignored. Therefore, the output energy across a resistive load in a **closed-circuit** condition for the shell energy harvesters should be exploited. Axisymmetric cylindrical energy harvester in the closed-circuit condition was

H. (HS) Tzou, *Piezoelectric Shells*, Solid Mechanics and Its Applications 247,
https://doi.org/10.1007/978-94-024-1258-1_11

proposed and the voltage output across the resistive load was estimated (Rao, Li, Tzou, 2011). And based on a circular ring-type energy harvester, voltage output and power across an external resistive load were also derived (Zhang, Hu, Tzou, 2011). However, these energy harvesting mechanisms of specific shell structures are limited in their applications. Accordingly, this study focuses on a development of a generic piezoelectric shell energy harvesting mechanism that can be simplified and applied to many linear/nonlinear shell and non-shell distributed energy harvesters.

In this chapter, the generic shell energy harvester is made of a double-curvature shell laminated with a piezoelectric patch defined by its corner coordinates. Based on the linear or nonlinear strains generated by the harmonic shell vibration and a closed-circuit current model, one can estimate the power generation of the shell energy harvester. Once the power generation and energy equations are established, with Lamé parameters and radii of curvature of a selected distributed harvester design, one can simplify the shell equations to account for many linear/nonlinear shell and non-shell-type energy harvesters in engineering applications. This procedure is demonstrated in a beam model, a circular ring harvester and a conical shell harvester in case studies.

11.1 Dynamic Responses of Shells

Power and energy generation is based on dynamic strains in the piezoelectric patch. This section focuses on dynamic strain generation of a generic double-curvature piezoelectric energy harvester. Since the harvester is mostly excited at resonance to maximize the energy output, a harmonic excitation is considered and its steady-state dynamic response is defined. Power and energy generation is discussed in the next section.

11.1.1 Dynamic Strains of Double-Curvature Shells

As shown in Fig. 11.1, a generic double-curvature shell energy harvester is defined in a tri-orthogonal curvilinear coordinate system with α_1 and α_2 defining the shell neutral surface and α_3 the normal direction, similar to Chaps. 2 and 4. R_1 and R_2 are the curvature radii of the α_1 and α_2 axes, respectively; $F_3(\alpha_1, \alpha_2)$ is a transverse excitation. The generic double-curvature shell, which has the uniform thickness h, is laminated with a piezoelectric patch (thickness is h^e). It is assumed the patch is perfectly laminated on the shell surface and its stiffness and mass effects are ignored since the patch thickness h^e is much thinner than the shell thickness h. The patch location and size is generically defined by its four-corner coordinates, i.e., the segmentation technique, Fig. 11.1. The energy harvester patch can be assumed to be an equivalent circuit where C_p is the equivalent capacitance of the piezoelectric patch. The energy harvester is used to power a resistive load R_L in the closed-circuit condition.

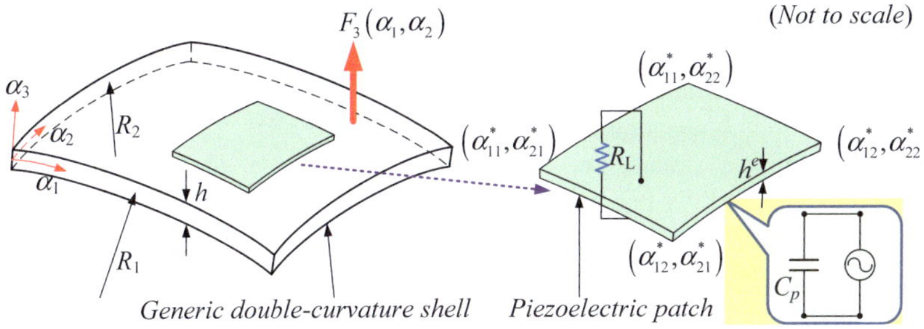

Fig. 11.1 Model of a generic piezoelectric shell energy harvester

Since the piezoelectric patch is laminated perfectly on the surface of the double-curvature shell, the strain in the piezoelectric patch is assumed to be consistent with the local strain on the generic double-curvature shell. Based on the thin shell theory and Kirchhoff–Love assumptions, normal strain and shear strains in the transverse direction can be neglected, i.e., $S_{33} = S_{13} = S_{23} = 0$, Chap. 4. Thus, the normal and shear strains in the piezoelectric patch in the α_1 and α_2 directions are defined as functions of shell displacements (Tzou, 1993)

$$\begin{aligned} S_{11} = & \frac{1}{A_1}\frac{\partial u_1}{\partial \alpha_1} + \frac{u_2}{A_1 A_2}\frac{\partial A_1}{\partial \alpha_2} + \frac{u_3}{R_1} \\ & + r_1^e\left[\left(\frac{1}{A_1}\frac{\partial}{\partial \alpha_1}\left(\frac{u_1}{R_1} - \frac{1}{A_1}\frac{\partial u_3}{\partial \alpha_1}\right) + \frac{1}{A_1 A_2}\left(\frac{u_2}{R_2} - \frac{1}{A_2}\frac{\partial u_3}{\partial \alpha_1}\right)\frac{\partial A_1}{\partial \alpha_2}\right)\right], \end{aligned} \tag{11.1.1}$$

$$\begin{aligned} S_{22} = & \frac{1}{A_2}\frac{\partial u_2}{\partial \alpha_2} + \frac{u_1}{A_1 A_2}\frac{\partial A_2}{\partial \alpha_1} + \frac{u_3}{R_2} \\ & + r_2^e\left[\left(\frac{1}{A_2}\frac{\partial}{\partial \alpha_2}\left(\frac{u_2}{R_2} - \frac{1}{A_2}\frac{\partial u_3}{\partial \alpha_2}\right) + \frac{1}{A_1 A_2}\left(\frac{u_1}{R_1} - \frac{1}{A_1}\frac{\partial u_3}{\partial \alpha_1}\right)\frac{\partial A_2}{\partial \alpha_1}\right)\right], \end{aligned} \tag{11.1.2}$$

$$\begin{aligned} S_{12} = & \frac{1}{A_2}\frac{\partial u_1}{\partial \alpha_2} + \frac{1}{A_1}\frac{\partial u_2}{\partial \alpha_1} - \frac{u_1}{A_1 A_2}\frac{\partial A_1}{\partial \alpha_2} - \frac{u_2}{A_1 A_2}\frac{\partial A_2}{\partial \alpha_1} \\ & + r_6^e\left[\frac{1}{A_2}\frac{\partial}{\partial \alpha_2}\left(\frac{u_1}{R_1} - \frac{1}{A_1}\frac{\partial u_3}{\partial \alpha_1}\right) + \frac{1}{A_1}\frac{\partial}{\partial \alpha_1}\left(\frac{u_2}{R_2} - \frac{1}{A_2}\frac{\partial u_3}{\partial \alpha_2}\right)\right. \\ & \left. - \frac{1}{A_1 A_2}\left(\frac{u_1}{R_1} - \frac{1}{A_1}\frac{\partial u_3}{\partial \alpha_1}\right)\frac{\partial A_1}{\partial \alpha_2} - \frac{1}{A_1 A_2}\left(\frac{u_2}{R_2} - \frac{1}{A_2}\frac{\partial u_3}{\partial \alpha_2}\right)\frac{\partial A_2}{\partial \alpha_1}\right]. \end{aligned} \tag{11.1.3}$$

where u_i (i = 1,2,3) is the displacement in the i direction; A_1 and A_2 are the Lamé parameters in the α_1 and α_2 directions, respectively; r_i^e (i = 1, 2, 6) is the moment arm measured from the shell's neutral surface to the middle plane of the piezoelectric patch, which are assumed to be the same as $r^e = (h + h^e)/2$ because of the uniform thickness of the generic double-curvature shell and the piezoelectric patch. Note that the terms after the moment arms r_i^e in the bracket are related to the bending strain component and those before the bracket are the membrane strain component. However, if the shell experiences a large transverse deformation u_3, *the von Karman nonlinearity* can be adopted. Based on *the von Karman geometrical non-linearity*, a quadratic effect of the non-linear transverse displacement is considered in the membrane strains (Howard, Chai, Tzou, 2001; Tzou and Bao, 1997). Thus, the nonlinear strains, i.e., Eqs. (11.1.1)–(11.1.3), of the piezoelectric harvester patch can be rewritten as

$$\begin{aligned} S_{11} &= \frac{1}{A_1}\frac{\partial u_1}{\partial \alpha_1} + \frac{u_2}{A_1A_2}\frac{\partial A_1}{\partial \alpha_2} + \frac{u_3}{R_1} + \frac{1}{2}\left(\frac{1}{A_1}\frac{\partial u_3}{\partial \alpha_1}\right)^2 \\ &\quad + r_1^e\left[\left(\frac{1}{A_1}\frac{\partial}{\partial \alpha_1}\left(\frac{u_1}{R_1} - \frac{1}{A_1}\frac{\partial u_3}{\partial \alpha_1}\right) + \frac{1}{A_1A_2}\left(\frac{u_2}{R_2} - \frac{1}{A_2}\frac{\partial u_3}{\partial \alpha_2}\right)\frac{\partial A_1}{\partial \alpha_2}\right)\right], \end{aligned} \tag{11.1.4}$$

$$\begin{aligned} S_{22} &= \frac{1}{A_2}\frac{\partial u_2}{\partial \alpha_2} + \frac{u_1}{A_1A_2}\frac{\partial A_2}{\partial \alpha_1} + \frac{u_3}{R_2} + \frac{1}{2}\left(\frac{1}{A_2}\frac{\partial u_3}{\partial \alpha_2}\right)^2 \\ &\quad + r_2^e\left[\left(\frac{1}{A_2}\frac{\partial}{\partial \alpha_2}\left(\frac{u_2}{R_2} - \frac{1}{A_2}\frac{\partial u_3}{\partial \alpha_2}\right) + \frac{1}{A_1A_2}\left(\frac{u_1}{R_1} - \frac{1}{A_1}\frac{\partial u_3}{\partial \alpha_1}\right)\frac{\partial A_2}{\partial \alpha_1}\right)\right], \end{aligned} \tag{11.1.5}$$

$$\begin{aligned} S_{12} &= \frac{1}{A_1}\frac{\partial u_1}{\partial \alpha_2} + \frac{1}{A_1}\frac{\partial u_2}{\partial \alpha_1} - \frac{u_1}{A_1A_2}\frac{\partial A_1}{\partial \alpha_2} - \frac{u_2}{A_1A_2}\frac{\partial A_2}{\partial \alpha_1} + \frac{1}{A_1}\frac{1}{A_2}\frac{\partial u_3}{\partial \alpha_1}\frac{\partial u_3}{\partial \alpha_2} \\ &\quad + r_6^e\left[\frac{1}{A_2}\frac{\partial}{\partial \alpha_2}\left(\frac{u_1}{R_1} - \frac{1}{A_1}\frac{\partial u_3}{\partial \alpha_1}\right) + \frac{1}{A_1}\frac{\partial}{\partial \alpha_1}\left(\frac{u_2}{R_2} - \frac{1}{A_2}\frac{\partial u_3}{\partial \alpha_2}\right)\right. \\ &\quad \left. - \frac{1}{A_1A_2}\left(\frac{u_1}{R_1} - \frac{1}{A_1}\frac{\partial u_3}{\partial \alpha_1}\right)\frac{\partial A_1}{\partial \alpha_2} - \frac{1}{A_1A_2}\left(\frac{u_2}{R_2} - \frac{1}{A_2}\frac{\partial u_3}{\partial \alpha_2}\right)\frac{\partial A_2}{\partial \alpha_1}\right]. \end{aligned} \tag{11.1.6}$$

Accordingly, the energy conversion based on the linear (i.e., small oscillation) and nonlinear (i.e., large oscillation) shell oscillations can be accommodated. Note that, for simplicity, later derivations are based on the linear shell theory. With the additions of the quadratic nonlinearity, the linear harvester derivations can be easily extended to the nonlinear ones.

With the modal expansion, the dynamic response u_i can be assumed as the sum of all participating modes multiplied by their corresponding modal participation factors η_{mn} (Tzou, 1993)

$$u_i(\alpha_1, \alpha_2, t) = \sum_{m=1}^{\infty}\sum_{n=1}^{\infty} \eta_{mn}(t)U_{imn}(\alpha_1, \alpha_2), \quad (i = 1, 2, 3) \tag{11.1.7}$$

where m, n are the mode numbers in the α_1 and α_2 directions, respectively; U_{imn} is the mode shape function and η_{mn} is the modal participation factor of the (m, n)th mode. The mode shape function U_{imn} denotes the spatial distribution for the mode (m, n), while the modal participation factor η_{mn} indicates the temporal contribution of each mode to the total dynamic response during the vibration. Since the harvester is mostly excited at resonance to maximize the energy output, a harmonic excitation is considered and its steady-state dynamic response is defined next.

11.1.2 Harmonic Excitation and Modal Response

As discussed previously, a harmonic external loading $F_i(\alpha_1, \alpha_2, t)$ with the excitation frequency ω in the i-th direction (i = 1, 2, 3) is applied to the shell (Soedel, 1993)

$$F_i(\alpha_1, \alpha_2, t) = F_i^*(\alpha_1, \alpha_2)\mathrm{e}^{\mathrm{j}\omega t}, \quad (i = 1, 2, 3) \tag{11.1.8}$$

where $F_i^*(\alpha_1, \alpha_2)$ denotes the magnitude; $\mathrm{j} = \sqrt{-1}$ is the imaginary unit; and $\mathrm{e}^{\mathrm{j}\omega t}$ is the harmonic function. This equation also denotes the separation of the spatial coordinate and the time-dependent external harmonic loading. Usually, a transverse harmonic excitation is used to excite the shell at one of its natural frequencies, i.e., $F_i^*(\alpha_1, \alpha_2) = F_3^*(\alpha_1, \alpha_2)$. Modal participation factor (or the (m, n)th modal response) η_{mn} is determined in order to evaluate the amount of participation of each mode in the total dynamic response. By considering Love's shell equations, eigenvalue analysis and orthogonality of mode shape functions, the modal response η_{mn} for each mode can be solved from the modal equation (Tzou, 1993)

$$\ddot{\eta}_{mn} + 2\zeta_{mn}\omega_{mn}\dot{\eta}_{mn} + \omega_{mn}^2\eta_{mn} = \hat{F}_{mn}(t), \tag{11.1.9}$$

where ω_{mn} denotes the (m, n)th natural frequency; $\hat{F}_{mn}(t)$ represents the modal force for the (m, n)th mode; ζ_{mn} is the modal damping ratio which is related to the equivalent damping constant c and the natural frequency ω_{mn}, i.e., $\zeta_{mn} = c/(2\rho h\omega_{mn})$ where ρ is shell's mass–density. With the harmonic excitation, the harmonic response occurs. When only considering the steady-state response, the modal force and the modal response can also be written in the harmonic forms:

$$\hat{F}_{mn} = \hat{F}_{mn}^*\mathrm{e}^{\mathrm{j}\omega t} \quad \text{and} \quad \eta_{mn} = \eta_{mn}^*\mathrm{e}^{\mathrm{j}\omega t}. \tag{11.1.10}$$

Thus, the modal response η_{mn} can be solved from Eq. (11.1.9) as

$$\eta_{mn} = \eta_{mn}^*\mathrm{e}^{\mathrm{j}\omega t} = \frac{\hat{F}_{mn}^*}{(\omega_{mn}^2 - \omega^2) + 2\mathrm{j}\zeta_{mn}\omega_{mn}\omega}\mathrm{e}^{\mathrm{j}\omega t}. \tag{11.1.11}$$

The modal force is related to the external loading $F_i(\alpha_1, \alpha_2, t)$ applied to the shell and the amplitude $\hat{F}_{mn}^*$ of the modal force is defined as (Tzou, 1993)

$$\hat{F}_{mn}^* = \frac{1}{\rho h N_{mn}} \int_{\alpha_1}\int_{\alpha_2} \left\{ \sum_{i=1}^{3} F_i^* U_{imn} \right\} A_1 A_2 \, \mathrm{d}\alpha_1 \mathrm{d}\alpha_2, \tag{11.1.12}$$

where N_{mn} is a constant defined by the mode shape functions

$$N_{mn} = \int_{\alpha_1}\int_{\alpha_2} \left\{ \sum_{i=1}^{3} U_{imn}^2 \right\} A_1 A_2 \, \mathrm{d}\alpha_1 \mathrm{d}\alpha_2. \qquad (11.1.13)$$

With the harmonic loading and mode shape functions, the modal response can be determined to estimate the contribution of each mode to the total dynamic response and subsequently the strains leading to energy generations of the piezoelectric energy harvester presented next.

11.2 Double-Curvature Shell Energy Harvester

With the dynamic response of the generic double-generic shell, a generic piezoelectric energy harvesting mechanism of the double-curvature shell is discussed in this section.

11.2.1 Shell Piezoelectric Coupling Mechanism

Based on the direct piezoelectric effect, the strains (either linear or nonlinear) imposed on the piezoelectric patch can generate electric charges or voltages. Since the transverse output is the most important component in thin shell and non-shell harvesters, the transverse electric displacement is defined by

$$D_3 = e_{31}S_{11} + e_{32}S_{22} + e_{36}S_{12} + \varepsilon_{33}E_3, \qquad (11.2.1)$$

where e_{3i} is the piezoelectric stress constant in the i direction; ε_{33} is the dielectric constant; E_3 is the electric filed in the transverse direction. With the electric displacement in the transverse direction, the total electric charge can be obtained by integrating the electric displacement over the whole effective electrode. Then, the current i in the closed-circuit condition is the time derivative of charge Q (Meeker, 1996)

$$i(t) = \frac{\mathrm{d}Q}{\mathrm{d}t} = \frac{\partial}{\partial t}\int_{S^e} D_3 \, \mathrm{d}S^e = \frac{\partial}{\partial t}\int_{S^e} (e_{31}S_{11} + e_{32}S_{22} + e_{36}S_{12} + \varepsilon_{33}E_3)\mathrm{d}S^e, \qquad (11.2.2)$$

where S^e is the area of the effective electrode. Note that, in reality, $i(t) \equiv i(\alpha_1, \alpha_2, t)$ and $Q(t) \equiv Q(\alpha_1, \alpha_2, t)$ although assumed instantaneously averaged on the electrode. Since the focus is on the dynamic response of the harvester, for simplification, only the time-dependent characteristics are emphasized in later derivations. So the voltage across the resistive load V_{R_L} becomes

$$V_{R_L}(t) = i(t)R_L = R_L \frac{\mathrm{d}}{\mathrm{d}t} \int\limits_{S^e} (e_{31}S_{11} + e_{32}S_{22} + e_{36}S_{12} + \varepsilon_{33}E_3)\mathrm{d}S^e. \qquad (11.2.3)$$

Based on the equivalent current model of the piezoelectric energy harvester, an equivalent capacitance of the piezoelectric patch C_p connects with the resistive load R_L in parallel. Thus, the voltage across the equivalent capacitance of the piezoelectric patch C_p is equal to the voltage across the resistive load R_L. With Maxwell's equation, the transverse electric field and voltage has the relationship $E_3 = -V_{R_L}/h^e$. Thus, the above equation can be further rewritten as

$$V_{R_L}(t) = R_L \frac{\mathrm{d}}{\mathrm{d}t} \int\limits_{S^e} (e_{31}S_{11} + e_{32}S_{22} + e_{36}S_{12} - \varepsilon_{33}V_{R_L}(t)/h^e)\mathrm{d}S^e. \qquad (11.2.4)$$

The voltage across the resistive load is a function of time. Thus, from Eq. (11.2.4) or the Kirchhoff's current law, the electromechanical coupling equation of the voltage across the resistive load can be rewritten as

$$\frac{\varepsilon_{33}S^e}{h^e}\frac{\mathrm{d}V_{R_L}(t)}{\mathrm{d}t} + \frac{V_{R_L}(t)}{R_L} = \frac{\mathrm{d}}{\mathrm{d}t}\left(\int\limits_{S^e} (e_{31}S_{11} + e_{32}S_{22} + e_{36}S_{12})\mathrm{d}S^e\right), \qquad (11.2.5a)$$

or

$$\tau\frac{\mathrm{d}V_{R_L}(t)}{\mathrm{d}t} + V_{R_L}(t) = R_L\frac{\mathrm{d}}{\mathrm{d}t}\left(\int\limits_{S^e} (e_{31}S_{11} + e_{32}S_{22} + e_{36}S_{12})\mathrm{d}S^e\right), \qquad (11.2.5b)$$

where $\varepsilon_{33}S^e/h^e = C_p$; the time constant τ of the first-order voltage equation $\tau\frac{\mathrm{d}V_{R_L}(t)}{\mathrm{d}t} + V_{R_L}(t) = R_L\frac{\mathrm{d}}{\mathrm{d}t}\left(\int_{S^e} (e_{31}S_{11} + e_{32}S_{22} + e_{36}S_{12})\mathrm{d}S^e\right)$ is $\tau = R_L\varepsilon_{33}S^e/h^e$ or $\tau = R_LC_p$. Substituting Eqs. (11.1.1)–(11.1.3) and Eq. (11.1.7) into Eq. (11.2.5) and simplifying yields

$$\frac{\varepsilon_{33}S^e}{h^e}\frac{\mathrm{d}V_{R_L}(t)}{\mathrm{d}t} + \frac{V_{R_L}(t)}{R_L} = \frac{\varepsilon_{33}S^e}{h^e}\sum_{m=1}^{\infty}\sum_{n=1}^{\infty}\phi_{mn}\dot{\eta}_{mn}, \qquad (11.2.6a)$$

or

$$\tau\frac{\mathrm{d}V_{R_L}(t)}{\mathrm{d}t} + V_{R_L}(t) = \frac{R_L\varepsilon_{33}S^e}{h^e}\sum_{m=1}^{\infty}\sum_{n=1}^{\infty}\phi_{mn}\dot{\eta}_{mn} = \tau\sum_{m=1}^{\infty}\sum_{n=1}^{\infty}\phi_{mn}\dot{\eta}_{mn}, \qquad (11.2.6b)$$

where "·" denotes the time derivative; ϕ_{mn} is the modal voltage across the resistive load for the mode (m, n), expressed as (Tzou, 1993; Sect. 4.2 of Chap. 4)

$$
\begin{aligned}
\phi_{mn} = \frac{h^e}{\varepsilon_{33}S^e}\int\limits_{S^e} \Bigg\{ e_{31}\Bigg\{ & \frac{1}{A_1}\frac{\partial}{\partial\alpha_1}U_{1mn}(\alpha_1,\alpha_2)+\frac{1}{A_1A_2}U_{2mn}(\alpha_1,\alpha_2)\frac{\partial A_1}{\partial\alpha_2} \\
& + \frac{1}{R_1}U_{3mn}(\alpha_1,\alpha_2) \\
& + r^e\Bigg[\frac{1}{A_1}\frac{\partial}{\partial\alpha_1}\left(\frac{U_{1mn}(\alpha_1,\alpha_2)}{R_1}-\frac{1}{A_1}\frac{\partial U_{3mn}(\alpha_1,\alpha_2)}{\partial\alpha_1}\right) \\
& + \left(\frac{1}{A_1A_2}\left(\frac{U_{2mn}(\alpha_1,\alpha_2)}{R_2}-\frac{1}{A_2}\frac{\partial U_{3mn}(\alpha_1,\alpha_2)}{\partial\alpha_2}\right)\frac{\partial A_1}{\partial\alpha_2}\right)\Bigg]\Bigg\} \\
+ e_{32}\Bigg\{ & \frac{1}{A_2}\frac{\partial}{\partial\alpha_2}U_{2mn}(\alpha_1,\alpha_2)+\frac{1}{A_1A_2}U_{1mn}(\alpha_1,\alpha_2)\frac{\partial A_2}{\partial\alpha_1}+\frac{1}{R_2}U_{3mn}(\alpha_1,\alpha_2) \\
& + r^e\Bigg[\frac{1}{A_2}\frac{\partial}{\partial\alpha_2}\left(\frac{U_{2mn}(\alpha_1,\alpha_2)}{R_2}-\frac{1}{A_2}\frac{\partial U_{3mn}(\alpha_1,\alpha_2)}{\partial\alpha_2}\right) \\
& + \left(\frac{1}{A_1A_2}\left(\frac{U_{1mn}(\alpha_1,\alpha_2)}{R_1}-\frac{1}{A_1}\frac{\partial U_{3mn}(\alpha_1,\alpha_2)}{\partial\alpha_1}\right)\frac{\partial A_2}{\partial\alpha_1}\right)\Bigg]\Bigg\} \\
+ e_{36}\Bigg\{ & \frac{1}{A_2}\frac{\partial U_{1mn}(\alpha_1,\alpha_2)}{\partial\alpha_2}+\frac{1}{A_1}\frac{\partial U_{2mn}(\alpha_1,\alpha_2)}{\partial\alpha_1}-\frac{U_{2mn}(\alpha_1,\alpha_2)}{A_1A_2}\frac{\partial A_2}{\partial\alpha_1} \\
& - \frac{U_{1mn}(\alpha_1,\alpha_2)}{A_1A_2}\frac{\partial A_1}{\partial\alpha_2} \\
& + r^e\Bigg[\frac{1}{A_2}\frac{\partial}{\partial\alpha_2}\left(\frac{U_{1mn}(\alpha_1,\alpha_2)}{R_1}-\frac{1}{A_1}\frac{\partial U_{3mn}(\alpha_1,\alpha_2)}{\partial\alpha_1}\right) \\
& + \frac{1}{A_1}\frac{\partial}{\partial\alpha_1}\left(\frac{U_{2mn}(\alpha_1,\alpha_2)}{R_2}-\frac{1}{A_2}\frac{\partial U_{3mn}(\alpha_1,\alpha_2)}{\partial\alpha_2}\right) \\
& - \frac{1}{A_1A_2}\left(\frac{U_{1mn}(\alpha_1,\alpha_2)}{R_1}-\frac{1}{A_1}\frac{\partial U_{3mn}(\alpha_1,\alpha_2)}{\partial\alpha_1}\right)\frac{\partial A_1}{\partial\alpha_2} \\
& - \frac{1}{A_1A_2}\left(\frac{U_{2mn}(\alpha_1,\alpha_2)}{R_2}-\frac{1}{A_2}\frac{\partial U_{3mn}(\alpha_1,\alpha_2)}{\partial\alpha_2}\right)\frac{\partial A_2}{\partial\alpha_1}\Bigg]\Bigg\}\Bigg\}\mathrm{d}S^e.
\end{aligned}
\tag{11.2.7}
$$

It shows that the modal voltage of each mode only depends on intrinsic structural parameters and mode shape functions U_{jmn}, $j=1,2,3$ and it is independent of external excitations. Once the structural geometry is specified, its distinct Lamé parameters A_i, i = 1, 2 and radii of curvature R_i, i = 1, 2 are known (Tzou and Bao, 1997). Substituting these four parameters into Eqs. (11.2.6) and (11.2.7) and simplifying, one obtains the distributed electromechanical coupling equation for the selected geometry, Sect. 4.5 of Chap. 4. The mode shape function is defined based on specified boundary conditions. Thus, the modal voltage across the resistive load is unique with a given size/location of the piezoelectric patch laminated on the shell surface. With the modal voltage and modal response for each mode, voltage output/ power across the resistive load can be estimated next.

11.2.2 Voltage and Power Output

With a harmonic excitation, the voltage across the resistive load R_L is also harmonic, which can be expressed as $V_{R_L} = V_{R_L}^* e^{j\omega t}$. With the known modal response η_{mn} (Eq. 11.1.11) and modal voltage ϕ_{mn} for the mode (m, n), the voltage V_{R_L} can be determined from Eq. (11.2.6) as

$$\begin{aligned} V_{R_L} &= \sum_{m=1}^{\infty}\sum_{n=1}^{\infty} \frac{\phi_{mn} R_L \varepsilon_{33} S^e}{h^e} \frac{j\omega \hat{F}_{mn}^*}{(1 + j\omega R_L \varepsilon_{33} S^e/h^e)\left[(\omega_{mn}^2 - \omega^2) + 2j\zeta_{mn}\omega_{mn}\omega\right]} e^{j\omega t} \\ &= \sum_{m=1}^{\infty}\sum_{n=1}^{\infty} \frac{(\phi_{mn}\hat{F}_{mn}^* \omega R_L \varepsilon_{33} S^e/h^e) e^{j(\omega t - \varphi*)}}{\left\{\left[\omega_{mn}^2 - \omega^2(1 + 2\zeta_{mn}\omega_{mn} R_L \varepsilon_{33} S^e/h^e)\right]^2 + \left[(\omega_{mn}^2 - \omega^2)\omega R_L \varepsilon_{33} S^e/h^e + 2\zeta_{mn} er \omega_{mn}\omega\right]^2\right\}^{1/2}} \end{aligned} \tag{11.2.8}$$

where φ^* is the phase angle lagging behind the excitation and it is defined as

$$\varphi^* = \arctan\left(\frac{\omega_{mn}^2 - \omega^2(1 + 2\zeta_{mn}\omega_{mn} R_L \varepsilon_{33} S^e/h^e)}{(\omega_{mn}^2 - \omega^2)\omega R_L \varepsilon_{33} S^e/h^e + 2\zeta_{mn}\omega_{mn}\omega}\right). \tag{11.2.9}$$

To maximize the energy output, the excitation frequency is assumed to be equal to the (m, n)th natural frequency, i.e., resonance. Thus, the vibration response of the (m, n)th mode dominates and the output reaches the maximum. The generic shell resonant voltage response, Eq. (11.2.7), is degraded to a single (m, n) mode for the steady-state response. The independent voltage output across the resistive load is expressed as

$$\begin{aligned} (V_{R_L})_{mn} &= (V_{R_L})_{mn}^* e^{j(\omega t - \varphi*)} \\ &= \frac{(\phi_{mn}\hat{F}_{mn}^* \omega R_L \varepsilon_{33} S^e/h^e) e^{j(\omega t - \varphi*)}}{\left\{\left[\omega_{mn}^2 - \omega^2(1 + 2\zeta_{mn}\omega_{mn} R_L \varepsilon_{33} S^e/h^e)\right]^2 + \left[(\omega_{mn}^2 - \omega^2)\omega R_L \varepsilon_{33} S^e/h^e + 2\zeta_{mn}\omega_{mn}\omega\right]^2\right\}^{1/2}} \end{aligned} \tag{11.2.10}$$

In order to explicitly define the shell's voltage and power outputs ready for simplified shell/non-shell engineering geometries and structures, substituting the shell modal voltage Eq. (11.2.7) into Eq. (11.2.10) and restricting to a single resonant (m, n)th mode yields the voltage output magnitude

$$
\begin{aligned}
(V_{R_L})^*_{mn} = R_L\hat{F}^*_{mn}\omega \int_{S^e} \Bigg\{ & e_{31}\left\{\frac{1}{A_1}\frac{\partial}{\partial\alpha_1}U_{1mn}(\alpha_1,\alpha_2) + \frac{1}{A_1A_2}U_{2mn}(\alpha_1,\alpha_2)\frac{\partial A_1}{\partial\alpha_2}\right. \\
& + \frac{U_{3mn}(\alpha_1,\alpha_2)}{R_1} \\
& \quad + r^e\left[\frac{1}{A_1}\frac{\partial}{\partial\alpha_1}\left(\frac{U_{1mn}(\alpha_1,\alpha_2)}{R_1} - \frac{1}{A_1}\frac{\partial U_{3mn}(\alpha_1,\alpha_2)}{\partial\alpha_1}\right)\right. \\
& \quad \left.\left. + \left(\frac{1}{A_1A_2}\left(\frac{U_{2mn}(\alpha_1,\alpha_2)}{R_2} - \frac{1}{A_2}\frac{\partial U_{3mn}(\alpha_1,\alpha_2)}{\partial\alpha_2}\right)\frac{\partial A_1}{\partial\alpha_2}\right)\right]\right\} \\
& + e_{32}\left\{\frac{1}{A_2}\frac{\partial}{\partial\alpha_2}U_{2mn}(\alpha_1,\alpha_2) + \frac{1}{A_1A_2}U_{1mn}(\alpha_1,\alpha_2)\frac{\partial A_2}{\partial\alpha_1} + \frac{1}{R_2}U_{3mn}(\alpha_1,\alpha_2)\right. \\
& \quad + r^e\left[\frac{1}{A_2}\frac{\partial}{\partial\alpha_2}\left(\frac{U_{2mn}(\alpha_1,\alpha_2)}{R_2} - \frac{1}{A_2}\frac{\partial U_{3mn}(\alpha_1,\alpha_2)}{\partial\alpha_2}\right)\right. \\
& \quad \left.\left. + \left(\frac{1}{A_1A_2}\left(\frac{U_{1mn}(\alpha_1,\alpha_2)}{R_1} - \frac{1}{A_1}\frac{\partial U_{3mn}(\alpha_1,\alpha_2)}{\partial\alpha_1}\right)\frac{\partial A_2}{\partial\alpha_1}\right)\right]\right\} \\
& + e_{36}\left\{\frac{1}{A_2}\frac{\partial U_{1mn}(\alpha_1,\alpha_2)}{\partial\alpha_2} + \frac{1}{A_1}\frac{\partial U_{2mn}(\alpha_1,\alpha_2)}{\partial\alpha_1} - \frac{U_{2mn}(\alpha_1,\alpha_2)}{A_1A_2}\frac{\partial A_2}{\partial\alpha_1}\right. \\
& - \frac{U_{1mn}(\alpha_1,\alpha_2)}{A_1A_2}\frac{\partial A_1}{\partial\alpha_2} \\
& \quad + r^e\left[\frac{1}{A_2}\frac{\partial}{\partial\alpha_2}\left(\frac{U_{1mn}(\alpha_1,\alpha_2)}{R_1} - \frac{1}{A_1}\frac{\partial U_{3mn}(\alpha_1,\alpha_2)}{\partial\alpha_1}\right)\right. \\
& \quad + \frac{1}{A_1}\frac{\partial}{\partial\alpha_1}\left(\frac{U_{2mn}(\alpha_1,\alpha_2)}{R_2} - \frac{1}{A_2}\frac{\partial U_{3mn}(\alpha_1,\alpha_2)}{\partial\alpha_2}\right) \\
& \quad - \frac{1}{A_1A_2}\left(\frac{U_{1mn}(\alpha_1,\alpha_2)}{R_1} - \frac{1}{A_1}\frac{\partial U_{3mn}(\alpha_1,\alpha_2)}{\partial\alpha_1}\right)\frac{\partial A_1}{\partial\alpha_2} \\
& \quad \left.\left.\left. - \frac{1}{A_1A_2}\left(\frac{U_{2mn}(\alpha_1,\alpha_2)}{R_2} - \frac{1}{A_2}\frac{\partial U_{3mn}(\alpha_1,\alpha_2)}{\partial\alpha_2}\right)\frac{\partial A_2}{\partial\alpha_1}\right]\right\}\right\}\mathrm{d}S^e \\
\Bigg/ \Big\{ & \left[\omega_{mn}^2 - \omega^2(1 + 2\zeta_{mn}\omega_{mn}R_L\varepsilon_{33}S^e/h^e)\right]^2 \\
& + \left[(\omega_{mn}^2 - \omega^2)\omega R_L\varepsilon_{33}S^e/h^e + 2\zeta_{mn}\omega_{mn}\omega\right]^2\Big\}^{1/2}.
\end{aligned}
\tag{11.2.11}
$$

Note that this is an explicit voltage expression with an arbitrary excitation ω and it is readily simplified to other shell and non-shell geometries if two Lamé parameters and two curvature radii are specified, Sect. 4.5 of Chap. 4. Considering the steady-state harmonic response, the root-mean-square (RMS) of the power output across the resistive load for the (m, n)th mode is

$$
\begin{aligned}
&(P^*_{mn})_{\mathrm{RMS}} = \left((V_{R_L})^*_{mn} \Big/ \sqrt{2} \right)^2 \Big/ R_L \\
&= R_L \Bigg(\hat{F}^*_{mn} \omega \int_{S^e} \Bigg\{ e_{31} \Bigg\{ \frac{1}{A_1} \frac{\partial}{\partial \alpha_1} U_{1mn}(\alpha_1, \alpha_2) + \frac{1}{A_1 A_2} U_{2mn}(\alpha_1, \alpha_2) \frac{\partial A_1}{\partial \alpha_2} + \frac{U_{3mn}(\alpha_1, \alpha_2)}{R_1} \\
&\qquad + r^e \Bigg[\frac{1}{A_1} \frac{\partial}{\partial \alpha_1} \left(\frac{U_{1mn}(\alpha_1, \alpha_2)}{R_1} - \frac{1}{A_1} \frac{\partial U_{3mn}(\alpha_1, \alpha_2)}{\partial \alpha_1} \right) \\
&\qquad + \left(\frac{1}{A_1 A_2} \left(\frac{U_{2mn}(\alpha_1, \alpha_2)}{R_2} - \frac{1}{A_2} \frac{\partial U_{3mn}(\alpha_1, \alpha_2)}{\partial \alpha_2} \right) \frac{\partial A_1}{\partial \alpha_2} \right) \Bigg] \Bigg\} \\
&\quad + e_{32} \Bigg\{ \frac{1}{A_2} \frac{\partial}{\partial \alpha_2} U_{2mn}(\alpha_1, \alpha_2) + \frac{1}{A_1 A_2} U_{1mn}(\alpha_1, \alpha_2) \frac{\partial A_2}{\partial \alpha_1} + \frac{1}{R_2} U_{3mn}(\alpha_1, \alpha_2) \\
&\qquad + r^e \Bigg[\frac{1}{A_2} \frac{\partial}{\partial \alpha_2} \left(\frac{U_{2mn}(\alpha_1, \alpha_2)}{R_2} - \frac{1}{A_2} \frac{\partial U_{3mn}(\alpha_1, \alpha_2)}{\partial \alpha_2} \right) \\
&\qquad + \left(\frac{1}{A_1 A_2} \left(\frac{U_{1mn}(\alpha_1, \alpha_2)}{R_1} - \frac{1}{A_1} \frac{\partial U_{3mn}(\alpha_1, \alpha_2)}{\partial \alpha_1} \right) \frac{\partial A_2}{\partial \alpha_1} \right) \Bigg] \Bigg\} \\
&\quad + e_{36} \Bigg\{ \frac{1}{A_2} \frac{\partial U_{1mn}(\alpha_1, \alpha_2)}{\partial \alpha_2} + \frac{1}{A_1} \frac{\partial U_{2mn}(\alpha_1, \alpha_2)}{\partial \alpha_1} - \frac{U_{2mn}(\alpha_1, \alpha_2)}{A_1 A_2} \frac{\partial A_2}{\partial \alpha_1} \\
&\qquad - \frac{U_{1mn}(\alpha_1, \alpha_2)}{A_1 A_2} \frac{\partial A_1}{\partial \alpha_2} + r^e \Bigg[\frac{1}{A_2} \frac{\partial}{\partial \alpha_2} \left(\frac{U_{1mn}(\alpha_1, \alpha_2)}{R_1} - \frac{1}{A_1} \frac{\partial U_{3mn}(\alpha_1, \alpha_2)}{\partial \alpha_1} \right) \\
&\qquad + \frac{1}{A_1} \frac{\partial}{\partial \alpha_1} \left(\frac{U_{2mn}(\alpha_1, \alpha_2)}{R_2} - \frac{1}{A_2} \frac{\partial U_{3mn}(\alpha_1, \alpha_2)}{\partial \alpha_2} \right) \\
&\qquad - \frac{1}{A_1 A_2} \left(\frac{U_{1mn}(\alpha_1, \alpha_2)}{R_1} - \frac{1}{A_1} \frac{\partial U_{3mn}(\alpha_1, \alpha_2)}{\partial \alpha_1} \right) \frac{\partial A_1}{\partial \alpha_2} \\
&\qquad - \frac{1}{A_1 A_2} \left(\frac{U_{2mn}(\alpha_1, \alpha_2)}{R_2} - \frac{1}{A_2} \frac{\partial U_{3mn}(\alpha_1, \alpha_2)}{\partial \alpha_2} \right) \frac{\partial A_2}{\partial \alpha_1} \Bigg] \Bigg\} \Bigg\} dS^e \Bigg)^2 \\
&\Big/ 2 \Big\{ \left[\omega_{mn}^2 - \omega^2 (1 + 2\zeta_{mn} \omega_{mn} R_L \varepsilon_{33} S^e / h^e) \right]^2 \\
&+ \left[(\omega_{mn}^2 - \omega^2) \omega R_L \varepsilon_{33} S^e / h^e + 2\zeta_{mn} \omega_{mn} \omega \right]^2 \Big\}
\end{aligned}
\qquad (11.2.12)
$$

Usually, the energy harvester works at resonance, i.e., $\omega = \omega_{mn}$ for each shell mode to maximize the output. Thus, the voltage and power outputs Eqs. (11.2.10)–(11.2.12) at resonance become

$$(V_{R_L})_{mn} = (V_{R_L})^*_{mn} e^{j(\omega_{mn}t-\varphi*)} = \frac{\left(\phi_{mn}\hat{F}^*_{mn}\omega_{mn}R_L\varepsilon_{33}S^e/h^e\right)e^{j(\omega_{mn}t-\varphi*)}}{\left(2\zeta_{mn}\omega^2_{mn}\right)\sqrt{1+\left(R_L\omega_{mn}\varepsilon_{33}S^e/h^e\right)^2}}, \tag{11.2.13}$$

and its resonance voltage magnitude $(V_{R_L})^*_{mn}$ and RMS power $\left(P^*_{mn}\right)_{\mathrm{RMS}}$ are respectively

$$\begin{aligned}
(V_{R_L})^*_{mn} = R_L\hat{F}^*_{mn}\omega_{mn} \int\limits_{S^e} \Bigg\{ & e_{31}\Bigg\{\frac{1}{A_1}\frac{\partial}{\partial\alpha_1}U_{1mn}(\alpha_1,\alpha_2) + \frac{1}{A_1A_2}U_{2mn}(\alpha_1,\alpha_2)\frac{\partial A_1}{\partial\alpha_2} \\
& + \frac{U_{3mn}(\alpha_1,\alpha_2)}{R_1} + r^e\left[\frac{1}{A_1}\frac{\partial}{\partial\alpha_1}\left(\frac{U_{1mn}(\alpha_1,\alpha_2)}{R_1} - \frac{1}{A_1}\frac{\partial U_{3mn}(\alpha_1,\alpha_2)}{\partial\alpha_1}\right)\right. \\
& \left.+\left(\frac{1}{A_1A_2}\left(\frac{U_{2mn}(\alpha_1,\alpha_2)}{R_2} - \frac{1}{A_2}\frac{\partial U_{3mn}(\alpha_1,\alpha_2)}{\partial\alpha_2}\right)\frac{\partial A_1}{\partial\alpha_2}\right)\right]\Bigg\} \\
& + e_{32}\Bigg\{\frac{1}{A_2}\frac{\partial}{\partial\alpha_2}U_{2mn}(\alpha_1,\alpha_2) + \frac{1}{A_1A_2}U_{1mn}(\alpha_1,\alpha_2)\frac{\partial A_2}{\partial\alpha_1} + \frac{1}{R_2}U_{3mn}(\alpha_1,\alpha_2) \\
& + r^e\left[\frac{1}{A_2}\frac{\partial}{\partial\alpha_2}\left(\frac{U_{2mn}(\alpha_1,\alpha_2)}{R_2} - \frac{1}{A_2}\frac{\partial U_{3mn}(\alpha_1,\alpha_2)}{\partial\alpha_2}\right)\right. \\
& \left.+\left(\frac{1}{A_1A_2}\left(\frac{U_{1mn}(\alpha_1,\alpha_2)}{R_1} - \frac{1}{A_1}\frac{\partial U_{3mn}(\alpha_1,\alpha_2)}{\partial\alpha_1}\right)\frac{\partial A_2}{\partial\alpha_1}\right)\right]\Bigg\} \\
& + e_{36}\Bigg\{\frac{1}{A_2}\frac{\partial U_{1mn}(\alpha_1,\alpha_2)}{\partial\alpha_2} + \frac{1}{A_1}\frac{\partial U_{2mn}(\alpha_1,\alpha_2)}{\partial\alpha_1} - \frac{U_{2mn}(\alpha_1,\alpha_2)}{A_1A_2}\frac{\partial A_2}{\partial\alpha_1} \\
& - \frac{U_{1mn}(\alpha_1,\alpha_2)}{A_1A_2}\frac{\partial A_1}{\partial\alpha_2} + r^e\left[\frac{1}{A_2}\frac{\partial}{\partial\alpha_2}\left(\frac{U_{1mn}(\alpha_1,\alpha_2)}{R_1} - \frac{1}{A_1}\frac{\partial U_{3mn}(\alpha_1,\alpha_2)}{\partial\alpha_1}\right)\right. \\
& + \frac{1}{A_1}\frac{\partial}{\partial\alpha_1}\left(\frac{U_{2mn}(\alpha_1,\alpha_2)}{R_2} - \frac{1}{A_2}\frac{\partial U_{3mn}(\alpha_1,\alpha_2)}{\partial\alpha_2}\right) \\
& - \frac{1}{A_1A_2}\left(\frac{U_{1mn}(\alpha_1,\alpha_2)}{R_1} - \frac{1}{A_1}\frac{\partial U_{3mn}(\alpha_1,\alpha_2)}{\partial\alpha_1}\right)\frac{\partial A_1}{\partial\alpha_2} \\
& \left.- \frac{1}{A_1A_2}\left(\frac{U_{2mn}(\alpha_1,\alpha_2)}{R_2} - \frac{1}{A_2}\frac{\partial U_{3mn}(\alpha_1,\alpha_2)}{\partial\alpha_2}\right)\frac{\partial A_2}{\partial\alpha_1}\right]\Bigg\}\Bigg\} dS^e \\
& \Bigg/ \left(2\zeta_{mn}\omega^2_{mn}\right)\sqrt{1+\left(R_L\omega_{mn}\varepsilon_{33}S^e/h^e\right)^2}.
\end{aligned} \tag{11.2.14}$$

$$
\begin{aligned}
\left(P_{mn}^*\right)_{\mathrm{RMS}} &= \left(\left(V_{R_L}\right)_{mn}^* \Big/ \sqrt{2}\right)^2 \Big/ R_L \\
&= R_L \left(\hat{F}_{mn}^* \omega_{mn} \int_{S^e} \left\{ e_{31} \left\{ \frac{1}{A_1} \frac{\partial U_{1mn}(\alpha_1, \alpha_2)}{\partial \alpha_1} + \frac{U_{2mn}(\alpha_1, \alpha_2)}{A_1 A_2} \frac{\partial A_1}{\partial \alpha_2} + \frac{U_{3mn}(\alpha_1, \alpha_2)}{R_1} \right. \right. \right. \\
&+ r^e \left[\frac{1}{A_1} \frac{\partial}{\partial \alpha_1} \left(\frac{U_{1mn}(\alpha_1, \alpha_2)}{R_1} - \frac{1}{A_1} \frac{\partial U_{3mn}(\alpha_1, \alpha_2)}{\partial \alpha_1} \right) \right. \\
&+ \left. \left. \left(\frac{1}{A_1 A_2} \left(\frac{U_{2mn}(\alpha_1, \alpha_2)}{R_2} - \frac{1}{A_2} \frac{\partial U_{3mn}(\alpha_1, \alpha_2)}{\partial \alpha_2} \right) \frac{\partial A_1}{\partial \alpha_2} \right) \right] \right\} \\
&+ e_{32} \left\{ \frac{1}{A_2} \frac{\partial U_{2mn}(\alpha_1, \alpha_2)}{\partial \alpha_2} + \frac{U_{1mn}(\alpha_1, \alpha_2)}{A_1 A_2} \frac{\partial A_2}{\partial \alpha_1} + \frac{U_{3mn}(\alpha_1, \alpha_2)}{R_2} \right. \\
&+ r^e \left[\frac{1}{A_2} \frac{\partial}{\partial \alpha_2} \left(\frac{U_{2mn}(\alpha_1, \alpha_2)}{R_2} - \frac{1}{A_2} \frac{\partial U_{3mn}(\alpha_1, \alpha_2)}{\partial \alpha_2} \right) \right. \\
&+ \left. \left. \left(\frac{1}{A_1 A_2} \left(\frac{U_{1mn}(\alpha_1, \alpha_2)}{R_1} - \frac{1}{A_1} \frac{\partial U_{3mn}(\alpha_1, \alpha_2)}{\partial \alpha_1} \right) \frac{\partial A_2}{\partial \alpha_1} \right) \right] \right\} \\
&+ e_{36} \left\{ \frac{1}{A_2} \frac{\partial U_{1mn}(\alpha_1, \alpha_2)}{\partial \alpha_2} + \frac{1}{A_1} \frac{\partial U_{2mn}(\alpha_1, \alpha_2)}{\partial \alpha_1} - \frac{U_{2mn}(\alpha_1, \alpha_2)}{A_1 A_2} \frac{\partial A_2}{\partial \alpha_1} - \frac{U_{1mn}(\alpha_1, \alpha_2)}{A_1 A_2} \frac{\partial A_1}{\partial \alpha_2} \right. \\
&+ r^e \left[\frac{1}{A_2} \frac{\partial}{\partial \alpha_2} \left(\frac{U_{1mn}(\alpha_1, \alpha_2)}{R_1} - \frac{1}{A_1} \frac{\partial U_{3mn}(\alpha_1, \alpha_2)}{\partial \alpha_1} \right) + \frac{1}{A_1} \frac{\partial}{\partial \alpha_1} \left(\frac{U_{2mn}(\alpha_1, \alpha_2)}{R_2} \right. \right. \\
&- \left. \frac{1}{A_2} \frac{\partial U_{3mn}(\alpha_1, \alpha_2)}{\partial \alpha_2} \right) + \frac{1}{A_1} \frac{\partial}{\partial \alpha_1} \left(\frac{U_{2mn}(\alpha_1, \alpha_2)}{R_2} - \frac{1}{A_2} \frac{\partial U_{3mn}(\alpha_1, \alpha_2)}{\partial \alpha_2} \right) \\
&- \frac{1}{A_1 A_2} \left(\frac{U_{1mn}(\alpha_1, \alpha_2)}{R_1} - \frac{1}{A_1} \frac{\partial U_{3mn}(\alpha_1, \alpha_2)}{\partial \alpha_1} \right) \frac{\partial A_1}{\partial \alpha_2} - \frac{1}{A_1 A_2} \left(\frac{U_{2mn}(\alpha_1, \alpha_2)}{R_2} \right. \\
&- \left. \left. \left. \left. \left. \frac{1}{A_2} \frac{\partial U_{3mn}(\alpha_1, \alpha_2)}{\partial \alpha_2} \right) \frac{\partial A_2}{\partial \alpha_1} \right] \right\} \right\} dS^e \right)^2 \Big/ 2\left(2\zeta_{mn}\omega_{mn}^2\right)^2 \left[1 + \left(R_L \omega_{mn} \varepsilon_{33} S^e / h^e \right)^2 \right].
\end{aligned}
\tag{11.2.15}
$$

Again, these are explicit voltage and power expressions ready to be simplified to account for other shell and non-shell geometries. These procedures are demonstrated in case studies. (Note that similar procedures were used in Chaps. 3 and 4.) On the other hand, the maximal power output across the resistive load is expressed as

$$
\left(P_{mn}^*\right)_{\max} = 2\left(P_{mn}^*\right)_{\mathrm{RMS}}. \tag{11.2.16}
$$

The $(m,\ n)$th modal voltage output and power across the resistive load are derived based on the generic double-curvature shell harvester with linear or nonlinear oscillations and they are explicitly expressed in design and geometric parameters, i.e., harvester geometry (defined by Lamé parameters A_i, radii of curvature R_i), shell/piezoelectric material properties, frequency, damping ratio, load resistance R_L, etc. Again, with appropriate material selections for a selected linear/nonlinear shell or non-shell harvester geometry, one can simplify the voltage and power equations to estimate the specified distributed harvester outputs and efficiency during the design stage, e.g., Sect. 4.5, Chap. 4. Electromechanical voltage and output characteristics of a beam harvester, a ring harvester and a conical shell harvester simplified from the generic double-curvature shell harvester are demonstrated in case studies presented next.

11.3 Various Energy Harvesters

In practical applications, the piezoelectric material is usually symmetrical hexagonal and the piezoelectric constant e_{36} is assumed zero, i.e., no in-plane twist effect, Chap. 4. This indicates that the shear strain S_{12} makes no contribution to the output signal. Thus, the electric energy is only related to the normal strains in the α_1 and α_2 directions. Accordingly, the $(m,\ n)$th modal voltage induced by the hexagonal piezoelectric patch laminated on a nonlinear shell can be simplified to

$$
\begin{aligned}
\phi_{mn} = \frac{h^e}{\varepsilon_{33} S^e} \int_{S^e} \Bigg\{ & e_{31} \Bigg\{ \frac{1}{A_1} \frac{\partial}{\partial \alpha_1} U_{1mn}(\alpha_1, \alpha_2) + \frac{1}{A_1 A_2} U_{2mn}(\alpha_1, \alpha_2) \frac{\partial A_1}{\partial \alpha_2} \\
& + \frac{1}{R_1} U_{3mn}(\alpha_1, \alpha_2) + \frac{1}{2} \left(\frac{1}{A_1} \frac{\partial U_{3mn}(\alpha_1, \alpha_2)}{\partial \alpha_1} \right)^2 \\
& + r^e \left[\frac{1}{A_1} \frac{\partial}{\partial \alpha_1} \left(\frac{U_{1mn}(\alpha_1, \alpha_2)}{R_1} - \frac{1}{A_1} \frac{\partial U_{3mn}(\alpha_1, \alpha_2)}{\partial \alpha_1} \right) \right. \\
& \left. + \left(\frac{1}{A_1 A_2} \left(\frac{U_{2mn}(\alpha_1, \alpha_2)}{R_2} - \frac{1}{A_2} \frac{\partial U_{3mn}(\alpha_1, \alpha_2)}{\partial \alpha_2} \right) \frac{\partial A_1}{\partial \alpha_2} \right) \right] \Bigg\} \\
& + e_{32} \Bigg\{ \frac{1}{A_2} \frac{\partial}{\partial \alpha_2} U_{2mn}(\alpha_1, \alpha_2) + \frac{1}{A_1 A_2} U_{1mn}(\alpha_1, \alpha_2) \frac{\partial A_2}{\partial \alpha_1} + \frac{1}{R_2} U_{3mn}(\alpha_1, \alpha_2) \\
& + \frac{1}{2} \left(\frac{1}{A_2} \frac{\partial U_{3mn}(\alpha_1, \alpha_2)}{\partial \alpha_2} \right)^2 + r^e \left[\frac{1}{A_2} \frac{\partial}{\partial \alpha_2} \left(\frac{U_{2mn}(\alpha_1, \alpha_2)}{R_2} - \frac{1}{A_2} \frac{\partial U_{3mn}(\alpha_1, \alpha_2)}{\partial \alpha_2} \right) \right. \\
& \left. + \left(\frac{1}{A_1 A_2} \left(\frac{U_{1mn}(\alpha_1, \alpha_2)}{R_1} - \frac{1}{A_1} \frac{\partial U_{3mn}(\alpha_1, \alpha_2)}{\partial \alpha_1} \right) \frac{\partial A_2}{\partial \alpha_1} \right) \right] \Bigg\} \mathrm{d}S^e.
\end{aligned}
\tag{11.2.17}
$$

Note that the quadratic terms are related the von Karman geometric nonlinearity. For linear shell oscillations, these quadratic terms can be neglected. In this section, the generic piezoelectric energy harvesting mechanism of the double-curvature shell is simplified to account for three selected distributed piezoelectric harvesters limited to linear oscillations. Recall that simplification procedures were presented in Sect. 3.1 (Chap. 3) and Sect. 4.5 (Chap. 4).

11.3.1 A Cantilever Beam Harvester

For a beam harvester (as shown in Fig. 11.2), x and z are used to denote the α_1 and α_2 directions, respectively. The distinct Lamé parameters $A_1 = A_x = A_2 = A_y = 1$, and the curvature radii $R_1 = R_x = R_2 = R_y = \infty$ for the beam structure.

The electromechanical coupling equation (11.2.16) can be simplified to

$$\frac{\varepsilon_{33} S^e}{h^e} \frac{\mathrm{d}V_{R_L}(t)}{\mathrm{d}t} + \frac{V_{R_L}(t)}{R_L} = \frac{\varepsilon_{33} S^e}{h^e} \sum_{m=1}^{\infty} \phi_m \dot{\eta}_m, \tag{11.3.1}$$

or

$$\tau \frac{\mathrm{d}V_{R_L}(t)}{\mathrm{d}t} + V_{R_L}(t) = \frac{\varepsilon_{33} R_L S^e}{h^e} \sum_{m=1}^{\infty} \phi_m \dot{\eta}_m. \tag{11.3.2}$$

Here m is the mode number in the x direction. The vibration in the width direction is ignored, i.e., $\partial(\bullet)/\partial y = 0$. When only considering the transverse vibration of the beam, the displacement u_3 in the z direction dominates the vibration. Substituting the Lamé parameters and curvature radii into Eq. (11.2.17), neglecting the nonlinear quadratic terms and only considering the transverse deformation yields the modal voltage across the resistive load as

$$\phi_m(x) = -\frac{h^e e_{31}}{\varepsilon_{33}\left(x_2^* - x_1^*\right)} \int_{x_1^*}^{x_2^*} r^e \frac{\mathrm{d}^2}{\mathrm{d}x^2} U_{3m}(x) \mathrm{d}x, \tag{11.3.3}$$

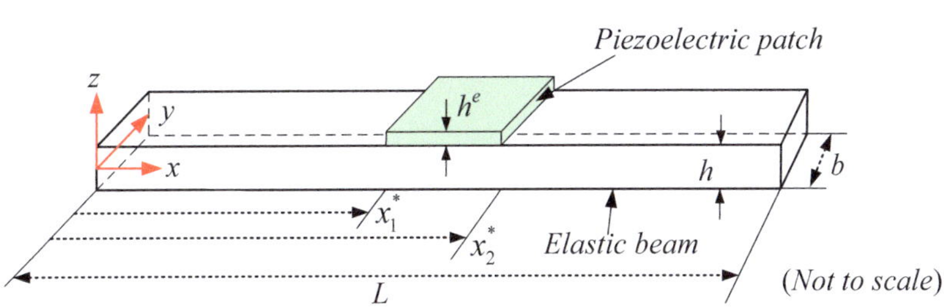

Fig. 11.2 Energy harvester based on a cantilever beam

where x_1^* and x_2^* denote the two ends of the piezoelectric patch. Thus, the output voltage magnitude across the resistor can be simplified from Eq. (11.2.10) as

$$(V_{R_L})_m^* = \frac{\phi_m \hat{F}_m^* R_L \omega\, \varepsilon_{33} S^e/h^e}{\sqrt{\left[\omega_m^2 - \omega^2(1 + 2\zeta_m\omega_m R_L\, \varepsilon_{33} S^e/h^e)\right]^2 + \left[(\omega_m^2 - \omega^2)\omega R_L\, \varepsilon_{33} S^e/h^e + 2\zeta_m\omega_m\omega)\right]^2}}. \tag{11.3.4}$$

and the maximal output power across the resistor is simplified from Eqs. (11.2.12) and (11.2.16) as

$$(P_m^*)_{\max} = \frac{R_L\left(\phi_m \hat{F}_m^*\, \omega\, \varepsilon_{33} S^e/h^e\right)^2}{\left\{\left[\omega_m^2 - \omega^2(1 + 2\zeta_m\omega_m R_L\, \varepsilon_{33} S^e/h^e)\right]^2 + \left[(\omega_m^2 - \omega^2)\omega R_L\, \varepsilon_{33} S^e/h^e + 2\zeta_m\omega_m\omega)\right]^2\right\}} \tag{11.3.5}$$

Note that Eqs. (11.3.4) and (11.3.5) are valid for distributed cantilever beam harvesters subjected to generic harmonic excitations. When the cantilever beam is subjected to a transverse harmonic base excitation $z(t) = Z^* e^{j\omega t}$ where Z^* is the amplitude of the excitation, the transverse distributed loading force F_3 to the beam is

$$F_3 = -\frac{m^*}{b}\frac{\partial^2 z(t)}{\partial t^2} = \frac{m^*}{b} Z^* \omega^2 e^{j\omega t} = F_3^* e^{j\omega t}, \tag{11.3.6}$$

where m^* is the mass per unit length, i.e., $m^* = \rho b h$; b is the width of the beam; F_3^* is the loading amplitude, i.e., $F_3^* = Z^* \omega^2 m^*/b$. With the mode shape function $U_{3m} = \sqrt{\frac{1}{m^* L}}\left[\cosh\frac{\lambda_m}{L}x - \cos\frac{\lambda_m}{L}x - \frac{\sinh\lambda_m - \sin\lambda_m}{\cosh\lambda_m + \cos\lambda_m}\left(\sinh\frac{\lambda_m}{L}x - \sin\frac{\lambda_m}{L}x\right)\right]$ of the cantilever beam, where λ_m is the characteristic value of the beam, the term

$$N_m = \int_0^L U_{3m}^2\, dx = \frac{1}{m^*}. \tag{11.3.7}$$

Then, the amplitude of the modal force Eq. (11.1.12) becomes

$$\hat{F}_m^* = \frac{1}{\rho h N_m}\int_x F_3^* U_{3m} dx = m^* \omega^2 Z^* \frac{2}{\lambda_m}\sqrt{\frac{L}{m^*}}\frac{\sinh\lambda_m - \sin\lambda_m}{\cosh\lambda_m + \cos\lambda_m}, \tag{11.3.8a}$$

$$\text{or} \quad \hat{F}_m^* = m^* \omega^2 Z^* \gamma_m, \tag{11.3.8b}$$

where $\gamma_m = \int_0^L U_{3m} dx = \frac{2}{\lambda_m}\sqrt{\frac{L}{m^*}}\frac{\sinh\lambda_m - \sin\lambda_m}{\cosh\lambda_m + \cos\lambda_m}$. Thus, with the modal force, the voltage output magnitude across the resistive load can be simplified from Eq. (11.3.4) as

$$(V_{R_L})^*_m = \frac{\phi_m m^* \omega^3 Z^* \gamma_m R_L \, \varepsilon_{33} S^e / h^e}{\sqrt{\left[\omega_m^2 - \omega^2(1 + 2\zeta_m \omega_m R_L \, \varepsilon_{33} S^e / h^e)\right]^2 + \left[(\omega_m^2 - \omega^2)\omega R_L \, \varepsilon_{33} S^e / h^e + 2\zeta_m \omega_m \omega)\right]^2}}. \tag{11.3.9}$$

And the maximal power output across the resistive load is simplified from Eq. (11.3.5) as

$$(P^*_m)_{\max} = \frac{R_L(\phi_m m^* \omega^3 Z^* \gamma_m \varepsilon_{33} S^e / h^e)^2}{\left[\omega_m^2 - \omega^2(1 + 2\zeta_m \omega_m R_L \varepsilon_{33} S^e / h^e)\right]^2 + \left[(\omega_m^2 - \omega^2)\omega R_L \varepsilon_{33} S^e / h^e + 2\zeta_m \omega_m \omega)\right]^2} \tag{11.3.10}$$

In (Erturk and Inman, 2008), piezoelectric ceramic PZT was used to harvest the energy from a cantilever beam subjected to base excitations. Using the notations defined in this study and $\tau = R_L \varepsilon_{33} S^e / h^e$, one defines the modal output voltage and power across the resistive load.

$$\begin{aligned} (V_{R_L})^*_m &= (\phi_m m^* \omega^3 Z^* \gamma_m R_L \varepsilon_{33} S^e / h^e) / \Big\{ \left[\omega_m^2 - \omega^2(1 + 2\zeta_m \omega_m R_L \varepsilon_{33} S^e / h^e)\right]^2 \\ &\quad + \left[(\chi_m \phi_m + \omega_m^2 - \omega^2)\omega R_L \varepsilon_{33} S^e / h^e + 2\zeta_m \omega_m \omega\right]^2 \Big\}^{1/2}, \end{aligned} \tag{11.3.11}$$

$$\begin{aligned} (P^*_m)_{\max} &= R_L(\phi_m m^* \omega^3 Z^* \gamma_m \varepsilon_{33} S^e / h^e)^2 \Big/ \Big\{ \left[\omega_m^2 - \omega^2(1 + 2\zeta_m \omega_m R_L \varepsilon_{33} S^e / h^e)\right]^2 \\ &\quad + \left[(\chi_m \phi_m + \omega_m^2 - \omega^2)\omega R_L \varepsilon_{33} S^e / h^e + 2\zeta_m \omega_m \omega\right]^2 \Big\}, \end{aligned} \tag{11.3.12}$$

where χ_m is the secondary coupling term (Erturk and Inman, 2008). When the PZT material is replaced by polyvinylidene fluoride (PVDF) polymer and PVDF's stiffness and mass effects can be neglected, the coupling term χ_m is very small and ignored in Eqs. (11.3.11) and (11.3.12). Thus, the derived voltage and power equations (11.3.11) and (11.3.12) are identical to the expressions in Eqs. (11.3.9) and (11.3.10) directly simplified from the generic shell formulations. Thus, the piezoelectric energy harvesting mechanism of double-curvature shell can be simplified to account for the cantilever beam case.

11.3.2 A Circular Ring Harvester

The generic piezoelectric shell energy harvesting theory can be applied to not only simple non-shell structures (i.e., beams and plates), but also other thin shell structures, e.g., cylindrical shells, conical shells, parabolic shells, etc. In this

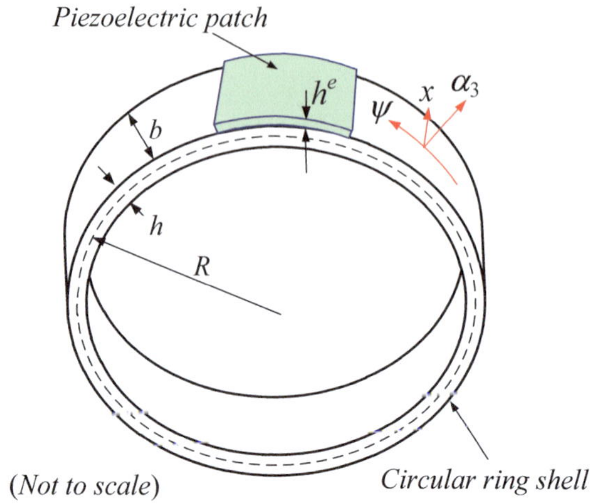

Fig. 11.3 Energy harvester based on a circular ring

section, the generic piezoelectric shell energy harvesting theory is simplified to a piezoelectric circular ring harvester. As shown in Fig. 11.3, x and ψ respectively define the width and circumferential directions of ring shell's neutral surface. The Lamé parameters A_1 and A_2 are defined as $A_1 = A_x = 1$ and $A_2 = A_\psi = R$; two curvature radii $R_1 = R_x = \infty$ and $R_2 = R_\psi = R$, where R is the radius of the ring, Chap. 8. The strain and displacement in the width direction is neglected for the ring.

The electromechanical coupling equation can be simplified from Eq. (11.2.6) as

$$\frac{\varepsilon_{33}S^e}{h^e}\frac{\mathrm{d}V_{R_L}(t)}{\mathrm{d}t} + \frac{V_{R_L}(t)}{R_L} = \frac{\varepsilon_{33}S^e}{h^e}\sum_{n=2}^{\infty}\phi_n\dot{\eta}_n, \tag{11.3.13a}$$

or

$$\tau\frac{\mathrm{d}V_{R_L}(t)}{\mathrm{d}t} + V_{R_L}(t) = \frac{\varepsilon_{33}R_L S^e}{h^e}\sum_{n=2}^{\infty}\phi_n\dot{\eta}_n. \tag{11.3.13b}$$

Here modes $n = 0$ and $n = 1$ are neglected, since $n = 0$ is a breathing mode and $n = 1$ is a rigid-body motion mode. The modal voltage is simplified from Eq. (11.2.17) as

$$\phi_n = \frac{h^e}{\varepsilon_{33}S^e}\int_{S^e} e_{32}\left[\frac{1}{R}\left(\frac{\partial U_{\psi n}(\psi)}{\partial\psi} + U_{3n}(\psi)\right) + \frac{r^e}{R^2}\left(\frac{\partial U_{\psi n}(\psi)}{\partial\psi} - \frac{\partial^2 U_{3n}(\psi)}{\partial\psi^2}\right)\right]\mathrm{d}S^e. \tag{11.3.14}$$

The voltage and power output magnitude can be simplified from Eqs. (11.2.11) and (11.2.12) as

$$(V_{R_L})_n^* = R_L \hat{F}_n^* \omega \int_{S^e} e_{32} \left[\frac{1}{R} \left(\frac{\partial U_{\psi n}(\psi)}{\partial \psi} + U_{3n}(\psi) \right) + \frac{r^e}{R^2} \left(\frac{\partial U_{\psi n}(\psi)}{\partial \psi} - \frac{\partial^2 U_{3n}(\psi)}{\partial \psi^2} \right) \right] \mathrm{d}S^e$$
$$/ \left\{ \left[\omega_n^2 - \omega^2 (1 + 2\zeta_n \omega_n R_L \varepsilon_{33} S^e / h^e) \right]^2 + \left[(\omega_n^2 - \omega^2) \omega R_L \varepsilon_{33} S^e / h^e + 2\zeta_n \omega_n \omega) \right]^2 \right\}^{1/2}, \tag{11.3.15}$$

$$(P_n^*)_{RMS} = R_L \left(\hat{F}_n^* \omega \int_{S^e} e_{32} \left[\frac{\partial U_{\psi n}(\psi)}{R \partial \psi} + \frac{U_{3n}(\psi)}{R} + \frac{r^e}{R^2} \left(\frac{\partial U_{\psi n}(\psi)}{\partial \psi} - \frac{\partial^2 U_{3n}(\psi)}{\partial \psi^2} \right) \right] \mathrm{d}S^e \right)^2$$
$$/2 \left\{ \left[\omega_n^2 - \omega^2 (1 + 2\zeta_n \omega_n R_L \varepsilon_{33} S^e / h^e) \right]^2 + \left[(\omega_n^2 - \omega^2) \omega R_L \varepsilon_{33} S^e / h^e + 2\zeta_n \omega_n \omega) \right]^2 \right\}. \tag{11.3.16}$$

The spatially distributed modal force $\hat{F}_n^*(\psi)$ is defined by Eqs. (11.1.12) and (11.1.13), which can be simplified to the circular ring as

$$\hat{F}_n^* = \frac{1}{\rho h N_n} \int_x \int_\psi \sum_i^3 F_i^*(\psi) U_{in} R \mathrm{d}x \mathrm{d}\psi, \tag{11.3.17}$$

$$\text{where} \qquad N_n = \int_x \int_\psi \sum_i^3 U_{in}^2 R \, \mathrm{d}x \mathrm{d}\psi. \tag{11.3.18}$$

11.3.3 A Conical Shell Harvester

In this section, the generic piezoelectric shell energy harvesting theory is simplified to a piezoelectric conical shell harvester illustrated in Fig. 11.4.

As shown in Fig. 11.4, x and ψ respectively define the longitudinal and circumferential axes of shell's neutral surface, Sect. 3.3.1.3, Chap. 3. The Lamé parameters

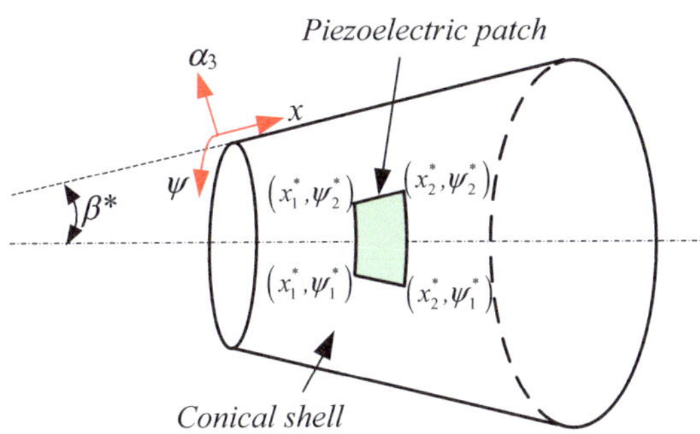

Fig. 11.4 Energy harvester model based on a conical shell

A_1 and A_2 are defined as $A_1 = A_x = 1$ and $A_2 = A_\psi = x \sin \beta^*$; two curvature radii $R_1 = R_x = \infty$ and $R_2 = R_\psi = x \tan \beta^*$ where * is the half-apex angle and x is the distance measured from the apex (Chai, Smithmaitrie, Tzou, 2004) x_i^* and ψ_i^* (i = 1, 2) denote the corner coordinates (i.e., location) of the piezoelectric harvester patch.

The electromechanical coupling equation can be simplified from Eq. (11.2.6) as

$$\frac{\varepsilon_{33} S^e}{h^e} \frac{\mathrm{d} V_{R_L}(t)}{\mathrm{d} t} + \frac{V_{R_L}(t)}{R_L} = \frac{\varepsilon_{33} S^e}{h^e} \sum_{m=1}^{\infty} \sum_{n=0}^{\infty} \phi_{mn} \dot{\eta}_{mn}, \tag{11.3.19a}$$

or

$$\tau \frac{\mathrm{d} V_{R_L}(t)}{\mathrm{d} t} + V_{R_L}(t) = \frac{\varepsilon_{33} R_L S^e}{h^e} \sum_{m=1}^{\infty} \sum_{n=0}^{\infty} \phi_{mn} \dot{\eta}_{mn}. \tag{11.3.19b}$$

Here, $n = 0$ denotes the axisymmetric vibration of the circumferential direction. According to the Donnel-Mushtari-Vlasov assumptions, the in-plane displacements u_x and u_ψ are usually neglected in the bending strains; only the transverse displacement dominants the bending strains (Tzou, Chai, Wang, 2003). The modal voltage is simplified from Eq. (11.2.17) as

$$\begin{aligned} \phi_{mn} = \frac{h^e}{\varepsilon_{33} S^e} \int_{S^e} \Bigg\{ & e_{31} \left(\frac{\partial}{\partial x} U_{xmn}(x,\psi) - r^e \frac{\partial^2}{\partial x^2} U_{3mn}(x,\psi) \right) \\ & + e_{32} \Bigg[\frac{1}{x \sin \beta^*} \frac{\partial U_{\psi mn}(x,\psi)}{\partial \psi} + \frac{U_{xmn}(x,\psi)}{x} + \frac{U_{3mn}(x,\psi)}{x \tan \beta^*} \\ & + r^e \left(-\frac{1}{x^2 \sin^2 \beta^*} \frac{\partial^2 U_{3mn}(x,\psi)}{\partial \psi^2} - \frac{1}{x} \frac{\partial U_{3mn}(x,\psi)}{\partial x} \right) \Bigg] \Bigg\} \mathrm{d}S^e. \end{aligned} \tag{11.3.20}$$

The voltage output amplitude $(V_{R_L})^*_{mn}$ and the root-mean-square power $(P^*_{mn})_{RMS}$ of the linear conical shell harvester at resonance are simplified from Eqs. (11.2.14) and (11.2.15) as

$$\begin{aligned} (V_{R_L})^*_{mn} = R_L \hat{F}^*_{mn} \omega_{mn} \int_{S^e} \Bigg\{ & e_{31} \left(\frac{\partial}{\partial x} U_{xmn}(x,\psi) - r^e \frac{\partial^2}{\partial x^2} U_{3mn}(x,\psi) \right) \\ & + e_{32} \Bigg[\frac{1}{x \sin \beta^*} \frac{\partial U_{\psi mn}(x,\psi)}{\partial \psi} + \frac{U_{xmn}(x,\psi)}{x} + \frac{U_{3mn}(x,\psi)}{x \tan \beta^*} \\ & + r^e \left(-\frac{1}{x^2 \sin^2 \beta^*} \frac{\partial^2 U_{3mn}(x,\psi)}{\partial \psi^2} - \frac{1}{x} \frac{\partial U_{3mn}(x,\psi)}{\partial x} \right) \Bigg] \Bigg\} \mathrm{d}S^e \\ & / \left(2 \zeta_{mn} \omega_{mn}^2 \right) \sqrt{1 + (R_L \omega_{mn} \varepsilon_{33} S^e / h^e)^2}. \end{aligned} \tag{11.3.21}$$

$$\begin{aligned}(P^*_{mn})_{RMS} = R_L \Bigg(\hat{F}^*_{mn}\omega_{mn} \int_{S^e} \Bigg\{ e_{31}\left(\frac{\partial}{\partial x}U_{xmn}(x,\psi) - r^e\frac{\partial^2}{\partial x^2}U_{3mn}(x,\psi)\right) \\ + e_{32}\Bigg[\frac{1}{x\sin\beta^*}\frac{\partial U_{\psi mn}(x,\psi)}{\partial\psi} + \frac{U_{xmn}(x,\psi)}{x} + \frac{U_{3mn}(x,\psi)}{x\tan\beta^*} \\ + r^e\left(-\frac{1}{x^2\sin^2\beta^*}\frac{\partial^2 U_{3mn}(x,\psi)}{\partial\psi^2} - \frac{1}{x}\frac{\partial U_{3mn}(x,\psi)}{\partial x}\right)\Bigg]\Bigg\} dS^e\Bigg)^2 \\ /2\left(2\zeta_{mn}\omega^2_{mn}\right)^2\left[1 + (R_L\omega_{mn}\varepsilon_{33}S^e/h^e)^2\right].\end{aligned} \quad (11.3.22)$$

Again, the spatially distributed modal force $\hat{F}^*_{mn}(x,\psi)$ defined by Eqs. (11.1.12) and (11.1.13) and applied to the conical shell becomes

$$\hat{F}^*_{mn}(x,\psi) = \frac{1}{\rho h N_{mn}} \int_x \int_\psi \sum_i^3 F^*_i(x,\psi) U_{imn} x \sin\beta^* dx d\psi, \quad (11.3.23)$$

where $$N_{mn}(x,\psi) = \int_x \int_\psi \sum_i^3 U^2_{imn} x \sin\beta^* dx d\psi. \quad (11.3.24)$$

Note that $\hat{F}^*_{mn}(x,\psi)$ can be either distributed, line or point excitation. U_{imn} is the mode shape functions defined by shell's boundary conditions. With specified boundary conditions and loading distributions, the detailed voltage and power output across the resistive load can be established. Thus, the piezoelectric energy harvesting mechanism based on common shell and non-shell structures can be established using the intrinsic Lamé parameters and the curvature radii of the selected structure. Note that only linear cases are considered in the simplified geometries in case studies. The nonlinear quadratic terms can be added to those expressions if the von Karman geometric nonlinearity is considered.

11.4 Ring Harvester Experiments

Laboratory experiments are conducted in this section to validate the energy harvesting mechanism of the piezoelectric ring energy harvester. Voltage outputs across the resistive load will be measured to compare with the predictions.

Table 11.1 Material properties and geometry dimensions

Properties	Ring	PVDF
Radius, R (m)	43.03×10^{-3}	–
Thickness, h (m)	1.32×10^{-3}	40×10^{-6}
Width, b (m)	9.74×10^{-3}	10×10^{-3}
Mass density, ρ (kg/m^3)	907	1800
Poisson's ratio, μ	0.4438	0.29
Young's modulus, Y (N/m^2)	1.556×10^9	1.6×10^9
Dielectric constant, ε_{33} (F/m)	–	8.85×10^{-11}
Piezoelectric stress constant, e_{31} (C/m^2)	–	9.6×10^{-3}
Piezoelectric strain constant, d_{31} (C/N)	–	6×10^{-12}

11.4.1 Material Properties and Ring Shell Harvester

Elastic ring model and piezoelectric PVDF material are used in the experiments. The material properties and the geometry dimensions are the same as listed in the

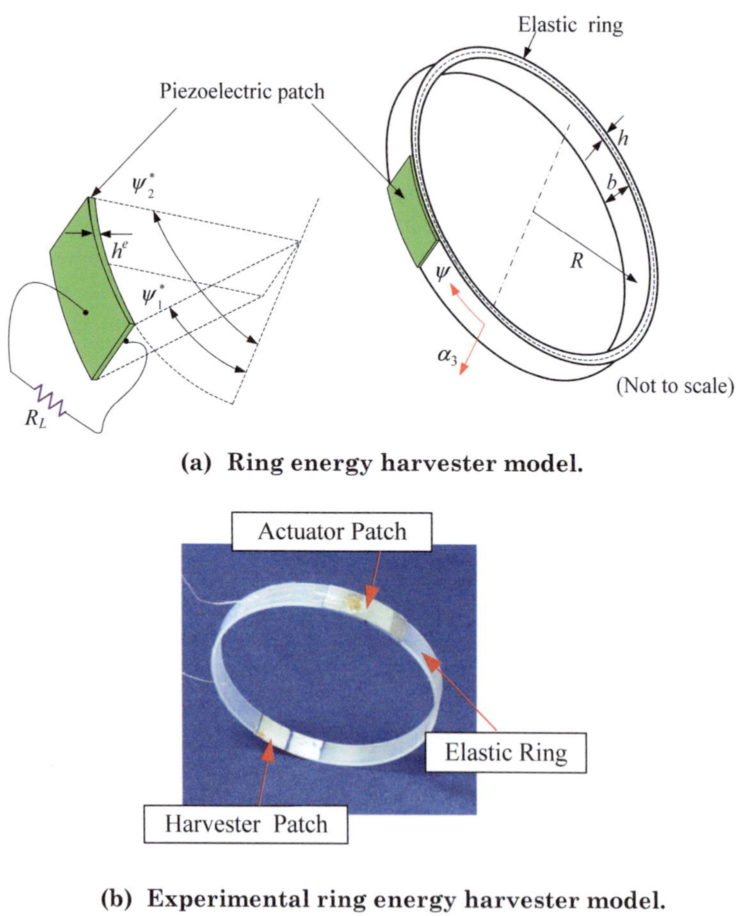

Fig. 11.5 Energy harvester model of ring shells

Table 11.1. The experimental energy harvester model is shown in Fig. 11.5a, b. The actuator patch and the harvester patch are both $30 \times 10\ \mathrm{mm}^2$, laminated on the symmetric location of the ring by considering the optimal loading location in the case studies. The experimental natural frequency and the damping ratio for each mode are tested firstly in order to carry out the energy harvesting experiments.

11.4.2 Natural Frequency and Damping Ratio

Based on the dynamic vibration of the circular ring shell theory, Chap. 8, the theoretical natural frequency can be calculated for each mode. The experimental natural frequency is measured with the resonant dynamics, and the experiment setup is shown in Fig. 11.6. The experimental natural frequency is estimated by the frequency spectrum and listed in Table 11.2.

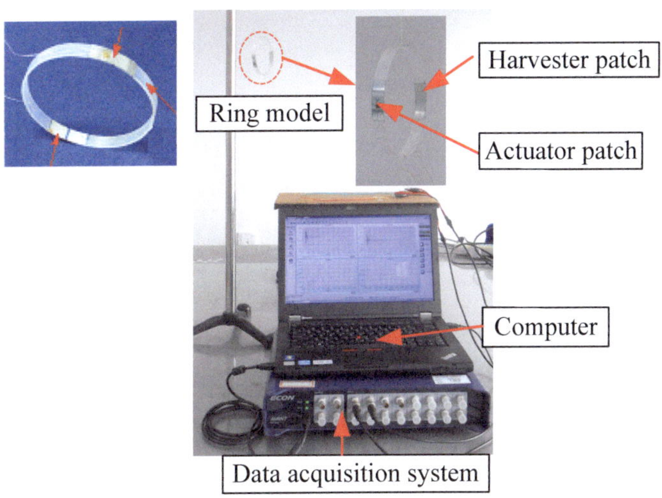

Fig. 11.6 Setup of testing the natural frequency

Table 11.2 Comparison of theoretical and experimental natural frequencies

Mode number n	Theory frequency (Hz)	Experiment frequency (Hz)	Error (%)**	Damping ratio (ζ_n)
2	128.50	120	−6.61	0.013
3	363.46	338	−7.00	0.007
4	696.88	662	−5.01	0.005
5	1127.01	1069.5	−5.10	
6	1653.29	1579.5	−4.46	

**The error is calculated by the equation: (experiment frequency-theory frequency)/theory frequency * 100

The damping ratios for the three modes are estimated by the time-domain history—the logarithmic decrement method with the experimental natural frequency for each mode.

11.4.3 Voltage and Power Outputs

With the experimental natural frequency and the damping ratio for each mode, the output voltage across the resistive load is measured in this section. The setup of the energy harvesting experiment is shown in Fig. 11.7, and the flow diagram of the energy harvesting experiment is shown in the Fig. 11.8. The actuator patch is subjected to a harmonic excitation.

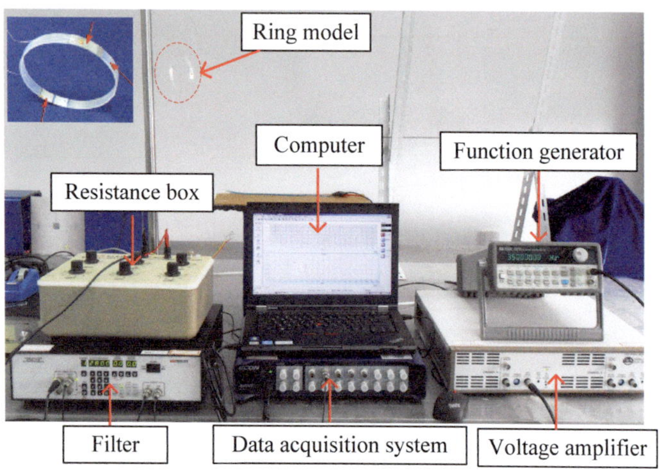

Fig. 11.7 Setup of the energy harvesting experiment

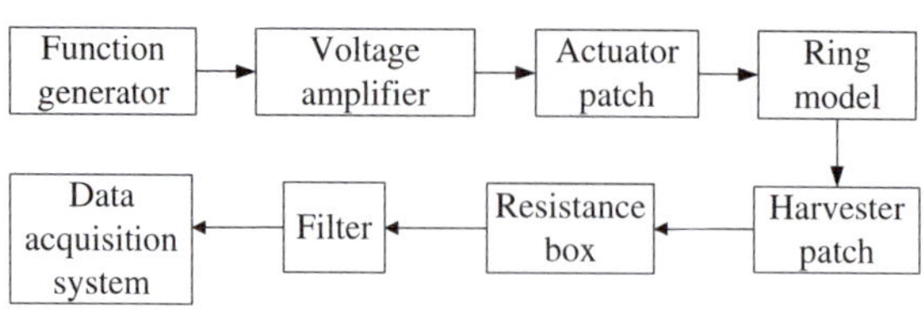

Fig. 11.8 Flow diagram of the energy harvesting experiment

The signal of the function generator is 1 V, and the excitation frequency is equal to the experimental natural frequency for each mode, i.e., $\omega = \omega_n$. The excitation actuator voltage on the actuator patch can be 200 V by magnifying 200 times by the high voltage amplifier, which can be converted to control force signal on the elastic ring. The harvester patch is connected to the resistance box. Then the signal across the resistance connects to the data acquisition system through the filter. The filter is used here to filter the signals of other frequencies. Because even though the function generator set at a single frequency, the ring response is composed of other frequencies. The filter with band pass can reject unwanted frequencies and keep signal around the natural frequency. Then the voltage across the resistance can be obtained from the data acquisition system. Energy harvesting experiments from mode 2 to mode 4 are carried out next. Note that mode 1 is a rigid body ring mode.

Figures 11.9, 11.10, and 11.11 present the experimental (dotted-line) and analytical (solid-line) output voltages and powers for ring modes 2–4, in which part

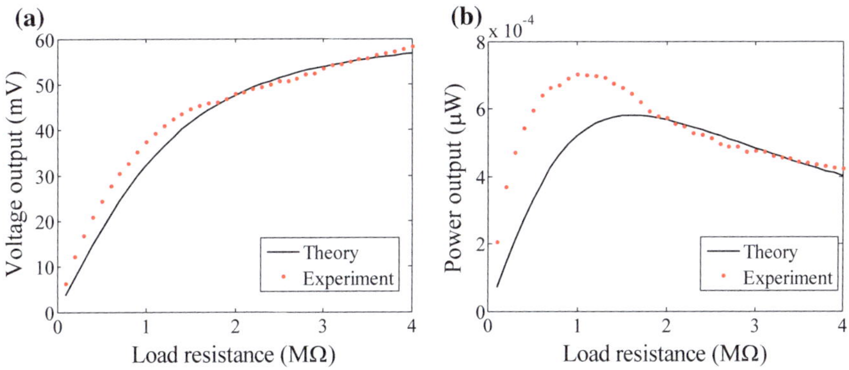

Fig. 11.9 Voltage/power output across the resistance for mode 2. Excitation frequency is 120 Hz and the filter frequency band is 96–150 Hz

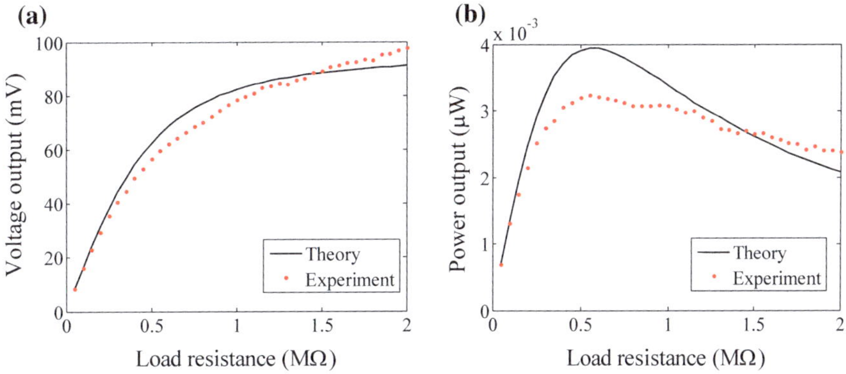

Fig. 11.10 Output voltage/power across the resistance for mode 3. Excitation frequency is 340 Hz and filter frequency band is 270–425 Hz

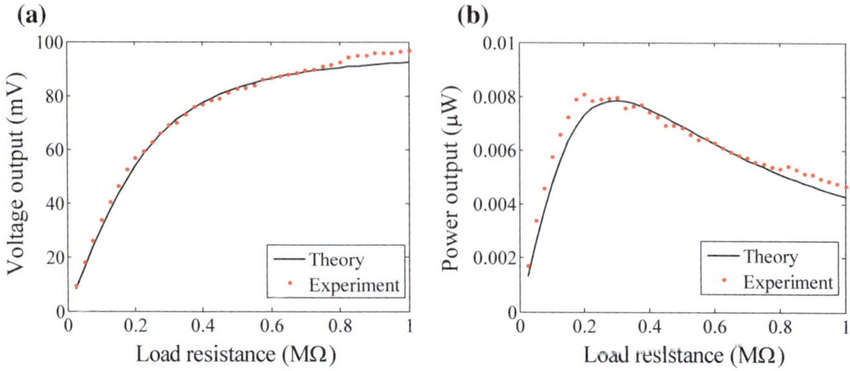

Fig. 11.11 Output voltage/power across the resistance for mode 4. Excitation frequency is 662 Hz and the filter frequency band is 530–828 Hz

(a) shows the output **voltage** across the resistance and part (b) denotes the **power** output across the resistance.

Energy harvesting experiments are conducted at the experimental frequency for each mode. The band-pass frequency setup of the filter is based on the impendence of the filter, from $0.8f_n$ to $1.25f_n$, where f_n is the experimental natural frequency for each mode.

The voltage and power output changing with the resistance is consistent with the theory. The voltage output increases with the resistance increasing. When dynamic response of the ring is specified, the current in the closed-circuit condition is constant. Thus, the voltage output is direct proportion to the load resistance. The errors between the experimental and theoretical output voltages are within 10%.

The experimental power output can't be measured directly. It is calculated by the relation $P = V^2/2R_L$, which shows the root-mean-square of the power output. The results show that the power output achieves the maximum at a specified resistance, i.e. an optimal resistance, for both the experiment and the theory. The errors for the power output between the experiment and the theory are larger than the voltages, since the power is calculated with the voltage output measured directly and the resistance, and other affecting factors. The experiments show the practical uses of the energy harvesting mechanism about the output voltage and power of the ring harvester.

11.5 Summary

Energy harvesting based on distributed piezoelectric laminated structures has been proposed and extensively investigated for over a decade. The objective of this study is to develop a generic distributed piezoelectric shell energy harvester theory based on a generic linear/nonlinear double-curvature shell structure, which can be simplified to account for many linear/nonlinear shell and non-shell type distributed energy harvesters. Distributed electromechanical coupling mechanism of the energy harvester was discussed; voltage and power output across the external resistive load of the shell energy harvester were evaluated. Those equations were explicitly expressed in terms of design parameters and modes. Once the intrinsic Lamé parameters and the curvature radii of the selected host structure are specified, one can simplify the piezoelectric shell energy harvesting equations to account for common shell and non-shell structures.

To demonstrate the simplifications, the generic piezoelectric shell energy harvesting mechanism was applied to a cantilever beam, a circular ring and a conical shell in cases studies. Again, the generic piezoelectric energy harvesting formulations derived from a double-curvature shell can be applied to many shell, e.g., ring shells, cylindrical shell, conical shells, paraboloidal shells, etc., and non-shell, e.g., plates, beams, etc., structures using two Lamé parameters and two curvature radii of the specified structures. Besides, these shell and non-shell structures can be either linear or nonlinear with the von Karman geometric nonlinearity. With given boundary conditions and external loading forces, generated voltage and power across the resistive load in the closed-circuit condition can be estimated for the distributed piezoelectric laminated harvester structure.

Acknowledgements The research was supported by the National Natural Science Foundation of China (11472241 & 11172262), the Nanjing University of Aeronautics and Astronautics Foundation (NUAA-NP2016203) and the State Key Laboratory of Mechanics and Control of Mechanical Structures (NUAA-MCMS-0516G01).

References

Anton, S.R. and Sodano, H.A., A review of power harvesting using piezoelectric materials (2003–2006), *Smart Materials & Structures* 16 (2007) 1–21.

Beeby, S.P., Tudor, M.J. and White, N.M., Energy harvesting vibration sources for microsystems applications, *Measurement Science & Technology* 17 (2006) 175–195.

Chai, W.K., Smithmaitrie, P. and Tzou, H.S., Neural potentials and micro-signals of non-linear deep and shallow conical shells, *Mechanical Systems and Signal Processing* 18 (2004) 959–975.

Erturk, A. and D. J. Inman, D.J., A distributed parameter electromechanical model for cantilevered piezoelectric energy harvesters, *Journal of Vibration and Acoustics* 130 (2008) 0410021–04100215.

Erturk, A. and Inman, D.J., An experimentally validated bimorph cantilever model for piezoelectric energy harvesting from base excitations, *Smart Materials and Structures* 18 (2009) 025009–025027.

Howard, R.V., Chai, W.K. and Tzou, H.S., Modal voltages of linear and non-linear structures using distributed artificial neurons, *Mechanical Systems and Signal Processing* 15 (2001) 629–640.

Hu, S.D., Li, H. and Tzou, H.S., Distributed neural signals on parabolic cylindrical shells, *Journal of Sound and Vibration* 332 (2013) 2984–3001.

Hu, S.D., H. Li, H. and Tzou, H.S., Comparison of flexoelectric and piezoelectric dynamic signal responses on flexible rings, *Journal of Intelligent Material Systems and Structures* 25 (2014) 832–844.

Huang, S.L., Chu, C.C., Chang, C.C. and Tzou, H.S., Spatial electrostrictive actuations of flexible circular tubes, *Journal of Pressure Vessel Technology* 134 (2012) 0212081–02120812.

Karami, M.A., Bilgen, O.,Inman, D.J. and Friswell, M.I., Experimental and analytical parametric study of single-crystal unimorph beams for vibration energy harvesting, *IEEE Transactions on Ultrasonics, Ferroelectrics, and Frequency Control* 58 (2011) 1508–1520.

Kim, S. Piezoelectric energy harvesting with a clamped circular plate: analysis, *Journal of Intelligent Material Systems and Structures* 16 (2005a) 847–854.

Kim, S. Piezoelectric energy harvesting with a clamped circular plate: experimental study, *Journal of Intelligent Material Systems and Structures* 16 (2005b) 855–863.

Kim, H.S., Kim, J.H. and Kim, J., A review of piezoelectric energy harvesting based on vibration, *International Journal of Precision Engineering and Manufacturing* 12 (2011) 1129–1141.

Li, H., Hu, S.D and Tzou, H.S., Z. B. Chen, Optimal vibration control of conical shells with collocated helical sensor/actuator pairs, *Journal of Theoretical and Applied Mechanics* 50 (2012) 769–784.

Liu, X.J. and Chen, R.W., Current situation and developing trend of piezoelectric vibration energy harvesters, *Journal of Vibration and Shock* 31 (2012) 169–176.

Meeker, T., ANSI/IEEE Standard on piezoelectricity, *IEEE Transactions on Ultrasonics Ferroelectrics and Frequency Control* 43 (1996) 717–772.

Rao, Z., Li, H. and Tzou, H.S., Breathing cylindrical piezoelectric energy harvesters, Proceedings of the 2011 Symposium on Piezoelectricity, Acoustic Waves and Device Applications, pp. 506–509, 2011 SPAWDA, Shenzhen, 9–11 Dec. 2011.

Smithmaitrie, P. and Tzou, H.S., Micro-control actions of actuator patches laminated on hemispherical shells, *Journal of Sound and Vibration*, 277 (2004) 691–710.

Sodano, H.A., Inman, D.J. and Park, G., A review of power harvesting from vibration using piezoelectric materials, *Shock and Vibration Digest* 36 (2004) 197–205.

Soedel, W. *Vibrations of Shells and Plates*, Marcel Dekker, New York, 1993.

Tzou, H.S., *Piezoelectric Shells: Distributed Sensing & Control*, Kluwer Academic Publishers, London, 1993.

Tzou, H.S. and Bao, Y., Nonlinear piezothermoelasticity and multi-field actuations, part1: nonlinear anisotropic piezothermoelastic shell laminates, *Journal of Vibration and Acoustics* 119 (1997) 374–381.

Tzou, H.S., Bao, Y. and Venkayya, V.R., Parametric study of segmented transducers laminated on cylindrical shells, Part 2: Actuator patches, *Journal of Sound and Vibration* 197 (1996a) 225–249.

Tzou, H.S., Bao, Y. and Venkayya, V.B., Parametric study of segmented transducers laminated on cylindrical shells, Part 1: Sensor patches, *Journal of Sound and Vibration* 197 (1996b) 207–224.

Tzou, H.S., Chai, W.K. and Wang, D.W., Modal voltages and micro-signal analysis of conical shells of revolution, *Journal of Sound and Vibration* 260 (2003) 589–609.

Tzou, H.S., Zhong, J.P. and Hollkamp, J.J., Spatially distributed orthogonal piezoelectric shell actuators: theory and applications, *Journal of Sound and Vibration*, 177 (1994) 363–378.

Zhang, X.F.. Hu, S.D. and Tzou, H.S., Electromechanical coupling and energy harvesting of circular rings, Proceedings of the 2011 Symposium on Piezoelectricity, Acoustic Waves and Device Applications, pp. 514–517, 2011 SPAWDA, Shenzhen, 9–11 Dec. 2011.

Zhang, X.F.. Hu, S.D. and Tzou, H.S., A generic double-curvature piezoelectric shell energy harvester: Linear/nonlinear theory and applications, *Journal of Sound and Vibration,* 333 (2014) 7286–7298.

Chapter 12
Tubular Shell Energy Harvester

Harvesting energy from ambient vibrations by using piezoelectric materials has been investigated over the years (Sodano, et al., 2004, 2005; Lefeuvre, et al., 2005; Baddel, et al., 2006; Erturk and Inman, 2009; Poulin, et al., 2004; Lu, et al., 2004; Anton and Sodano, 2007). Conventionally, piezoelectric energy harvesting is based on zero curvature structures (Poulin, et al., 2004; Kim, William and Wang, 2005), but little has been reported on harvesting energy using curvature structures such as shells. This study investigates a circular cylindrical shell (or tubular shell) laminated with segmented piezoelectric energy harvester patches. Figure 12.1 illustrates a piezoelectric energy harvesting system, including a shell structure, signal rectifiers, signal comparators, multiplexers (MUX), and an energy storage (with inductor L and capacitor C.) Using the dynamic characteristics and distributed sensing or control of a circular cylindrical shell (Pelicano, 2007; Poore, et al., 2007; Guo, et al., 2001; Tzou, Bao and Venkayya, 1996; Tzou and Zhang, 1996; Callahan and Baruh, 1999; Tzou and Yang, 2000; DeHaven, Han, and Tzou, 2007; Huang, et al., 2008), one can further investigate the distributed modal energy generation of piezoelectric energy harvester patches laminated on the circular cylindrical shell.

Piezoelectric power and energy generation is based on dynamic strains in the piezoelectric patch due to the direct piezoelectric effect when subjected to external mechanical excitations. Energy and power generations based on the double-curvature shell experiencing linear and the von Karman nonlinear oscillations were evaluated and detailed formulations were presented in Chap. 11. With *the von Karman geometrical non-linearity*, a quadratic effect of the non-linear transverse displacement was considered in the membrane strains of oscillations (Howard, Chai, Tzou, 2001; Tzou & Bao, 1997).

Tubular shells (or circular cylindrical shells) are commonly found in piping or tubing systems, e.g., oil and gas plants, drug deliveries, blood vessels, etc., for liquid and gas deliveries. Flow induced kinetic energies can be converted to electric energies or power in piezoelectric patches laminated on the tubular shells.

H. (HS) Tzou, *Piezoelectric Shells*, Solid Mechanics and Its Applications 247,
https://doi.org/10.1007/978-94-024-1258-1_12

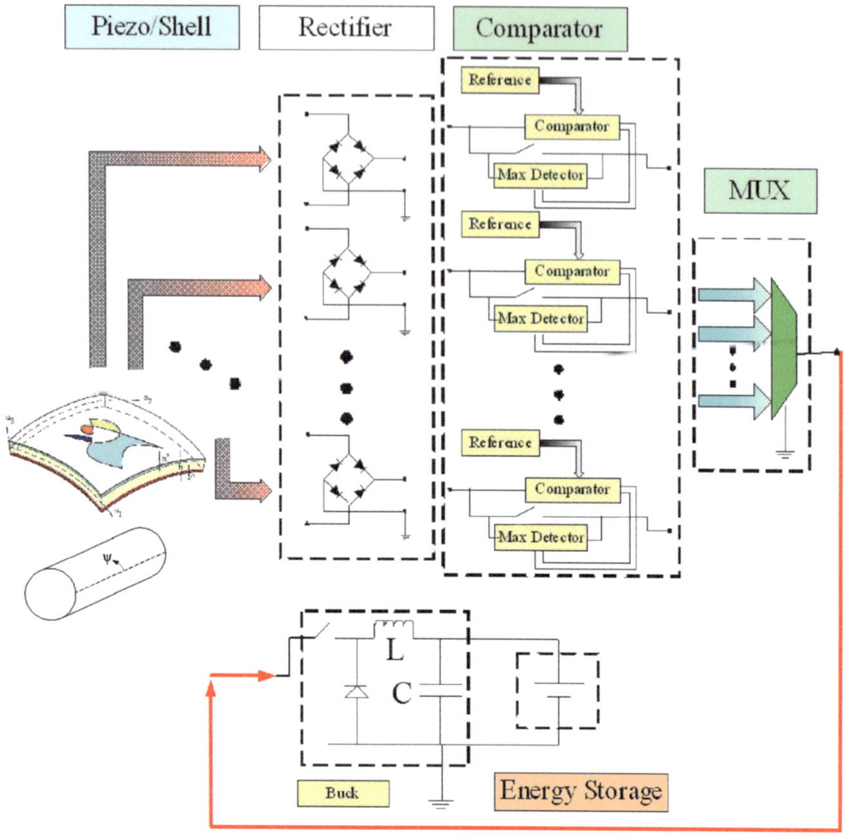

Fig. 12.1 A generic piezoelectric energy harvesting system

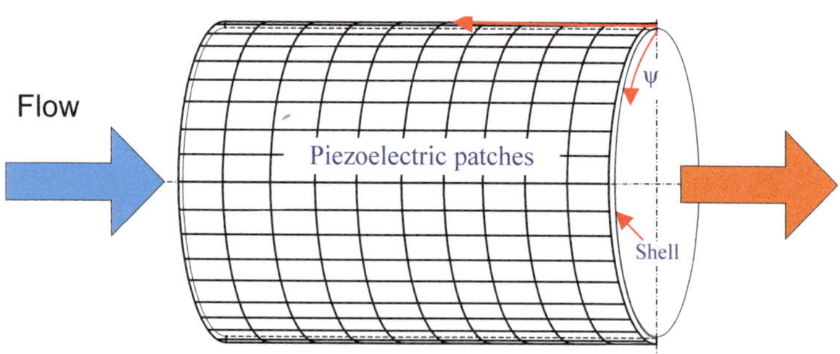

Fig. 12.2 A piezoelectric tubular shell energy harvester

Figure 12.2 illustrates a tubular shell laminated with piezoelectric energy harvesting patches. Liquids or gasses flowing through the tube could induce structural vibrations and consequently the piezoelectric patches generate signals and energies due to the direct piezoelectric effect.

A tubular shell physical model of energy harvesting system is presented first, followed by the generation of the modal energy. Three energy harvester patch sizes (i.e., (1 mm,3.6°) in Case 1, (10 mm,30°) in Case 2, and (20 mm,60°) in Case 3) are evaluated and compared in case studies. Design parameters and shell's geometry (i.e., shell radius, shell thickness, and energy harvester thickness) are also investigated in parametric studies. Laboratory experiments of a circular cylindrical shell model laminated with piezoelectric patches are carried out and power generations of experimental data are compared with those of analytical predictions. The objective of this chapter is to investigate the spatial distribution characteristics of the modal energies, and to evaluate the effects of energy harvester patch size and design parameters on the generation of the modal energies.

12.1 Tubular Shell Energy Generations

Since the harvester is mostly excited at resonances to maximize the energy output, a harmonic excitation is considered and its steady-state dynamic response is used to evaluate energy efficiency. This section focuses on dynamic strain generation of a generic double-curvature piezoelectric energy harvester. Since the harvester is mostly excited at resonance to maximize the energy output, a harmonic excitation is considered and its steady-state dynamic response is defined.

12.1.1 Model of a Circular Cylindrical Shell Tube

The physical model of the energy harvesting system of a simply-supported circular cylindrical tubular shell laminated with segmented piezoelectric energy harvesters is shown in Fig. 12.3. Each energy harvester is uniformly thin compared with the elastic shell. Thus, the Kirchhoff—Love assumptions can be used. An energy harvester patch is defined from x_1^* to x_2^* in the longitudinal direction, and ψ_1^* to ψ_2^* in the circumferential direction. h_p denotes the energy harvester's thickness; h represents the elastic shell's thickness; and R is the radius of the elastic shell.

12.1.2 Generation of the Modal Energy

The dynamic equations of a generic isotropic circular cylindrical (or tubular) shell are expressed as:

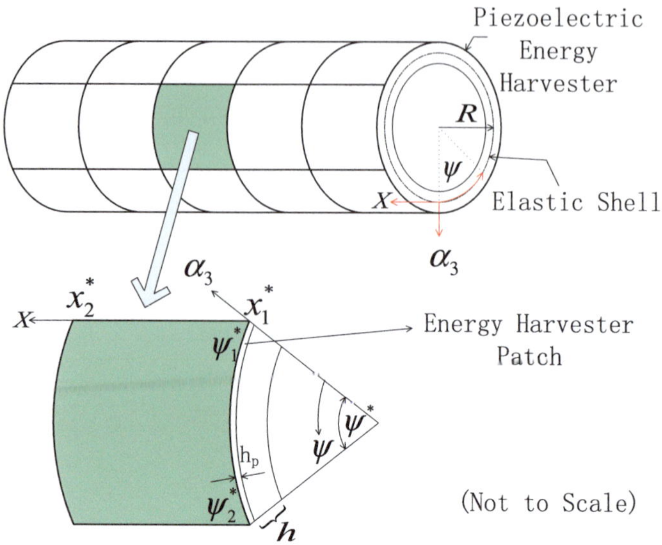

Fig. 12.3 A simply-supported circular cylindrical shell laminated with piezoelectric energy harvester patches

$$K\left[R\frac{\partial^2 u_x}{\partial x^2}+\frac{1}{2}\frac{\partial^2 u_\psi}{\partial x\partial\psi}+\frac{1}{2R}\frac{\partial^2 u_x}{\partial\psi^2}+\mu\left(\frac{1}{2}\frac{\partial^2 u_\psi}{\partial x\partial\psi}+\frac{\partial u_3}{\partial x}-\frac{1}{2R}\frac{\partial^2 u_x}{\partial\psi^2}\right)\right]-R\rho h\ddot{u}_x=0, \tag{12.1.1}$$

$$\begin{aligned}&K\left[\frac{R}{2}\frac{\partial^2 u_\psi}{\partial x^2}+\frac{1}{2}\frac{\partial^2 u_x}{\partial x\partial\psi}+\frac{1}{R}\frac{\partial^2 u_\psi}{\partial\psi^2}+\frac{1}{R}\frac{\partial u_3}{\partial\psi}+\mu\left(\frac{1}{2}\frac{\partial^2 u_x}{\partial x\partial\psi}-\frac{R}{2}\frac{\partial^2 u_\psi}{\partial x^2}\right)\right]\\&+D\left(\frac{1}{2R}\frac{\partial^2 u_\psi}{\partial x^2}-\frac{1}{R}\frac{\partial^3 u_3}{\partial\psi\partial x^2}+\frac{1}{R^3}\frac{\partial^2 u_\psi}{\partial\psi^2}-\frac{1}{R^3}\frac{\partial^3 u_3}{\partial\psi^3}-\frac{\mu}{2R}\frac{\partial^2 u_\psi}{\partial x^2}\right)-R\rho h\ddot{u}_\psi=0,\end{aligned} \tag{12.1.2}$$

$$\begin{aligned}&-K\left(\frac{1}{R}\frac{\partial u_\psi}{\partial\psi}+\frac{u_3}{R}+\mu\frac{\partial u_x}{\partial x}\right)+D\left(-R\frac{\partial^4 u_3}{\partial x^4}-\frac{2}{R}\frac{\partial^4 u_3}{\partial x^2\partial\psi^2}\right.\\&\left.-\frac{1}{R^3}\frac{\partial^4 u_3}{\partial\psi^4}+\frac{1}{R^3}\frac{\partial^3 u_\psi}{\partial\psi^3}+\frac{1}{R}\frac{\partial^3 u_\psi}{\partial\psi\partial x^2}\right)-R\rho h\ddot{u}_3=0,\end{aligned} \tag{12.1.3}$$

where x, ψ, α_3 denote the longitudinal, circumferential, and transverse directions, respectively; u_i is the displacement in the ith direction; R is the radius; h is the shell thickness; ρ is the density; μ is Poisson's ratio of the elastic shell; $K = Yh/(1-\mu^2)$ is the membrane stiffness; and $D = Yh^3/12(1-\mu^2)$ is the bending stiffness.

12.1.3 Energy Equation

The modal energy generated by the distributed energy harvester patches can be calculated according to the following formula:

$$\mathscr{E} = \frac{1}{2} C_p (\phi_p)^2 = \frac{1}{2} \frac{\varepsilon_{33} S_p}{h_p} (\phi_p)^2, \tag{12.1.4}$$

where $\mathscr{E}$ represents the energy (Units: *Joule*); ϕ_p (Units: *V*) is the voltage generated by the energy harvester patch. $C_p = \varepsilon_{33} S_p / h_p$ is the capacitance of an energy harvester patch.

12.1.4 Voltage Generation

The voltage ϕ_p in the transverse direction induced by the modal strains for an open-circuit is defined as (Tzou 1993):

$$\phi_p = \frac{h_p}{S_p} \int_\psi \int_x \left[h_{31} S_{11}^p + h_{32} S_{22}^p + h_{36} S_{12}^p \right] R dx d\psi, \tag{12.1.5}$$

where h_{3i} are the piezoelectric constants; S_{ij}^p are induced strains; S_p is the electrode area of an energy harvester patch:

$$S_p = \int_{x_1^*}^{x_2^*} \int_{\psi_1^*}^{\psi_2^*} R dx d\psi = R(x_2^* - x_1^*)(\psi_2^* - \psi_1^*). \tag{12.1.6}$$

Neglecting the effect of the in-plane shear strains, the voltage represented by displacements u_i becomes:

$$\begin{aligned} \phi_p = \frac{h_p}{S_p} \int_\psi \int_x h_{31} \left(R \frac{\partial u_x}{\partial x} - r_p R \frac{\partial^2 u_3}{\partial x^2} \right) \\ + h_{32} \left[\left(\frac{\partial u_\psi}{\partial \psi} + u_3 \right) + r_p \left(\frac{1}{R} \frac{\partial u_\psi}{\partial \psi} - \frac{1}{R} \frac{\partial^2 u_3}{\partial \psi^2} \right) \right] dx d\psi, \end{aligned} \tag{12.1.7}$$

where $r_p = (h_p + h)/2$. The transverse mode shape function of a circular cylindrical shell is (Soedel, 1981):

$$\mathrm{U}_{3mn} = C_{mn} \sin\left(\frac{m\pi x}{L} \right) \cos[n(\psi - \phi)], \tag{12.1.8}$$

where m denotes the longitudinal mode number and n denotes the circumferential mode number. Assume there is no phase shift, i.e., $\phi = 0$, and set the modal amplitudes equal to unity (i.e., $C_{mn} = 1$) for generic analysis. Note that, the longitudinal and circumferential shell oscillations are negligible, since they are usually much smaller than the transverse displacement. The transverse voltage generated by the energy harvester patch is:

$$\begin{aligned}\phi_p &= \frac{h_p e_{31}}{S_p \varepsilon_{33}} \iint_{S_p} \left[\frac{u_3}{R} - r_p \left(\frac{\partial^2 u_3}{\partial x^2} + \frac{\partial^2 u_3}{R^2 \partial \psi^2} \right) \right] R dx d\psi \\ &= \frac{h_p e_{31}}{S_p \varepsilon_{33}} \sum_{m=1}^{\infty} \sum_{n=1}^{\infty} \eta_{mn} \iint_{S_p} \left[\frac{U_{3mn}}{R} - r_p \left(\frac{\partial^2 U_{3mn}}{\partial x^2} + \frac{\partial^2 U_{3mn}}{R^2 \partial \psi^2} \right) \right] R dx d\psi,\end{aligned} \tag{12.1.9}$$

where the modal participation factor $\eta_{\mathrm{mn}}(t)$ is assumed harmonic $\eta_{\mathrm{mn}}(t) = e^{j\omega_{mn}t}$ in free modal oscillation; ω_{mn} is the (m, n)th natural frequency; ε_{33} is the permittivity; and e_{31} is the piezoelectric stress constant.

12.1.5 Modal Energy Generation

The modal energy of an energy harvester patch can be obtained by substituting Eq. (12.1.9) into Eq. (12.1.4). The total energy consists of three components of modal energy generation of an energy harvester patch.

$$(\mathscr{E}_p)_{\psi\psi,mem} = \frac{1}{2} C_p \left(\frac{h_p e_{31}}{S_p \varepsilon_{33}} \sum_{m=1}^{\infty} \sum_{n=1}^{\infty} \eta_{mn}(t) \iint_{S_p} U_{3mn} dx d\psi \right)^2, \tag{12.1.10a}$$

$$(\mathscr{E}_p)_{xx,bend} = \frac{1}{2} C_p \left(\frac{h_p e_{31}}{S_p \varepsilon_{33}} \sum_{m=1}^{\infty} \sum_{n=1}^{\infty} \eta_{mn}(t) \iint_{S_p} -r_p \frac{\partial^2 U_{3mn}}{\partial x^2} R dx d\psi \right)^2, \tag{12.1.10b}$$

$$(\mathscr{E}_p)_{\psi\psi,bend} = \frac{1}{2} C_p \left(\frac{h_p e_{31}}{S_p \varepsilon_{33}} \sum_{m=1}^{\infty} \sum_{n=1}^{\infty} \eta_{mn}(t) \iint_{S_p} -r_p \frac{\partial^2 U_{3mn}}{R^2 \partial \psi^2} R dx d\psi \right)^2. \tag{12.1.10c}$$

Thus, the total modal energy $(\mathscr{E}_p)_{total}$ is contributed by three energy components: 1) the circumferential membrane strain induced energy $(\mathscr{E}_p)_{\psi\psi,mem}$, 2) the longitudinal bending strain induced energy $(\mathscr{E}_p)_{xx,bend}$, and 3) the circumferential bending strain induced energy $(\mathscr{E}_p)_{\psi\psi,bend}$. For a specific mode (m, n), they are expressed as:

$$(\mathscr{E}_p)_{\psi\psi,mem} = \frac{1}{2} C_p \left(\frac{-h_p e_{31}}{S_p \varepsilon_{33}} \frac{L}{nm\pi} \cos\left(\frac{m\pi x}{L}\right) \Big|_{x_1^*}^{x_2^*} \sin(n\psi) \Big|_{\psi_1^*}^{\psi_2^*} \right)^2, \tag{12.1.11a}$$

$$\left(\mathscr{E}_p\right)_{xx,bend} = \frac{1}{2}C_p\left(\frac{-h_p e_{31}}{S_p \varepsilon_{33}}\frac{R(h_p+h)}{2n}\frac{m\pi}{L}\cos\left(\frac{m\pi x}{L}\right)\Big|_{x_1^*}^{x_2^*}\sin(n\psi)\Big|_{\psi_1^*}^{\psi_2^*}\right)^2, \quad (12.1.11b)$$

$$\left(\mathscr{E}_p\right)_{\psi\psi,bend} = \frac{1}{2}C_p\left(\frac{-h_p e_{31}}{S_p \varepsilon_{33}}\frac{L(h_p+h)}{2m\pi}\frac{n}{R}\cos\left(\frac{m\pi x}{L}\right)\Big|_{x_1^*}^{x_2^*}\sin(n\psi)\Big|_{\psi_1^*}^{\psi_2^*}\right)^2, \quad (12.1.11c)$$

$$\begin{aligned}\left(\mathscr{E}_p\right)_{total} = \frac{1}{2}C_p\Bigg(&\frac{-h_p e_{31}}{S_p \varepsilon_{33}}\frac{L}{nm\pi}\cos\left(\frac{m\pi x}{L}\right)\Big|_{x_1^*}^{x_2^*}\sin(n\psi)\Big|_{\psi_1^*}^{\psi_2^*}\\ &+\frac{-h_p e_{31}}{S_p \varepsilon_{33}}\frac{R(h_p+h)}{2n}\frac{m\pi}{L}\cos\left(\frac{m\pi x}{L}\right)\Big|_{x_1^*}^{x_2^*}\sin(n\psi)\Big|_{\psi_1^*}^{\psi_2^*} \qquad (12.1.11d)\\ &+\frac{-h_p e_{31}}{S_p \varepsilon_{33}}\frac{L(h_p+h)}{2m\pi}\frac{n}{R}\cos\left(\frac{m\pi x}{L}\right)\Big|_{x_1^*}^{x_2^*}\sin(n\psi)\Big|_{\psi_1^*}^{\psi_2^*}\Bigg)^2.\end{aligned}$$

The spatial distributions of the modal energies are investigated next. Experimental demonstrations are presented later.

12.2 Evaluation of Energy Generations

The piezoelectric material used in this study (model shown in Fig. 12.3) is Polymeric Polyvinylidene Fluoride (PVDF). In order to investigate the effect of energy harvester patch size on the modal energy distribution, three patch sizes are evaluated in this section. They are (1 mm,3.6°) for Case 1, (10 mm,30°) for Case 2,

Table 12.1 Parameters of the energy harvesting system

Cylindrical shell	Geometry	Units
Length, L	100	mm
Radius, R	25, 50, 100	mm
Elastic shell thickness, h	0.5, 1, 2	mm
Harvester patch area, S^e	(1,3.6°), (10,30°), (20,60°)	mm^2
Harvester thickness, h^e	12, 25, 50	μm

Table 12.2 Material properties

Properties	Steel	PVDF	Units
Density, ρ	7.8×10^3	1800	kg/m^3
Young's modulus, Y	2.1×10^{11}	1.6×10^9	N/m^2
Poisson's ratio, μ	0.3	0.29	
Electric permittivity, ε_{33}		1.06×10^{-10}	F/m
Piezoelectric strain const., $d_{31} = d_{32}$		6.0×10^{-12}	C/N
Piezoelectric stress const., $e_{31} = e_{32}$		9.6×10^{-3}	C/m^2

and (20 mm,60°) for Case 3, respectively. Material properties and parameters of the physical model of these three cases are listed in Tables 12.1 and 12.2, respectively.

12.2.1 Case 1: Energy Harvester Patch Size of (1 mm,3.6°)

Since the patch size is small, modal energies generated by energy harvester patches are fine enough to display the spatially distributed microscopic modal energy generation behaviors. The modal energy distributions of the (1,1), (1,2), (2,1), and (2,2) modes are plotted in Figs. 12.4, 12.5, 12.6, and 12.7. The top-left is the *circumferential membrane* energy component; the top-right is the *longitudinal bending* energy component; the bottom-left is the *circumferential bending* energy component; the bottom-right is the **total** modal energy. Simulation results suggest that the two bending energy components are much smaller than that of the circumferential membrane energy, but they increase as the mode number increases, due to increased bending behaviors.

Figure 12.8 shows the maximum magnitude of the modal energy changes with the mode number in Case 1. It is observed that: 1) the circumferential membrane energy remains almost constant when m or n changes; 2) the longitudinal bending energy increases when m increases; 3) the circumferential bending energy increases when n increases; 4) the total modal energy increases when m or n increases.

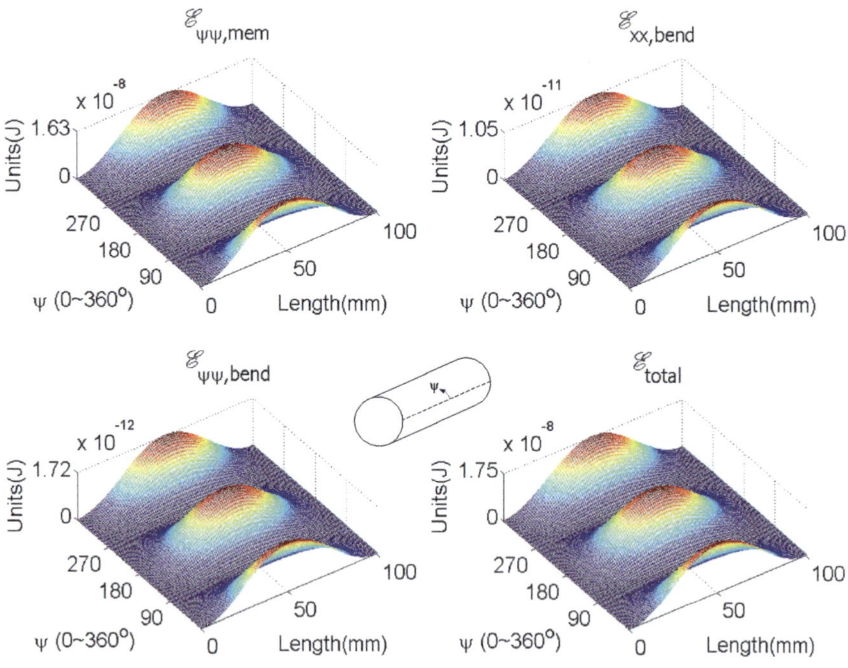

Fig. 12.4 Distributed modal energy magnitude of the (1,1) mode, case 1

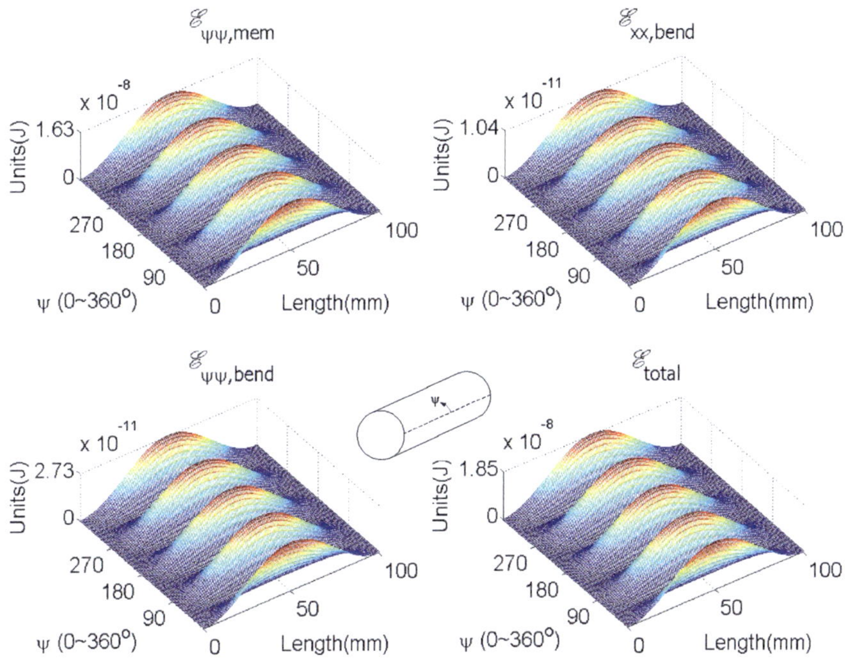

Fig. 12.5 Distributed modal energy magnitude of the (1,2) mode, case 1

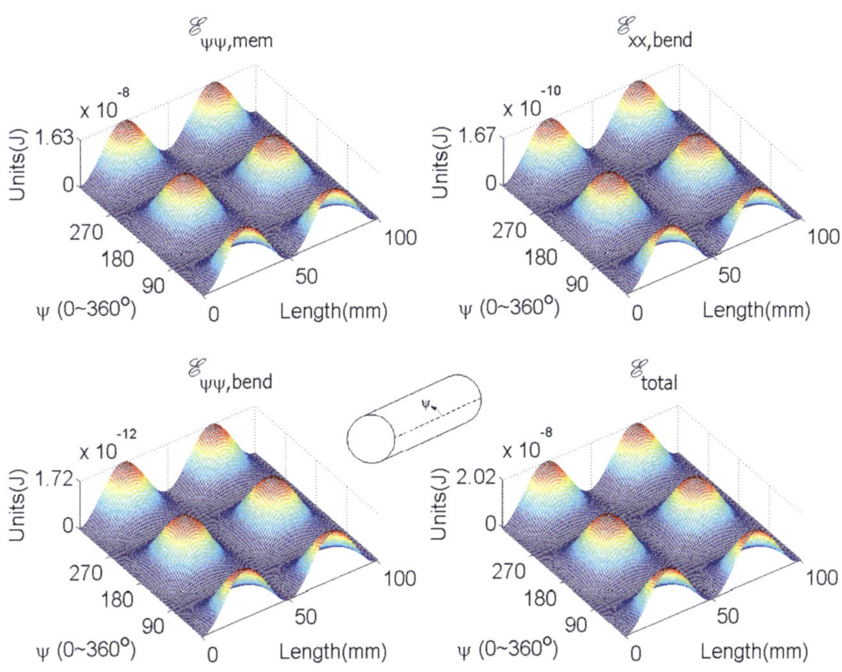

Fig. 12.6 Distributed modal energy magnitude of the (2,1) mode, case 1

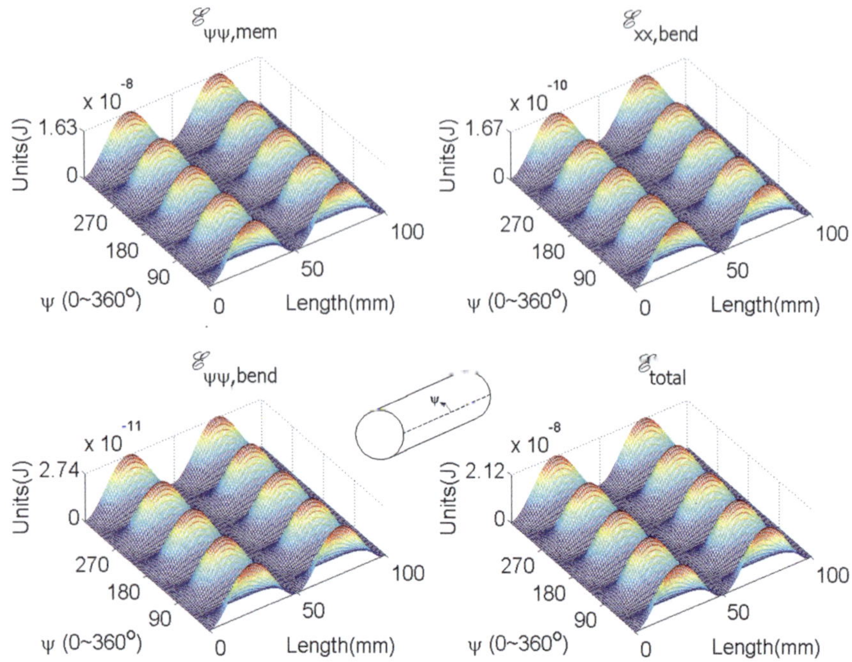

Fig. 12.7 Distributed modal energy magnitude of the (2,2) mode, case 1

12.2.2 Case 2: Energy Harvester Patch Size of (10 mm,30°)

In this case, a bigger energy harvester patch size is considered, Fig. 12.9 shows the spatial distribution of the modal energy of the (1,1) mode and only the (1,1) mode is demonstrated in these two cases. The magnitude of spatially distributed modal energy becomes slightly higher than that in Case 1. Accordingly, there exists the signal averaging effect on any energy harvester patch at various modes. The bending induced energies are still much lower than the circumferential membrane energy.

Figure 12.10 illustrates the variation of the maximum magnitude of the modal energy of different modes. The modal energy changes apparently when n changes, and the modal energy becomes zero when $n = 6$. This phenomenon can be illustrated in Fig. 12.11, which shows that there exist positive and negative voltages in each energy harvester patch in the circumferential direction, thus, resulting in the cancellation effect on a specific patch.

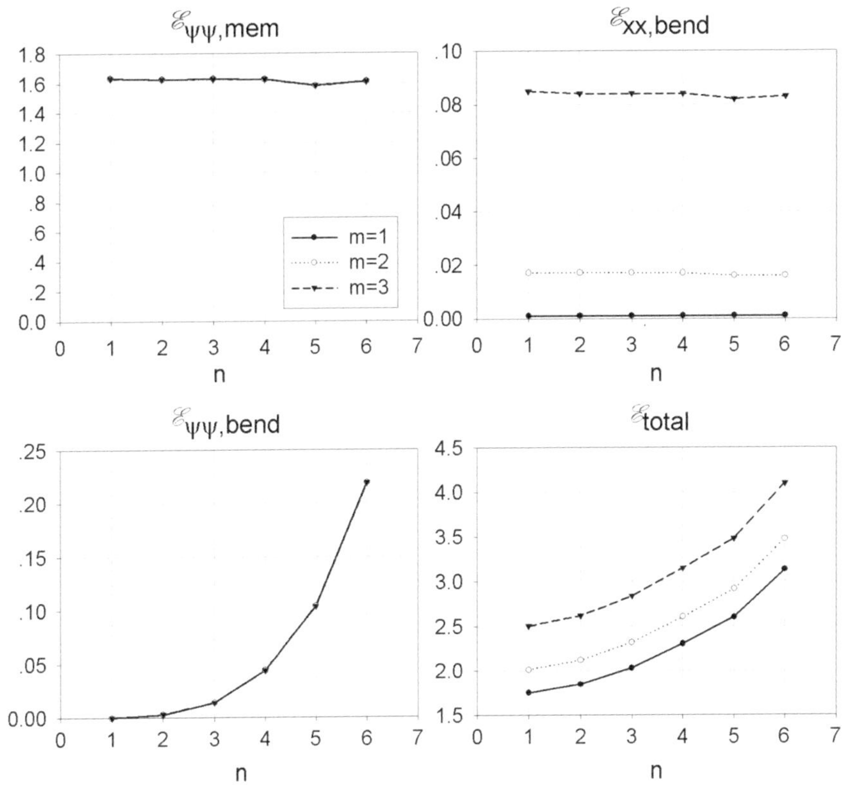

Fig. 12.8 Maximum magnitude of modal energy, case 1 (magnification: 10^8)

12.2.3 Case 3: Energy Harvester Patch Size of (20 mm,60°)

Finally, a patch size of $\Delta L = 20$ mm in the longitudinal direction and $\Delta\psi = 60°$ in the circumferential direction is considered. Figure 12.12 shows the distribution of the modal energy of (1,1) mode. It is observed that the modal energy become slightly higher than that in Case 2 and Case 1, but the larger the energy harvester patch size is, the more significant surface averaging effect reveals on each piezoelectric patch. It should be noted that the value at the coordinate $(x, \Delta\psi)$ represents the energy generation of the patch $(x + \Delta L, \psi + \Delta\psi)$ in the plots of Fig. 12.12. Figure 12.13 shows the maximum magnitude of the modal energy of various modes, and becomes zero when $n = 3$, 6, or 12, etc. This phenomenon can be illustrated in Fig. 12.14, which also shows that there exist positive and negative voltages on each energy harvester patch in the circumferential direction, but the number of energy harvester patches in the circumferential direction decreased compared with that in Case 2, due to the patch size enlarged.

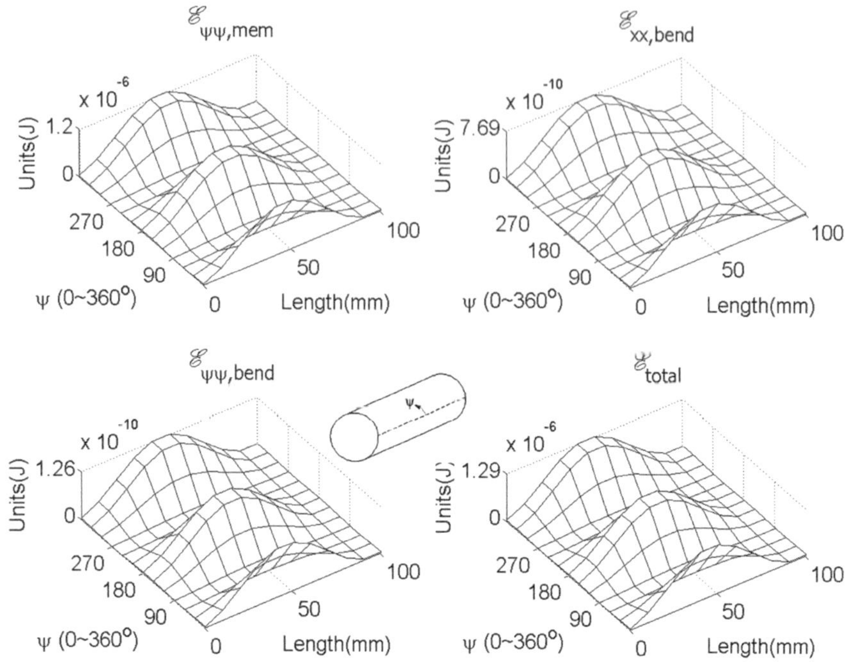

Fig. 12.9 Distributed modal energy magnitude of the (1,1) mode, case 2

12.3 Studies of Design Parameters

Modal energy generation of piezoelectric energy harvester patches with different design parameters and shell's geometry (e.g., shell radius, shell thickness, and energy harvester thickness) are studied in this section. Shell's radius effect is reported first, followed by the shell thickness effect, and the energy harvester thickness.

12.3.1 Shell Radius

Effect of the shell radius to the modal energy generation is investigated in this section. Figure 12.15 summarizes the maximum magnitudes of the circumferential membrane energy (1st row), the longitudinal bending energy (2nd row), the circumferential bending energy (3rd row), and the total modal energy (4th row) with respect to three shell radii, i.e., (a) R = 25 mm, (b) R = 50 mm, (c) R = 100 mm plotted in three columns, of the (m = 1–3, n = 1–6) shell modes. It is observed that: 1) the circumferential membrane energy decreases when the shell radius increases; 2) the longitudinal bending energy almost unchanged when the shell radius

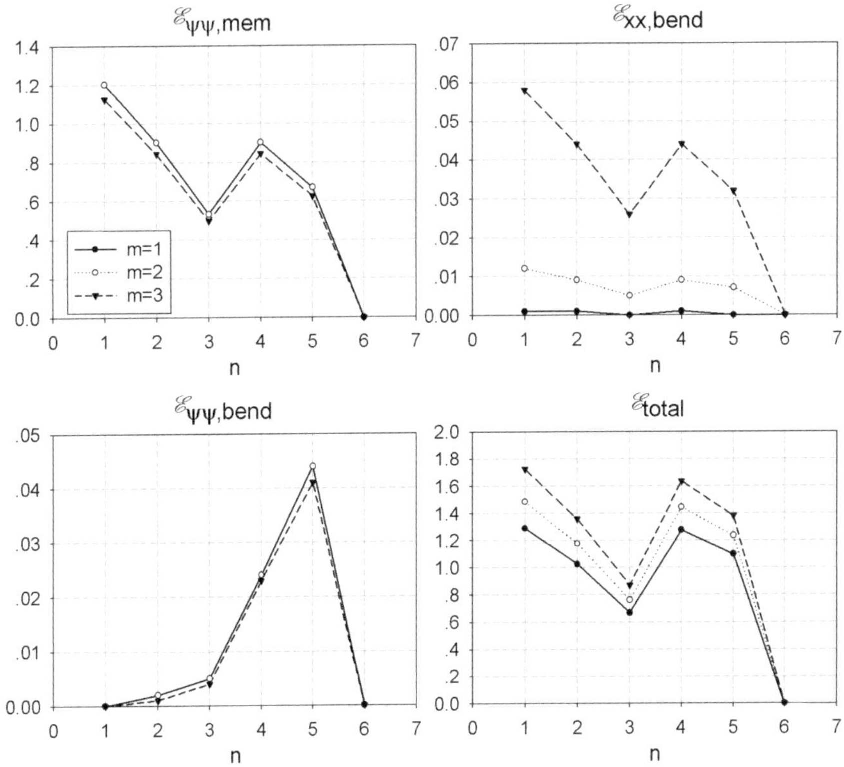

Fig. 12.10 Maximum magnitude of modal energy, case 2 (magnification: 10^6)

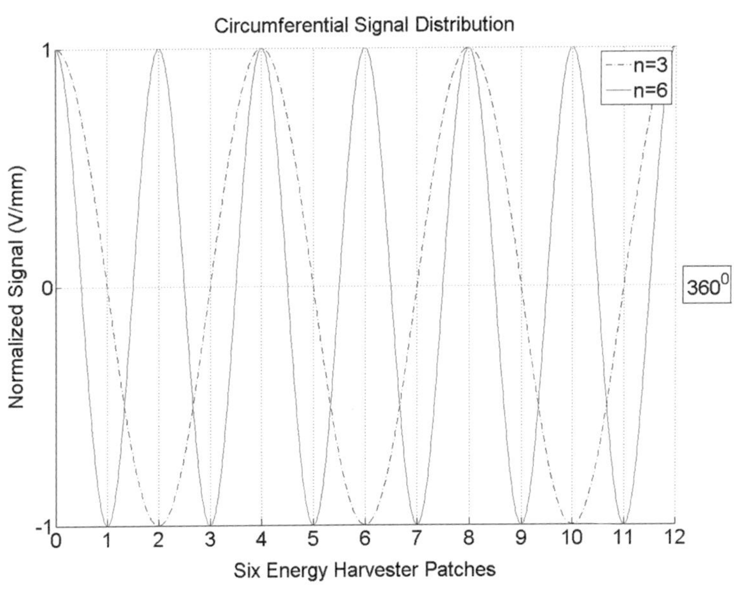

Fig. 12.11 Circumferential signal distribution on the twelve energy harvester patches, case 2

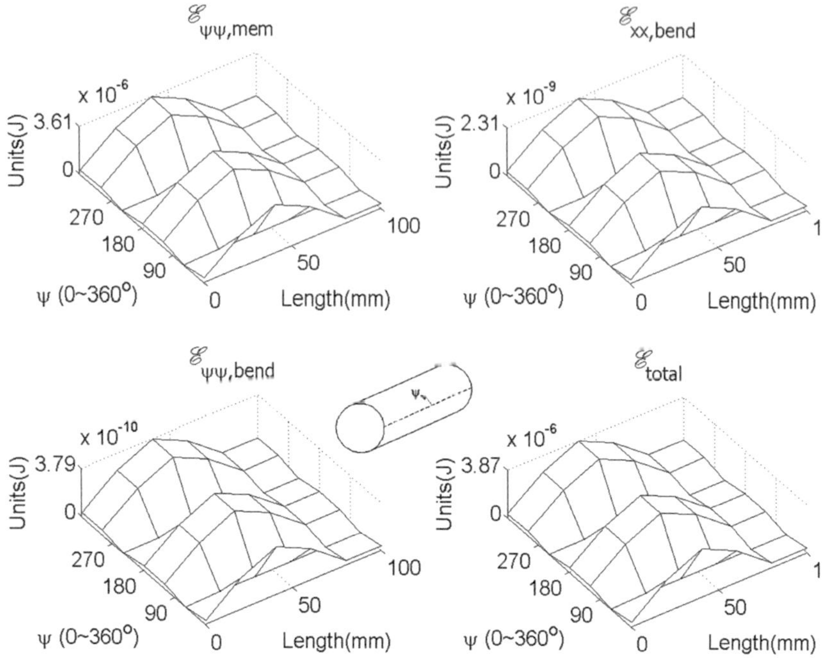

Fig. 12.12 Distributed modal energy magnitude of (1,1) mode, case 3 (magnification: 10^6)

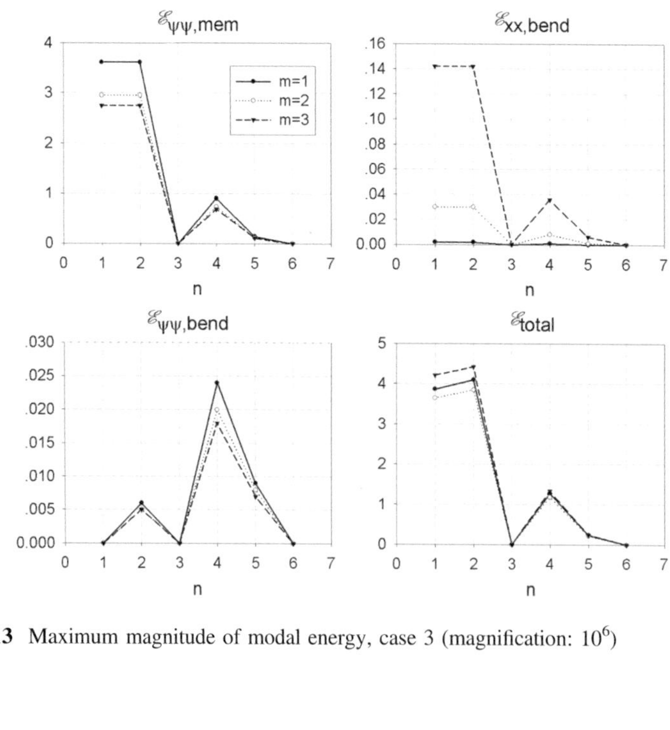

Fig. 12.13 Maximum magnitude of modal energy, case 3 (magnification: 10^6)

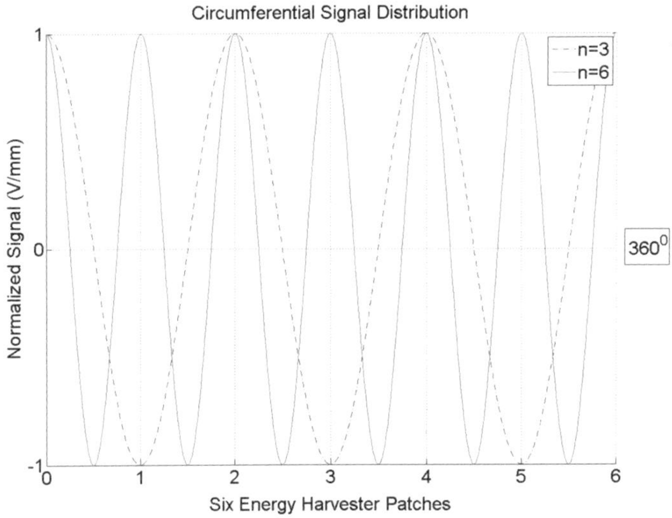

Fig. 12.14 Circumferential signal distribution on the six energy harvester patches, case 3

increases; 3) the circumferential bending energy decreases rapidly when the shell radius increases; 4) the total modal energy decreases when the shell radius increases, due to reduced strain levels when tube's circumference becomes larger.

12.3.2 Shell Thickness

Effect of the shell thickness to the modal energy generation is investigated in this section. Figure 12.16 summarizes the maximum magnitudes of the modal energy with respect to three shell thicknesses, i.e., a) h = 0.5 mm, b) h = 1 mm, c) h = 2 mm plotted in three columns, of the (m = 1–3, n = 1–6) shell modes. It is observed that: 1) the circumferential membrane energy doesn't change with shell thickness; 2) the longitudinal bending energy is proportional to the shell thickness; 3) the circumferential bending energy is also proportional to the shell thickness; and 4) the total modal energy increases when the shell thickness increases, due to increased bending behavior when shell becomes thicker.

12.3.3 Harvester Patch Thickness

Effect of the energy harvester's thickness to the modal energy generation is investigated in this section. Figure 12.17 summarizes the maximum magnitudes of

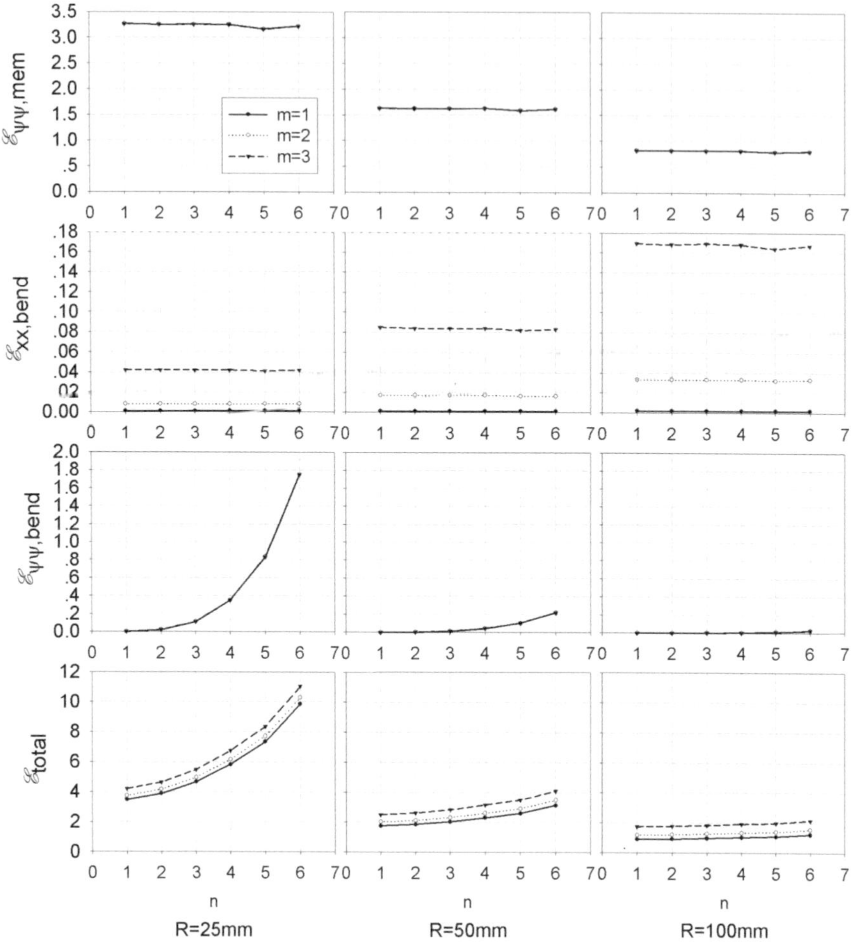

Fig. 12.15 Maximum modal energy at different shell radius, case 1 (m = 1–3, n = 1–6, magnification: 10^8)

the modal energy with respect to three energy harvester thicknesses, i.e., (a) h_p = 12 μm, (b) h_p = 25 μm, (c) h_p = 50 μm plotted in three columns, of the (m = 1–3, n = 1–6) shell modes.

It is observed that: 1) the circumferential membrane energy increases when the energy harvester thickness increases; 2) both the longitudinal bending energy and the circumferential bending energy increase rapidly with the energy harvester thickens; and 3) the total modal energy increases with the energy harvester thickens.

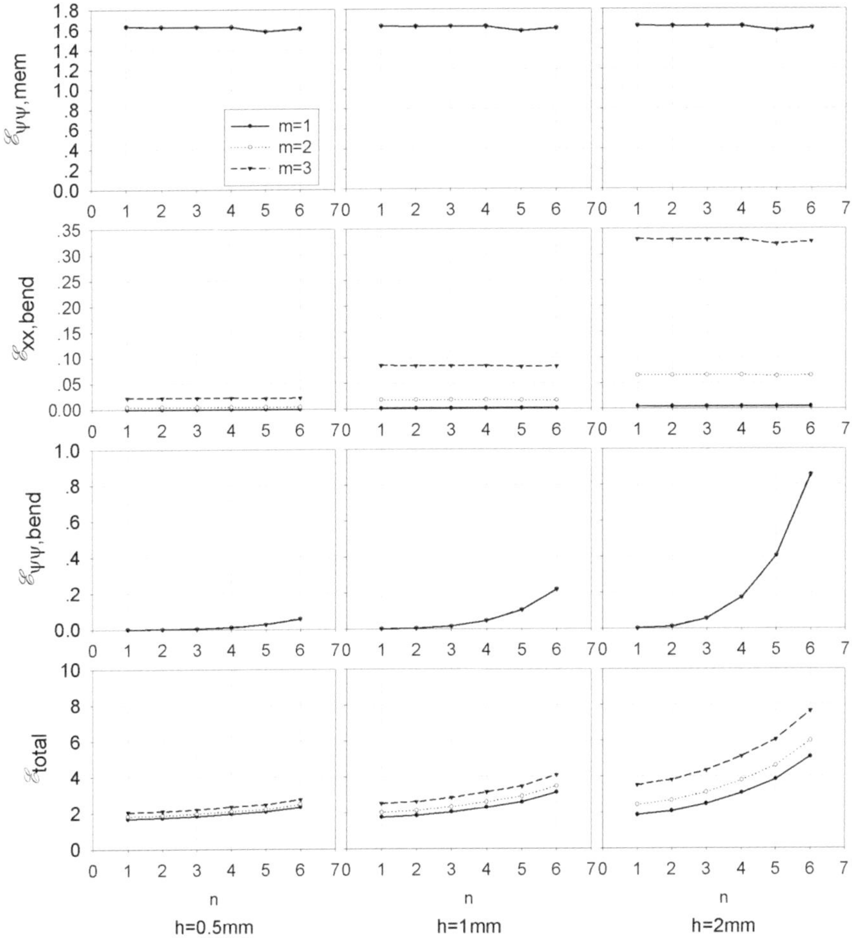

Fig. 12.16 Maximum modal energy at different shell thickness, case 1 ($m = 1–3$, $n = 1–6$, magnification: 10^8)

12.4 Laboratory Experiments

Based on the theoretical analysis, parametric studies and performance evaluations, a tubular shell energy harvesting system is setup and tested in the StrucTronics and Control Laboratory at Zhejiang University. Dimensions and properties of the shell model and piezoelectric patches are summarized in Tables 12.3 and 12.4. The experimental setup, including a tubular shell model mounted on a shaker driven by a wave generator, an external loading resistor circuit, etc., is shown in Fig. 12.18.

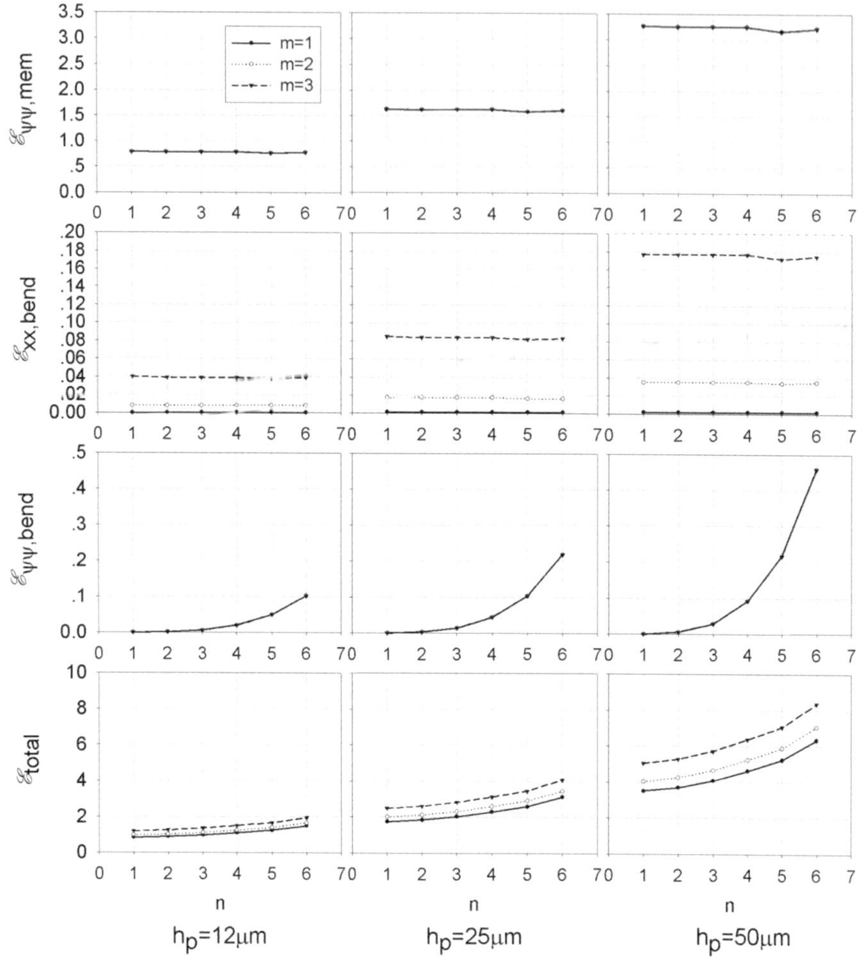

Fig. 12.17 Maximum modal energy at different energy harvester thickness in case 1 (m = 1–3, n = 1–6, magnification: 10^8)

Table 12.3 Dimensions of tubular shell and piezoelectric patch

Parameter	Dimensions	Units
Length, L	121	mm
Radius, R	42	mm
Shell thickness, h	0.37	mm
Piezoelectric patch area, S_p^e	20 × 20	mm^2
Piezoelectric patch thickness, h^e	50	μm

Natural frequencies and mode shapes are evaluated first and its first three modes are summarized in Table 12.5. It suggests that the experimental data are compared with analytical and finite element results favorably.

Table 12.4 Material properties

Properties	Plastic shell	PVDF	Units
Density, ρ	1.307×10^3	1.78×10^3	kg/m^3
Young's modulus, Y	4×10^9	2.5×10^9	N/m^2
Poisson ratio, μ	0.4	0.29	/
Dielectric constant, ε_{33}	/	8.41×10^{-11}	F/m
Piezoelectric constant, d_{31}	/	17×10^{-12}	C/N
Piezoelectric constant, d_{32}	/	6×10^{-12}	C/N

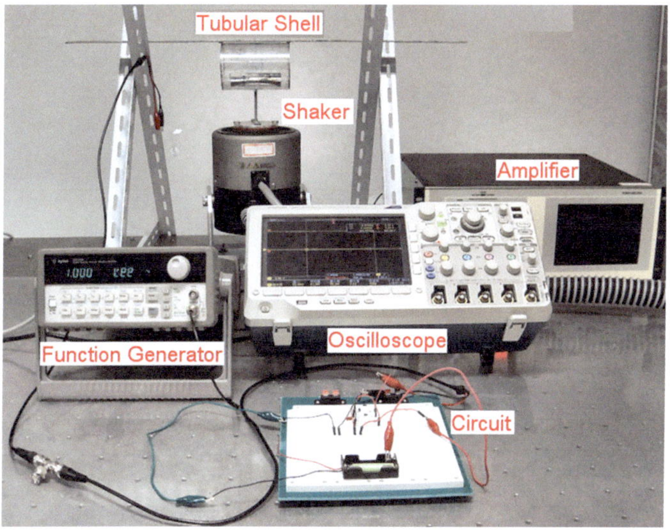

Fig. 12.18 A tubular shell energy harvester setup

Comparing the three modes indicates that the (1.3) mode is the lowest mode. Adjusting the external loading resistors R_L and recording the voltage V (i.e., $V = V_{pp}/2$), one can estimate the effective power output P, i.e., $P = \left(V/\sqrt{2}\right)^2 \Big/ R_L$. Note that since the oscilloscope impedance is 1M Ohm. the true external loading resistance is $R_L = (R_0 \times 10^6)/(R_0 + 1.0 \times 10^6)$ where R_o is the added resistance. Figure 12.19 shows the power output and the external loading resistance relationships (i.e., both theoretical and experimental data) of the (1,3) mode at 779.8 Hz. In general, the theoretical predictions and experimental data are compared favorably. The experimental maximal power appears at R_L = 200 K while the theoretical value is R_L = 230 K. With the same procedures, one can proceed to test the (1,2) and (1,1) shell modes. Figure 12.20 shows the power output and external

Table 12.5 Natural frequencies and mode shapes

Modes (m, n)	Natural frequency			Mode shape	
	Theory	Finite element	Lab data (difference%)		
(1,1)	2683.48	2683.5	2573 (−4.12)		
(1,2)	1339.82	1339.8	1360 (1.51)	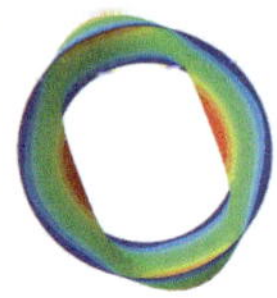	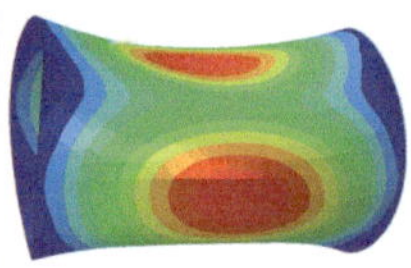
(1,3)	744.64	744.66	779.8 (4.72)	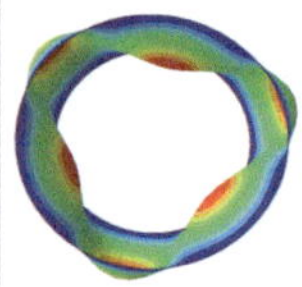	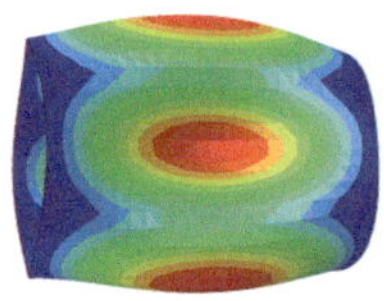

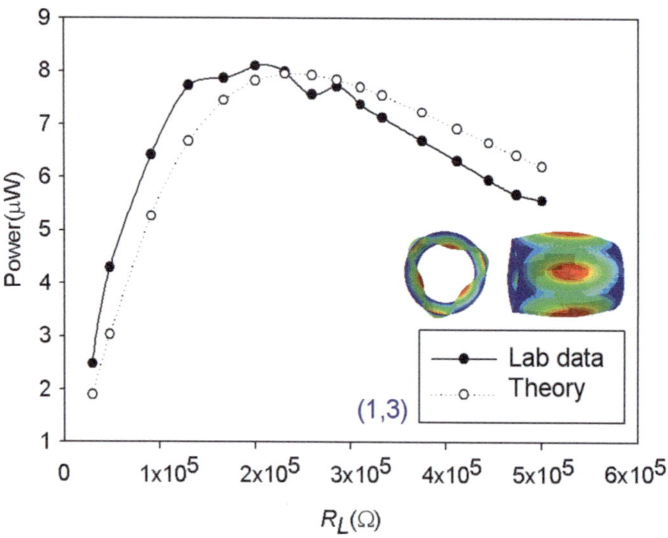

Fig. 12.19 Power versus loading resistance, (1,3) mode

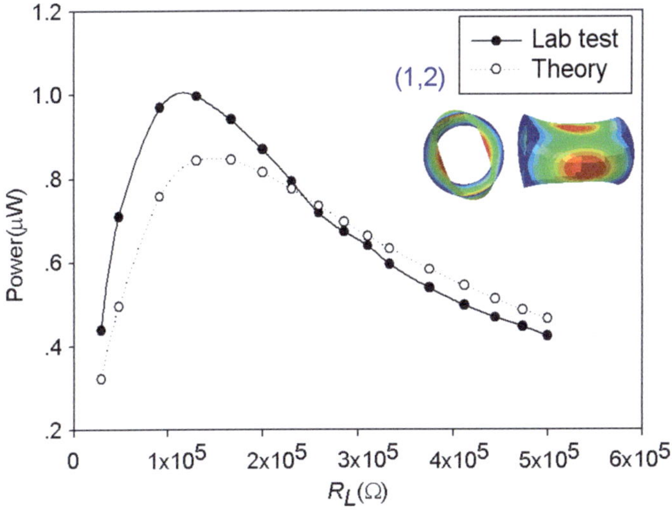

Fig. 12.20 Power versus loading resistance, (1,2) mode

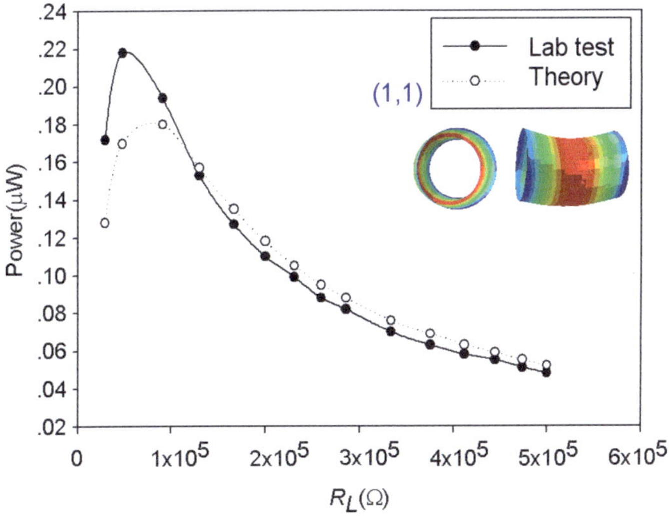

Fig. 12.21 Power versus loading resistance, (1,1) mode

loading resistance relationship of the (1,2) mode at 1339.8 Hz and Fig. 12.21 shows the theoretical and experimental power output and external loading resistance relationships of the (1,1) mode at 2573 Hz. Both figures show similar behaviors that there is an optimal loading resistance leading to the maximal power output.

12.5 Summary

This chapter involved energy harvesting of a simply supported tubular (circular) cylindrical shell laminated with piezoelectric patches. The distributed modal energy generations using different energy harvester patch sizes (i.e., (1 mm,3.6°) in Case 1, (10 mm,30°) in Case 2, and (20 mm,60°) in Case 3) at various mode numbers were evaluated in case studies. Analytical and simulation results suggest that the maximum magnitude of the spatially distributed modal energies changes at various modes in two cases, due to the patch size enlarged or the number of energy harvester patches in the circumferential direction decreased. It should be noted that the signal averaging effects on energy harvester patches become more significant when the patch size continuously increasing. Additionally, the bending energy components are much smaller than the circumferential membrane energy component, and they increase when mode number increases due to increased bending behaviors. Furthermore, the maximum magnitude of the (m, n)th modal energy, in general, increases when energy harvester's thickness h_p or shell's thickness h increases, but decreases when the shell radius R increases.

A tubular shell energy harvesting system was designed and tested in the StrucTronics and Control Laboratory at Zhejiang University. Experimental results suggest that there is an optimal external loading resistance leading to the maximal power output. Both analytical predictions and experimental data were compared favorably. These data evaluated in this study can be used as guidelines to design the optimum piezoelectric energy harvester in practical engineering applications.

Acknowledgements The research was supported by the National Natural Science Foundation of China (11472241 & 11172262), the Nanjing University of Aeronautics and Astronautics Foundation (NUAA-NP2016203), the State Key Laboratory of Mechanics and Control of Mechanical Structures (NUAA-MCMS-0516G01), and the State Key Laboratory of Fluid Power Transmission and Control (No.SKLoFP_CX_001) at Zhejiang University.

References

Anton, S.R. and Sodano, H.A., A review of power harvesting using piezoelectric materials (2003–2006), *Smart Material & Structure*. **16**, R1–R21, 2007.

Badel, A., Guyomar, D. and Richard, C., Piezoelectric energy harvesting using a synchronized switch technique, *Journal of Intelligent Material Systems and Structures*. **17(8)**, 831–839, 2006.

Callahan, J. and Baruh, H., Modal sensing of circular cylindrical shells using segmented piezoelectric elements, *Smart Material and Structure*. **8**, 125–135, 1999.

DeHaven, J.G. Han, Y. and Tzou, H.S., Transition of membrane/bending neural signals on transforming adaptive shells, *Journal of Vibration and Control*. **13(7)**, 1007–1029, 2007.

Erturk, A. and D. J. Inman, An experimentally validated bimorph cantilever model for piezoelectric energy harvesting from base excitations, *Smart Mater. Struct.* **18**, 1–18, 2009.

Guo, D. Chu, and F.L. Zheng, Z.C., The influence of rotation on vibration of a thick cylindrical shell, *Journal of Sound and Vibration*. **242(3)**, 487–505, 2001.

Howard, R.V., Chai, W.K. and Tzou, H.S., Modal voltages of linear and non-linear structures using distributed artificial neurons, *Mechanical Systems and Signal Processing* 15 (2001) 629–640.

Huang, S.L. Chu, C.C. Chang, C.C. and Tzou, H.S., Electromechanical coupling and signal distribution of a circular cylindrical shell coupled with segmented sensors, *ASME Conf. Proc.1*, 145–154(IMECE2008-68016), 2008.

Kim, S. William, W.C. and Wang, Q., Piezoelectric energy harvesting with a clamped circular plate: analysis, *Journal of Intelligent Material Systems and Structures*. **16**, 847, 2005.

Lefeuvre, E., Badel, A. and Guyomar, D, Piezoelectric energy harvesting device optimization by synchronous electric charge extraction, *Journal of Intelligent Material Systems and Structures*. **16(10)**, 865–876, 2005.

Lu, F. Lee, and Lim, H.P., Modeling and analysis of micro piezoelectric power generators for micro-electromechanical-systems Applications, *Smart Material and Structure*. **13**, 57–63, 2004.

Pellicano, F., Vibrations of circular cylindrical shells: theory and experiments, *Journal of Sound and Vibration*. **303**, 154–170, 2007.

Poore, A.L., Barut, and A. Madenci, E., Free vibrations of laminated cylindrical shells with a circular cutout, *Journal of Sound and Vibration*. **303**, 154–170, 2007.

Poulin, G. Sarraute, and E. Costa, F., Generation of electrical energy for portable devices comparative study of an electromagnetic and a piezoelectric system, *Sensors and Actuators A: Physical*. **116**, 461–471, 2004.

Sodano, H.A., Inman, D.J. and Park, G., A review of power harvesting from vibration using piezoelectric materials, *Shock & Vibration Digest* 36 (2004) 197–205.

Sodano, H.A., Inman, D.J. and Park, G., Generation and storage of electricity from power harvesting devices, *Journal of Intelligent Material Systems and Structures*. **16(1)**, 67–75, 2005.

Soedel, W., Vibration of shells and plates, *Marcel Dekker, Inc, New York. Basel*, 1981.

Tzou, H.S., Piezoelectric shells—distributed sensing and control of continua, *Kluwer Academic Publishers, Boston/Dordrecht*, 1993.

Tzou, H.S. and Bao, Y., Nonlinear piezothermoelasticity and multi-field actuations, part1: nonlinear anisotropic piezothermoelastic shell laminates, *Journal of Vibration and Acoustics* 119 (1997) 374–381.

Tzou, H.S., Bao, Y.M, and Venkayya, V.B., Parametric study of segmented transducers laminated on cylindrical shells, part 1: sensor patches, *Journal of Sound and Vibration*. **117(2)**, 207–224, 1996.

Tzou, H.S. and Yang, R.J., Nonlinear piezo-thermoelastic shell theory applied to control of variable-geometry shells, *Theoretical and Applied Mechanics*. **38(3)**, 623–644, 2000.

Tzou, H.S. and Zhong, J.P., Spatially filtered vibration control of cylindrical shells, *Shock and Vibration*. **3(4)**, 269–278, 1996.

Chapter 13
Finite Element Formulation and Analyses

The rapid development of high-speed computers has facilitated the use of computational and numerical techniques in a variety of engineering analyses and applications. The finite element method is one of the most popular and powerful techniques in modern engineering design and analysis of complicated structures, system and multi-field problems. In recent years, however, researches on active piezoelectric structures have focused primarily on experimental and theoretical studies; there has been little general-purpose piezoelectric finite element development (Tzou & Tseng, 1988a & b; Allik & Hughes, 1979; Nailon, et al., 1983).

In general, experimental models are limited by sizes, cost, noises, and many other laboratory unknowns. Theoretical models can be more general, however, analytical solutions are restricted to relatively simple geometries and boundary conditions. When the geometries and/or boundary conditions become complicated, both theoretical and experimental techniques encounter technical difficulties. Thus, finite element development becomes very important in modeling and analysis of elastic/piezoelectric coupled distributed parameter systems (DPS's).

A typical active elastic/piezoelectric structure is composed of a (master or base) elastic continuum (e.g., shell or plate) coupled or embedded with piezoelectric (or other smart-material based) sensors and actuators. The thickness of the master structure can be about two to three orders thicker than that of the piezoelectric layer. Thus, it would be very inefficient and time-consuming if the entire structure were modeled by isoparametric hexahedron or tetrahedral solid elements. Thus, development of new "thin" piezoelectric finite elements applied to piezoelectric laminated shell structures is essential for investigating distributed sensing and control of active piezoelectric structures. Piezoelectric solid finite elements were formulated and applied to piezoceramic transducer and oscillator designs (Allik & Hughes, 1979; Nailon, et al., 1983). Tzou and Tseng (1990) formulated a new thin piezoelectric solid finite element and applied it to the modeling and analysis of flexible continua—shells and plates with distributed piezoelectric sensors/actuators. State variable transformation of the dynamic equation has also been presented (Tzou & Tseng, 1991a & b; Tseng, 1989).

H. (HS) Tzou, *Piezoelectric Shells*, Solid Mechanics and Its Applications 247,
https://doi.org/10.1007/978-94-024-1258-1_13

In this chapter, development of the new piezoelectric thin hexahedron solid element with internal degrees of freedom is presented. Distributed sensing and control of continua are formulated. High-precision manipulation, distributed structural identification (modal voltage), and control of DPS's—elastic continua—are investigated in case studies.

13.1 Piezoelectric Finite Element Formulation

In this section, a new thin piezoelectric finite solid element with three internal degrees of freedom (DOF's) is formulated using the variation method and Hamilton's principle. The system matrix equation and control strategies are derived.

13.1.1 Variation Equation and Piezoelectric Element

The Lagrangian $\mathcal{L}$ of a bounded piezoelectric volume V is defined by a sum of a kinetic energy $\mathcal{K}$ and a potential energy $\mathcal{U}$ (including strain and electric energies and excluding externally applied work) (Tzou & Tseng, 1990):

$$\begin{aligned}\mathcal{L} &= \int_V (\mathcal{K} - \mathcal{U})\mathrm{dV} \\ &= \int_V \left(\tfrac{1}{2}\rho\{\dot{u}\}^t\{\dot{u}\} - \tfrac{1}{2}\left[\{S\}^t\{T\} - \{E\}^t\{D\}\right]\right)\mathrm{dV}\end{aligned}, \tag{13.1.1}$$

where $\dot{u}$ is the velocity (the time derivative of displacement u), and V is the piezoelectric volume. The virtual work $\delta\mathcal{W}$ done by the external forces P_x's (x = b (body), s (surface), c (concentrated)) and the applied surface charge Q is

$$\delta\mathcal{W} = \int_V \{\delta u\}^t\{P_b\}\mathrm{dV} + \int_{S_1} \{\delta u\}^t\{P_S\}\mathrm{dS_1} + \{\delta u\}^t\{P_c\} - \int_{S_2} \delta\phi Q \mathrm{dS_2}, \tag{13.1.2}$$

where $\{P_b\}$ is the body force; $\{P_s\}$ is the surface force; $\{P_c\}$ is the concentrated load; S_i is the surface area; ϕ is the electric potential; $\boldsymbol{Q}$ is the surface charge. Based on the Lagrangian and the virtual work defined above, one can derive the dynamic equations of a piezoelectric structure by using Hamilton's principle:

$$\delta \int_{t_0}^{t_1} (\mathcal{L} + \mathcal{W})\mathrm{dt} = 0, \tag{13.1.3}$$

where t_0 and t_1 define the time interval, and all variations must vanish at t = t_0 and t = t_1. The linear constitutive piezoelectric equation can be expressed as

$$\{T\} = [c]\{S\} - [e]^t\{E\} \quad (13.1.4a)$$

$$\{D\} = [e]\{S\} + [\varepsilon]\{E\} \quad (13.1.4b)$$

where $\{T\}$ is the stress vector; [c] is the elasticity matrix; $\{S\}$ is the strain vector; $[e]^t$ is the transpose of the piezoelectric constant matrix; $\{E\}$ is the electric field vector; $\{D\}$ is the electric displacement vector; $[\varepsilon]$ is the dielectric constant matrix.

Substituting all strain, electric displacement, and virtual work terms into Hamilton's equation and taking the variation, one can derive

$$\int_{t_0}^{t_1} \Big\{\int_V \big[\rho\{\delta\dot{u}\}^t\{\dot{u}\} - \{\delta S\}^t[c]\{S\} + \{\delta S\}^t[e]\{E\} - \{\delta E\}^t[e]\{S\} - \{\delta E\}^t[\varepsilon]\{E\} + \{\delta u\}^t\{P_b\}\big]dV + \int_{S_1} \{\delta u\}^t\{P_S\}dS_1 - \int_{S_2} \delta\phi Q dS_2 + \{\delta u\}^t\{P_c\}\Big\}dt = 0 \quad (13.1.5)$$

To derive the piezoelastic matrix relationship for an **eight-node hexahedron isoparametric piezoelectric solid** finite element **(FE)**, the displacement $\{u\}$ and electric potential ϕ are defined in terms of i nodal variables via the shape function matrices $[N_u]$ and $[N_\phi]$:

$$\{u\} = [N_u]\{u_i\}, \quad (13.1.6a)$$

$$\phi = [N_\phi]\{\phi_i\}, \quad (13.1.6b)$$

where

$$\{u_i\} = \sum_{i=1}^{8} [\bar{u}_i \quad \bar{v}_i \quad \bar{w}_i]^t. \quad (13.1.7a)$$

$$\{\phi_i\} = \sum_{i=1}^{8} [\bar{\phi}_i]^t, \quad (13.1.7b)$$

$$\{N_u\} = \sum_{i=1}^{8} \begin{bmatrix} N_i & 0 & 0 \\ 0 & N_i & 0 \\ 0 & 0 & N_i \end{bmatrix}, \quad (13.1.7c)$$

$$\{N_\phi\} = \sum_{i=1}^{8} [N_i] = \lfloor N_1\, N_2\, N_3\, N_4\, N_5\, N_6\, N_7\, N_8 \rfloor. \quad (13.1.7d)$$

Note that the subscript "i" denotes the nodal variables, and $\lfloor \cdots \rfloor$ denotes a row vector. Equation (13.1.7d) gives an example for the eight nodal variables. The strains $\{S\}$ are defined by the first derivative of displacement vector $\{u\}$ using a differential operator matrix $[L_u]$ (Bathe & Wilson, 1976). The electric field vector $\{E\}$ is defined by the electric potential ϕ using a gradient operator ∇, i.e.,

$\{E\} = -\nabla\phi$, where $\nabla = \{\partial/\partial x, \partial/\partial y, \partial/\partial z\}^t$. Writing the strain {S} and {E} in nodal variables, one can obtain

$$\{S\} = [B_u]\{u_i\}, \tag{13.1.8}$$

$$\{E\} = -[B_\phi]\{\phi_i\}, \tag{13.1.9}$$

where

$$[B_u] = [L_u][N_u] = \sum_{i=1}^{8} \begin{bmatrix} \partial N_i/\partial x & 0 & 0 \\ 0 & \partial N_i/\partial y & 0 \\ 0 & 0 & \partial N_i/\partial z \\ 0 & \partial N_i/\partial z & \partial N_i/\partial y \\ \partial N_i/\partial z & 0 & \partial N_i/\partial x \\ \partial N_i/\partial y & \partial N_i/\partial x & 0 \end{bmatrix} \tag{13.1.10a}$$

$$[B_\phi] = \nabla[N_\phi] = \sum_{i=1}^{8} \begin{bmatrix} \partial N_i/\partial x \\ \partial N_i/\partial y \\ \partial N_i/\partial z \end{bmatrix}. \tag{13.1.10b}$$

Note that the differential operator $[L_u]$ is defined as

$$[L_u] = \begin{bmatrix} \partial/\partial x & 0 & 0 \\ 0 & \partial/\partial y & 0 \\ 0 & 0 & \partial/\partial z \\ 0 & \partial/\partial z & \partial/\partial y \\ \partial/\partial z & 0 & \partial/\partial x \\ \partial/\partial y & \partial/\partial x & 0 \end{bmatrix}. \tag{13.1.10c}$$

Detailed derivations of the element are presented in the next section.

13.1.2 New Piezoelectric Element with Internal DOF's

Conventional isoparametric solid elements have significant deficiencies in "thin" structural applications. If the element thickness is thin compared with the element span, excessive shear strain energy is stored in the thickness direction. Accordingly, the stiffness coefficients in the thickness direction become much higher than those in the planar directions. This leads to poor estimations and inaccurate results (Bathe & Wilson, 1976; Cook, 1974). An important technique for improving the behavior of isoparametric elements is to introduce internal degrees of freedom (DOF's).

The original piezoelectric hexahedron solid element was an eight-node hexahedron element (Tzou & Tseng, 1988a). The added three internal nodal DOF's are numbered from 9 to 11. By adding these internal DOF's to the dependent variables, one can derive the displacement {u} vector as

$$\{u\} = [N_u]\{u_i\} + [X_u]\{\hat{u}_j\} \tag{13.1.11}$$

where $[N_u]$ is the displacement shape function matrix for nodal displacements $\{u_i\}$, and $[X_u]$ is the extra mode shape function matrix for the added internal DOF's $\{\hat{u}_j\}$. $[X_u]$ and $\{\hat{u}_j\}$ are represented as

$$[X_u] = \sum_{j=9}^{11} \begin{bmatrix} N_j & 0 & 0 \\ 0 & N_j & 0 \\ 0 & 0 & N_j \end{bmatrix}, \tag{13.1.12a}$$

$$\{\hat{u}_j\} = \sum_{j=9}^{11} [\,\bar{u}_i \quad \bar{v}_i \quad \bar{w}_i\,]^t. \tag{13.1.12b}$$

Note that the added DOF's $\{\hat{u}_j\}$ are not physical displacements, but can be regarded as generalized coordinates, or as "displacements" relative to the nodal displacements. These "displacements" vanish at all element edges, so that these DOF's are internal and have no physical effect on inter-element compatibility. The strain-displacement equation can be written as

$$\{S\} = [B_u]\{u_i\} + [Y_u]\{\hat{u}_j\}, \tag{13.1.13}$$

where

$$[B_u] = [L_u][N_u], \tag{13.1.14a}$$

$$[Y_u] = \sum_{j=9}^{11} \begin{bmatrix} \partial N_j/\partial x & 0 & 0 \\ 0 & \partial N_j/\partial y & 0 \\ 0 & 0 & \partial N_j/\partial z \\ 0 & \partial N_j/\partial z & \partial N_j/\partial y \\ \partial N_j/\partial z & 0 & \partial N_j/\partial x \\ \partial N_j/\partial y & \partial N_j/\partial x & 0 \end{bmatrix}. \tag{13.1.14b}$$

Note $[L_u]$ is a differential operator matrix. Using the variation principle, one can derive the electric enthalpy $\mathcal{H}$ (see Sect. 2.1 of Chap. 2) of the piezoelectric finite element with the internal DOF's as

$$\begin{aligned}\delta\mathcal{H} = &\int_V \left[\left\{\begin{matrix}[B_u]^t\\ [Y_u]^t\end{matrix}\right\}[c]\{[B_u][Y_u]\}\begin{bmatrix}\{u_i\}\\ \{\hat{u}_j\}\end{bmatrix}dV\right]^t \cdot \delta\begin{bmatrix}\{u_i\}\\ \{\hat{u}_j\}\end{bmatrix}\\ &-\int_V \left[\left\{\begin{matrix}[B_u]^t\\ [Y_u]^t\end{matrix}\right\}[e]^t[B_\phi]\{\phi_i\}dV\right]^t \cdot \delta\begin{bmatrix}\{u_i\}\\ \{\hat{u}_j\}\end{bmatrix}\\ &-\int_V \left[[B_\phi]^t[e]\{[B_u][Y_u]\}\begin{bmatrix}\{u_i\}\\ \{\hat{u}_j\}\end{bmatrix}dV\right]^t \cdot \delta\{\phi_i\}\\ &-\int_V \left[[B_\phi]^t[\varepsilon][B_\phi]\{\phi_i\}dV\right]^t \cdot \delta\{\phi_i\}\end{aligned}, \qquad (13.1.15)$$

where [c] is the elasticity matrix; V denotes the integration volume; $[e]^t$ is a transpose of the piezoelectric constant matrix; $[B_\phi] = \nabla[N_\phi]$, Eq. (13.1.10b); $[\varepsilon]$ is the dielectric constant matrix. The static homogeneous system equations of the piezoelectric finite element are derived as

$$\begin{bmatrix}[k_{uu}] & [k_{u\hat{u}}]\\ [k_{\hat{u}u}] & [k_{\hat{u}\hat{u}}]\end{bmatrix}\begin{bmatrix}\{u_i\}\\ \{\hat{u}_j\}\end{bmatrix} + \begin{bmatrix}[k_{u\phi}]\\ [k_{\hat{u}\phi}]\end{bmatrix}\{\phi_i\} = 0, \qquad (13.1.16)$$

$$\begin{bmatrix}[k_{\phi u}] & [k_{\phi\hat{u}}]\end{bmatrix}\begin{bmatrix}\{u_i\}\\ \{\hat{u}_j\}\end{bmatrix} + [k_{\phi\phi}]\{\phi_i\} = 0; \qquad (13.1.17)$$

where the elemental matrices are defined as:

$$[k_{uu}] = \int_V [B_u]^t[c][B_u]dV, \qquad (13.1.18a)$$

$$[k_{u\hat{u}}] = \int_V [B_u]^t[c][Y_u]dV, \qquad (13.1.18b)$$

$$[k_{u\phi}] = \int_V [B_u]^t[e]^t[B_\phi]dV, \qquad (13.1.18c)$$

$$[k_{\hat{u}u}] = [k_{u\hat{u}}]^t, \qquad (13.1.18d)$$

$$[k_{\hat{u}\hat{u}}] = \int_V [Y_u]^t[c][Y_u]dV, \qquad (13.1.18e)$$

$$[k_{\phi\hat{u}}] = \int_V [B_\phi]^t[e][Y_u]dV, \qquad (13.1.18f)$$

$$[k_{\phi u}] = [k_{u\phi}]^t, \qquad (13.1.18g)$$

$$\left[k_{\hat{u}\phi}\right] = \left[k_{\phi\hat{u}}\right]^t, \tag{13.1.18h}$$

$$\left[k_{\phi\phi}\right] = -\int_V \left[B_\phi\right]^t [\varepsilon] \left[B_\phi\right] dV. \tag{13.1.18}$$

Note the *lower-case* letters represent the element matrices (e.g., stiffness, damping, charge, force, etc.) and the *upper-case* letters denote the corresponding assembled global matrices. (The upper case letters will be introduced later.)

13.1.3 Condensation of Internal DOF's

Since the internal DOF's have no physical significance, they can be condensed from the system equations in order to improve the computation efficiency. Employing Guyan's reduction method (Guyan, 1965), one can obtain the modified element static matrix equations as

$$\left[k^*_{uu}\right]\{u_i\} + \left[k^*_{u\phi}\right]\{\phi_i\} = \{0\}, \tag{13.1.19}$$

$$\left[k^*_{\phi u}\right]\{u_i\} + \left[k^*_{\phi\phi}\right]\{\phi_i\} = \{0\}; \tag{13.1.20}$$

where

$$\left[k^*_{uu}\right] = \left[k_{uu}\right] - \left[k_{u\hat{u}}\right]\left[k_{\hat{u}\hat{u}}\right]^{-1}\left[k_{\hat{u}u}\right], \tag{13.1.21a}$$

$$\left[k^*_{u\phi}\right] = \left[k_{u\phi}\right] - \left[k_{u\hat{u}}\right]\left[k_{\hat{u}\hat{u}}\right]^{-1}\left[k_{\hat{u}\phi}\right], \tag{13.1.21b}$$

$$\left[k^*_{\phi\phi}\right] = \left[k_{\phi\phi}\right] - \left[k_{\phi\hat{u}}\right]\left[k_{\hat{u}\hat{u}}\right]^{-1}\left[k_{\hat{u}\phi}\right]. \tag{13.1.21c}$$

Note that $\left[k^*_{uu}\right]$ is the modified elastic stiffness matrix; $\left[k^*_{u\phi}\right]$ is the piezoelectric stiffness matrix; $\left[k^*_{\phi\phi}\right]$ is the electric potential stiffness matrix. The modified dynamic equations are derived accordingly:

$$[m_{uu}]\{\ddot{u}\} + [c^*]\{\dot{u}\} + \left[k^*_{uu}\right]\{u\} + \left[k^*_{u\phi}\right]\{\phi\} = \{f_i\}, \tag{13.1.22}$$

$$\left[k^*_{\phi u}\right]\{u\} + \left[k^*_{\phi\phi}\right]\{\phi\} = \{q_i\}, \tag{13.1.23}$$

where $[m_{uu}]$ is the element mass matrix, $[c^*]$ is the element proportional damping matrix.

$$[m_{uu}] = \int_V \rho [N_u]^t [N_u] dV, \tag{13.1.24a}$$

$$[c^*] = \alpha[m_{uu}] + \beta\left[k^*_{uu}\right], \tag{13.1.24b}$$

where α and β are Rayleigh's coefficients. $\{f_i\}$ is the mechanical excitation and $\{q_i\}$ is the electric excitation defined as

$$\{f_i\} = \int_V [N_u]^t \{P_b\} dV + \int_{S_1} [N_u]^t \{P_S\} dS_1 + [N_u]^t \{P_c\}, \tag{13.1.25a}$$

$$\{q_i\} = -\int_{S_2} \left[N_\phi\right]^t Q dS_2. \tag{13.1.25b}$$

Note that the modal damping ratio of the k-th mode, with a natural frequency ω_k, is defined as

$$\zeta_k = \frac{\alpha + \beta\omega_k^2}{2\omega_k}. \tag{13.1.26}$$

The corresponding Rayleigh's coefficients α and β, Eq. (13.1.24b), can be estimated accordingly. If the piezoelectric properties are not considered in the equation, the element is reduced to a conventional thin elastic solid element.

13.1.4 System Equations

Assembling the elemental matrices gives the global system matrices. The resulting dynamic equation, in matrix form, becomes

$$[M_{uu}]\{\ddot{u}\} + [C]\{\dot{u}\} + [K_{uu}]\{u\} + \left[K_{u\phi}\right]\{\phi\} = \{F\}, \tag{13.1.27}$$

$$\left[K_{\phi u}\right]\{u\} + \left[K_{\phi\phi}\right]\{\phi\} = \{Q\}, \tag{13.1.28}$$

where the global matrices are assembled from the elemental matrices, i.e.,

$$[M_{uu}] = \sum^{\wedge} [m_{uu}] \tag{13.1.29a}$$

$$[C] = \sum^{\wedge} [c^*], \tag{13.1.29b}$$

$$[K_{uu}] = \sum^{\wedge} \left[k^*_{uu}\right], \tag{13.1.29c}$$

$$\left[K_{u\phi}\right] = \hat{\sum} \left[K^*_{u\phi}\right], \tag{13.1.29d}$$

$$\left[K_{\phi u}\right] = \left[k_{u\phi}\right]^t, \tag{13.1.29e}$$

$$\left[K_{\phi\phi}\right] = \hat{\sum} \left[K^*_{\phi\phi}\right], \tag{13.1.29f}$$

$$\{F\} = \hat{\sum} \{f_i\}, \tag{13.1.29g}$$

$$\{Q\} = \hat{\sum} \{q_i\}. \tag{13.1.29h}$$

Note that $\hat{\sum}$ denotes the matrix assemblage. The combined matrix representation becomes

$$\begin{bmatrix} [M_{uu}] & 0 \\ 0 & 0 \end{bmatrix} \begin{bmatrix} \{\ddot{u}\} \\ \{\ddot{\phi}\} \end{bmatrix} + \begin{bmatrix} [C] & 0 \\ 0 & 0 \end{bmatrix} \begin{bmatrix} \{\dot{u}\} \\ \{\dot{\phi}\} \end{bmatrix} + \begin{bmatrix} [K_{uu}] & [K_{u\phi}] \\ [K_{\phi u}] & [K_{\phi\phi}] \end{bmatrix} \begin{bmatrix} \{u\} \\ \{\phi\} \end{bmatrix} = \begin{Bmatrix} F \\ Q \end{Bmatrix}. \tag{13.1.30}$$

Note that except for the static couplings, ϕ has no contribution to the higher derivative components, e.g., damping and acceleration, in this formulation. However, component couplings can change when the closed-loop feedback schemes are employed, e.g., the velocity feedback.

13.2 Eigenvalue Analysis

Natural frequencies and mode shapes can be obtained by reducing the assembled global matrices into a standard eigenvalue form. For an eigenvalue analysis, the undamped homogeneous system matrices are used, i.e.,

$$[M]\{\ddot{u}\} + [K_{uu}]\{u\} + \left[K_{u\phi}\right]\{\phi\} = 0, \tag{13.2.1}$$

$$\left[K_{\phi u}\right]\{u\} + \left[K_{\phi\phi}\right]\{\phi\} = 0. \tag{13.2.2}$$

(Note that all external forces are zero in the eigenvalue analysis.) To improve the computational efficiency, the unspecified potentials can be condensed from the system matrices. Thus, Eqs. (13.2.1) and (13.2.2) can be reduced to a standard eigenvalue equation:

$$\left(\left[\hat{K}^{*}\right]-\omega^{2}[M]\right)\{u\}=0, \qquad (13.2.3)$$

where

$$\left[\hat{K}^{*}\right]=[K_{uu}]-\left[K_{u\phi}\right]\left[K_{\phi\phi}\right]^{-1}\left[K_{\phi u}\right], \qquad (13.2.4)$$

where $[\cdot]^{-1}$ denotes the inverted matrix. Eigenvalues and eigenvectors (mode shapes) can be calculated and defined accordingly.

13.3 Structural Sensing and Control

Piezoelectric elements can be used as sensing elements for structural identification via the direct piezoelectric effect, and as active control elements—actuators—via the converse piezoelectric effect. Figure 13.1 illustrates a generic elastic DPS (continuum) coupled with a distributed piezoelectric sensing layer and a piezoelectric actuator layer. The corresponding governing structural sensing and control equations are formulated in this section.

Assembling all element matrices and applying Guyan's reduction technique, one can decouple the displacement and electric potential equations as:

$$[M]\{\ddot{u}\}+[C]\{\dot{u}\}+\left[\hat{K}^{*}\right]\{u\}=\{F\}+\{F_{e}\}, \qquad (13.3.1)$$

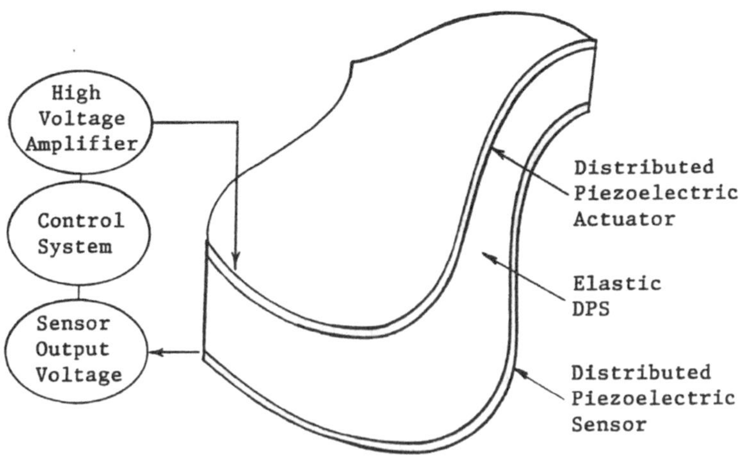

Fig. 13.1 A generic DPS with distributed piezoelectric sensor and actuator

$$\{\phi\} = \left[\mathrm{K}_{\phi\phi}\right]^{-1}\left(\{Q\} - \left[\mathrm{K}_{\phi \mathrm{u}}\right]\{\mathrm{u}\}\right); \tag{13.3.2}$$

with

$$[\mathrm{K}^*] = [\mathrm{K}_{\mathrm{uu}}] - \left[\mathrm{K}_{\mathrm{u}\phi}\right]\left[\mathrm{K}_{\phi\phi}\right]^{-1}\left[\mathrm{K}_{\phi \mathrm{u}}\right], \tag{13.3.3}$$

$$\{\mathrm{F_e}\} = -\left[\mathrm{K}_{\mathrm{u}\phi}\right]\left[\mathrm{K}_{\phi\phi}\right]^{-1}\{\mathrm{Q}\}; \tag{13.3.4}$$

where [M], [C], $[\mathrm{K}_{\mathrm{uu}}]$, and $\left[\hat{\mathrm{K}}^*\right]$ are (n × n) matrices; $\left[\mathrm{K}_{\mathrm{u}\phi}\right]$, $\left[\mathrm{K}_{\phi\phi}\right]$, and $\left[\mathrm{K}_{\phi \mathrm{u}}\right]$ are, respectively, (n × j), (j × j), and (j × n) matrices; {F}, $\{\mathrm{F_e}\}$, $\{Q\}$, and {u} are, respectively, (n × 1), (n × 1), (j × 1), and (n × 1) vectors. Note that there are two force components, {F} and $\{\mathrm{F_e}\}$, in Eq. (13.3.1): the former is a mechanical excitation force and the latter is a feedback (electric) control force. Moreover, the sensor output is contributed by two components: 1) a feedback $\{Q\}$ and 2) a displacement {u} in the general expression, Eq. (13.3.2). However, usually no feedback input is injected into the sensor layer, i.e., $\{Q\} = 0$, in practical applications. Consequently, the sensor signal is contributed only by mechanical displacements.

The electric potentials calculated in Eq. (13.3.2) are regarded as the output signals of piezoelectric sensors for structural identification; they can be processed to provide feedback signals to the piezoelectric actuators for active vibration controls. Note that the sensor output is contributed only by the mechanical displacement {u} because no external charge, i.e., $\{Q\} = 0$, is applied to the distributed piezoelectric sensor layers. Note that the (nodal) sensor signal is an averaged output contributed by all adjacent elements in which a finite separation of surface electrodes is required.

It is assumed that the direct effect in the piezoelectric actuator is negligible because the feedback voltage is much higher than the self-generated voltage. Thus, separating the actuator DOF's from the sensor DOF's, one can further define the feedback control force and can partition the matrix as

$$\begin{pmatrix} \{\mathrm{F_e}\}_\mathrm{a} \\ \{0\} \end{pmatrix} = -\begin{pmatrix} \left[\mathrm{K}_{\mathrm{u}\phi}\right]_\mathrm{a}\left[\mathrm{K}_{\phi\phi}\right]_\mathrm{a}^{-1} & [0] \\ [0] & \left[\mathrm{K}_{\mathrm{u}\phi}\right]_\mathrm{s}\left[\mathrm{K}_{\phi\phi}\right]_\mathrm{s}^{-1} \end{pmatrix}\begin{pmatrix} \{Q\}_\mathrm{a} \\ \{0\} \end{pmatrix}, \tag{13.3.5}$$

where $\left[\mathrm{K}_{\mathrm{u}\phi}\right]_\mathrm{a}$, $\left[\mathrm{K}_{\phi\phi}\right]_\mathrm{a}$, $\left[\mathrm{K}_{\mathrm{u}\phi}\right]_\mathrm{s}$, and $\left[\mathrm{K}_{\phi\phi}\right]_\mathrm{s}$ are, respectively, (n − k) × (j − p), (j − p) × (j − p), (k × p), and (p × p) matrices. $\{\mathrm{F}\}_\mathrm{a}$ and $\{Q\}_\mathrm{a}$ are, respectively, (n − k) × 1, and (j − p) × 1 vectors. Note the subscript "a" denotes the DOF's associated with the piezoelectric actuator layer, and "s" the sensor DOF's. (Note that these subscripts were used as superscripts in previous chapters. Conventionally, the location for superscripts is reserved for symbols such as matrix transpose "t" and matrix inverse "−1" in matrix representations. Thus, subscripts are adopted in this chapter.)

The sign and magnitude of the feedback is of importance to the velocity feedback control. Thus, the external charge $\{Q\}_a$ applied to the actuator needs to be properly controlled in order to produce adequate and desirable feedback control forces and moments. Using Eq. (13.3.2), one can define the induced electric potential due to mechanical displacements:

$$\begin{pmatrix} \{\phi\}_a \\ \{\phi\}_s \end{pmatrix} = \begin{pmatrix} [K_{\phi\phi}]_a^{-1} & [0] \\ [0] & [K_{\phi\phi}]_s^{-1} \end{pmatrix} \left\{ \begin{pmatrix} \{Q\}_a \\ \{Q\}_s \end{pmatrix} - \begin{bmatrix} [K_{\phi u}]_a & [0] \\ [0] & [K_{\phi u}]_s \end{bmatrix} \begin{pmatrix} \{u\}_a \\ \{u\}_s \end{pmatrix} \right\}, \tag{13.3.6}$$

where $\{\phi\}_a$ and $\{\phi\}_a$ are, respectively, (j − p) × 1, and (p × 1) vectors. As discussed previously, the subscript "a" denotes actuator DOF's and "s" denotes the sensor DOF's.

13.3.1 Sensor Signal

Since the charge applied to the sensor layer is zero, the voltage from the sensor layer can be written as

$$\{\phi\}_s = -[K_{\phi\phi}]_s^{-1}[K_{\phi u}]_s\{u\}_s. \tag{13.3.7}$$

Again, the (nodal) sensor signal is an averaged output contributed by all adjacent elements which require finite (electrical) separations of the surface electrodes, Sect. 4.2 of Chap. 4.

13.3.2 Distributed Actuators and Vibration Controls

In order to provide proper velocity information to the distributed piezoelectric actuators, the voltage induced in the sensor layer (or elements) is differentiated and fed back. Moreover, the magnitude of feedback voltage is chosen as large as possible, but it is kept below the *breakdown voltage*. Accordingly, a feedback control gain is used to enhance the sensor signal and also to change its sign before the voltage is injected into the piezoelectric actuators. Note that the feedback gains and capacitances for actuators need not to be identical. Assume $\mathcal{G}_1$, $\mathcal{G}_2$, ⋯ $\mathcal{G}_g$ represent the feedback control gains and C_1, C_2, ⋯ C_g the capacitances. Multiplying the feedback voltage by a capacitance, one can derive the feedback charge injected into the i-th actuator. Substituting the feedback charge into Eq. (13.3.5) and multiplying a feedback gain, one can obtain the equivalent feedback force generated by the i-th actuator. Two control algorithms, the *constant amplitude negative velocity feedback control* and the *constant gain negative velocity feedback control*, are

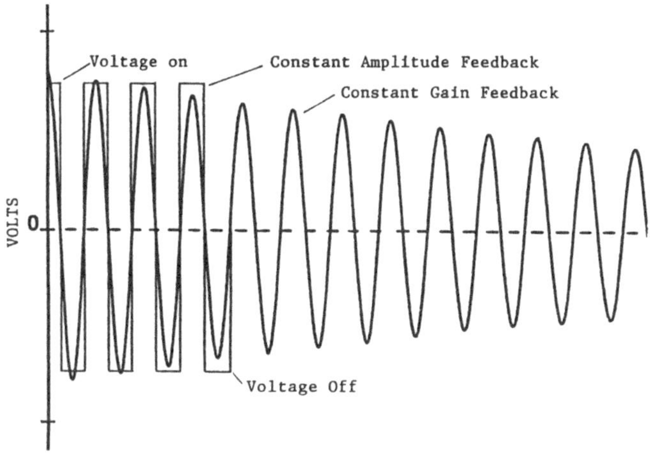

Fig. 13.2 Feedback of the constant-gain and constant-amplitude controls

formulated (Tzou & Tseng, 1990 & 1991a). Figure 13.2 shows the feedback voltages of the two control algorithms. Note that the velocity is opposite to the control voltage.

13.3.2.1 Constant Amplitude Negative Velocity Feedback Control

In the *constant amplitude negative velocity feedback control*, the feedback amplitude is constant and the sign is opposite to the velocity,

$$\{F_e\}_{a,i} = -\mathcal{G}_i \cdot C_i \cdot [K_{u\phi}]_{a,i} [K_{\phi\phi}]_{a,i}^{-1} [K_{\phi\phi}]_{s,i}^{-1} [K_{\phi u}]_{s,i} \mathrm{sgn}\left(\{\dot{u}\}_{s,i}\right), \tag{13.3.8}$$

where "sgn" is a signum function:

$$\mathrm{sgn}[u] = \begin{cases} -1, & \text{if } u < 0; \\ 0, & \text{if } u = 0; \\ +1, & \text{if } u > 0. \end{cases} \tag{13.3.9}$$

Note that the subscript ",i" indicates the i-th sensor or actuator if multiple sensors and actuators are used in the DPS's.

13.3.2.2 Constant Gain Negative Velocity Feedback Control

In the *constant gain negative velocity feedback control*, the feedback control force is calculated by

$$\{F_e\}_{a,i} = -\mathcal{G}_i\, \omega_{mn} \cdot C_i \cdot [K_{u\phi}]_{a,i} [K_{\phi\phi}]_{a,i}^{-1} [K_{\phi\phi}]_{s,i}^{-1} [K_{\phi u}]_{s,i} \{\dot{u}\}_{s,i}. \qquad (13.3.10)$$

Note that the *constant amplitude negative velocity feedback control* is nonlinear and discontinuous. The *constant gain negative velocity feedback control* is linear and continuous; the feedback control voltage decreases as the vibration velocity decays.

In the finite element analysis, the time-history responses of the elastic/ piezoelectric system are calculated using a time-domain direct integration algorithm—the modified Wilson-θ method and a pseudo-force method (Tzou & Schiff, 1987) to accommodate the control force derived from the applied surface charges.

13.3.3 Time-Domain Integration—The Modified Wilson-θ Method

Time-history responses of the structronic system are integrated in the time domain using *the modified Wilson-θ method* with *a pseudo force approximation* (Tzou & Schiff, 1987). The original method is an extension of a linear acceleration method, in which the acceleration is assumed linear from time t to (t + Δt). In the Wilson-θ method, the acceleration is assumed linear from time t to (t + θΔt) where θ is constant. Note that if θ = 1.0, the method is reduced to the linear acceleration method. The system equation at time (t + θΔt) can be expressed as

$$[M]\{\ddot{u}(t+\theta \cdot \Delta t)\} + [C]\{\dot{u}(t+\theta \cdot \Delta t)\} + \left[\hat{K}^*\right]\{u(t+\theta \cdot \Delta t)\} = \left\{\tilde{F}^{\circ}(t+\theta \cdot \Delta t)\right\}, \qquad (13.3.11)$$

where θ is a constant and is taken as 1.4 to guarantee an unconditional stability in the time-domain integration. $\left\{\tilde{F}^{\circ}(t+\theta \cdot \Delta t)\right\}$ is a linearly extrapolated load vector and can be expressed as

$$\left\{\tilde{F}^{\circ}(t+\theta \cdot \Delta t)\right\} = \left\{\tilde{F}^{\circ}(t)\right\} + \theta \cdot \left(\left\{\tilde{F}^{\circ}(t+\Delta t)\right\} - \left\{\tilde{F}^{\circ}(t)\right\}\right). \qquad (13.3.12)$$

Considering additional control forces are generated in feedback controls, one can rewrite the extrapolated force vector as

$$\left\{\tilde{\tilde{F}}^{\circ}(t+\theta \cdot \Delta t)\right\} = \left\{\tilde{\tilde{F}}^{\circ}(t)\right\} + \theta \cdot \left(\left\{\tilde{F}^{\circ}(t+\Delta t)\right\} + \{F_e(t+\Delta t)\} - \left\{\tilde{F}^{\circ}(t)\right\}\right), \qquad (13.3.13)$$

where $\left\{\tilde{\tilde{F}}^{\circ}(t)\right\}$ is the resultant force, including the mechanical and feedback forces, at the time t. Thus, the system equation can be written as

$$\begin{aligned}&[M]\{\ddot{u}(t+\theta\cdot\Delta t)\}+[C]\{\dot{u}(t+\theta\cdot\Delta t)\}+\left[\hat{K}^*\right]\{u(t+\theta\cdot\Delta t)\}\\&=\left\{\tilde{F}^\circ(t)\right\}+\theta\cdot\left(\left\{\tilde{F}^\circ(t+\Delta t)\right\}+\{F_e(t+\Delta t)\}-\left\{\tilde{F}^\circ(t)\right\}\right)\end{aligned}. \quad (13.3.14)$$

It should be noted that since the control forces are estimated by a linear extrapolation, a relatively small integration time-step, i.e., at least, 1/10 of the lowest interested frequency, is required to assure the numerical stability (Tzou, 1983).

13.4 State Variable Transformation

State variable representation of structural system equations is highly desirable in design and analysis of control systems. In this section, the derived piezoelectric/elastic system equations are transferred into the state space (Tzou & Tseng, 1991b). (Note that bold-faced symbols are used in the state variable transformations.) Define state variables $\mathbf{x}_1$ and $\mathbf{x}_2$ as

$$\mathbf{x}_1=\{u\}, \quad (13.4.1)$$

$$\mathbf{x}_2=\{\dot{u}\}. \quad (13.4.2)$$

For an **open-loop** system, the system is represented by a set of two first order differential equations in terms of state variables $\mathbf{x}_1$ and $\mathbf{x}_2$,

$$\dot{\mathbf{x}}_1=\mathbf{x}_2, \quad (13.4.3)$$

$$\dot{\mathbf{x}}_2=-[M]^{-1}\left[\hat{K}^*\right]\mathbf{x}_1-[M]^{-1}[C]\mathbf{x}_2+[M]^{-1}\{F\}, \quad (13.4.4)$$

where the superscript "−1" denotes the matrix inversion. Writing in a matrix form gives

$$\begin{pmatrix}\dot{\mathbf{x}}_1\\ \dot{\mathbf{x}}_2\end{pmatrix}=\begin{bmatrix}[0] & [I]\\ -[M]^{-1}\left[\hat{K}^*\right] & -[M]^{-1}[C]\end{bmatrix}\begin{pmatrix}\mathbf{x}_1\\ \mathbf{x}_2\end{pmatrix}+\begin{pmatrix}[0]\\ [M]^{-1}\end{pmatrix}\{F\}, \quad (13.4.5)$$

$$\{\phi\}_s=\left[\left([0],-[K_{\phi\phi}]_s^{-1}[K_{\phi u}]_s\right)\quad [0]\right]\begin{pmatrix}\mathbf{x}_1\\ \mathbf{x}_2\end{pmatrix}; \quad (13.4.6)$$

or simply

$$\dot{\mathbf{x}}=\mathbf{Ax}+\mathbf{Br}, \quad (13.4.7)$$

$$\mathbf{y}=\mathbf{Cx}; \quad (13.4.8)$$

where **x** is a (2n × 1) state vector; **A** is a (2n × 2n) system matrix; **B** is a (2n × n) control matrix; **r** is an (n × 1) input vector; **y** is a (p × 1) output vector; **C** is a (p × 2n) matrix. They are respectively defined as

$$\mathbf{x} = \begin{pmatrix} \mathbf{x}_1 \\ \mathbf{x}_2 \end{pmatrix}, \tag{13.4.9a}$$

$$\mathbf{A} = \begin{bmatrix} [0] & [I] \\ -[M]^{-1}\left[\hat{K}^*\right] & -[M]^{-1}[C] \end{bmatrix}, \tag{13.4.9b}$$

$$\mathbf{B} = \begin{pmatrix} [0] \\ [M]^{-1} \end{pmatrix}, \tag{13.4.9c}$$

$$\mathbf{r} = \{F\}, \tag{13.4.9d}$$

$$\mathbf{y} = \{\phi\}_s, \tag{13.4.9e}$$

$$\mathbf{C} = \left[\left([0], -[K_{\phi\phi}]_s^{-1}[K_{\phi u}]_s\right) \quad [0] \right]. \tag{13.4.9f}$$

Note that the electromechanical force $\{F_e\}$ does not appear in the above equations; this is because $\{F_e\}$ is used as a feedback control force in a closed-loop control system.

Considering a closed-loop system with u representing the control force with a dimension of (n × 1), one can derive the state equation as

$$\dot{\mathbf{x}} = \mathbf{A}\mathbf{x} + \mathbf{B}\mathbf{u}, \tag{13.4.10}$$

$$\mathbf{y} = \mathbf{C}\mathbf{x}. \tag{13.4.11}$$

With the conventional feedback control law, the feedback force is

$$\mathbf{u} = \mathbf{r} - \mathbf{K}^t\mathbf{x}, \tag{13.4.12}$$

where $\mathbf{K}^t$ is the feedback gain matrix and it is defined by the feedback algorithms.

1) Constant Amplitude Negative Velocity Feedback Control:

$$K^t = \left[[0] \quad \left([0], \sum \left(\mathcal{G}_i \cdot C_i \cdot [K_{u\phi}]_{a,i}[K_{\phi\phi}]_{a,i}^{-1}[K_{\phi\phi}]_{s,i}^{-1}[K_{\phi u}]_{s,i}\mathrm{sgn}(\cdot)\right)\right) \right] \begin{pmatrix} \mathbf{x}_1 \\ \mathbf{x}_2 \end{pmatrix}. \tag{13.4.13}$$

2) Constant Gain Negative Velocity Feedback Control:

$$K^t = \left[[0] \quad \left([0], \sum \left(\mathcal{G}_i \cdot C_i \cdot [K_{u\phi}]_{a,i} [K_{\phi\phi}]_{a,i}^{-1} [K_{\phi\phi}]_{s,i}^{-1} [K_{\phi u}]_{s,i} \right) \right) \right] \begin{pmatrix} \mathbf{x}_1 \\ \mathbf{x}_2 \end{pmatrix}. \tag{13.4.14}$$

13.5 Matrix Condensation and Reduction

Assembling the element matrices yields a global system matrix equation, Sect. 13.1.4. As far as calculating the system eigenvalues and evaluating distributed sensing and controls, it is not necessary to employ the reduction scheme (Guyan, 1965) and reduce the overall system order in the analysis. However, in order to improve the computational efficiency, it is best to employ Guyan's reduction scheme to reduce the system order. In this section, the matrix reduction scheme is presented (Tseng, 1989).

A congruent transformation matrix is derived from the static equivalent equation. The static equation is written as

$$\begin{bmatrix} [K_{uu}] & [K_{u\phi}] \\ [K_{\phi u}] & [K_{\phi\phi}] \end{bmatrix} \begin{bmatrix} \{u\} \\ \{\phi\} \end{bmatrix} = \begin{bmatrix} \{F\} \\ \{Q\} \end{bmatrix}. \tag{13.5.1}$$

(Note the capital letters denote the global matrices.) Partitioning the static equation in terms of retained DOF's $\{u_1\}$ and reduced DOF's $\{u_2\}$ and $\{\phi\}$ gives

$$\begin{bmatrix} [K_{uu}^{11}] & [K_{uu}^{12}] & [K_{u\phi}^{1}] \\ [K_{uu}^{21}] & [K_{uu}^{22}] & [K_{u\phi}^{2}] \\ [K_{\phi u}^{1}] & [K_{\phi u}^{2}] & [K_{\phi\phi}] \end{bmatrix} \begin{pmatrix} \{u_1\} \\ \{u_2\} \\ \{\phi\} \end{pmatrix} = \begin{bmatrix} \{F_1\} \\ \{F_2\} \\ \{Q\} \end{bmatrix}. \tag{13.5.2}$$

Note that the superscripts do not imply the "power" relationship and the potential; $\{\phi\}$ is also condensed from the time-domain integration. Solving the condensed displacement and potential DOF's in terms of $\{u_1\}$ gives

$$[\bar{K}_{uu}]\{u_1\} = \{\bar{F}_1\}, \tag{13.5.3}$$

where the modified stiffness matrix $[\bar{K}_{uu}]$ and force vector $\{\bar{F}_1\}$ are defined as

$$[\bar{K}_{uu}] = [K_{uu}^{11}] - \left\{ [K_{uu}^{12}] \quad [K_{u\phi}^{1}] \right\} \begin{bmatrix} [K_{uu}^{22}] & [K_{u\phi}^{2}] \\ [K_{\phi u}^{2}] & [K_{\phi\phi}] \end{bmatrix}^{-1} \begin{bmatrix} [K_{uu}^{21}] \\ [K_{\phi u}^{1}] \end{bmatrix}, \tag{13.5.4}$$

$$\{\bar{F}_1\} = \{F_1\} - \left\{\left[K_{uu}^{12}\right]\left[K_{u\phi}^{1}\right]\right\}\begin{bmatrix}\left[K_{uu}^{22}\right] & \left[K_{u\phi}^{2}\right] \\ \left[K_{\phi u}^{2}\right] & \left[K_{\phi\phi}\right]\end{bmatrix}^{-1}\begin{bmatrix}\{F_2\} \\ \{Q\}\end{bmatrix}. \tag{13.5.5}$$

The condensed system mass matrix becomes

$$[\bar{M}] = \left[T_g\right]^t\begin{bmatrix}\left[M^{11}\right] & \left[M^{12}\right] & [0] \\ \left[M^{21}\right] & \left[M^{22}\right] & [0] \\ [0] & [0] & [0]\end{bmatrix}\left[T_g\right], \tag{13.5.6}$$

where $[T_g]$ is a congruent transformation matrix defined as

$$\left[T_g\right] = \begin{bmatrix} [I] \\ -\begin{bmatrix}\left[K_{uu}^{22}\right] & \left[K_{u\phi}^{2}\right] \\ \left[K_{\phi u}^{2}\right] & \left[K_{\phi\phi}\right]\end{bmatrix}^{-1}\begin{bmatrix}\left[K_{uu}^{21}\right] \\ \left[K_{\phi u}^{1}\right]\end{bmatrix}\end{bmatrix}. \tag{13.5.7}$$

where [I] is an identity matrix with the dimension equal to the retained DOF's. Note that a consistent mass formulation is used in this derivation. Following the same procedure, one can derive the condensed damping matrix $[\bar{C}]$ as

$$[\bar{C}] = \left[T_g\right]^t\begin{bmatrix}\left[C^{11}\right] & \left[C^{12}\right] & [0] \\ \left[C^{21}\right] & \left[C^{22}\right] & [0] \\ [0] & [0] & [0]\end{bmatrix}\left[T_g\right]. \tag{13.5.8}$$

Hence, the condensed dynamic system matrix equation (open-loop) can be written as

$$[\bar{M}]\{\ddot{\bar{u}}\} + [\bar{C}]\{\dot{\bar{u}}\} + [\bar{K}]\{\bar{u}\} = \{\bar{F}\}, \tag{13.5.9}$$

where $\{\bar{u}\} = \{u_1\}$ which denote the displacement of the retained master DOF's, and $\{\bar{F}\} = \{\bar{F}_1\}$, as derived previously.

In order to improve the computational efficiency, a condensation technique is implemented in the finite element code. The theoretical derivation of the condensation technique was discussed previously. Pre-multiplied by an $[M]^{-1}$, a transformation matrix $[T_g]$ can be formulated, which is similar to that discussed previously. Considering the closed-loop case, one obtains the system equation:

$$[\bar{M}]\{\ddot{\bar{u}}\} + [\bar{C}]\{\dot{\bar{u}}\} + \left[\bar{K}^*\right]\{\bar{u}\} = \left\{\bar{F}^*\right\}\left\{\bar{F}_e^*\right\}, \tag{13.5.10}$$

$$\{\phi\} = \left[K_{\phi\phi}\right]^{-1}\left(\{Q\} - \left[K_{\phi u}\right]\left[T_g\right]\{\bar{u}\}\right); \tag{13.5.11}$$

with

$$\{u\} = [T_g]^t\{\bar{u}\}, \tag{13.5.12a}$$

$$[\bar{M}] = [T_g]^t[T_g], \tag{13.5.12b}$$

$$[\bar{C}] = [T_g]^t[M]^{-1}[C][T_g], \tag{13.5.12c}$$

$$[\bar{K}^*] = [T_g]^t[M]^{-1}[\hat{K}^*][T_g], \tag{13.5.12d}$$

$$\{\bar{F}^*\} = [T_g]^t[M]^{-1}[F], \tag{13.5.12e}$$

$$\{\bar{F}_e^*\} = [T_g]^t[M]^{-1}[F_e]; \tag{13.5.12f}$$

where the overhead bar represents the condensed matrices or vectors.

In the state space representation, $z_1 = \{\bar{u}\}$, $z_2 = \{\dot{\bar{u}}\}$, Eqs. (13.5.10) and (13.5.11) are written as

$$\dot{\mathbf{z}} = \mathbf{Fz} + \mathbf{Gu}, \tag{13.5.13}$$

$$\mathbf{y} = \mathbf{Hz}, \tag{13.5.14}$$

where the vector and matrices are defined as

$$\mathbf{z} = \begin{pmatrix} z_1 \\ z_2 \end{pmatrix}, \tag{13.5.15a}$$

$$\mathbf{F} = \begin{bmatrix} [0] & [I] \\ -[\bar{M}]^{-1}[\bar{K}^*] & -[\bar{M}]^{-1}[\bar{C}] \end{bmatrix}, \tag{13.5.15b}$$

$$\mathbf{G} = \begin{bmatrix} [0] \\ [\bar{M}]^{-1}[T_g]^t[M]^{-1} \end{bmatrix}, \tag{13.5.15c}$$

$$\mathbf{H} = \left[\left([0], -[K_{\phi\phi}]_s^{-1}[K_{\phi u}]_s \right) \quad [0] \right] \begin{bmatrix} [T_g] & [0] \\ [0] & [T_g] \end{bmatrix}; \tag{13.5.15d}$$

with the same feedback control law as

$$\mathbf{u} = \mathbf{r} - \mathbf{K^t x}, \tag{13.5.16}$$

where $\mathbf{r}$ and $\mathbf{K^t}$ are defined previously in Sect. 13.4.

13.6 System Aggregation for Control of Large-Scale Systems

The principle of aggregation technique is to use a lower-order model which preserves the dominant modes in a control analysis. The concept is similar to the matrix condensation method which is usually used to reduce the computational effort in finite element analyses. In this section, a combination of the condensation and aggregation techniques is presented to derive a reduced order system (Tseng, 1989; Tzou & Tseng, 1991a).

Aggregation techniques involve replacing the initial system model S^i by a reduced-order model S^r which preserves the dominant system characteristics, Fig. 13.3, In general, S^i can be represented by a state vector **x** of dimension 2n and S^r by a state vector **z** of dimension 2 m (m < < n). In this section, the analyses are restricted by the assumptions that: 1) S^i and S^r are both linear dynamic models, and 2) the aggregation is linear, i.e., **z** is a linear function in **x**.

Thus, the aggregation technique transforms the state x to z by

$$\mathbf{z} = \mathbf{L}\mathbf{x}, \tag{13.6.1}$$

where **L** is so called the **aggregation matrix** of dimension (2m × 2n). The state variable representation of the initial model S^i can be written as

$$\dot{\mathbf{x}} = \mathbf{A}\mathbf{x} + \mathbf{B}\mathbf{u},, \tag{13.6.2}$$

$$\mathbf{y}^i = \mathbf{C}\mathbf{x}. \tag{13.6.3}$$

A reduced model ($\mathbf{y}^r$) takes the form as

$$\dot{\mathbf{z}} = \mathbf{F}\mathbf{z} + \mathbf{G}\mathbf{u} \tag{13.6.4}$$

$$\mathbf{y}^r = \mathbf{H}\mathbf{z}. \tag{13.6.5}$$

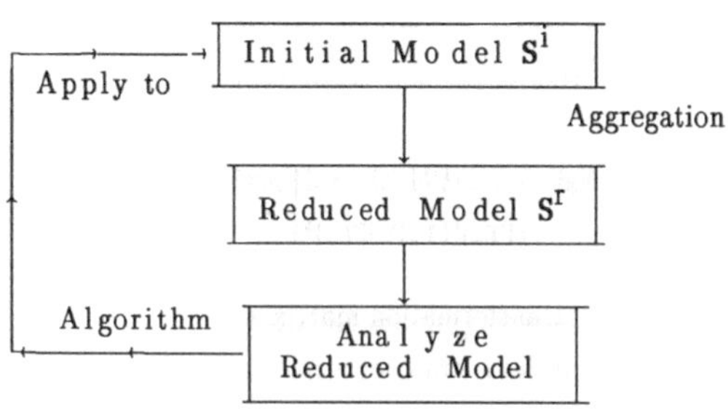

Fig. 13.3 Control system aggregation

Note that Eqs. (13.6.4)–(13.6.5) represent the aggregated model for the initial model S^i. There are still two problems which need to be addressed: 1) the determination of the aggregation matrix **L**, and 2) the determination of control strategies for the reduced model S^r suitable for the initial model S^i. It can be proved that (Tseng, 1989)

$$\mathbf{z} = \begin{bmatrix} \left([T_g]^t[T_g]\right)^{-1}[T_g]^t & [0] \\ [0] & \left([T_g]^t[T_g]\right)^{-1}[T_g]^t \end{bmatrix} \mathbf{x}, \tag{13.6.6}$$

where $[T_g]$ is the congruent transformation matrix and

$$\mathbf{F} = \mathbf{LA}, \tag{13.6.7}$$

$$\mathbf{G} = \mathbf{LB}; \tag{13.6.8}$$

where

$$\mathbf{L} = \begin{bmatrix} \left([T_g]^t[T_g]\right)^{-1}[T_g]^t & [0] \\ [0] & \left([T_g]^t[T_g]\right)^{-1}[T_g]^t \end{bmatrix}. \tag{13.6.9}$$

Thus, the reduced-order model can sufficiently represent the initial system model in the feedback controls. Control algorithms derived previously are still applicable in the reduced system. Detailed discussion and proofs are presented in a reference (Tseng, 1989). Control spill-over problems and system stabilities for the reduced order model are also discussed (Tseng, 1989).

13.7 Case Studies

To demonstrate the usefulness of the derived finite element scheme, three case studies are presented in this section: 1) an active piezoelectric bimorph beam (validated by theoretical and experimental data), 2) a laminated piezoceramics/steel beam (compared with published finite-difference data), and 3) a square plate laminated with distributed sensor and actuator layers. The first case—a bimorph structure—can be applied to a number of engineering applications, e.g., a micro-positioning device, a piezoelectric gripper, a high-precision actuator, a flexible robot manipulator, etc. Thus, two aspects of the bimorph structure are investigated: 1) micro-displacement actuation, and 2) distributed structural identification.

The second case is a steel beam laminated with a lead–titanate–zirconate (PZT) ceramic layer. Distributed control effectiveness of the steel/PZT laminated beam via the two feedback algorithms is calculated and compared with published results.

Plate structures or components are very common in many mechanical and aerospace systems. The third case presents a study on distributed structural sensing, identification and vibration control of an elastic plate with integrated piezoelectric sensor/actuator layers. Distributed ***modal voltages*** of the first three modes are illustrated and control effectiveness of the two control algorithms evaluated.

It should be noted that assumptions are made in the finite element analysis: 1) the resistance of deposited metal electrode on each side of the piezoelectric material is negligible, so that the applied voltage can quickly reach a steady state; 2) material properties of the electrodes are not considered; 3) the induced strain contraction and expansion settles very fast without any time delay; 4) a velocity information can be obtained and used instantaneously as a feedback signal.

13.7.1 A Piezoelectric Bimorph Beam

The piezoelectric bimorph beam is made of two layers of piezoelectric polymeric polyvinylidene fluoride (PVDF) with opposite polarity. Material properties are summarized in Table 13.1.

When an external voltage is applied across the beam thickness, the induced internal stresses induce a resultant bending moment which forces the bimorph beam bend. The first study is a static deflection test in which a strong voltage is applied across the thickness and the induced deflection is studied using theoretical, experimental, and finite element techniques. The second study is a distributed structural identification in which the distributed voltage along the beam is calculated and illustrated using the developed finite element code.

The bimorph beam model is discretized into ten piezoelectric finite elements, five on each layer. (Note the aspect ratio of each element.) One end of the bimorph beam is assumed fixed. The physical dimension and polarity are illustrated in Fig. 13.4; the experimental setup is shown in Fig. 13.5.

Table 13.1 Material properties of the piezoelectric PVDF

Properties	Value	Units
Poisson ratio	0.2900	
Mass density	0.1800×10^4	Kg/m^3
Modulus	0.2000×10^{10}	N/m^2
Piezoelectric constants		
e_{31}	0.0460	C/m^2
e_{32}	0.0460	C/m^2
Dielectricity		
ε_{11}	0.1062×10^{-9}	F/m
ε_{22}	0.1062×10^{-9}	F/m
ε_{33}	0.1062×10^{-9}	F/m

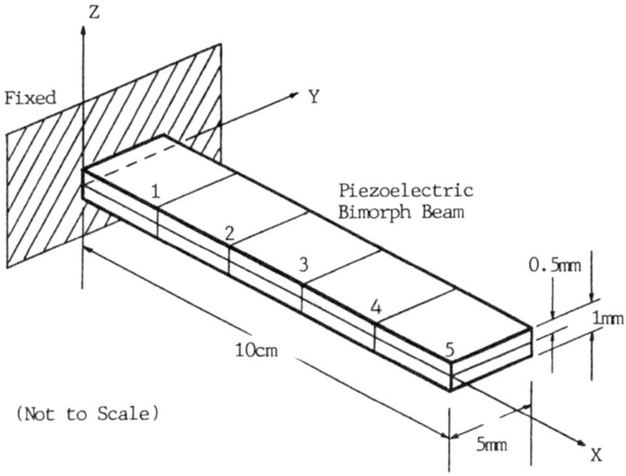

Fig. 13.4 A piezoelectric polymeric bimorph beam

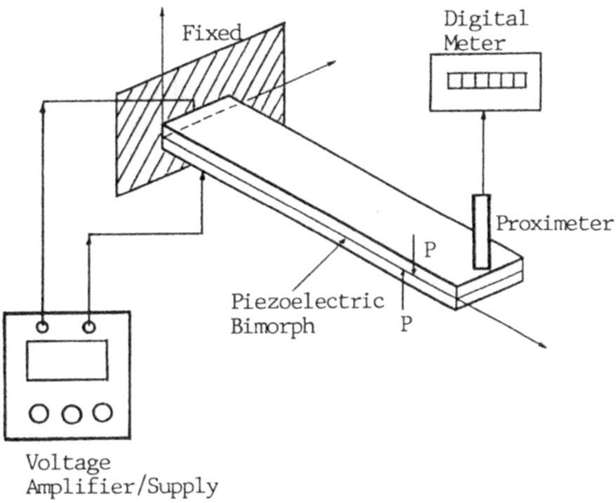

Fig. 13.5 Experimental apparatus

13.7.1.1 Micro-Displacement Actuation

A unit voltage (1 V) is applied across the thickness and the static deflections of five nodes are calculated analytically and by the finite element method. A physical model with the same dimension is also tested in laboratory. The results are

Table 13.2 Static deflection of the piezoelectric bimorph beam (10^{-7}m)

Node	1	2	3	4	5
Theory	0.138	0.552	1.24	2.21	3.45
FEM	0.124	0.508	1.16	2.10	3.30
Error (%)	10	8.0	6.2	5.1	4.4

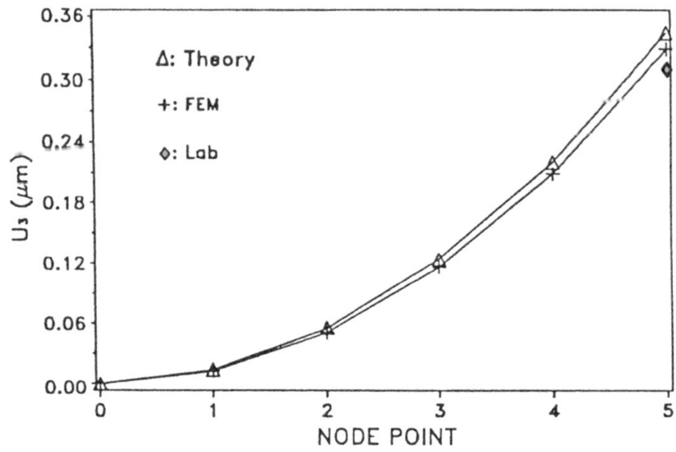

Fig. 13.6 Nodal deflection of the piezoelectric bimorph beam

summarized in Table 13.2 and plotted in Fig. 13.6. Note that only the tip deflection is measured in laboratory experiments, Node 5 in Fig. 13.6.

The error at the fixed end is more significant than that at the free end. This is because the discretized finite element is relatively "rigid" near the fixed end; the model can be improved by using more elements. It is also observed that the experimental results are lower than the other two calculations; this could be caused by: 1) non-perfect bonding, 2) voltage leakage, 3) energy dissipations, etc. Furthermore, it should be noted that there are only **five** finite elements along the beam and this proves the effectiveness of the new "thin" piezoelectric hexahedron element with internal degree-of-freedom (DOF).

13.7.1.2 Distributed Structural Sensing and Identification

The piezoelectric bimorph beam is studied for its voltage distribution in dynamic oscillations. A snap-back test with an initial tip displacement of 1 cm is performed and the voltage response is calculated using the developed finite element code. Figure 13.7 shows the displacement response starting from the lower-right fixed-end corner, and Fig. 13.8 the voltage distribution along the beam. Note that

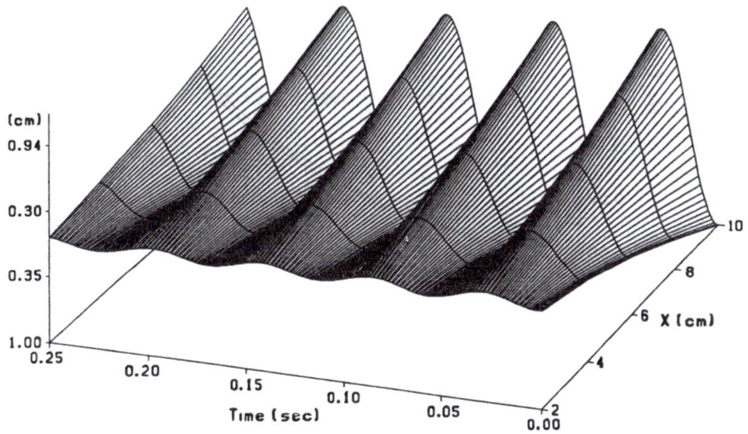

Fig. 13.7 Snap-back response of the piezoelectric bimorph beam

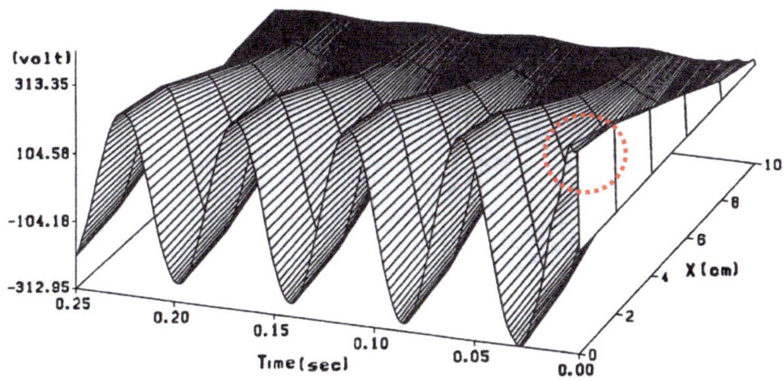

Fig. 13.8 Distributed voltage of the piezoelectric bimorph beam

only the span from 2 to 10 cm is plotted in Fig. 13.8 because relatively fine response curves occurring near the fixed end, see Sect. 4.2 of Chap. 4.

It is observed that the displacement profile exhibits a first mode shape of the bimorph beam. Thus, the voltage distribution should also represent the ***modal voltage*** (Tzou & Tseng, 1990 & 1991b) of the first mode. However, the transient voltage response (encircled) is observed in the beginning of the snap-back response. It should be pointed out that a finite separation of surface electrodes is required to develop the surface averaged voltage distribution presented here.

13.7.2 Vibration Control of a Laminated Steel/PZT Beam

A steel beam is sandwiched between two layers of piezoelectric ceramics, lead–titanate–zirconate (PZT), Fig. 13.9. One of the layers serves as a distributed sensor and the other a distributed actuator. The sensor output is used in feedback controls based on the control algorithms proposed previously: 1) the *constant amplitude negative velocity feedback control* and 2) the *constant gain negative velocity feedback control*.

A snap-back test with an initial tip velocity of 1 m/sec is studied by using the finite element technique, and the time responses with and without feedback controls are compared with the finite difference solutions (Obal, 1986) presented in 13.9 Appendix appended at the end of this chapter. Figure 13.10 shows a free tip response with a 2% natural damping.

Figure 13.11a, b show the controlled responses in which a) is from the **constant amplitude** feedback control and b) is from the **constant gain** feedback control. The major difference between the finite element and finite difference models should be noted. The finite difference model (Obal, 1986) did not consider the effect of piezoceramic layers; this led to a softer beam with a first natural frequency of 5.2 Hz. However, because of the new piezoelectric finite element, these two thin piezoceramic layers were easily modeled and included in the analysis; this should provide more accurate results. The finite element solutions show a first natural frequency of 5.4 Hz and natural damping of 2%; these are identical to the analytical solution and assumed viscous damping, respectively.

In the constant amplitude feedback control (Fig. 13.11a), the control voltage was turned on at t = 0 and turned off at t = 4 s. In the constant gain feedback control, gain = 0.036, (Fig. 13.11b), the feedback voltage was always proportional to the

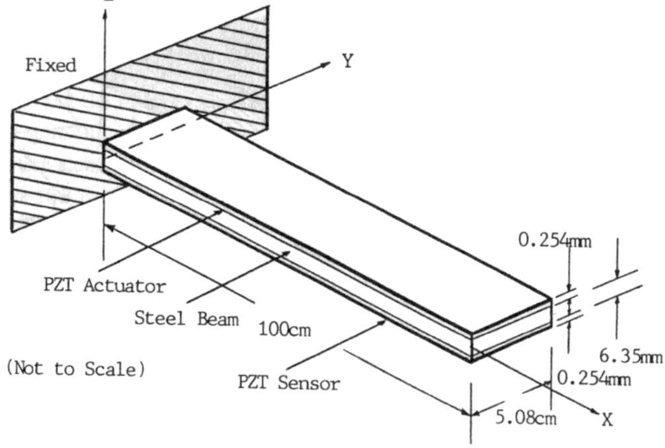

Fig. 13.9 A steel beam laminated with distributed PZT sensor and actuator layers

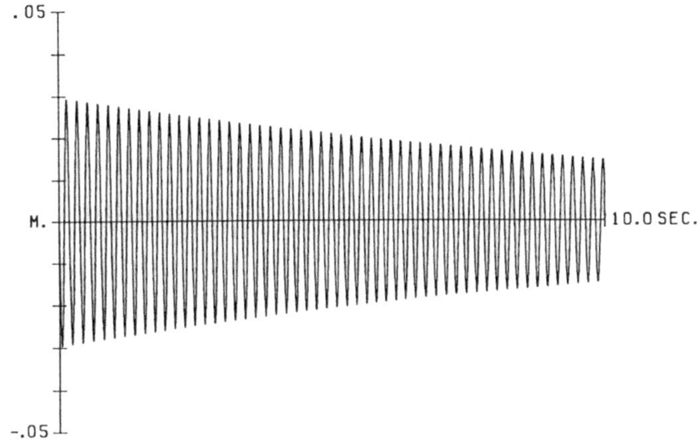

Fig. 13.10 Free response of the steel/PZT cantilever beam

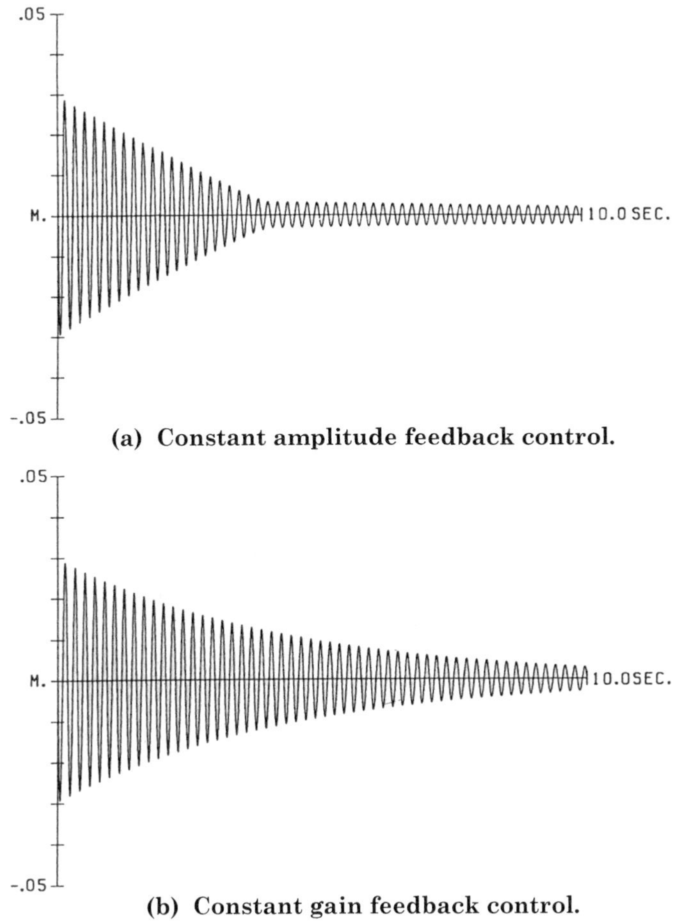

Fig. 13.11 Active distributed vibration control of the steel/PZT beam

negative velocity and it was smaller when the oscillation reduced. Thus, the constant amplitude feedback control more effectively controls the beam oscillation. Note that the voltage amplitude in the constant amplitude feedback was kept about the same as the maximum of that in the second control algorithm. Consequently, the energy used in the former is greater than that in the latter.

13.7.3 Distributed Vibration Identification and Control of a Plate

In this section, distributed structural sensing, identification and vibration control of a zero-curvature shell—a plate—are studied. The physical model is a plexiglas cantilever plate (10 cm × 10 cm × 0.31 cm) with a top distributed piezoelectric PVDF actuator layer (40 μm) and a bottom distributed sensor layer. The plate with the distributed sensor/actuator is divided into seventy-five elements, twenty-five for each layer, modeled by the piezoelectric finite elements developed earlier. Figure 13.12 shows the cantilever plate finite element model.

Distributed ***modal voltage*** measured by the distributed sensor is demonstrated in the first three modes. The distributed piezoelectric actuator can be made either **mono-axially** oriented, aligned with the x-axis, or **bi-axially** oriented, effective in both x and y axes. (The effects of material orientation are evaluated in distributed vibration controls of the plate.)

As discussed previously, the distributed piezoelectric sensing layer should respond to the plate oscillation and generate an electric voltage representing the distributed dynamic state of the plate. This distributed sensing phenomenon is

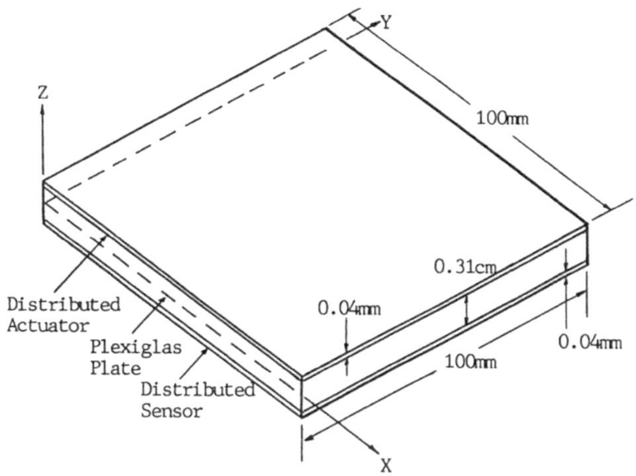

Fig. 13.12 Finite element modeling of a laminated plate

demonstrated in an eigenvalue analysis. The active distributed vibration suppression and control of the plate is studied and evaluated in a snap-back analysis in which an initial displacement (first mode) is imposed.

13.7.3.1 Distributed Structural Identification: *Modal Voltages*

The output signals of each node on the distributed piezoelectric sensor layer can be calculated as a function of displacements, ultimately local strains. (Note that the $\{Q\}$ vector is zero in an eigenvalue analysis.) After the nodal voltage is calculated, the overall voltage distribution of the plate can be plotted by connecting all nodal voltage amplitudes. (Note that each nodal voltage is contributed by all surrounding elements. Finite separations of electrode surfaces are required.) Thus, for a given mode, the ***modal voltage*** distribution (distributed sensing phenomena) can be observed. The first three plate mode shapes and ***modal voltage*** distributions are illustrated in Figs. 13.13, 13.14 and 13.15.

It is observed that the first mode is a bending mode, the second a torsion mode, and the third a warping mode. The sensitivities of the modes will vary because the tension and compression vary in different modes. Voltage drops at two corners, V (0,0) and V(0,100) are observed in all three figures. This is because that the strains are (numerically) smaller at these two boundary nodes in the finite element calculation. According to the theory discussed previously, the output amplitude is

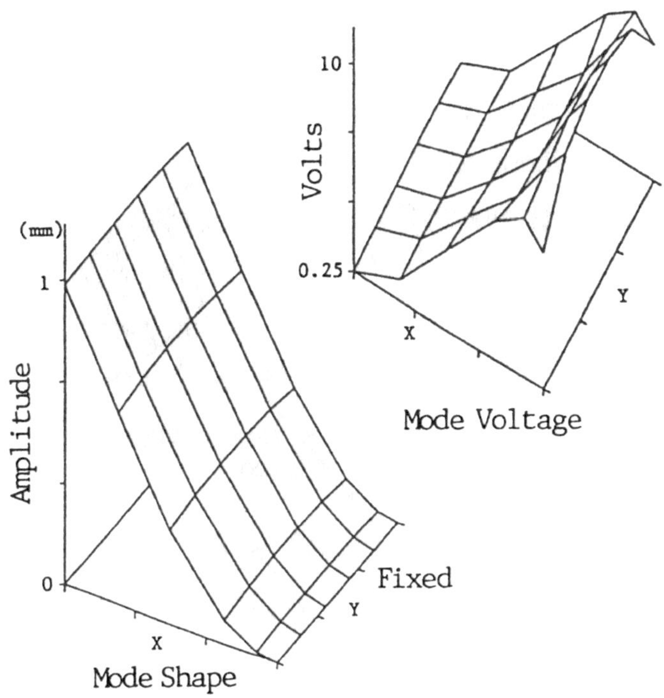

Fig. 13.13 First mode shape and distributed modal voltage

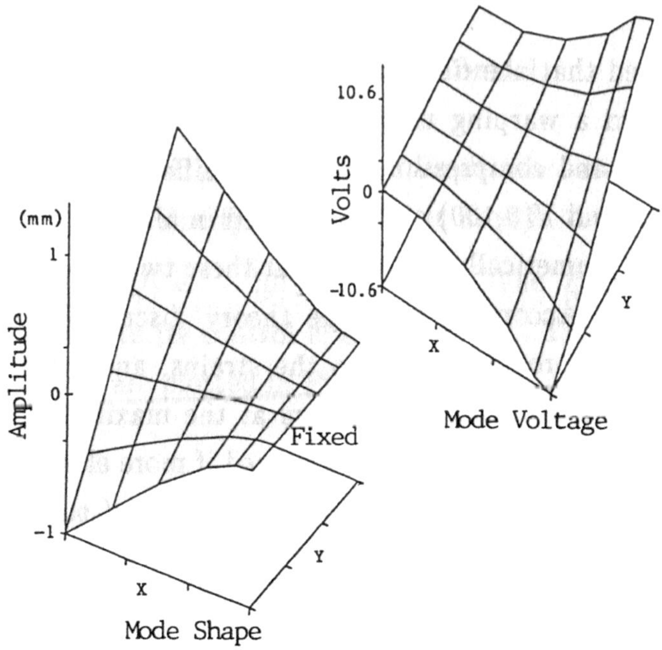

Fig. 13.14 Second mode shape and distributed modal voltage

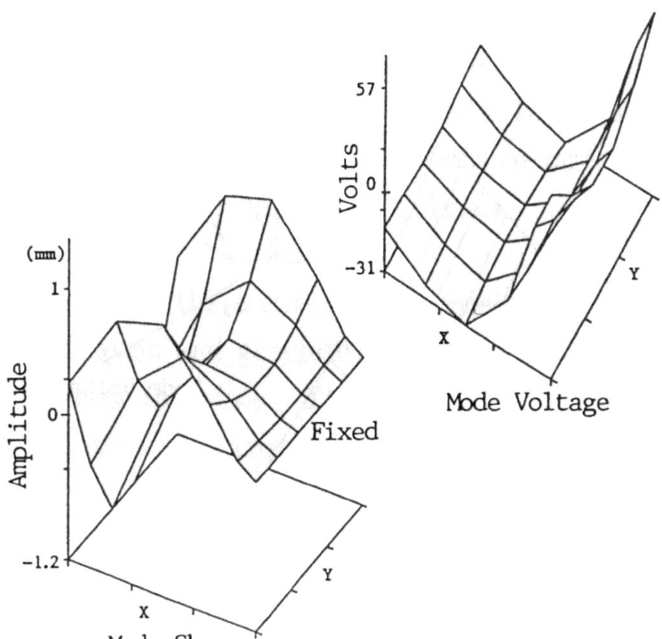

Fig. 13.15 Third mode shape and distributed modal voltage

inversely proportional to the strains, and ultimately displacements, i.e., the maximum positive voltage occurs at the maximum negative strain. The ***modal voltage*** distribution could also be refined if more elements and finer mesh are used in the finite element modeling such that *modal voltages* of higher natural modes can be observed.

13.7.3.2 Distributed Vibration Controls

The distributed piezoelectric actuator on the top surface of the plate contracts or expands depending on the signs of the feedback voltages (based on the converse piezoelectric effect). In general, a feedback voltage is selected to counteract the motion of the plate.. There are two distributed piezoelectric actuators evaluated in the finite element analyses: 1) a mono-axial actuator whose piezoelectric orientation is aligned with the x-axis, and 2) a bi-axial actuator whose orientation is aligned with both x and y axes.

An initial displacement (first mode) is imposed, and then the plate is set free. The vibration amplitude decays, depending on the modal damping and the feedback. An initial modal damping is assumed to be 0.9% (based on a laboratory experiment); the damping ratio variations are evaluated when comparing the control effectiveness of the distributed piezoelectric actuator with different control algorithms. It should be noted that a tip velocity (node P) is used in feedback controls, since in practical applications, this is the simplest way to implement it.

1) Negative Velocity, Constant Gain Feedback Control

In the first case, the feedback voltage amplitude varies with respect to the negative velocity. Since the velocity at note P is used in the feedback, the original equation is modified as $\{Q\} = -\left[\mathcal{G}\right]^* \cdot \{\dot{u}_p\}$. The gain matrix $\left[\mathcal{G}\right]^*$ is also changed in order to evaluate its control effectiveness. The time histories without and with feedback control are plotted in Fig. 13.16.

The vibration amplitudes are suppressed by the distributed piezoelectric actuator, and the damping ratio changes are calculated and plotted versus feedback gains. Figure 13.17 shows that the damping ratio increases when the feedback gain increases. The mono-axial actuator is more effective on first mode vibration control.

2) Negative Velocity, Constant Amplitude Feedback Control

In the second case, although the feedback amplitude is constant, the feedback voltage changes sign when the nodal velocity $\dot{u}_p$ changes its direction, i.e., $\{Q\} = -\left[\mathcal{G}\right]^* \mathrm{sgn}\left[\{\dot{u}_p\}\right]$ (the nonlinear bang-bang control). The controlled and uncontrolled time histories are plotted in Fig. 13.18; the damping ratios are studied for different feedback voltages in Fig. 13.19.

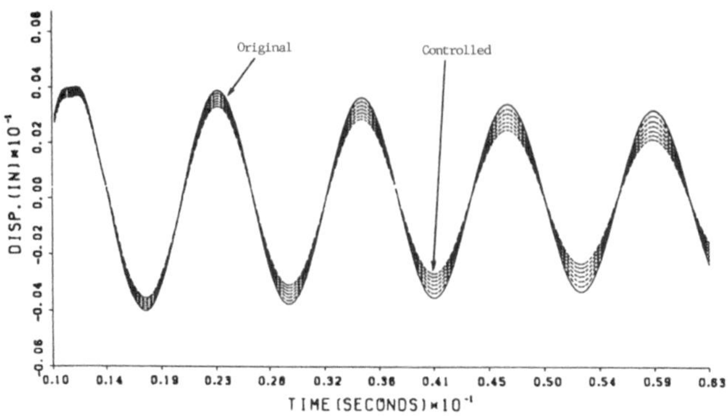

Fig. 13.16 Time histories of controlled and uncontrolled responses

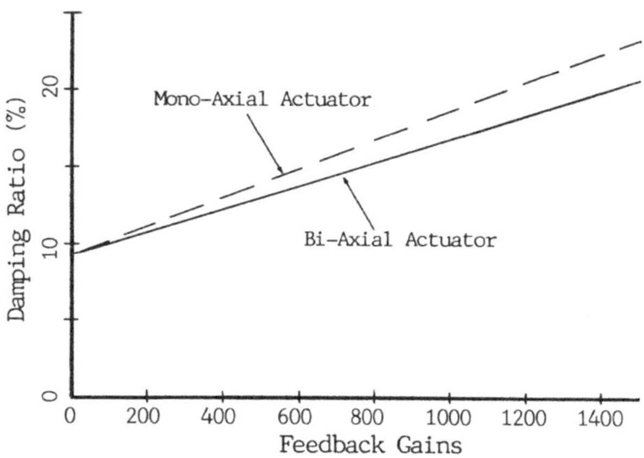

Fig. 13.17 Damping calculation at constant gain feedback controls

Note that a single node velocity was used in the feedback controls, even though the distributed voltage was discussed in the previous section. This single signal was amplified and fed back to all nodes of the actuator. In practice, one could use multiple nodes or areas as the feedback reference. It is observed that a higher gain leads to a higher damping matrix in the system equation, so that the plate oscillations are damped much faster. It is also noted that the mono-axially oriented piezoelectric actuator is better than the bi-axially oriented actuator for controlling the first plate mode. (It is anticipated that the bi-axial actuator would be more effective if mode shapes involve relatively significant motions in the y direction, however, this is not demonstrated in this study.)

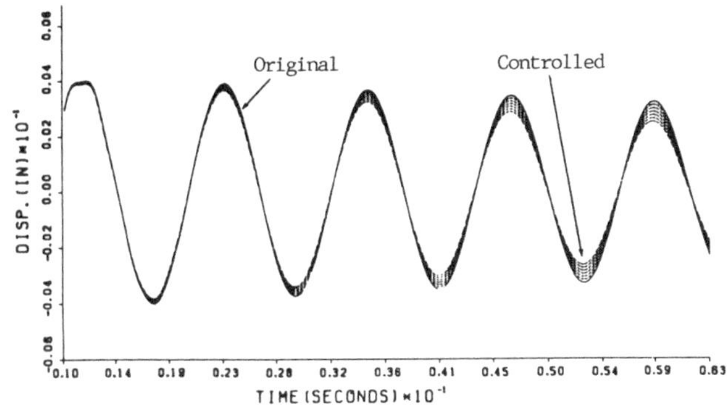

Fig. 13.18 Time histories of controlled and uncontrolled responses

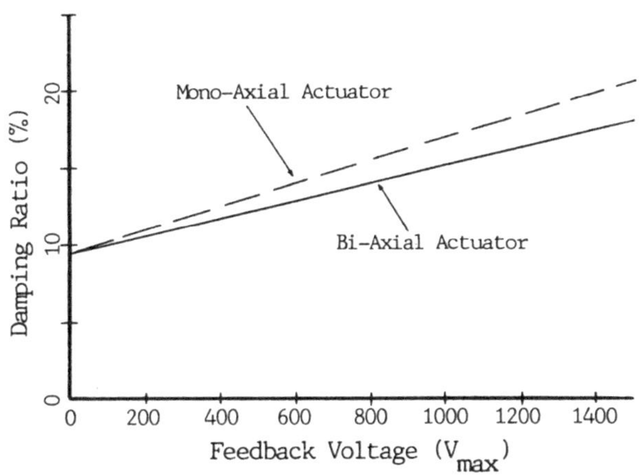

Fig. 13.19 Damping calculation at constant amplitude feedback controls

13.8 Summary

Conventional elastic structures are "passive" in nature, i.e., they do not possess any inherent self-sensation and action/reaction capabilities. Thus, development of new-generation active structures with integrated sensors, actuators, and control electronics, i.e., so called the **structronic system**, has received an increasing attention and interest in recent years (Tzou & Anderson, 1992). This chapter presents a finite element development and analysis of integrated distributed piezoelectric sensor/actuator structures—active distributed parameter systems (DPS's) or **structronic systems**.

Conventional isoparametric hexahedron piezoelectric elements are very inefficient and undesirable in the modeling and analysis of "**thin**" piezoelectric/elastic active continua. Thus, a new "thin" piezoelectric hexahedron element with internal degree-of-freedom (DOF) was formulated using the variation principle, Hamilton's principle, and linear piezoelectricity theory. System electromechanical equations of the piezoelectric/elastic coupled DPS's were derived and then transformed into the state equations. Two control algorithms, namely, a *constant-amplitude* and a *constant-gain (negative velocity) feedback controls* were proposed and integrated into the finite element formulations. Control of the original large-scale DPS's was reduced to a control of an aggregated (or reduced) DPS's using Guyan's reduction scheme. A modified Wilson-θ method was implemented to carry out the direct time-domain integrations. Three active piezoelectric/elastic laminated structures: 1) a piezoelectric polyvinylidene fluoride (PVDF) bimorph beam, 2) a steel/PZT laminated beam and 3) a laminated zero-curvature shell—a plate—sandwiched between two piezoelectric sensor/actuator layers were investigated in case studies.

In Case 1, micro-positioning capabilities (static deflections) of the bimorph beam were calculated using the newly developed finite element code, and results were compared closely with the analytical solutions and experimental data. This proved the dependability of the finite element code. Time-history voltage distributions along the beam were also calculated, which showed the ***potential map*** representing the dynamic state of the beam. A transient phenomenon was also observed in the beginning of the snap-back response. Applications of this bimorph configuration include: 1) micro-positioning, 2) high-precision actuation, 3) robot fingers, 4) flexible robot manipulators, etc.

In Case 2, distributed piezoelectric PZT layers were laminated on a steel beam and were used as a distributed sensor and a distributed actuator. Voltage outputs from the distributed sensor were processed and fed back to the distributed actuator using two control algorithms: 1) *the constant amplitude negative velocity feedback* and 2) *the constant gain negative velocity feedback*. It was shown that the constant amplitude feedback control is superior to the constant gain feedback control for controlling the first bending mode. Finite element results were also closely compared with the finite difference solutions.

Distributed modal identification and vibration control of a PVDF laminated plate was studied and evaluated in Case 3. The distributed piezoelectric sensor responded to modal oscillations and generated distributed ***modal voltages***. The first three *modal voltages* were plotted along with the corresponding mode shapes. Output signals from the sensor were processed, amplified, and fed back to the distributed actuator via two control algorithms: 1) the constant amplitude negative velocity feedback and 2) the constant gain negative velocity feedback. It was shown that the mono-axial sensitive actuator works better than that of the bi-axial sensitive actuator for the first mode vibration controls.

The use of distributed piezoelectric sensors and actuators for distributed vibration control and identification of DPS's was demonstrated to be an effective technique. The piezoelectric finite element and control techniques were also proved to be suitable for modeling and analysis of active piezoelectric/elastic coupled electromechanical DPS's. In practical applications, however, some potential problems, such as 1) bonding techniques, 2) resistance of electrodes, 3) voltage leakages, 4) temperature effects, 5) time delay associated with the control electronics, 6) segmenting and/or shaping electrodes and signal transmission, etc. still need further investigations.

Appendix

Uncontrolled and control responses of the steel/PZT laminated beam were calculated by the finite difference method (Obal, 1986) and time-history responses are plotted in Figs. 13.20, 13.21 and 13.22.

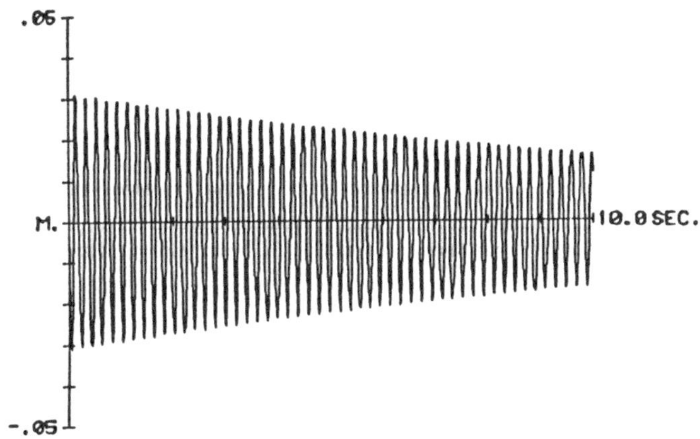

Fig. 13.20 Free vibration of a steel/PZT laminated plate

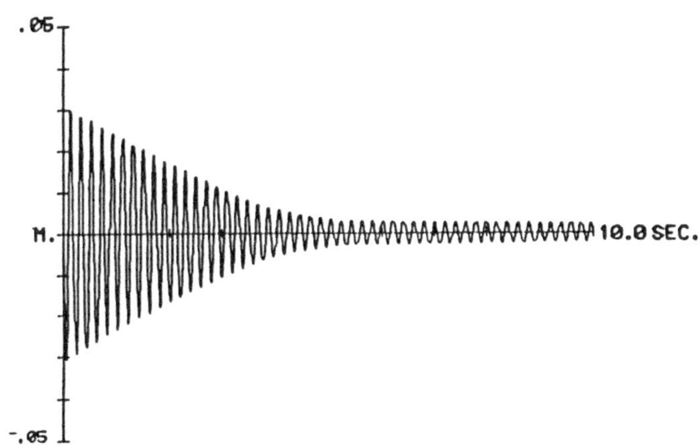

Fig. 13.21 Controlled response of a steel/PZT beam (Constant amplitude)

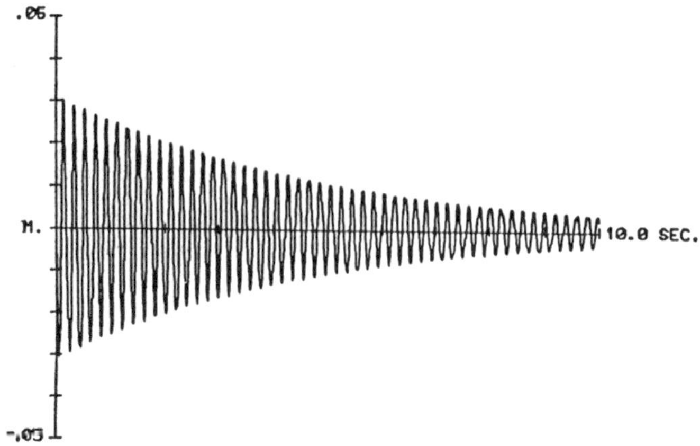

Fig. 13.22 Controlled response of a steel/PZT beam (Constant gain)

References

Allik H. and Hughes, T.J., 1979, "Finite element method for piezoelectric vibration," *Int. J. of Numerical Methods Eng.*, Vol.2, 1979, pp. 151–168.

Bathe, K. J. and Wilson, E.L., 1976, *Numerical Method in Finite Element Analysis*, Prentice Hall, London, 1976.

Cook, R. D. 1974, *Concepts and Applications of Finite Element Analysis*, John Wiley & Sons, Inc., New York, 1974.

Guyan, R. J., 1965, "Reduction of Stiffness and Mass Matrices," *AIAA Journal*, Vol.3, No.2, 1965, p. 380.

Nailon, M, Coursant, R.H., and Besnier, F., 1983, "Analysis of piezoelectric structures by a finite element method," *ACTA Electronica*, Vol.25, No.4, pp. 341–362.

Obal, M.W., 1986, *Vibration Control of Flexible Structures Using Piezoelectric Devices as Sensors and Actuators*, Ph.D. Thesis, Georgia Institute of Technology.

Tseng, C.I., 1989, *Electromechanical Dynamics of a Coupled Piezoelectric/Mechanical System Applied to Vibration Control and Distributed Sensing*, Ph.D. Thesis, Department of Mechanical Engineering, University of Kentucky, December 1989.

Tzou, H.S., 1983, *A Nonlinear Dynamic Analysis Program Applied to Elevator Counterweight Systems*, Ph.D. Thesis, School of Mechanical Engineering, Purdue University, December 1983.

Tzou, H.S. & Anderson, G.L., 1992, *Intelligent Structural Systems*, Kluwer Academic Pub., August 1992.

Tzou, H.S. and Schiff, A.J., 1987, "Development and Evaluation of a Pseudo–Force Approximation Applied to Nonlinear Contacts and Viscoelastic Damping," *Intl. J. Comptr. _ Structures*, Vol.26, No.3, pp. 481–493.

Tzou, H. S. Tzou and Tseng, C. I., 1988a, "Active Vibration Control of Distributed Parameter Systems by Finite Element Method," *Computers in Engineering 1988*, Vol.(3), 1988, pp. 599–604.

Tzou, H.S. and Tseng, C.I., 1988b, "Development of a Thin Piezoelectric Finite Element Applied to Distributed Sensing and Active Vibration Controls," 88–ASME/CIE–2, ASME Winter Annual Meeting, Chicago, Ill, 1988.

Tzou H.S. and Tseng, C.I., 1990, "Distributed piezoelectric sensor/actuator design for dynamic measurement/control of distributed parameter systems: a finite element approach," *Journal of Sound and Vibration*, Vol.(138), No.(1), pp. 17–34.

Tzou, H.S., and Tseng, C.I., 1991a, "Distributed Vibration Control and Identification of Coupled Elastic/Piezoelectric Systems," Mechanical Systems and Signal Processing, Vol.(5), No.3, pp. 215–231, May 1991.

Tzou, H.S., and Tseng, C.I., 1991b, "Distributed Modal Identification and Vibration Control of Continua: Piezoelectric Finite Element Formulation and Analysis," ASME *Journal of Dynamic Systems, Measurements, and Control*, 113(3), pp. 500–505, September 1991. (10/PzFem. BkPzSm5).

Appendix
Linear Piezoelectricity Theory

The linear *piezoelectricity* theory couples the electric and mechanical fields, respectively represented by the *direct* piezoelectric effect and the *converse* piezoelectric effect. In Appendix, the linear piezoelectricity theory is reviewed, and the relations among piezoelectricity related constants are defined. Note that the *pyroelectric* effect, i.e., a temperature differential induced signal generation, is not considered.

Definitions

The relationship between the conventional **Timoshenko** notation and the **IEEE piezoelectricity** notation needs to be defined first. Double and single subscripts of the stress and strain components are related as follows:

11 by 1; 22 by 2; 33 by 3; 23 by 4; 13 by 5; 12 by 6.

Thus, the stress and strain components defined in the two notations can be related as

$$\begin{bmatrix} \sigma_{xx} & \tau_{xy} & \tau_{xz} \\ \tau_{xy} & \sigma_{yy} & \tau_{yz} \\ \tau_{xz} & \tau_{yz} & \sigma_{zz} \end{bmatrix} \equiv \begin{bmatrix} T_{11} & T_{12} & T_{13} \\ T_{12} & T_{22} & T_{23} \\ T_{13} & T_{23} & T_{33} \end{bmatrix} \equiv \begin{bmatrix} T_1 & T_6 & T_5 \\ T_6 & T_2 & T_4 \\ T_5 & T_4 & T_3 \end{bmatrix}, \tag{A.1.1}$$

$$\begin{bmatrix} \varepsilon_{xx} & \gamma_{xy} & \gamma_{xz} \\ \gamma_{xy} & \varepsilon_{yy} & \gamma_{yz} \\ \gamma_{xz} & \gamma_{yz} & \varepsilon_{zz} \end{bmatrix} \equiv \begin{bmatrix} S_{11} & S_{12} & S_{13} \\ S_{12} & S_{22} & S_{23} \\ S_{13} & S_{23} & S_{33} \end{bmatrix} \equiv \begin{bmatrix} S_1 & S_6/2 & S_5/2 \\ S_6/2 & S_2 & S_4/2 \\ S_5/2 & S_4/2 & S_3 \end{bmatrix}. \tag{A.1.2}$$

Note that ε_{xx}, ε_{yy}, and ε_{zz} are the normal strains; γ_{ij} are the shear strains in the Timoshenko notation. However, $[\varepsilon_{ij}]$ denotes the dielectric constant matrix in the

H. (HS) Tzou, *Piezoelectric Shells*, Solid Mechanics and Its Applications 247,
https://doi.org/10.1007/978-94-024-1258-1

IEEE piezoelectricity notation, adopted in this book. Accordingly, the stress is defined by {T} and the strain is defined by {S} and they are represented in column vector forms:

$$\{T\} = [T_1 \quad T_2 \quad T_3 \quad T_4 \quad T_5 \quad T_6]^t, \tag{A.1.3}$$

$$\{S\} = [S_1 \quad S_2 \quad S_3 \quad S_4 \quad S_5 \quad S_6]^t, \tag{A.1.4}$$

where the superscript "t", i.e., $[\cdot]^t$, denotes the vector or matrix transpose. The electric field {E} and the electric displacement (flux density) {D} can also be written in column vector forms:

$$\{E\} = [E_1 \quad E_2 \quad E_3]^t, \tag{A.1.5}$$

$$\{D\} = [D_1 \quad D_2 \quad D_3]^t. \tag{A.1.6}$$

Linear Piezoelectricity Relations

The linear piezoelectricity theory is based on a quasi-static assumption in which the electric field is balanced with the elastic field so that these two fields can be decoupled at a given time instant. The direct and converse piezoelectric effects are respectively written as

$$\{D\} = [e]\{S\} + [\varepsilon^S]\{E\}, \tag{A.2.1a}$$

$$\{T\} = [c^E]\{S\} - [e]^t\{E\}; \tag{A.2.1b}$$

where {T} is the stress tensor (second order); {S} is the strain tensor (second order); {E} is the electric field vector; {D} is the electric displacement vector; $[c^E]$ is the elasticity tensor (4th order) evaluated at a constant electric field; [e] is a tensor of piezoelectric stress coefficients (3rd order); $[e]^t$ is the transposed tensor of [e]; $[\varepsilon^S]$ is the dielectric tensor (2nd order) evaluated at a constant strain. If the piezoelectric tensor [e] is set to zero, Eqs. (A.2.1a & b) become the conventional *dielectric* and *Hooke's* equations.

There are a number of ways to write the elastic, piezoelectric, and dielectric governing equations in which 1) stress and electric field, 2) strain and electric displacement, 3) strain and electric field are respectively used as dependent and/or independent variables.

$$\{T\} = [c^D]\{S\} - [h]^t\{D\}, \tag{A.2.2a}$$

$$\{E\} = -[h]\{S\} + [\beta^S]\{D\}; \tag{A.2.2b}$$

or

$$\{S\} = [s^E]\{T\} + [d]^t\{E\}, \tag{A.2.3a}$$

$$\{D\} = [d]\{T\} + [e^T]\{E\}; \tag{A.2.3b}$$

or

$$\{S\} = [s^D]\{T\} - [g]^t\{D\}, \tag{A.2.4a}$$

$$\{E\} = -[g]\{T\} + [\beta^T]\{D\}; \tag{A.2.4b}$$

where $[\beta^T]$ is the impermeability matrix obtained from the inverse of the dielectric matrix [ε]. Note that the superscript "T" (inside the matrix) denotes the properties measured at the constant stress T. (Recall that $[\cdot]^t$ denotes a vector or matrix transpose.) [d] is the piezoelectric strain constant matrix; [e] is the piezoelectric stress constant matrix; [g] matrix relates the open-circuit voltage at a given stress; [h] matrix relates the open-circuit voltage at a given strain. The four piezoelectric coefficient matrices [d], [e], [g], and [h] are also related and their relations are derived next.

Elastic, Piezoelectric, and Dielectric Relations

An adiabatic condition ensures that no heat is added or removed from a given space or volume. The elastic constants are defined at the adiabatic conditions. The piezoelectric and dielectric constants of a piezoelectric material are defined at the isothermal and adiabatic conditions. (Note that the piezoelectric material is assumed non-pyroelectric.) To derive the relations among [d], [e], [g], and [h] matrices, one can pre-multiply Eq. (A.2.3b) by $[\beta^T]$, i.e.,

$$[\beta^T]\{D\} = [\beta^T][d]\{T\} + [\beta^T][\varepsilon^T]\{E\}, \tag{A.3.1}$$

where $[\beta^T]$ is a tensor of free dielectric impermeability evaluated at the constant stress, and is defined by

$$[\beta^T] = [\varepsilon^T]^{-1}. \tag{A.3.2}$$

Note $[\cdot]^{-1}$ denotes the matrix inverse. Substituting Eq. (A.3.2) into Eq. (A.3.1) yields

$$\{E\} = [\beta^T]\{D\} - [[\beta^T][d]]\{T\}. \tag{A.3.3}$$

Setting the term inside parentheses equal to [g], i.e., $[g]=[\beta^T][d]$, and substituting Eq. (A.3.3) into Eq. (A.2.3), one can derive

$$\{S\} = [s^D]\{T\} + [g]^t\{D\}, \tag{A.3.4a}$$

$$\{E\} = -[g]\{T\} + [\beta^T]\{D\}, \tag{A.3.4b}$$

with

$$[s^D] = [s^E] - [d]^t[g], \tag{A.3.5}$$

where $[s^D]$ is the compliance matrix defined at the constant dielectric displacement. Note that Eqs. (A.3.4a & b) are identical to Eq. (A.2.4a & b). Following similar procedures, one can derive Eqs. (A.2.1a & b) and (A.2.2a & b). Four different forms of the elastic, dielectric, and piezoelectric equations can be defined and their relations are summarized in Tables A.1 and A.2.

Table A.1 Four sets of piezoelectric equations

No.	Elastic relationship	Electric relationship
1	$\{S\} = [s^E]\{T\} + [d]^t\{E\}$	$\{D\} = [d]\{T\} + [\varepsilon^T]\{E\}$
2	$\{S\} = [s^D]\{T\} + [g]^t\{D\}$	$\{E\} = -[g]\{T\} + [\beta^T]\{D\}$
3	$\{T\} = [c^E]\{S\} - [e]^t\{E\}$	$\{D\} = [e]\{S\} + [\varepsilon^S]\{E\}$
4	$\{T\} = [c^D]\{S\} - [h]^t\{D\}$	$\{E\} = -[h]\{S\} + [\beta^S]\{D\}$

Note that the superscripts T, S, E, D denote the matrix/tensor defined at the constant stress, strain, electric field, and dielectric displacement, respectively.

Table A.2 Relations among piezoelectric, dielectric, and elastic constants

No.	Elastic	Piezoelectric	Dielectric
1	$[s^D] = [s^E] - [d]^t[g]$	$[g] = [\beta^T][d]$	$[\beta^T] = [\varepsilon^T]^{-1}$
2	$[c^E] = [s^E]^{-1}$	$[e] = [d][c^E]$	$[\varepsilon^S] = [\varepsilon^T] - [e][d]^t$
3	$[c^D] = [c^E] - [e]^t[h]$	$[h] = [\beta^S][e]$	$[\beta^S] = [\beta^T] + [h][g]^t$
4	$[c^D] = [s^D]^{-1}$	$[h] = [g][c^D]$	$[\beta^S] = [\varepsilon^S]^{-1}$

Note that the superscripts T, S, E, D denote the matrix/tensor defined at the constant stress, strain, electric field, and dielectric displacement, respectively.

Index

H. (HS) Tzou, *Piezoelectric Shells*, Solid Mechanics and Its Applications 247,
https://doi.org/10.1007/978-94-024-1258-1

MIX
Papier aus verantwortungsvollen Quellen
Paper from responsible sources
FSC® C105338

FSC
www.fsc.org

If you have any concerns about our products,
you can contact us on
ProductSafety@springernature.com

In case Publisher is established outside the EU,
the EU authorized representative is:
Springer Nature Customer Service Center GmbH
Europaplatz 3, 69115 Heidelberg, Germany

Printed by Libri Plureos GmbH
in Hamburg, Germany